Antioxidants and Functional Foods for Diabetic Neuropathy

Manish Kumar Gautam
Mayank Gangwar • Dileep Kumar
Editors

Antioxidants and Functional Foods for Diabetic Neuropathy

Prevention and Cure

Springer

Editors
Manish Kumar Gautam
Faculty of Pharmaceutical Science
Assam down town University
Guwahati, Assam, India

Mayank Gangwar
Department of Health Research
Indian Council of Medical Research
New Delhi, Delhi, India

Dileep Kumar
Department of Entomology and Nematology
University of California Davis
Davis, CA, CA, USA

ISBN 978-981-95-9325-5 ISBN 978-981-95-9326-2 (eBook)
https://doi.org/10.1007/978-981-95-9326-2

This Springer imprint is published by the registered company Springer Nature Singapore Pte Ltd.
The registered company address is: 152 Beach Road, #21-01/04 Gateway East, Singapore 189721, Singapore

Each science only opens doors to us of knowledge—but far away lies the One who knows it all. The more we discover, the more we realize how much is left undiscovered. That is the glory of science.

—Swami Vivekananda

World's most respected Vedanta thinker (1863–1902)

Lectures from Colombo to Almora, Advaita Ashrama, 2004, p. 128.

Foreword

It is with great pleasure and scholarly enthusiasm that I pen this foreword to the latest academic endeavor, *Antioxidants and Functional Foods for Diabetic Neuropathy: Prevention and Cure*. This comprehensive volume serves as a timely and essential contribution to the evolving field of diabetic complications, particularly focusing on the growing interest in nutritional and antioxidant-based interventions for diabetic neuropathy.

The book is thoughtfully structured into six major categories—*Introduction to Diabetic Neuropathy*, *Role of Antioxidants in Diabetic Neuropathy*, *Functional Foods for Diabetic Neuropathy*, *Specific Antioxidants and Their Efficacy*, *Integrative Approaches to Prevention and Cure of Diabetic Neuropathy*, and *Case Studies and Future Perspectives*. Each section is meticulously curated, offering readers a rich blend of foundational knowledge and cutting-edge research, which not only informs but also inspires future investigations in this vital area of biomedical science.

At the heart of this academic masterpiece is *Dr. Manish Kumar Gautam*, a distinguished academician and researcher currently serving as Associate Professor and Associate Dean, Faculty of Pharmaceutical Sciences at Assam down town University, Guwahati, India.

This book, under his expert stewardship, is not merely a compilation—it is a clarion call for integrated, evidence-based solutions for one of the most debilitating complications of diabetes. I am confident this work will serve as a beacon for clinicians, researchers, and policymakers committed to improving the quality of life for individuals with diabetic neuropathy.

I extend my sincerest gratitude to *Dr. Manish Kumar Gautam*, *Dr. Mayank Gangwar*, *Dr. Dileep Kumar*, and all the contributing authors for their dedication and expertise in bringing this seminal work to fruition. It is my hope that this book will serve as a catalyst for continued dialogue, collaboration, and action in the global fight against diabetic neuropathy.

Dean, Faculty of Pharmaceutical Science
Assam down town University,
Guwahati, Assam, India

Saikat Sen

Preface

Diabetic neuropathy remains one of the most persistent and life-altering complications of diabetes, often manifesting as chronic pain, sensory loss, and increased vulnerability to infections. While conventional therapies offer symptomatic relief, they often fail to address the underlying oxidative stress and metabolic disturbances that contribute to nerve damage. This book, *Antioxidants and Functional Foods for Diabetic Neuropathy: Prevention and Cure*, is an earnest attempt to bridge this crucial gap through scientific exploration of natural and nutritional interventions.

The inspiration for this work stems from a growing body of research suggesting that antioxidants and functional foods can play a vital role in preventing and mitigating diabetic neuropathy. The book is organized into six comprehensive sections: *Introduction to Diabetic Neuropathy*, *Role of Antioxidants in Diabetic Neuropathy*, *Functional Foods for Diabetic Neuropathy*, *Specific Antioxidants and Their Efficacy*, *Integrative Approaches to Prevention and Cure of Diabetic Neuropathy*, and *Case Studies and Future Perspectives*. Each chapter aims to present a balance of fundamental principles, recent advances, and potential clinical applications.

Drawing from the fields of pharmacology, nutrition, molecular biology, and traditional medicine, this book provides a multidisciplinary perspective, making it a valuable resource for researchers, clinicians, students, and policymakers. It also serves as a guide for further academic inquiry and innovation in managing diabetic complications through integrative and natural means.

This book is intended for a wide audience: students seeking to understand the interplay between nutrition and disease; researchers aiming to identify novel targets and therapeutic compounds; clinicians interested in integrative therapies; and policymakers shaping future healthcare strategies. I believe that the convergence of traditional wisdom and modern science holds the key to solving many of our most persistent health challenges—and diabetic neuropathy is no exception.

I express my deepest gratitude to Assam down town University, all contributors, collaborators, and mentors who made this endeavor possible. I hope this book encourages further research, innovation, and dialogue toward safer, accessible, and more effective solutions for managing diabetic complications.

Guwahati, Assam, India — Manish Kumar Gautam
New Delhi, India — Mayank Gangwar
Pune, Maharashtra, India — Dileep Kumar

Contents

Editors and Contributors

About the Editors

Manish Kumar Gautam is a distinguished academician and researcher currently serving as an Associate Professor and Associate Dean in Faculty of Pharmaceutical Sciences, Assam down town University, Guwahati, India.

He earned his PhD in Pharmacology from Banaras Hindu University, one of India's most prestigious institutions. His early research contributions were shaped at premier national institutes including CSIR–National Botanical Research Institute (NBRI), Lucknow, and the Central Council for Research in Ayurvedic Sciences (CCRAS), Ministry of AYUSH, where he successfully regulated and executed several intramural and extramural research projects.

His global impact is highlighted by two international travel grants awarded by the Indian Council of Medical Research (ICMR) and Science and Engineering Research Board (SERB, DST) for paper presentations in London, UK, and Malaysia.

His expertise has been recognized at the policy level, having contributed as a Technical Support Expert in the development of "General Guidelines for Safety/Toxicity Evaluation of Ayurvedic Formulations," published by CCRAS, Ministry of AYUSH, Government of India.

A prolific writer, he has authored 9 books and more than 80 peer-reviewed research papers published in high-impact journals. His research spans across molecular pharmacology, natural product drug discovery, diabetic complications, and Ayurvedic pharmacology, marking him as a versatile and impactful scientist.

He continues to inspire students and researchers alike through his academic excellence, innovative thinking, and unwavering dedication to advancing pharmacological science at both national and international levels.

Mayank Gangwar is currently serving as Scientist-D in Clinical Studies and Trials Unit (INTENT), Division of Development Research, ICMR, Headquarters, New Delhi. His extraordinary research odyssey commenced in 2011 where he embarked on a relentless pursuit of understanding infectious microorganisms during doctoral studies. With an unwavering passion for science, he not only mastered the intricacies of molecular assays but also showcased his brilliance on national and international scientific stages through impactful presentations. His academic journey is a testament to his unwavering commitment, culminating in a Doctorate in Pharmacology from Banaras Hindu University, complemented by a Master of Pharmacy in Pharmacology. Beyond academia, his imprint on the scientific community is indelible. His prolific contributions to research are encapsulated in over 50 publications, one book with Springer bearing witness to his profound influence with an impressive citation count of 1590, an h-index of 21, and an i10-index of 40. Recognitions have been his companions, with the CSIR-Direct SRF, DBT, and ICMR International Travel Grant among his crowning achievements. He served 5 years as Scientist in VRDL, IMS, BHU toward research management and quality standards, augmented by ISO and NABL accreditation training. His impact extends beyond the laboratory, as he has contributed as a board editor and held editorial roles in esteemed national and international journals, including a tenure as a Research Topic Editor in the *Frontiers in Pharmacology* journal. Additionally, his commitment to excellence is underscored by his certification as a Registered Pharmacist with the Pharmacy Council of Lucknow, India. Dr. Gangwar's multifaceted contributions reach their zenith with his involvement in the Society for Bacteriophage Research and Therapy, where he has held office, further solidifying his role in shaping the future of research. Currently, he is engaged in planning, managing, and executing clinical trials related to medical devices and therapeutics at INTENT,

ICMR. His expertise in clinical trials management, quality assurance, and regulatory compliance is instrumental in ensuring the successful execution of clinical trials.

Dileep Kumar was born in the small hamlet of Jehanabad. He earned his B.Pharm from the Manipal College of Pharmaceutical Sciences, Manipal University, and his M.Pharm from S.G.S.I.T.S., Indore, Madhya Pradesh. During his academic and research career, he published numerous research and review articles in peer-reviewed journals. He holds 10 granted patents and has filed over 40 patents, including several in the United States. He has been the recipient of several prestigious fellowships and awards, including a Junior Research Fellowship from the National Medicinal Plant Board (NMPB), New Delhi, a Senior Research Fellowship from the University Grants Commission (UGC), a Teaching Assistantship, and an Institute Post-Doctoral Fellowship from IIT (BHU), Varanasi. Currently, he serves as an Assistant Professor at the Manipal College of Pharmaceutical Sciences, Manipal, Karnataka. He is also an editor for esteemed journals such as Current Topics in Medicinal Chemistry (CTMC), Current Drug Targets, Combinatorial Chemistry, and Molecules. In addition, he has served as a visiting professor in the Department of Entomology and at the Comprehensive Cancer Center, University of California, Davis. In his spare time, Dr. Kumar enjoys Sanskrit poetry, astrology, photography, and horse riding.

Contributors

Rabinarayan Acharya Central Council for Research in Ayurvedic Sciences (Ministry of Ayush, Government of India), New Delhi, India

Raja Chakraborty Institute of Pharmacy, Assam Don Bosco University, Tepesia, Assam, India

Md Sayeed Akhtar Department of Clinical Pharmacy, College of Pharmacy, King Khalid University, Alfara, Abha, Saudi Arabia

Nihal AlMuraikhi Department of Pharmacology, Vision College, Riyadh, Saudi Arabia

S. Amatha Department of Pharmaceutical Regulatory Affairs and Management, Manipal College of Pharmaceutical Sciences, Manipal, Udupi, Karnataka, India

K N Anitha Department of Pharmacology, Government College of Pharmacy, Bangalore, Karnataka, India

Pooja Arora Swami Devi Dyal Institute of Pharmacy, Barwala, Panchkula, Haryana, India

Madhu Katyayani Balijepalli Department of Pharmacology, Faculty of Medicine, Jenjarom, Selangor, Malaysia

Vijay Kumar Bansal Lachoo Memorial College of Science and Technology, Pharmacy Wing, Jodhpur, Rajasthan, India

Partha Pratim Dutta Faculty of Pharmaceutical Science, Assam down town University, Panikhaiti, Guwahati, Assam, India

Pervej Alom Barbhuiya Centre for Research on Ethnomedicine, Faculty of Pharmaceutical Science, Assam down town University, Guwahati, Assam, India

Swati Barche RVSKVV – College of Agriculture Indore, Indore, Madhya Pradesh, India

Arghadeep Bhattachrjee Faculty of Pharmaceutical Science, Assam down town University, Guwahati, Assam, India

Gourab Biswas Department of Pharmaceutical Technology, Brainware University, Kolkata, West Bengal, India

Binita Chakraborty Faculty of Pharmaceutical Science, Assam down town University, Guwahati, Assam, India

Rajeshwari Chandrakar Department of Pharmacy, Institute of Technology and Management, GIDA, Gorakhpur, India

Prashant Chandra Department of Pharmacy Practice, Amrita School of Pharmacy, Amrita Vishwa Vidyapeetham, Kochi, Kerala, India

Ananta Choudhury Faculty of Pharmaceutical Science, Assam down town University, Guwahati, Assam, India

Sejuti Ray Chowdhury Eminent College of Pharmaceutical Technology, Barasat, Jagannathpur, West Bengal, India

Prasanth Damarasingu Department of Pharmacology, School of Pharmacy and Life Sciences, Centurion University of Technology and Management, Balasore, Odisha, India

Prity Das Department of Pharmaceutical Technology, School of Health and Medical Sciences, Adamas University, Kolkata, West Bengal, India

Priyam Jyoti Das Faculty of Pharmaceutical Science, Assam down town University, Guwahati, Assam, India

Rakesh Das Department of Pharmaceutical Technology, School of Health and Medical Sciences, Adamas University, Kolkata, West Bengal, India

Ranjit Das Department of Microbiology, All India Institute of Medical Sciences (AIIMS), Kalyani, Nadia, West Bengal, India

Biswajit Dash Amity Institute of Pharmacy, Amity University Kolkata, Kolkata, West Bengal, India

Gauri Nilesh Deodhar School of Pharmacy, G.H. Raisoni Skilltech University, Nagpur, Maharashtra, India

Lalchand D. Devhare Nagpur College of Pharmacy, Wanadongri, Maharashtra, India

Arup Dey Faculty of Pharmaceutical Science, Assam down town University, Guwahati, Assam, India

Joyjeet Dey Faculty of Pharmaceutical Science, Assam down town University, Guwahati, Assam, India

S Umadevi Department of Pharmaceutics, School of Pharmaceutical Sciences Vels Institute of Science Technology and Advanced Studies, Pallavaram, Chennai, India

G. Dharmamoorthy Department of Pharmaceutical Analysis, MB School of Pharmaceutical Sciences, Mohan Babu University, Tirupati, Andhra Pradesh, India

Manish Dubey Vananchal College of Science (Pharmacy), Babu Dinesh Singh University, Garhwa, Jharkhand, India

Koustav Dutta Department of Pharmaceutical Technology, School of Health and Medical Sciences, Adamas University, Kolkata, West Bengal, India

Sadaf Farooqui Department of Clinical Pharmacy, College of Pharmacy, Prince Sattam Bin Abdulaziz University, Al-Kharj, Saudi Arabia

Hemendra Gautam Faculty of Pharmaceutical Sciences, Future University, Bareilly, UP, India

Manish Kumar Gautam Faculty of Pharmaceutical Science, Assam down town University, Guwahati, Assam, India

Ravi Kumar Gautam Faculty of Pharmaceutical Sciences, Future University, Bareilly, UP, India

Swetlana Gautam Institution of Pharmacy and Paramedical Sciences, Dr Bhimrao Ambedkar University, Agra, Uttar Pradesh, India

Neha Ghimire Regional College of Pharmaceutical Sciences, Kamrup, Assam, India

Anirbita Ghosh Department of Pharmacology, Central Ayurveda Research Institute (CARI), Kolkata, West Bengal, India

Anuradha Gupta School of Materials Science and Technology, Indian Institute of Technology, Banaras Hindu University, Varanasi, Uttar Pradesh, India

Mahesh Kumar Gupta Career Point School of Pharmacy, Career Point University, Kota, Rajasthan, India

Gopinath Halder Department of Chemical Engineering, National Institute of Technology Durgapur, Durgapur, West Bengal, India

Tanuj Hooda Department of Pharmaceutical Chemistry, Delhi Pharmaceutical Science and Research University, Delhi, New Delhi, India

Farid Jahan Faculty of Pharmaceutical Science, Assam down town University, Guwahati, Assam, India

Diptimayee Jena School of Pharmacy and Life Science, Centurion University of Technology & Management, Bhubaneswar, Odisha, India
Faculty of Pharmaceutical Science, Assam down town University, Guwahati, Assam, India

Ashish Jorasia School of Arts and Humanities, Career Point University, Kota, Rajasthan, India

Ritu Kashyap SVKM NMIMS School of Pharmacy, Shirpur, Dhule, Maharashtra, India

Ramchander Khatri ACOE, Delhi Pharmaceutical Sciences and Research University, New Delhi, India

Mrunalini Harish Kulkarni School of Pharmacy, Vishwakarma University, Pune, Maharashtra, India

Angesh Kumar Department of Pharmaceutical Technology, Meerut Institute of Engineering and Technology (MIET), Meerut, Uttar Pradesh, India

Bharat Kumar School of Pharmaceutical Sciences, Bahra University, Shimla Hills, Himachal Pradesh, India

Punam Kumari Faculty of Pharmaceutical Science, Assam down town University, Guwahati, Assam, India
Centre for Research on Ethnomedicine, Faculty of Pharmaceutical Science, Assam down town University, Guwahati, Assam, India
Department of Pharmacy, School of Health Science, Central University of South Bihar, Gaya, Bihar, India

Yamjala Ganesh Kumar Sree Dattha Institute of Pharmacy, Sheriguda, Ibrahimpatnam, Rangareddy, Telangana, India

Amit Lather Department of Pharmacology, Hindu College of Pharmacy, Sonipat, Haryana, India

Virendra S. Ligade Department of Pharmaceutical Regulatory Affairs and Management, Manipal College of Pharmaceutical Sciences, Manipal, Udupi, Karnataka, India

Geeta N. Lodhi Taywade College of Pharmacy, Koradi Nagpur, Maharashtra, India

Puttaswamy Madhusudan Department of Civil and Environmental Engineering, Environmental Materials Laboratory, Hanyang University, Seoul, South Korea

Kit-Kay Mak Department of Pharmaceutical Chemistry, School of Pharmacy, IMU University, Kuala Lumpur, Malaysia

Manikandan Department of Pharmaceutical Analysis, SRM College of Pharmacy, SRM Institute of Science and Technology, Chennai, Tamil Nadu, India

Sreejan Manna Department of Pharmaceutical Technology, Brainware University, Kolkata, West Bengal, India

Sangeeta Mathur School of Arts and Humanities, Career Point University, Kota, Rajasthan, India

Sangita Mazumder Central Ayurveda Research Institute (CARI), Kolkata, West Bengal, India

Kirtimaya Mishra School of Pharmacy and Life Science, Centurion University of Technology & Management, Bhubaneswar, Odisha, India

Manish Kumar Mishra Department of Pharmaceutical Chemistry, SGRR University, Dehradun, Uttarakhand, India

Kaushik Modak Faculty of Pharmaceutical Science, Assam down town University, Guwahati, Assam, India

Debashri Mondal Department of Zoology, Raiganj University, Uttar Dinajpur, Raiganj, West Bengal, India

Sandip Mondal Amity Institute of Pharmacy, Amity University Kolkata, Kolkata, West Bengal, India

Tamal Mondal Department of Botany, Hiralal Mazumdar Memorial College for Women, Dakshineswar, Kolkata, India

Rita Mourya School of Pharmacy, SAM Global University, Bhopal, Madhya Pradesh, India

Namrata A. Muddalwar School of Pharmacy, G.H. Raisoni Skilltech University, Nagpur, Maharashtra, India

T. Naga Aparna G. Pulla Reddy College of Pharmacy, Hyderabad, Telangana, India

Shalini Singh Negi Seth Vishambhar Nath Institute of Engineering and Technology (SVNIET), Barabanki, UP, India

Shivli Nomani SNS College of Pharmacy, Motihari, Bihar, India

Khalid Orayj Department of Clinical Pharmacy, College of Pharmacy, King Khalid University, Alfara, Abha, Saudi Arabia

Budumuru Padmasri Department of Pharmaceutics, Sri Venkateswara College of Pharmacy, Srikakulam, Andhra Pradesh, India

Vikas Pandey Department of Pharmacy, Mangalayatan University-Jabalpur, Jabalpur, Madhya Pradesh, India

Divyang Patel Institute of Pharmaceutical Sciences, Faculty of Pharmacy, Parul University, Vadodara, Gujarat, India

Minesh Kumar Patel Saraswati Institute of Pharmaceutical Sciences, Randesan, Gandhinagar, India

Sejal ben Gunvantbhai Patel Nootan Pharmacy College, Sankalchand Patel University, Mahesana, Gujarat, India

Manash Pratim Pathak Centre for Research on Ethnomedicine, Faculty of Pharmaceutical Science, Assam down town University, Guwahati, Assam, India

Prithu Pathak Sagar Institute of Pharmacy and Technology, Bhopal, MP, India

Sharad D. Pawar Central Ayurveda Research Institute (CARI), Kolkata, West Bengal, India

Mallikarjuna Rao Pichika Newcastle University Medicine Malaysia, Iskandar Puteri, Johor, Malaysia

Subhashree Pramanik Centurion University of Technology and Management, Gopalpur, Balasore, Odisha, India

Sukirti Priya Central University of South Bihar, Fatehpur, Bihar, India

Ramakant Faculty of Pharmaceutical Sciences, Future University, Bareilly, UP, India

Anagha Ranade Central Council for Research in Ayurvedic Sciences (Ministry of Ayush, Government of India), New Delhi, India

Nalanda Tulsiram Rangari KJEIs, Trinity College of Pharmacy, Pune, India

Pratibha Rathee Department of Pharmacology, MM College of Pharmacy Maharishi Markandeshwar (Deemed to be University), Mullana, Haryana, India

Anamika Saha The Neotia University, School of Pharmacy, Diamond Harbour, West Bengal, India

Tiyas Saha Radiant Institute of Pharmaceutical Sciences, Department of Pharmaceutics, Kishanganj, Bihar, India

Prasurjya Saikia Himalayan University, Itanagar, Arunachal Pradesh, India

Faculty of Pharmaceutical Science, Assam down town University, Guwahati, Assam, India

Tanay Samanta Department of Pharmaceutical Technology, School of Health and Medical Sciences, Adamas University, Kolkata, West Bengal, India

Biresh Kumar Sarkar Central Ayurveda Research Institute, CCRAS, Ministry of AYUSH, Kolkata, India

Nidhi Saxena Sagar Institute of Research and Technology Pharmacy (SIRT PHARMACY), Sanjeev Agrawal Global Educational University, Bhopal, MP, India

Saikat Sen Faculty of Pharmaceutical Science, Assam down town University, Guwahati, Assam, India

Rajesh Kumar Sharma Department of Pharmacy, Faculty of Health and Allied Science, KAAF University, Buduburam, Ghana

Vandana Sharma School of Pharmaceutical Sciences, Bahra University, Shimla Hills, Himachal Pradesh, India

Brijesh Shivhare Department of Botany, Faculty of Sciences, Baba Mastnath University, Rohtak, Haryana, India

Abhishek Shrivastava Sagar Institute of Research and Technology Pharmacy (SIRT PHARMACY), Sanjeev Agrawal Global Educational University, Bhopal, MP, India

Rajesh Shukla Department of Pharmacy, Mangalayatan University-Jabalpur, Jabalpur, Madhya Pradesh, India

Sangita H. Shukla Indubhai Patel College of Pharmacy and Research Centre, Anand, Gujarat, India

Balwinder Singh Department of Pharmaceutical Sciences, Amritsar Group of Colleges, Amritsar, Punjab, India

Satyendra Singh BPS Educational Institution of Pharmacy, Etah, UP, India

Shweta Singh Department of Pharmacy, Madhyanchal Professional University, Bhopal, MP, India

Virendra Singh Pharmaceutical Chemistry, Sainik Pharmacy College, Prayagraj, Uttar Pradesh, India

Priyanka Sonker Department of Pharmacy, Institute of Technology and Management, GIDA, Gorakhpur, India

Fauzia Tabassum Department of Pharmacology, Vision College, Riyadh, Saudi Arabia

Mohd Tabish Department of Pharmacy, Institute of Biomedical Education and Research, Mangalayatan University, Aligarh, Uttar Pradesh, India

Durga Prasad Thammisetty Department of Pharmacy Practice, Sri Padmavathi School of Pharmacy (Autonomous), Tirupati, Andhra Pradesh, India

Gayatri Thapa Faculty of Pharmaceutical Science, Assam down town University, Guwahati, Assam, India

Himalayan Pharmacy Institute, Majitar, Sikkim, India

Nirbhay Narayan Tiwari Department of Pharmacy, Tahira Institute of Medical Sciences, GIDA, Gorakhpur, UP, India

Swarnajeet Tripathy School of Pharmacy and Life Sciences, Centurion University of Technology and Management, Balangir, Odisha, India

Prabhat Kumar Upadhyay Institute of Pharmaceutical Research, GLA University, Mathura, Uttar Pradesh, India

Amit Verma School of Pharmacy, Chandigarh University, Lucknow, Uttar Pradesh, India

Priyanka Verma Kamla Raja Girls Government Post Graduate College, Gwalior, Madhya Pradesh, India

Vandita Vishwakarma Department of Pharmacy, Institute of Technology and Management, GIDA, Gorakhpur, India

Vivek Metro College of Health Sciences and Research, Greater Noida, India

Ajit Kumar Yadav R K Institute of Pharmacy, Bareilly, UP, India

Amarjeet Yadav Department of Pharmacy, Guru Ghasidas Vishwavidyalaya, Bilaspur, Chhattisgarh, India

Manoj Kumar Yadav Institute of Pharmacy & Paramedical Sciences, Dr Bhimrao Ambedkar University, Agra, UP, India

Saroj Yadav School of Medical and Allied Sciences, K.R. Mangalam University, Sohna, Gurugram, Haryana, India

Part I

Introduction to Diabetic Neuropathy

1 Diabetic Neuropathy and Oxidative Stress: Molecular Pathways and Emerging Interventions

Diptimayee Jena, Kirtimaya Mishra, Arup Dey, Binita Chakraborty, and Manish Kumar Gautam

1.1 Introduction

A collection of clinical symptoms that develop when the autonomic and peripheral nerve systems are compromised are the most common complications of diabetes. Neuropathy is a term for symptoms that up to 50% of persons with diabetes exhibit. They are caused by injury to the neurological system that is either localised or widespread (Ang et al. 2014; Gautam et al. 2015). High blood sugar that lasts a long time might damage any part of the nervous system. People with type 2 diabetes are more likely to have hyperglycemia that damages the autonomic, motor, and sensory neurones in their peripheral nervous system. If you have persistent hyperglycemia, it could hurt any part of your neurological system. When a person's blood sugar levels are excessively high, their peripheral nervous system motor, autonomic, and sensory neurones can have issues (Duby et al. 2004).

There have been a lot of new ways to classify diabetic neuropathy in the last few years. But everyone doesn't agree on one classification. A classification based on how the patient looks is the most useful in real life. Most of the groups that have been suggested are based on the aetiology, the condition, or the place (Thomas 2003). There are numerous ways to organise diabetic neuropathies, but in real life, it's better to classify them by how they affect people. Autonomic neuropathy and

D. Jena · K. Mishra
School of Pharmacy and Life Science, Centurion University of Technology & Management, Bhubaneswar, Odisha, India

A. Dey · B. Chakraborty · M. K. Gautam (✉)
Faculty of Pharmaceutical Science, Assam down town University, Guwahati, Assam, India
e-mail: manish.g@adtu.in

M. K. Gautam et al. (eds.), *Antioxidants and Functional Foods for Diabetic Neuropathy*, https://doi.org/10.1007/978-981-95-9326-2_1

Table 1.1 Clinical presentations of diabetic neuropathies

Focal neuropathies	Generalized neuropathies	Multifocal neuropathies
Peripheral nerve neuropathies (median, femoral and peroneal nerve) Other mononeuropathies (Phrenic mononeuropathy) Entrapment neuropathies Cranial neuropathies (Cranial nerves III,VII, VI, IV) Mononeuropathy multiplex	Autonomic neuropathy(cardiovascular, gastrointestinal, urogenital, sudomotor) Distal symmetric polyneuropathy Diabetic neuropathiccachexia Acute sensory neuropathy	Diabetic radiculoplexus neuropathies Lumbosacral radiculoplexus neuropathy Thoracic radiculoneuropathy Cervical radiculoplexus neuropathy

distal symmetric polyneuropathy are the two most common kinds. High blood sugar, advanced age, long-term diabetes, and heart disease warning signs including being obese, having hypertension, or having high cholesterol can all induce diabetic neuropathy (Table 1.1) (Dilek 2022).

1.2 Epidemiology

The study shows that between 10 and 20 percent of people who have diabetes also have DPN. People who have had diabetes for a long time are more likely to have DPN, according to research. 41% of people with diabetes get neuropathy after 10 years, and 26% get peripheral neuropathy after 5 years. Previous studies have shown that between 50% and 66% of people with diabetes are likely to develop DPN (Dyck et al. 1993). Diabetic peripheral neuropathy can happen to people with either type 1 or type 2 diabetes. However, type 2 diabetes is more common because it lasts longer and has more problems (Baum et al. 2021). Charcot neuroarthropathy is mostly caused by diabetes mellitus, which accounts for 0.1% to 0.4% of cases and up to 29% of cases in people with peripheral nerve damage (Hjelm 2022).

There are signs of sense problems on one side of almost half of the people with DPN who go to a centre. Diabetes is more likely to strike you if you are healthier or have a family history of the disease. A lot of the time, people with diabetes mellitus get sick because of two main things: autonomic neuropathies and peripheral neuropathies. The risk of death within 5 years is 2.5 times higher for diabetics with foot ulcers than for those without the condition (Table 1.2) (Gautam et al. 2012, 2014; Murthy et al. 2013).

Table 1.2 Risk factors associate with DPN

Unmodifiable risk factors	Modifiable risk factors
Advance in age. Height Diabetes duration	Poor glucose control; HbA1C Abdominal obesity Smoking Hypertension Obesity (Weight, BMI) Dyslipidemia (High LDL) Heavy alcohol intake

1.3 Pathophysiology

Excessive elevated level of sugar in the blood can affects the endoplasmic reticulum, cause dyslipidaemia, damage mitochondria, and make insulin less effective. These things make inflammation, reactive oxygen species (ROS), and cell damage worse. The cytokines and chemokines that invading macrophages release into peripheral nerve cells make inflammation and nerve fibre damage worse (Pop-Busui et al. 2017). Advanced glycation end products, problems with the protein kinase C and polymerase pathways, and hexosamine have all been associated to peripheral nerve damage and endothelial dysfunction (Gautam et al. 2014).

1.4 Toxicokinetics

Adequate restorative physiological processes and metabolic processes that govern blood sugar control usually tolerate temporary hyperglycemia and high serum glucose level can also lead the chances of neuropathy. For those with type 1 diabetes who are identified early, strict glucose management can reduce the incidence of diabetic peripheral neuropathy by 78% (Elafros et al. 2022). Strict glucose management, on the other hand, often only reduces the risk by 5% – 9% when long-standing hyperglycemia or type 2 diabetes was discovered later (Giri et al. 2018).

1.5 Clinical Manifestation

These include profound discomfort, scorching, intense stabbing feelings, paraesthesia, hyperaesthesia, and pain. Furthermore, if the neuropathy is not identified and treated right away, patients may develop unfavourable symptoms including numbness in their legs and feet, which could result in painless foot ulcers and eventual amputations.

1.6 Cellular Oxidative Damage

An imbalance between oxidants and antioxidants causes oxidative stress, which damages biomolecules and interferes with redox signalling (Sies and Jones 2007; Sies et al. 2017). A stable redox state is crucial for metabolic regulation, while disturbances trigger stress responses. Reactive species such as ROS, RNS, RSS, RHS, and RES play key roles in redox control by modifying proteins, lipids, carbohydrates, and nucleic acids. These processes, integrated with regulatory mechanisms like phosphorylation, form a complex network essential for maintaining cellular homeostasis (Sies and Jones 2020; Möller et al. 2019; Nakamura and Lipton 2020; Olson 2020; Parvez et al. 2018; Davies and Hawkins 2020; Santolini et al. 2019).

1.7 Mechanism of Oxidative Stress

In cellular signalling, ROS and reactive nitrogen species (RNS) are important secondary messengers, but when left unchecked, they can be dangerous (Burton and Jauniaux 2011; Liu et al. 2020a, b). When the body antioxidant defenses are overpowered by ROS generation then generation of oxidative stress. These extremely reactive species can harm essential biomolecules including DNA, proteins, lipids, and carbohydrates. They might be free radicals such as superoxide and nitric oxide or non-free radicals like hydrogen peroxide and peroxynitrite (Halliwell 1994; Mohammad and Charitra 2018). One such reactive species is nitric oxide (NO), a lipid-soluble gas generated by epithelial cells (Dröge 2002).

1.8 Link among Oxidative Stress and Diabetes

According to Pham-Huy et al. (2008), oxidative stress causing major contributor to generate vascular complications in type 2 diabetes. Diabetic individuals often exhibit elevated ROS levels due to increased production or impaired breakdown of antioxidants molecules such as superoxide dismutase (SOD), glutathione peroxidase (GSH-Px), and catalase (Lipinski 2001) and theses imbalance leads to oxidative stress, damaging tissues and contributing to diabetic complications. Free radicals generated by glucose oxidation, lipid peroxidation, and non-enzymatic protein glycation further disrupt cellular machinery and worsen insulin resistance (Maritim et al. 2003). Moussa (2008) reported that in diabetics, oxidative modification of low density lipoproteins especially via hydroxyl radicals that leads to cross-linking of apo-B monomers, enhancing vascular damage. Oxygen used during metabolism can form oxygen-free radicals such as O_2^-, which then convert into OH, $ONOO^-$, and H_2O_2 (Erejuwa 2012). While ROS/RNS support insulin's physiological functions, they can also impair insulin signalling, contributing to insulin resistance—a key factor in type 2 diabetes (Ayepola et al. 2014).

1.9 Impact of Oxidative Stress Upon Diabetes Complications

According to clinical data, endothelial function in diabetic vasculature is compromised by oxidative stress brought on by hyperglycemia (Giugliano et al. 1995). Furthermore, people with diabetes who have dyslipidaemia, hyperglycemia, or hyperinsulinemia are more likely to develop macroangiopathies, which exacerbate atherosclerosis and oxidative stress (Zhu et al. 2024). Free radicals damage proteins, lipids, and DNA, increasing the risk of advanced diabetic complications. Oxidative stress is also associated with conditions such as diabetes, cancer, and rheumatoid arthritis (Phillips et al. 2004). In diabetes mellitus, elevated oxidative stress results in retinopathy, nephropathy, neuropathy, heart disease, and stroke (Asfandiyarova et al. 2007; Ceriello 2006).

1.10 Impact of Oxidative Stress in Diabetic Neuropathy

1.10.1 Diabetic Neuropathy: Its Pathophysiology

Disrupted insulin signalling, lipid metabolism, and chronic hyperglycemia are key contributors to diabetic neuropathy, though its exact causes remain unclear. Glial cells, neurovascular cells, perikaryons, and myelinated/unmyelinated axons are among the peripheral nervous system's anatomical and functional alterations brought on by these metabolic disorders. Axonal repair may be hampered by impaired insulin signalling, which can cause injured cells to undergo apoptosis (Lin et al. 2020). Dysfunction in peripheral motor-sensory neurones and glial cells is the main cause of DPN symptoms. Axons, Schwann cells (SCs), and cytoplasmic processes—essential parts of glial cells—make up peripheral nerves (Previtali 2021). Glial cells help to promote fast, saltatory conduction and insulate neurones (Hoke et al. 2021; Bosch-Queralt et al. 2023). Glial cells envelop myelinated axons, whereas Remak bundles are formed by smaller, unmyelinated axons (Salzer 2015). Through mechanical support and biochemical signalling, the polarised structure formed by the axon and myelin promotes axonal development (Monje 2018; Du et al. 2019). Hyperglycemia and hyperlipidaemia alter autophagy, metabolism, and mitochondrial integrity, impairing Schwann cell regulatory functions (Liu et al. 2020a, b). Due to the lack of protective glial support, sensory neurons—particularly those in the dorsal root ganglion—are especially vulnerable to oxidative stress, inflammation, and metabolic disruptions (Cernea and Raz 2021; Shen 2022; McDonald and Windebank 2002). Deranged metabolic pathways, oxidative stress, and mitochondrial dysfunction are the main causes of DPN development. Microvascular problems, insulin resistance, dyslipidaemia, and hyperglycemia are the main pathogenic causes. Due to inflammation, oxidative damage, and compromised cellular signalling, hyperglycemia and dyslipidaemia cause extensive neuronal damage by activating harmful pathways such PKC, polyol, AGE, hexosamine, and PARP.

1.11 Role of ROS in Nerve Damage

The loss of antioxidant defences, the breakdown of bioenergetics, the opening of ion channels, the rupture of microtubules, demyelination, neuroinflammation, the impairment of mitophagy, and the death of neurones through apoptosis are just a few of the ways that oxidative stress can lead to neurodegeneration (Salvemini et al. 2011; Ta et al. 2013; Ganesh et al. 2013). Medications that alter the Nrf2-NF-κB axis can help neurones maintain their redox balance. For this reason, scientists have assessed the effectiveness of various modulators in animal models of peripheral neuropathy (Fig. 1.1) (Chandrasekaran et al. 2015).

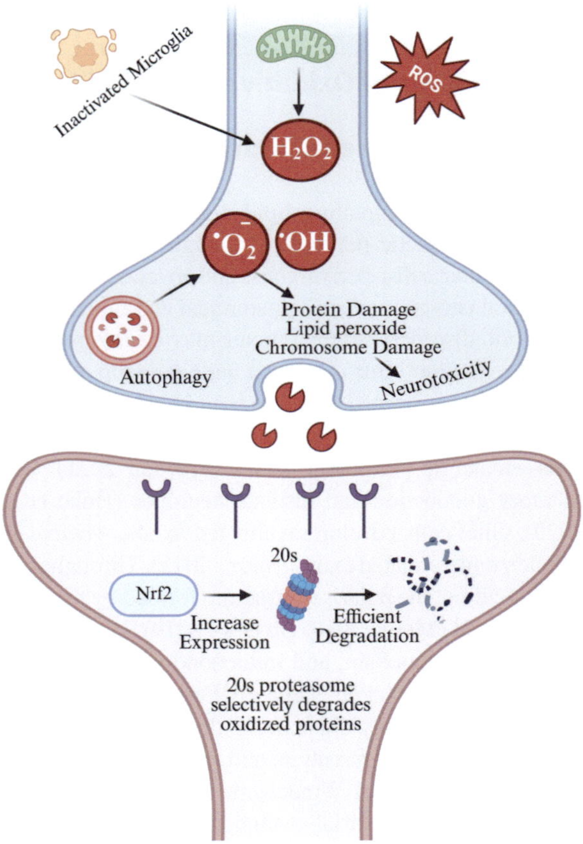

Fig. 1.1 Role of oxidative stress in neurotoxicity

1.12 Biomarkers Related to Oxidative Stress in Diabetic Neuropathy

In diabetic neuropathy (DN), oxidative stress indicators include

- *Lipid peroxidation products:* These include HNE; hydroxynonenal, TBARS; thiobarbituric acid reactive substances, MDA; malondialdehyde, and 8-iso-PGF2α. MDA indicates antioxidant status and oxidative stress.
- *ROS & RNS levels*: These levels imply oxidative stress.
- *Lipid & Protein modification*: The development of DN is aided by these changes.
- *Antioxidant enzyme activities*: These actions aid in the advancement of DN.
- *Pro-inflammatory cytokine analysis*: These cytokines aid in the development of DN.

Other biomarkers of DN include:

- *Genetic predisposition:* The term "genetic predisposition" describes inherited DNA changes that make a person more vulnerable to specific illnesses. Certain genetic variations may affect nerve growth, function, and the body's reaction to elevated blood sugar levels in diabetic neuropathy. Important genetic markers, including VEGF (Vascular Endothelial Growth Factor), NGF (Nerve Growth Factor), eNOS (Endothelial Nitric Oxide Synthase), TNF-α (Tumour Necrosis Factor-alpha), TCF7L2 (Transcription Factor 7-Like 2), IRS1 and IRS2 (Insulin Receptor Substrates), SOD2 (Superoxide Dismutase 2), and BDNF (Brain-Derived Neurotrophic Factor) provide information about disease mechanisms and help develop targeted treatments.
- *Mitochondrial SOD2:* Research suggests that recurring oxidative injury to Schwann cells and spinal axons might be the trigger of diabetic neuropathy (Chowdhury et al. 2011). Dorsal root ganglion (DRG) neurones in diabetic animals have lower mitochondrial DNA (mtDNA) levels, decreased respiration, and decreased mitochondrial membrane potential, particularly in early neuropathy (Vincent et al. 2004). Axons, Schwann cells, and DRG neurones all have mitochondrial dysfunction, which is thought to be a major cause of the illness. However, elevated blood sugar leads to an instability in the mitochondrial membrane potential, which increases the production of ROS and oxidative stress, which in turn leads to neuronal damage. By limiting the production of excessive ROS, proteins that alter the inner membrane potential may be able to mitigate these effects (Fig. 1.2) (Maniei et al. 2019).
- *CD31:* A cluster of differentiation proteins important in immunological control and cell adhesion is CD31, often referred to as Platelet Endothelial Cell Adhesion Molecule-1 (PECAM-1). It is expressed on a variety of cell types, including

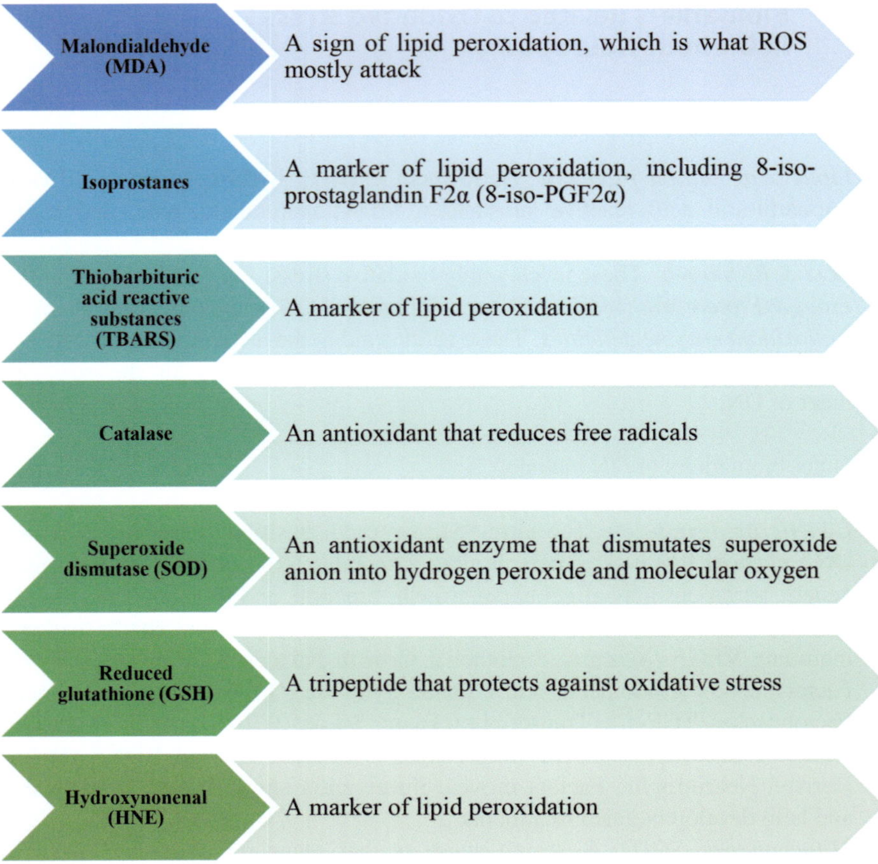

Fig. 1.2 Biomarkers of oxidative stress, lipid peroxidation and antioxidant defenses indicators in DN

endothelial cells, monocytes, platelets, neutrophils, macrophages, and some lymphocytes, and it is essential for the removal of senescent neutrophils (Kuang et al. 2013; Wilkinson et al. 2002). Leukocyte transmigration across the endothelium is facilitated by CD31, which also supports vascular integrity and immune surveillance by interacting with either different proteins (CD31-X) or identical molecules (CD31-CD31) on neighbouring cells (Laddha and Kulkarni 2020).

Transient receptor potential vanilloid type 1 (TRPV1): The primary mediators of pain and heat perception are TRPV1. Intraepidermal TRPV1 nerve fibre expression is frequently decreased or absent in painful neuropathic diseases, indicating that these sensory fibres may not be the source of pain (Stino and Smith 2017; Feldman-Billard et al. 2018). Tyrosine kinase, phosphatidylinositol, and protons are among the ligands that activate TRPV1, especially in peripheral and nociceptive neurones. Allodynia and hyperalgesia are important clinical characteristics of diabetic peripheral neuropathy (Sharopov et al. 2018; Shaqura et al. 2014). TRPV1 activation in peripheral nerves is primarily responsible for chronic pain in diabetes individuals (Hong et al. 2008). Increased TRPV1 expression may suppress peripheral sensory neurone activity through M-opioid receptor activation and improve reflex arc responses in diabetics. The dorsal root ganglion's restricted activity is partly caused by TRPV1 malfunction. As noted by Shaqura et al. (2014) and Lin et al. (2023), TRPV1 is not present in human embryonic kidney cells until it is stimulated by elevated PKA and PKC-mediated phosphorylation. One metabolic process that might cause oxidative damage in diabetic neuropathy is the polyol pathway. Glutathione levels may drop, F and reactive oxygen radicals may rise as a result of the polyol pathway being triggered by too much glucose in the body.

1.13 Types of Biomarkers

1.14 Therapeutic Strategies Targeting Oxidative Stress

Antioxidant supplements and exercise have been recommended as important treatments for people with diabetes through ROS-mediated mechanisms. Also, certain new ways to distribute drugs can make DPN work better and make antioxidants more available to the body (Pang et al. 2020).

1.15 Targeting Oxidative Stress

To address nerve dysfunction in diabetics, a variety of antioxidative stress-related tactics have been used, such as targeting mitochondria, specific oxidative stress pathways, or ROS directly (Negre-Salvayre et al. 2008).

1.16 ROS Targeting

Studies suggest that α-lipoic acid, taurine, melatonin, acetyl L-carnitine, N-acetylcysteine, and vitamins A, C, and E may help delay diabetic neuropathy. Among them, ALA shows promising symptom relief, though more research is needed on the role of vitamins (Feldan et al. 2019; Papanas and Ziegler 2014). According to Bertolotto and Massone (2012), ALA prevents oxidative stress by obstructing the hexosamine and AGE pathways. ALA is soluble in fat and water. When people with diabetic neuropathy took ALA and superoxide dismutase combined, their electroneurographic signals and symptoms improved (Shakher and Stevens 2011). A daily dose of 600 mg of ALA is permitted in Germany to treat painful diabetic neuropathy (Yagihashi et al. 2001).

1.16.1 Focussing on Distinct Oxidative Stress Pathways

This includes drugs that stop PKC, aldose reductase, and ageing. Aldose reductase inhibitors and anti-AGE drugs worked better than ruboxistaurin, a PKCβ inhibitor that didn't have a big effect in the clinic. Aldose reductase inhibitors slow down the metabolic conversion of glucose by reducing its admission into the polyol pathway. In diabetic people and mice, stopping aldose reductase has been shown to speed up the conduction of sensory and motor nerves in the sural nerve, delay the F-wave in the wrist and ankle, and ease neuropathic pain in diabetic neuropathy (Kawai et al. 2010; Morrison et al. 2010). But further research is needed to find out how well it works.

1.17 Pharmacological Interventions

In individuals with diabetic peripheral neuropathy (DPN), it has been demonstrated that a healthy diet and frequent exercise, such as strength and balance training, improve adherence, reduce discomfort, and encourage skin re-innervation (Smith et al. 2006; Bertelsmann et al. 1987). Nerve conduction, vibration sensitivity, and peripheral small fibre function are all improved by lowering HbA1C levels (Callaghan et al. 2012; Handelsman et al. 2011). An HbA1C target of less than 7% is advised for the majority of DPN patients; however, physicians must balance this against concerns including hypoglycemia and a shortened life expectancy (Martin et al. 2006; Albers et al. 2010). Extensive glycaemic control in type 1 diabetes can avoid DPN, and research supports the idea of "metabolic memory," which holds that glycaemic control in the past affects the risk of neuropathy in the future. The BARI 2D experiment shown that thiazolidinediones and insulin sensitisers such as metformin dramatically decreased the incidence of DPN, and intensive therapy improves tactile sensitivity in type 2 diabetes (Nathan et al. 2005).

1.18 Future Directions

The development and testing of novel antioxidant drugs that target the oxidative stress pathways linked to diabetic neuropathy should be the main focus of future research. It may be possible to prevent or repair nerve injury by discovering substances that may efficiently reduce the generation of reactive oxygen species (ROS) or strengthen the body's built-in antioxidant defences. Researchers ought to investigate the ways in which antioxidants interact with other forms of treatment, including blood sugar regulation, anti-inflammatory medications, and neuroprotective medications. Because it addresses multiple causes of nerve loss, this combination therapy approach may be a more comprehensive approach to treating diabetic neuropathy.

Early detection and treatment of diabetic neuropathy would be greatly facilitated by the discovery of trustworthy biomarkers for oxidative stress. Finding non-invasive markers that can monitor the effectiveness of antioxidant therapy and the extent of oxidative damage should be the main objective of future studies. People may react differently to oxidative stress depending on their genetic makeup. Future research should employ pharmacogenomic techniques to identify genetic markers that increase the risk of oxidative stress in diabetics. This could result in customised treatment regimens that optimise the effectiveness of antioxidant therapy for every patient.

1.19 Conclusion

The development and progression of diabetic neuropathy are significantly influenced by oxidative stress. The equilibrium between oxidative damage and antioxidant defence is upset by elevated blood glucose levels, which results in an overabundance of reactive oxygen species (ROS). In addition to causing inflammation and nerve dysfunction, these ROS harm neurones and blood vessels. This cascade exacerbates neuropathy and decreases nerve healing. Treatments based on antioxidants and better glycaemic management may help stop or slow nerve degeneration. Research on oxidative pathways must continue to create efficient treatments that can enhance results and make managing diabetes easier.

References

Albers JW, Herman WH, Pop-Busui R, et al. Effect of prior intensive insulin treatment during the Diabetes Control and Complications Trial (DCCT) on peripheral neuropathy in type 1 diabetes during the Epidemiology of Diabetes Interventions and Complications (EDIC) study. Diabetes Care. 2010;33:1090–6.

Ang L, Jaiswal M, Martin C, Pop-Busui R. Glucose control and diabetic neuropathy: lessons from recent large clinical trials. Curr Diab Rep. 2014;14:528.

Asfandiyarova N, Kolcheva N, Ryazantsev I, Ryazantsev V. Risk factors for stroke in type 2 diabetes mellitus. Diab Vasc Dis Res. 2007;3:57–60.

Ayepola OR, Brooks NL, Oguntibeju OO. Oxidative stress and diabetic complications: the role of antioxidant vitamins and flavonoids. Antioxidant-antidiabetic agents and human health. 2014. Available from: https://doi.org/10.5772/57029.

Baum P, Toyka KV, Blüher M, Kosacka J, Nowicki M. Inflammatory mechanisms in the pathophysiology of diabetic peripheral neuropathy (DN)-new aspects. Int J Mol Sci. 2021;22(19):10835.

Bertelsmann FW, Heimans JJ, Van RJC, et al. Peripheral nerve function in patients with painful diabetic neuropathy treated with continuous subcutaneous insulin infusion. J Neurosurg Psychiatry. 1987;50:1337–41.

Bertolotto F, Massone A. Combination of alpha lipoic acid and superoxide dismutase leads to physiological and symptomatic improvements in diabetic neuropathy. Drugs R D. 2012;12(1):29–34.

Bosch-Queralt M, Fledrich R, Stassart RM. Schwann cell functions in peripheral nerve development and repair. Neurobiol Dis. 2023;176:105952.

Burton GJ, Jauniaux E. Oxidative stress. Best Pract Res Clin Obstet Gynaecol. 2011;25(3):287–99.

Callaghan BC, Little AA, Feldman EL, et al. Enhanced glucose control for preventing and treating diabetic neuropathy. Cochrane Database Syst Rev. 2012;6:CD007543.

Ceriello A. Oxidative stress and diabetes-associated complications. Endocr Pract. 2006;12(l):60–2.

Cernea S, Raz I. Management of diabetic neuropathy. Metab Clin Exp. 2021;123:154867.

Chandrasekaran K, Anjaneyulu M, Inoue T, Choi J, Sagi AR, Chen C, et al. Mitochondrial transcription factor a regulation of mitochondrial degeneration in experimental diabetic neuropathy. Am J Physiol Endocrinol Metab. 2015;309(2):E132–41.

Chowdhury SK, Dobrowsky RT, Fernyhough P. Nutrient excess and altered mitochondrial proteome and function contribute to neurodegeneration in diabetes. Mitochondrion. 2011;11(6):845–54.

Davies MJ, Hawkins CL. The role of myeloperoxidase in biomolecule modification, chronic inflammation, and disease. Antioxid Redox Signal. 2020;32:957–81.

Dilek GY. Classification, risk factors, and clinical presentation diabetic neuropathy. Diabetic Neuropathy. 2022:1–9.

Dröge W. Free radicals in the physiological control of cell function. Physiol Rev. 2002;82:47–95.

Du W, Wang N, Li F, Jia K, An J, Liu Y, et al. STAT3 phosphorylation mediates high glucose-impaired cell autophagy in an HDAC1-dependent and independent manner in Schwann cells of diabetic peripheral neuropathy. FASEB J. 2019;33(7):8008–21.

Duby JJ, Campbell RK, Stephen MS, John RW, Kristin AR. Diabetic neuropathy: an intensive review. Am J Health Syst Pharm. 2004;61(2):160–73.

Dyck PJ, Kratz KM, Karnes JL, Litchy WJ, Klein R, Pach JM, Wilson DM, O'Brien PC, Melton LJ, Service FJ. The prevalence by staged severity of various types of diabetic neuropathy, retinopathy, and nephropathy in a population-based cohort: the Rochester Diabetic Neuropathy Study. Neurology. 1993;43(4):817–24.

Elafros MA, Andersen H, Bennett DL, Savelieff MG, Viswanathan V, Callaghan BC, Feldman EL. Towards prevention of diabetic peripheral neuropathy: clinical presentation, pathogenesis, and new treatments. Lancet Neurol. 2022;21(10):922–36.

Erejuwa OO. Oxidative stress in diabetes mellitus: is there a role for hypoglycemic drugs and/or antioxidants. In: Oxidative stress and disease; 2012. p. 217–46.

Feldan EL, Callaghan BC, Pop-Busui R, et al. Diabetic neuropathy. Nat Rev Dis Primers. 2019;5(41)

Feldman-Billard S, Larger É, Massin P. Early worsening of diabetic retinopathy after rapid improvement of blood glucose control in patients with diabetes. Diabetes Metab. 2018;44(1):4–14.

Ganesh YV, Negi G, Sharma SS, Kumar A. Potential therapeutic effects of the simultaneous targeting of the Nrf2 and NF-kB pathways in diabetic neuropathy. Redox Biol. 2013;1:394–7.

Gautam MK, Singh A, Rao CV, Goel RK. Toxicological evaluation of Murraya paniculata (L.) leaves extract on rodents. Am J Pharmacol Toxicol. 2012;7(2):62–7.

Gautam MK, Purohit V, Agarwal M, Singh A, Goel RK. In-vivo healing potential of Aegle marmelos in excision, incision, and dead space wound models. Sci World J. 2014;2014:740107.

Gautam MK, Gangwar M, Singh SK, Goel RK. Effects of *Azadirachta indica* on vascular endothelial growth factor and cytokines in diabetic deep wound. Planta Med. 2015;81(9):713–21.

Giri B, Dey S, Das T, Sarkar M, Banerjee J, Dash SK. Chronic hyperglycemia mediated physiological alteration and metabolic distortion leads to organ dysfunction, infection, cancer progression and other pathophysiological consequences: an update on glucose toxicity. Biomed Pharmacother. 2018;107:306–28.

Giugliano D, Ceriello A, Paolisso G. Diabetes mellitus, hypertension, and cardiovascular disease: which role for oxidative stress? Metabolism. 1995;44(3):363–8.

Halliwell B. Free radicals, antioxidants, and human disease: curiosity, cause, or consequence? Lancet. 1994;344:721–4.

Handelsman Y, Mechanick JI, Blonde L, et al. American Association of Clinical Endocrinologists Medical Guidelines for clinical practice for developing a diabetes mellitus comprehensive care plan. Endocr Pract. 2011;17(suppl 2):1–53.

Hjelm LR. Diabetes mellitus: An overview in relationship to Charcot Neuroarthropathy. Clin Podiatr Med Surg. 2022;39(4):535–42.

Hoke A, Cerri F, Fisgin A, Riva N, Quattrini A. Normal structure and pathological features in peripheral neuropathies. J Peripher Nerv Syst. 2021;26(Suppl 2):s11–20.

Hong S, Agresta L, Guo C, Wiley JW. The TRPV1 receptor is associated with preferential stress in large dorsal root ganglion neurons in early diabetic sensory neuropathy. J Neurochem. 2008;105(4):1212–22.

Kawai I, Takei M, et al. Effects of epalrestat, analdosereductase inhibitor, on diabetic peripheral neuropathyin patients with type 2 diabetes, in relation to suppression of Nε-carboxymethyl lysine. J Diabetes Complicat. 2010;24(6):424–32.

Kuang B, Wen X, Ding Y, Peng R, Cai P, Zhang M, et al. The prognostic value of platelet endothelial cell adhesion molecule-1 in non-small-cell lung cancer patients. Med Oncol. 2013;30(2):536.
Laddha AP, Kulkarni YA. NADPH oxidase: a membrane-bound enzyme and its inhibitors in diabetic complications. Eur J Pharmacol. 2020;881:173206.
Lin T, Gargya A, Singh H, Sivanesan E, Gulati A. Mechanism of peripheral nerve stimulation in chronic pain. Pain Med. 2020;21(Suppl 1):s6–s12.
Lin Q, Li K, Chen Y, Xie J, Wu C, Cui C, Deng B. Oxidative stress in diabetic peripheral neuropathy: pathway and mechanism-based treatment. Mol Neurobiol. 2023;60(8):4574–94.
Lipinski B. Pathophysiology of oxidative stress in diabetes mellitus. J Diabetes Complic. 2001;15(4):203–10.
Liu L, Cui H, Xu Y. Quantitative estimation of oxidative stress in cancer tissue cells through gene expression data analyses. Front Genet. 2020a;11:494.
Liu YP, Shao SJ, Guo HD. Schwann cells apoptosis is induced by high glucose in diabetic peripheral neuropathy. Life Sci. 2020b;248:117459.
Maniei M, Aghababaeian H, Karimi H, Firozzadeh K, Amirgholami N, Moghaddam AS, et al. The use of Aloe Vera after infrared therapy in the treatment of a diabetic foot ulcer: a case report. J Biochem Technol. 2019;10(2):43–9.
Maritim AC, Sanders RA, Watkins JB. Diabetes, oxidative stress, and antioxidants: a review. J Biochem Mol Toxicol. 2003;17(1):24–38.
Martin CL, Albers J, Herman WH, et al. Neuropathy among the diabetes control and complications trial cohort 8 years after trial completion. Diabetes Care. 2006;20:340–4.
McDonald ES, Windebank AJ. Cisplatin-induced apoptosis of DRG neurons involves bax redistribution and cytochrome c release but not fas receptor signaling. Neurobiol Dis. 2002;9:220–33.
Mohammad JM, Charitra MR. A gentle introduction to gasotransmitters with special reference to nitric oxide: biological and chemical implications. Rev Inorg Chem. 2018;38:193–220.
Möller MN, Rios N, Trujillo M, Radi R, Denicola A, Alvarez B. Detection and quantification of nitric oxide-derived oxidants in biological systems. J Biol Chem. 2019;294:14776–802.
Monje M. Myelin plasticity and nervous system function. Annu Rev Neurosci. 2018;41:61–76.
Morrison S, Colberg SR, Mariano M. Balance training reduces falls risk in older individuals with type 2 diabetes. Diabetes Care. 2010;33:748–50.
Moussa SA. Oxidative stress in diabetes mellitus. Rom J Biophys. 2008;18(3):225–36.
Murthy S, Gautam MK, Goel S, Purohit V, Sharma H, Goel RK. Evaluation of in-vivo wound healing activity of Bacopa monniera on different wound models in rats. BioMed Res Int. 2013;2013:972028.
Nakamura T, Lipton SA. Nitric oxide-dependent protein post-translational modifications impair mitochondrial function and metabolism to contribute to neurodegenerative diseases. Antioxid Redox Signal. 2020;32:817–33.
Nathan DM, Cleary PA, Backlund JC, et al. Intensive diabetes treatment and cardiovascular disease in patients with type 1 diabetes. N Engl J Med. 2005;535:2643–53.
Negre-Salvayre A, Coatrieux C, Ingueneau C, Salvayre R. Advanced lipid peroxidation end products in oxidative damage to proteins. Potential role in diseases and therapeutic prospects for the inhibitors. Br J Pharmacol. 2008;153(1):6–20.
Olson KR. Reactive oxygen species or reactive sulfur species: why we should consider the latter. J Exp Biol. 2020;223:jeb196352.
Pang L, Lian X, Liu H, Zhang Y, Li Q, Cai Y, Ma H, Yu X. Understanding diabetic neuropathy: focus on oxidative stress. Oxid Med Cell Longev. 2020;2020:9524635.
Papanas N, Ziegler D. Efficacy of α-lipoic acid in diabetic neuropathy. Expert Opin Pharmacother. 2014;15(18):2721–31.
Parvez S, Long MJC, Poganik JR, Aye Y. Redox signaling by reactive electrophiles and oxidants. Chem Rev. 2018;118:8798–888.
Pham-Huy LA, He H, Pham-Huy C. Free radicals, antioxidants in disease and health. Int J Biol Sci. 2008;4(2):89–96.
Phillips M, Cataneo RN, Cheema T, Greenberg J. Increased breath biomarkers of oxidative stress in diabetes mellitus. Clin Chim Acta. 2004;344(1–2):189–94.

Pop-Busui R, Boulton AJ, Feldman EL, Bril V, Freeman R, Malik RA, Sosenko JM, Ziegler D. Diabetic neuropathy: a position statement by the American Diabetes Association. Diabetes Care. 2017;40(1):136–54.

Previtali SC. Peripheral nerve development and the pathogenesis of peripheral neuropathy: the sorting point. Neurotherapeutics. 2021;18(4):2156–68.

Salvemini D, Little JW, Doyle T, Neumann WL. Roles of reactive oxygen and nitrogen species in pain. Free Radic Biol Med. 2011;51:951–66.

Salzer JL. Schwann cell myelination. Cold Spring Harb Perspect Biol. 2015;7(8):a020529.

Santolini J, Wootton SA, Jackson AA, Feelisch M. The redox architecture of physiological function. Curr Opin Physio. 2019;9:34–47.

Shakher J, Stevens MJ. Update on the management of diabetic polyneuropathies. Diabetes Metab Syndr Obes. 2011;4:289–305.

Shaqura M, Khalefa BI, Shakibaei M, Zöllner C, Khrasani M, Fürst S, et al. New insights into mechanisms of opioid inhibitory effects on capsaicin-induced TRPV1 activity during painful diabetic neuropathy. Neuropharmacology. 2014;85:142–50.

Sharopov BR, Gulak KL, Philyppov IB, Sotkis AV, Shuba YM. TRPV1 alterations in urinary bladder dysfunction in a rat model of STZ-induced diabetes. Life Sci. 2018;193:207–13.

Shen J. Plasticity of the central nervous system involving peripheral nerve transfer. Neuroplasticity. 2022;2022:5345269.

Sies H, Jones DP. Oxidative stress. In: Fink G, editor. Encyclopedia of stress, vol. 3. 2nd ed. Amsterdam: Elsevier; 2007. p. 45–8.

Sies H, Jones DP. Reactive oxygen species (ROS) as pleiotropic physiological signalling agents. Nat Rev Mol Cell Biol. 2020;21:363–83.

Sies H, Berndt C, Jones DP. Oxidative stress. Annu Rev Biochem. 2017;86:715–48.

Smith AG, Russell J, Feldman EL. Lifestyle intervention for pre-diabetic neuropathy. Diabetes Care. 2006;29:1294–9.

Stino AM, Smith AG. Peripheral neuropathy in prediabetes and the metabolic syndrome. J Diabetes Investig. 2017;8(5):646–55.

Ta LE, Schmelzer JD, Bieber AJ, Loprinzi CL, Sieck GC, Brederson JD, Low PA, Windebank AJ. A novel and selective poly (ADP-ribose) polymerase inhibitor ameliorates chemotherapy-induced painful neuropathy. PLoS One. 2013;8:e54161.

Thomas PK. Classification of the diabetic neuropathies. In: Textbook of diabetic neuropathy. Stuggart; 2003. p. 175–7.

Vincent AM, Olzmann JA, Brownlee M, Sivitz WI, Russell JW. Uncoupling proteins prevent glucose-induced neuronal oxidative stress and programmed cell death. Diabetes. 2004;53:726–34.

Wilkinson R, Lyons AB, Roberts D, Wong MX, Bartley PA, Jackson DE. Platelet endothelial cell adhesion molecule-1 (PECAM-1/CD31) acts as a regulator of B-cell development, B-cell antigen receptor (BCR)-mediated activation, and autoimmune disease. Blood. 2002;100(1):184–93.

Yagihashi S, Yamagishi SI, Wada R, et al. Neuropathy in diabetic mice overexpressing human aldose reductase and effects of aldose reductase inhibitor. Brain. 2001;124(12):2448–58.

Zhu J, Hu Z, Luo Y, Liu Y, Luo W, Du X, Luo Z, Hu J, Peng S. Diabetic peripheral neuropathy: pathogenetic mechanisms and treatment. Front Endocrinol. 2024;14:1265372.

Mechanistic Insights into Oxidative Stress and Diabetic Neuropathy

2

Ritu Kashyap, Brijesh Shivhare, Vijay Kumar Bansal,
Pooja Arora, Balwinder Singh, and Minesh Kumar Patel

2.1 Introduction

Both macrovascular issues such as cardiovascular diseases and microvascular issues like retinopathy, neuropathy, and nephropathy are the features of diabetes mellitus. As a result of a lesion to the somatosensory nerve system, one of the clinical syndromes associated with diabetes is diabetic neuropathy, which exhibits significant morbidity and discomfort. When neuropathy affects the hands and feet, the most common clinical pattern is a distal-to-proximal slope of severity. Right now, managing glucose levels and taking care of one's feet is the sole therapy for diabetic neuropathy (Vinik and Mehrabyan 2004). Furthermore, there is a significant correlation between diabetic peripheral neuropathy (DPN) and increased mortality, morbidity, and worse quality of life. About 30% of diabetes patients have neuropathy, and up to 50% of diabetics will undoubtedly acquire neuropathy at some point throughout their illness. Consequently, given the prevalence of DN, it is imperative to look into

R. Kashyap (✉)
SVKM NMIMS School of Pharmacy, Shirpur, Dhule, Maharashtra, India

B. Shivhare
Department of Botany, Faculty of Sciences, Baba Mastnath University,
Rohtak, Haryana, India

V. K. Bansal
Lachoo Memorial College of Science and Technology, Pharmacy Wing,
Jodhpur, Rajasthan, India

P. Arora
Swami Devi Dyal Institute of Pharmacy, Barwala, Panchkula, Haryana, India

B. Singh
Department of Pharmaceutical Sciences, Amritsar Group of Colleges, Amritsar, Punjab, India

M. K. Patel
Saraswati Institute of Pharmaceutical Sciences, Randesan, Gandhinagar, India

M. K. Gautam et al. (eds.), *Antioxidants and Functional Foods for Diabetic Neuropathy*, https://doi.org/10.1007/978-981-95-9326-2_2

the specifics of its pathogenesis and possible treatments. DN arises in the context of hyperglycaemia and related metabolic dysregulation, mostly due to oxidative stress (Callaghan et al. 2012).

2.2 Decoding Oxidative Stress

An imbalance involving ROS production and antioxidant systems leads to oxidative stress. Via several molecular signalling channels, it alters the way nerve cells operate. Due to poor breakdown of glucose, excess glucose is redirected to various metabolic or nonmetabolic methods, which is one significant way that diabetes contributes to oxidative stress. This results in an overabundance of nicotinic acid adenine dinucleotide phosphate (NADPH) and the build-up of harmful substances. Together, these increase aberrant changes on protein, lipid, and DNA as well as intracellular redox stress; this leads to mitochondrial damage and an excess of ROS being produced. Loss of Schwann cells, myelinated axons, and sensory neurones in the dorsal root ganglia are signs that this impairs the peripheral nervous system. As this is going on, inadequate mitochondrial energy generation impairs axon trafficking, which exacerbates axonal damage and leads to a variety of chronic degenerative illnesses, such as diabetic neuropathy (Pang et al. 2020).

2.3 Sources of ROS Production

The accumulation of free radicals and the reduced activity of antioxidant enzymes in diabetic mice served as the main evidence that oxidative stress plays a role in diabetic neuropathy. This impact reduced when receiving antioxidant therapy to relieve symptoms (Cameron et al. 1994). It is believed that poor mitochondrial oxidised glucose is the main source of ROS production in diabetes. A quick review of glucose metabolism is useful in understanding how hyperglycaemia raises the generation of ROS. In a normal state, intracellular glucose is taken in and transformed to glucose-6-phosphate. Glycolysis and oxidation then occur to create acetyl CoA and NADH. Protons are expelled from the mitochondrial matrix after the mitochondrial electron transport chain (ETC), and NADH-carried electrons are transferred to oxygen. ATP synthase uses the proton gradient that is produced to produce ATP. Only 0.2–2% of the electrons in the ETC leak out under normal circumstances to form ROS, and neurones contain enough natural cellular antioxidants to eliminate ROS (Turrens 2003). In contrast, impaired glucose metabolism in diabetics causes glucose or glycolysis intermediates to be redirected to other metabolic and nonmetabolic pathways. As a result, a neurone produces a lot of ROS and more protons have returned to the cell mitochondria without generating ATP. Axons have direct access to the nerve's blood supply and are widely distributed among mitochondria. Axons in hyperglycaemia are more susceptible to ROS-mediated injury since they rely on local mitochondria for energy. This is because the neurone is unable to detoxify the excess ROS and ATP production is insufficient. This ultimately causes axonal degeneration (Fernyhough and McGavock 2014).

2.4 Antioxidant Defence Systems

Interaction between antioxidants and antioxidative enzymes makes up the antioxidant system. Both endogenous and exogenous resources can be used as antioxidants. The most common endogenous antioxidant present in the majority of cell types is glutathione, also known as GSH in its reduced form. Examples of exogenous antioxidants that may be acquired by diet or supplementation include vitamins A, C, and E as well as antioxidant minerals (Cu, Zn). Catalase, glutathione reductase/peroxidase, superoxide dismutase, and other key antioxidative enzymes convert free radicals into harmless intermediates. Together, they function as a team through processes that repair de novo antioxidants in addition to scavenging and inhibiting free radicals before they may harm cells (Mondal et al. 2025).

Cells have produced anti-oxidative enzymes including catalase, GPx, and SOD in response to the continuous production of ROS during aerobic metabolic processes. Millimolar amounts of GSH are also present in cells, which sustains the sulfhydryl buffering ability. The metabolic conversion of O_2 to H_2O_2 is catalysed by extracellular SOD, copper/zinc (cytoplasmic), and manganese (mitochondrial). H_2O_2 is then transformed by catalase into O_2 and water. GSH/GSH reductase breaks down GSH, thioredoxin/thioredoxin and GSH/GSH systems employ NADPH, and GPx, which contains selenium, lowers organic peroxides.

The mechanisms that restore, eliminate, or swap out oxidatively damaged molecules are referred to as secondary antioxidant defences. These might contain lipid or DNA repair enzymes, as well as protease systems designed specifically to eliminate oxidatively damaged proteins. The majority of ROS along with free radicals can be eliminated by enzymes, but some can evade the body's defences and damage cellular macromolecules including fatty substances, proteins, and nucleic acids. These might represent the early phases of OS, which are followed by fibrosis, apoptosis, inflammation, and senescence. These conditions lead to redox alterations in neurones and ED in the blood vessels vasorum (Chun and Park 2020).

2.5 Link Between Oxidative Stress and Diabetes

Many theories can be described, even if the pathogenic process that underlies the development of diabetes mellitus (DN) is still unclear. It is believed that DN is primarily caused by the activation of polyol, diacylglycerol/protein kinase C (PKC), oxidative stress, advanced glycation end products (AGEs), hexosamine, nitric oxide, and inflammation. However, there is evidence that oxidative stress is connected to all of the previously stated pathways (Hosseini and Abdollahi 2013).

Figure 2.1 shows the suggested mechanism of OS involvement in DMT-2. Though not entirely supported, an appealing theory has appeared in the last few

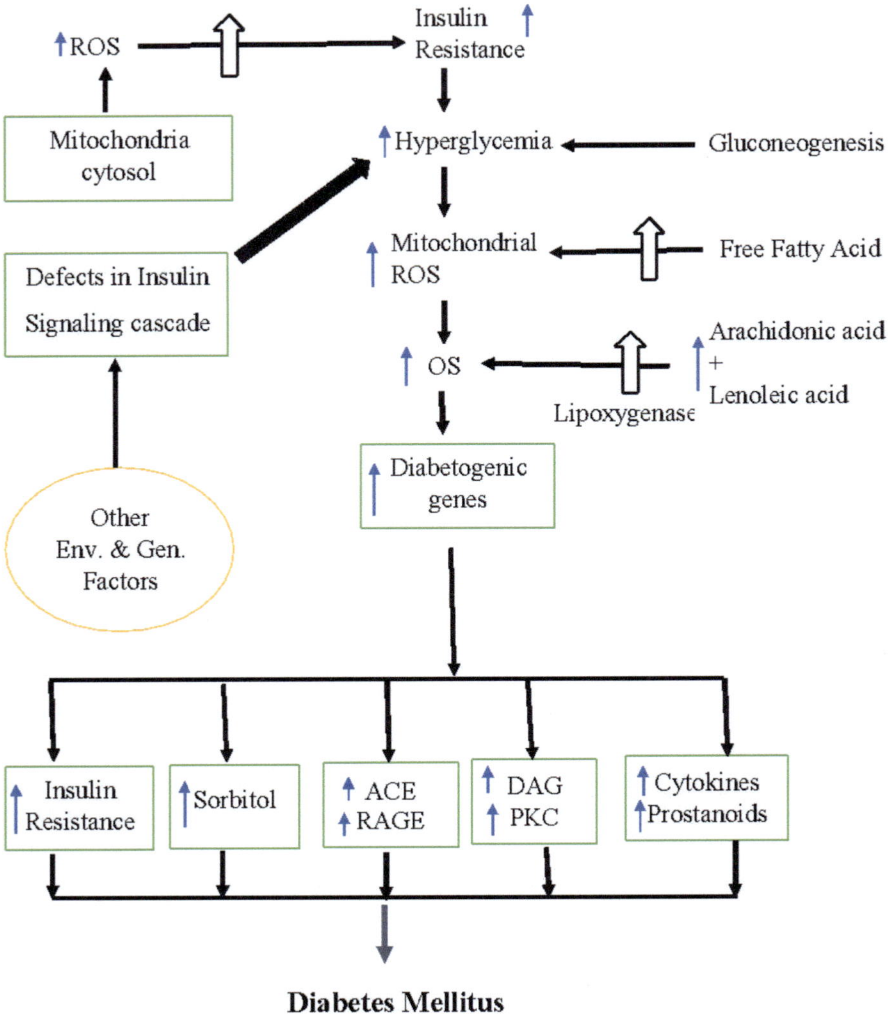

Fig. 2.1 Hypothesised mechanism of oxidative stress's role in diabetic mellitus

years suggesting that ROS and RNS might be crucial to the development of diabetic problems brought on by chronically inflamed OS and the onset of DMT-2, particularly through obesity and insulin resistance. It is believed that this chronic elevated OS is caused by a number of factors, such as increased AGE formation and a malfunction in the RAGE receptor, increased glucose diversion to sorbitol, hexosamine-stimulated secretion of several inflammatory cytokines, glucose autoxidation, and including tumour necrosis factor alpha (TNF-α) and several other IL (Gautam et al. 2015).

2.6 Oxidative Stress as Key Mediator

It is thought that OS is crucial to the pathophysiology of DN. Cell signalling is impacted by ROS and RNS, which are by-products of regular cellular metabolism. The mitochondrial enzymes xanthine reductase, NADPH oxidase, and superoxide anion (O_2^-) create the majority of ROS (hydrogen peroxide (H_2O_2), hydroxyl radical ($^{\cdot}OH$), and superoxide anion (ROS). NO synthase is responsible for producing RNS. Depending on the kind of cell, the environment, and the production of ROS and RNS, OS can cause a variety of cellular responses, including growth, division, and cell death, which have a substantial influence on the health of neuronal cells. While DNA damage and protein malfunction can result in variations in genes and neuronal cell death, the majority of ROS and RNS are associated with the formation of DPN. Moreover, they play advantageous roles in the normal cellular signalling pathways of senescence, apoptosis, necrosis, adhesion, differentiation, proliferation, and gene expression—all of which are primarily mediated by MAPKs. Furthermore, ROS can alter a number of stress-induced signalling pathways, including those of Sirt-1/AMPK, PI3K, Wnt, and NF-κB (Chun and Park 2020).

Numerous factors, including elevated hyperglycaemia, hypoxia, and AGEs, can cause cells to produce ROS. The most significant feature of neuropathy is the activation of MAPK in various Schwann and neuronal cells brought on by an excess of ROS. This activation subsequently causes MD, neuronal damage, and ultimately, the cell's demise. Diabetes also causes increased ER stress in other organs, such as brain cells. A major feature of diabetic tissues, AGE production plays a crucial role in the early stages of DN and is impacted by a number of pathways, including diminished NGF assist, OS, nerve hypoxia/ischemia, PKC, and enhanced polyol pathway activation. According to many studies, neuropathic morphologic nerve alterations, poor neurotrophic support, and neural dysfunction have all been associated with increased OS in peripheral nerves and the vasa nervorum (Purves et al. 2001).

When OS is induced by ROS and RNS with greater reactivities, cells respond to these changes by activating a range of coping mechanisms. One of the two possible causes of a rise in OS is a decrease in antioxidant defences or an increase when free radicals are produced. SOD, catalase, and GPx are always able to neutralise ROS chain reactions, however they are not able to prevent elevated OS in diabetic individuals (Gautam et al. 2012).

2.7 Pathological Effects of Oxidative Stress on Nerve Cells

2.7.1 General Effect on Central Nervous System

The hippocampal biochemical alterations brought on by oxidative stress play a major role in the function and connections between neurones. The hippocampus and dentate gyrus are both significantly impacted functionally by oxidative stress, regardless of the specific regions of the hippocampus that are affected. Chronic stress, particularly that caused by ROS/RNS, has been shown to cause dendritic constriction in medial PFC neurones and neural development in amygdala neurones. Neural connection is disrupted by these changes in dendritic shape. These investigations reveal the ways in which oxidative stress affects the memory area of the brain's biochemistry and structure, amygdala, and prefrontal cortex regions that are usually essential to the pathogenesis of various neurological diseases (Olufunmilayo et al. 2023).

2.7.2 Damage to Mitochondrial Cells

In addition to contributing to hyperglycaemia, the mitochondrial electron transport chain's generation of superoxide opens up a number of etiological routes. By increasing the reducing equivalents and electrochemical voltage across the inner mitochondrial membrane, hyperglycaemia promotes the production of superoxide. Superoxide facilitates the depletion of NADH+ through PARP, which starts the synthesis of AGE, an essential bodily function (Kumar et al. 2013). As a result of this inhibition, glycolytic intermediates build up and are subsequently targeted at cellular functions such as the formation of AGE, which activates the PKC pathway via dihydroxyacetone phosphate-mediated diacylglycerol. Increased fructose-6-phosphate flow and high glucose levels activate the polyol and hexosamine pathways, respectively. Ultimately, this causes the cells to experience osmotic stress, hastening necrotic cell death. The increased activity of Mn-SOD, a mitochondrial form of SOD, or upregulation of uncoupling proteins (UCP-1) in experimentally diabetic rats reduces oxidative stress-induced brain damage and delays the onset of vascular problems. One possible explanation for this neuroprotective effect is the removal of the notorious ROS from the cells and a reduction in the formation of ROS in the mitochondria (Brownlee 2001).

An important role play in the pathophysiology of diabetic neuropathy is played by oxidative stress linked with mitochondria and anomalies in the mitochondria, in addition to the aforementioned idea. Free radical production is seen to be enhanced and ATP synthesis compromised by malfunctioning components of the ETC chain. Nuclear and mitochondrial DNA are harmed by the free radicals produced, which exacerbates damage to the DNA of the mitochondria. Apoptosis or necrosis is the end result of the vicious cycle that has established inside the mitochondria, causing extreme oxidative stress. Diabetes has proven to impact ETC functional complexes' ability to breathe, which in turn modifies ATP synthesis. As electron leakage centres, complex I and III are the primary ones recognized to be impacted, that may increase the formation of ROS (Fernyhough et al. 2003).

2.7.3 Impact on Peripheral Nerves

It is well recognised that the high concentration of phospholipids, the mitochondria-rich axoplasm, and the poor antioxidant defences of mammals make them more vulnerable to oxidative stress. Furthermore, it has recently been noted that certain medicines' structural and functional damage increases the generation of free radicals within the mitochondria. This type of oxidative stress results in demyelination, mitochondrial malfunction, microtubular damage, and apoptosis, which physically harm neurones. Apoptosis is one of the mechanisms via which oxidative stress-induced neurodegeneration can occur, along with bioenergetic failure, ion channel activation, antioxidant defence depletion, neuroinflammation, biomolecular damage, microtubular disruption, demyelination, and impaired mitophagy. To remedy the redox imbalance generated in neuronal cells, the ratio of nuclear factor kappa light chain enhancer of B cells to nuclear erythroid factor-2 related factor can be changed pharmacologically (Nrf2-NF-κB axis). As a result, the effectiveness of these modulators has been evaluated in animal models of peripheral neuropathy (Areti et al. 2014).

2.7.4 Role in Neurodegeneration

The growing dysfunction and death of cells that often affect a specific brain system is the biological characteristic of neurodegenerative diseases. Clinically, these illnesses are distinguished by their chronic progression and gradual onset. In neurodegenerative diseases, it is referred to as selective neuronal susceptibility. Although many brain neurones are able to sustain a rise in oxidative stress, certain groups of neurones are more susceptible to these phenomena. The central nervous system (CNS) has neural populations that may vary in their sensitivity to stresses that cause cell loss or damage and neurodegeneration. Selective neuronal vulnerability is the term used to describe this phenomenon. Oxidative stress has a significant impact on neural cells, because (1) compared to other cells, their energy source is mostly oxidative phosphorylation; (2) despite making up just 5% of the body weight, they are subjected to high oxygen concentrations, using 20% of the oxygen breathed. Under normal circumstances, 1–2% of oxygen taken in is transformed into ROS, which causes oxidative stress; in the old brain, this proportion increases significantly; (3) they are rich in polyunsaturated fatty acids, which are oxidatively prone; (4) they are rich in ionised metals, which can act as a potent catalyst for the production of oxidative species and build up in brain tissue with age; and (5) they have relatively low concentrations of antioxidants and related enzymes. Lipids, proteins, and DNA oxidise as a result of oxidative stress in the neuronal microenvironment, which also produces a variety of by-products such peroxides, alcohols, aldehydes, ketones, and cholesterol oxide (Chen et al. 2012).

Studying the impacts of oxidative stress has made considerable use of in vitro models. When exposed to cadmium, the mouse neuroblastoma HT4 cell line shows enhanced cell protein ubiquitination and oxidative stress, which results in the breakdown of bound to another proteins. The ubiquitin peroxisome system may become overwhelmed by oxidative stress, leading to a build-up of some of these modified proteins within the neurones. The disruption of regular neuronal function resulting from these protein deposits could potentially lead to an ER stress state, as evidenced by elevated levels of GRP78 (78-kDa glucose-regulated protein), a chaperone protein implicated in the unfolded protein response (UPR) that ultimately triggers the autophagic pathway. Numerous diseases, especially neurological ones, are primarily caused by the unchecked and mismanaged build-up of mutated and misfolded proteins in the cytoplasm and endoplasm (Olufunmilayo et al. 2023).

2.8 Clinical Studies Linking Oxidative Stress and Neuropathy

Numerous clinical research projects have connected oxidative stress to a range of neuropathies. Research has indicated that oxidative stress is a crucial factor in the aetiology of conditions including DN, PDPN, and optic neuropathy. The following list includes several researches that have demonstrated the influence of elements such proinflammatory cytokines, antioxidant enzyme activity, and free radicals on neuropathic disorders.

The objective of one study was to compare DN patients to healthy persons in terms of oxidative stress biomarkers (SOD, CAT, GR, GPX) and inflammatory cytokines (IL-2, IL-4, IL-6). The activity of enzymatic antioxidants was drastically reduced in individuals with DN, highlighting the part oxidative stress plays in the pathophysiology of DN. In DN patients, higher levels of proinflammatory cytokines (IL-2, IL-4, and IL-6) were detected, suggesting a possible connection between inflammation and oxidative stress in the pathophysiology of DN (Devarapalli and Sudhakar 2024).

In another study, rats with experimentally generated sciatic nerve chronic constriction injury (CCI) had their peripheral nerve injury and behavioural pain responses analysed in relation to ROS and OX. Additionally, the effects of intraperitoneal N-acetyl-L-cysteine (NAC) treatment on rats suffering from neuropathic pain caused by CCI were investigated. This study shows that oxidative stress, the antioxidants decreased glutathione and superoxide dismutase, and the neuropathological and psychological effects of CCI-induced neuropathy are significant factors. NAC could also be a good way to alleviate neuropathic pain (Naik et al. 2006).

Furthermore, in an effort to comprehend PN pathophysiology better, a second systematic review was conducted with the goal of examining the relationship between oxidative stress and PN. The information collected from the included publications was evaluated and analysed in this study after an automated, systematic search of the PubMed database. 69 articles that satisfied the eligibility criterion were used for this review. Reduced endogenous antioxidants and elevated reactive oxygen species are linked to peripheral neuropathy. There might be a genetic component to oxidative damage susceptibility. Antioxidant treatment shows promise. It is evident that oxidative stress contributes to the pathophysiology of PN or is at least continuously present during it, even though further investigation is required to fully comprehend the underlying process (Mallet et al. 2020).

Additionally, Willems et al.'s study, which had 110 T1DM patients, found that individuals with greater illness duration had higher antioxidant levels on average.

Additionally, vitamin A levels were greater in PN patients than in individuals with other microvascular problems. These findings support the idea that longer disease duration is associated with neuropathy through an increase in oxidative stress. High level of vit. A in this situation may indicate a compensatory mechanism since increased oxidative stress may force the body to attempt to produce more of the antioxidant vitamin A (Willems et al. 1998).

Stress on the perineurium caused by damage to the brain may arise from vascular oxidative stress. The perineurium of biopsies obtained from 31 patients with distal symmetric neuropathy, whose aetiology ranged from DPN to other conditions, contained N(ε)- carboxymethyllisine (CML). This implies that circulating oxidative stress damages the perineurium, upsetting the endoneurial equilibrium environment and leading to neuropathy from oedema, as CML is an advanced glycation end product (Haslbeck et al. 2002).

Newly detected type I and type II diabetes patients also exhibited antioxidant depletion, as seen by elevated thiobarbituric acid reactive species and lower levels of SOD3 and glutathione (GSH) in comparison to healthy controls (Strom et al. 2017). Mendez et al. confirmed the GSH drop, indicating a faster depletion of endogenous reserves. This suggests that diabetes increases vulnerability to oxidative damage and gradually diminishes antioxidant defences. SOD3 levels were also associated with conduction velocities of the ulnar, medial, and sural sensory nerves along with the ulnar and peroneal motor nerves. SOD3 and distal symmetric polyneuropathy often had a negative correlation. All in all, our results highlight the crucial part that oxidative stress plays in the growth and course of different neuropathies, offering insightful information for prospective treatment approaches (Mendez et al. 2015).

2.9 Future Directions and Research

It is abundantly evident from the strong data shown by several research groups that oxidative stress-mediated neurodegeneration as well as the associated inflammatory responses are major players in the aetiology of diabetic neuropathy. Pharmacological drugs can modify these pathways to avoid the pathophysiological and functional abnormalities linked to peripheral neuropathy. This can also hasten the development of novel therapeutic approaches for diabetic neuropathy. Numerous preclinical and clinical investigations have reliably shown that oxidative stress is present in both animal models and human subjects suffering from a range of neurodegenerative illnesses in terms of symptoms and indicators. various studies have also consistently shown the complex and often reciprocal relationship between oxidative stress and various diseases; wherein oxidative stress promotes the development of the classic pathologic features of these diseases and inversely.

There is evidence that oxidative stress and neuroinflammation play a key role as pathophysiological initiators in a number of diabetes-related microvascular problems, such as diabetic neuropathy. It has been discovered that the sensory and functional abnormalities linked to diabetic neuropathy can be improved by using medications that target the oxidative stress-inflammatory pathways. The therapeutic efficacy of intracellular redox signalling mechanisms is still inadequate because of their complex nature and interactions with the cell genome, DNA, and epigenome. Because redox imbalances created in one route might trigger those in another, combining the several above-mentioned techniques may have greater positive benefits than doing them alone. The cellular damage caused by hyperglycaemia is believed to be initiated by mitochondrial malfunction; hence, in the near future, medications that target mitochondria will be more widely used to treat diabetic neuropathy. More study is needed to understand how swelling, malfunctioning mitochondria, and oxidative stress relate to the development of diabetic neuropathy.

Despite extensive research on oxidative stress and reactive oxygen species (ROS), there are few studies demonstrating ROS's role in chronic pain. The potential of PBN as a ROS scavenger in treating neuropathic pain was demonstrated when it was systemically delivered to reduce mechanical allodynia in a rat model of neuropathic pain. Antioxidant enzymes have a crucial role in neuropathic pain, despite the fact that these findings indicate the involvement of free radicals and oxidative stress in the condition (Naik et al. 2006). Numerous studies on antioxidant-based therapy approaches have been conducted, and many more are now underway.

2.10 Conclusion

These studies have produced results that are erratic and sometimes challenging to implement. The distinctions between preclinical and clinical study designs may help to explain this wide variation. Adopting the findings from such research might be challenging due to heterogeneity, which may also be explained by the fact that the majority of clinical investigations have included small study populations. The antioxidant properties of innovative synthetic and natural medicinal agents as well as their clinical applications in neurodegenerative disease states are being investigated. More comprehensive preclinical and clinical investigations of oxidative stress indicators and treatments will ultimately enhance the standard of life for people with neurodegenerative disease conditions. This will lead to more dependable and therapeutically useful outcomes.

References

Areti A, Yerra VG, Naidu VGM, Kumar A. Oxidative stress and nerve damage: role in chemotherapy induced peripheral neuropathy. Redox Biol. 2014;2:289–95.

Brownlee M. Biochemistry and molecular cell biology of diabetic complications. Nature. 2001;414(6865):813–20.

Callaghan BC, Cheng HT, Stables CL, Smith AL, Feldman EL. Diabetic neuropathy: clinical manifestations and current treatments. Lancet Neurol. 2012;11(6):521–34.

Cameron NE, Cotter MA, Archibald V, Dines KC, Maxfield EK. Anti-oxidant and pro-oxidant effects on nerve conduction velocity, endoneurial blood flow and oxygen tension in non-diabetic and streptozotocin-diabetic rats. Diabetologia. 1994;37:449–59.
Chen X, Guo C, Kong J. Oxidative stress in neurodegenerative diseases☆. Neural Regen Res. 2012;7(5):376–85.
Chun H, Park Y. Oxidative stress and diabetic neuropathy. In: Diabetes. Academic Press; 2020. p. 13–23.
Devarapalli P, Sudhakar G. Screening of oxidative stress biomarkers and inflammatory cytokines in the pathogenesis of diabetic neuropathy. Res J Biotechnol. 2024;19:7.
Fernyhough P, McGavock J. Mechanisms of disease: mitochondrial dysfunction in sensory neuropathy and other complications in diabetes. Handb Clin Neurol. 2014;126:353–77.
Fernyhough P, Huang TJ, Verkhratsky A. Mechanism of mitochondrial dysfunction in diabetic sensory neuropathy. J Peripher Nerv Syst. 2003;8(4):227–35.
Gautam MK, Gangwar M, Singh A, Rao CV, Goel RK. In-vitro antioxidant properties of Murraya paniculata (L.) leaves extract. Inventi Rapid Ethnopharmacol. 2012;4:1–3.
Gautam MK, Gangwar M, Singh SK, Goel RK. Effects of Azadirachta indica on vascular endothelial growth factor and cytokines in diabetic deep wound. Planta Med. 2015;81:713–21.
Haslbeck K, Schleicher E, Friess U, Kirchner A, Neundörfer B, Heuss D. N ε-Carboxymethyllysine in diabetic and non-diabetic polyneuropathies. Acta Neuropathol. 2002;104:45–52.
Hosseini A, Abdollahi M. Diabetic neuropathy and oxidative stress: therapeutic perspectives. Oxidative Med Cell Longev. 2013;2013(1):168039.
Kumar M, Gautam MK, Singh A, Goel RK. Healing effects of Musa sapientum var. paradisiaca in diabetic rats with co-occurring gastric ulcer: cytokines and growth factor by PCR amplification. BMC Complement Altern Med. 2013;13:305.
Mallet ML, Hadjivassiliou M, Sarrigiannis PG, Zis P. The role of oxidative stress in peripheral neuropathy. J Mol Neurosci. 2020;70:1009–17.
Mendez MM, Folgado J, Tormo C, Artero A, Ascaso M, Martinez-Hervás S, et al. Altered glutathione system is associated with the presence of distal symmetric peripheral polyneuropathy in type 2 diabetic subjects. J Diabetes Complicat. 2015;29(7):923–7.
Mondal NS, Islam A, Gautam MK, Mondal S, Ahmed M, Jamir SI, Khongwir LD, Mahajon B. Unraveling the mechanisms of diabetic wounds: insights into pathogenesis and advanced treatment strategies. Curr Diabetes Rev. 2025;22:e15733998368222.
Naik AK, Tandan SK, Dudhgaonkar SP, Jadhav SH, Kataria M, Prakash VR, Kumar D. Role of oxidative stress in pathophysiology of peripheral neuropathy and modulation by N-acetyl-L-cysteine in rats. Eur J Pain. 2006;10(7):573–9.
Olufunmilayo EO, Gerke-Duncan MB, Holsinger RD. Oxidative stress and antioxidants in neurodegenerative disorders. Antioxidants. 2023;12(2):517.
Pang L, Lian X, Liu H, Zhang Y, Li Q, Cai Y, et al. Understanding diabetic neuropathy: focus on oxidative stress. Oxidative Med Cell Longev. 2020;2020(1):9524635.
Purves T, Middlemas A, Agthong S, Jude EB, Boulton AJ, Fernyhough P, Tomlinson DR. A role for mitogen-activated protein kinases in the etiology of diabetic neuropathy. FASEB J. 2001;15(13):2508–14.
Strom A, Kaul K, Brüggemann J, Ziegler I, Rokitta I, Püttgen S, et al. Lower serum extracellular superoxide dismutase levels are associated with polyneuropathy in recent-onset diabetes. Exp Mol Med. 2017;49(11):e394.
Turrens JF. Mitochondrial formation of reactive oxygen species. J Physiol. 2003;552(2):335–44.
Vinik AI, Mehrabyan A. Diabetic neuropathies. Med Clin. 2004;88(4):947–99.
Willems D, Dorchy H, Dufrasne D. Serum antioxidant status and oxidized LDL in well-controlled young type 1 diabetic patients with and without subclinical complications. Atherosclerosis. 1998;137:S61–4.

Diagnostic Approaches in Diabetic Neuropathy & Differential Diagnosis: Distinguishing Diabetic Neuropathy from Other Conditions

3

Nalanda Tulsiram Rangari, Yamjala Ganesh Kumar, Lalchand D. Devhare, Gauri Nilesh Deodhar, Namrata A. Muddalwar, and G. Dharmamoorthy

3.1 Introduction

A major world health concern, diabetes impacts on 500 million people worldwide. Because of its alarming rate of global increase, its prevalence is expected to reach 700 million by 2045 (Saeedi et al. 2019). Type 2 diabetes is caused by environmental and behavioural factors such as obesity, an overly rich diet, and sedentary lifestyles, in addition to a family history of the disease, which is more prevalent in some ethnic groups. An integrated, international approach will be required for the early detection of diabetes and the management of its micro- and macrovascular effects if we are to witness a significant reduction in the massive amount of premature morbidity and mortality that diabetes causes (Zimmet et al. 2001).

Peripheral nerve damage and breakdown in diabetics is known as diabetic neuropathy. A progressive decline in nerve fibre function is a characteristic of neuropathies (Rosenberger et al. 2020). Diabetic neuropathy is a common side effect of diabetes mellitus (DM), affecting up to 51% of people with type 1 and type 2 DM

N. T. Rangari
KJEIs, Trinity College of Pharmacy, Pune, India

Y. G. Kumar
Sree Dattha Institute of Pharmacy, Sheriguda, Ibrahimpatnam, Rangareddy, Telangana, India

L. D. Devhare (✉)
Nagpur College of Pharmacy, Wanadongri, Maharashtra, India

G. N. Deodhar · N. A. Muddalwar
School of Pharmacy, G.H. Raisoni Skilltech University, Nagpur, Maharashtra, India

G. Dharmamoorthy
Department of Pharmaceutical Analysis, MB School of Pharmaceutical Sciences, Mohan Babu University, Tirupati, Andhra Pradesh, India

M. K. Gautam et al. (eds.), *Antioxidants and Functional Foods for Diabetic Neuropathy*, https://doi.org/10.1007/978-981-95-9326-2_3

(Hicks and Selvin 2019). Diabetic neuropathy can show up clinically as mild neuropathic pain and numbness or as severe pain and neuropathic pain. Diabetes-related neuropathy raises the risk of lower limb amputation, foot ulceration, and death (Sloan et al. 2021).

3.2 Clinical Presentation and Symptoms

Depending on the kind of neuropathy and the number and type of damaged nerves, diabetic neuropathy can have different symptoms. Diabetic mononeuropathy, or focal neuropathy: One nerve at a time is impacted by this kind of diabetic neuropathy, and the symptoms vary depending on which nerve is impacted. For example, it can affect the thoracic nerves in the chest, causing numbness and discomfort in the chest wall that feels like a heart attack, angina, or an appendix. Different kinds of focal neuropathy may result in discomfort in the thighs, severe discomfort in the flanks, stomach, or lower back, as well as in the chest, Eye pain behind the eyelids, difficulty focussing the eyes, dual vision, hearing issues, facial paralysis on one side.

Diabetic polyneuropathy (DPN) affects the peripheral sensory and motor nerves that extend from the spinal cord into the hands, arms, legs, and feet. The longest nerves, which extend from the spine to the feet, are often the most damaged. Paraesthesia, numbness, discomfort, muscular weakness, acute pains, excessive sensitivity to touch, insensitivity to pain, loss of balance, problems with coordination, and trouble walking on uneven surfaces are some of the symptoms (Deli et al. 2014).

Autonomic nerves supplying the cardiac system, gastrointestinal system, sexual organs, urinary system, and sweat glands are the main targets of diabetic autonomic neuropathy. Symptoms of this kind of diabetic polyneuropathy include persistent nausea and vomiting, constipation, diarrhoea, excessive sweating, dysfunction in sexual relationships, digestive issues, low blood pressure, a diminished ability to feel pain, and low blood sugar (Islam et al. 2025).

In addition to causing hyperalgesia and allodynia, spontaneous activity in the peripheral nociceptor system may also cause alterations in the central nervous system. Death by abrupt cardiac arrest, malignant arrhythmia, and cardiovascular autonomic neuropathy (CAN) may be associated. Disabled by gastrointestinal autonomic neuropathy, gastroparesis is the most harmful consequence. Neurogenic bladder and sexual dysfunction are the two possible outcomes of genitourinary autonomic neuropathy (Yasuda et al. 2003).

As an alternative, DN preventive and treatment techniques are recommended by national and international clinical recommendations, which draw upon the most up-to-date research. Prevention tactics aim to treat DN before symptoms occur or stop the illness from advancing, as opposed to therapy methods that address DN symptoms after they arise. These guidelines emphasise three primary preventive techniques that might be difficult to implement for various reasons: good foot care, lifestyle modifications, and glucose control. The significance of DN early diagnosis and its potential to enable the early implementation of strategies to stop the disease's progression have been highlighted; however, since the current invasive diagnostic evaluation tools have safety issues, research has to concentrate on creating non-invasive alternatives. Diabetic neuropathy is confirmed by a variety of diagnostic techniques, which also help to distinguish it from other illnesses that appear similarly (Smith et al. 2022).

3.3 Diagnostic Evaluation

3.3.1 Diagnosis of Pre-diabetes and Diabetes Mellitus

The American Diabetes Association recommends that HgbA1c, fasting plasma glucose, and 2-hour postprandial glucose levels be used to diagnose diabetes because neuropathy and pre-diabetes are related. Pre-diabetes is indicated by impaired glucose tolerance, which is characterised by 2-hour postprandial glucose levels of 7.8–11.0 mmol/l, or impaired fasting glucose. People who have an A1c between 5.7% and 6.4% are at a high risk of pre-diabetes, another name for diabetes, in the future (Mondal et al. 2025).

It is imperative to incorporate screening strategies into routine clinical practice due to the high prevalence of DPN and its severe consequences. This will help ensure that DPN pain is diagnosed at an early stage and treated promptly, preventing progression and the development of advanced complications, such as limb amputations and death. Every person should have their DPN evaluated at the time of type 2 diabetes diagnosis, five years after type 1 diabetes diagnosis, and then at least once a year after that (Pop-Busui et al. 2022).

3.3.2 Diagnosis of Sensorimotor Neuropathy

Every diabetic patient should undergo an annual screening for DPN, which includes testing ankle reflexes, temperature, vibration perception (using a 128-Hz tuning fork), pinprick sensation, and 10-g monofilament pressure sensation at the distal hallucices. Diagnosing DPN requires a comprehensive clinical examination. When many test combinations are used to detect DPN, the sensitivity is higher than 87% (Boulton et al. 2005).

In DPN, sensory problems frequently take precedence over motor nerve damage when diabetes duration or severity worsens. They travel proximally in a "stocking-glove" pattern after first appearing in the distal portions of the limbs. Larger nerve cells of this kind are frequently damaged later than smaller ones (Hartemann et al. 2011). Depending on the kind of fibre implicated, DPN might have different indications and symptoms. For example, big fibre illness can affect proprioception and light touch. Small-fibre illness causes paraesthesia, dysesthesia, and/or neuropathic pain by affecting the senses of pain and temperature. Only in the most extreme circumstances may distal weakening arise. Deep tendon reflexes, especially the Achilles tendon reflex, might be weak or absent, which is frequently a sign of moderate DPN with no other symptoms. Crucially, it's important to remember that every distinct type of nerve fibre has a distinct function and job, making focused assessments simple to carry out in the clinic when screening for DPN. Below is a step-by-step breakdown of a screening and diagnosing process.

3.3.2.1 Clinical Assessment

Medical History In order to gather information about a patient's specific medical history, it is either possible to quickly gather this information during a routine clinic visit or to look up specific risk factors from the electronic health record (EHR). These risk factors include obesity, metabolic syndrome, high height, high blood pressure, long-term diabetes, poor glycaemic control, and advanced age (>70 years). Additionally, a history of recent falls may reveal balance and gait problems that are directly related to the large-fibre dysfunction associated with DPN, provided that there are no other clear risk factors for falls (Khan et al. 2021).

Physical Examination Based on reflexes, autonomic function, and sensory and motor function, a comprehensive neurological examination should be performed. A typical finding in diabetic neuropathy is loss of feeling (pain, warmth, vibration), as well as diminished or missing ankle reflexes.

3.3.2.2 Neurological Testing

It is advised to include electrodiagnostic results in the work-up as they provide a higher level of specificity in the diagnosis of polyneuropathy. The most illuminating part of the electrodiagnostic test is nerve conduction studies (NCS), which are often conducted in combination with a needle electromyogram (EMG).

Nerve Conduction Studies (NCS) Peripheral nerves' electrical signal strength and speed are measured using nerve conduction studies (NCS). A decrease in nerve conduction velocity and amplitude in diabetic neuropathy is typically a sign of nerve damage. NCS is performed by applying small pads that monitor electric impulses and deliver mild electric shocks to the skin. Unlike whole EMG, which may involve inserting tiny needles into the muscles, NCS is relatively simple, painless, and time-efficient. Additionally, NCS has long been considered a minimal need or the gold standard test for validating the diagnosis of peripheral neuropathies due to its accuracy, consistency, and sensitivity in measuring peripheral nerve function. The place of recording, the kind of electrodes used, the temperature of the limb, the examiner's credentials, and other variables can all affect the results of NCS. It is necessary to measure and consider pertinent information, such as temperature, height, gender, age and weight, before reporting an NCS as normal or abnormal (Yang et al. 2018).

Electromyography (EMG) A diagnostic method for evaluating the health of motor neurones—nerve cells that control muscles is electromyography, or EMG. Muscle dysfunction, nerve-to-muscle, signalling issues, or both might be the cause.

Electrodes convert the electrical impulses sent by motor neurons which cause muscles to contract into graphs, sounds, or numerical data. To record a muscle's electrical activity, a needle electrode is inserted into it. EMG is frequently used in conjunction with NCS for a more thorough evaluation of nerve damage and aids in differentiating between neuropathic and myopathic disorders (England et al. 2005).

Quantitative Sensory Testing (QST) QST measures vibration, temperature, and pain thresholds to assess sensory nerve function. It aids in the identification of sensory deficiencies in diabetics. A 128-Hz tuning fork should be used to perform the vibration test. Ten grams of monofilament is advised for touch sensitivity. Several questionnaires have been created to assist professionals in the diagnosis of neuropathic pain. The DN4 questionnaire is particularly interesting since it is quick to complete, user-friendly, and has strong diagnostic performance—it can diagnose neuropathic pain with a sensitivity of 83% and a specificity of 90% for scores of ≥4/10 (Deli et al. 2014). Identifying prospective neuropathy patients is the primary benefit of screening methods, especially for non-specialists. These instruments, however, are unable to detect 10–20% of individuals with neuropathy as diagnosed by a physician, demonstrating that they cannot take the role of meticulous clinical judgement (Cruccu et al. 2010).

3.3.2.3 Autonomic Function Tests

In diabetic neuropathy, autonomic function testing is crucial, especially when there is autonomic nerve system dysfunction. These examinations assess the autonomic nerves that regulate involuntary bodily processes including blood pressure, heart rate, perspiration, and digestion. Heart Rate Variability (HRV) Testing is a frequently employed test that gauges the fluctuation in the duration between consecutive heartbeats, signifying the equilibrium between sympathetic and parasympathetic nervous system activity. Reduced HRV is indicative of autonomic dysfunction and is frequently observed in diabetic neuropathy. Another test that assesses the autonomic response to a controlled forced expiration against a closed airway is the Valsalva Manoeuvre. Abnormal blood pressure and heart rate responses can be signs of autonomic dysfunction. The Head-Up Tilt Test, also known as the Tilt Table Test, assesses the cardiovascular autonomic response to postural alterations. Orthostatic hypotension, a typical autonomic dysfunction in diabetic neuropathy, can be indicated by a considerable drop in blood pressure or an insufficient heart rate response upon tilting.

The function of the autonomic nerve-controlled sweat glands is evaluated by sudomotor function tests, such the Thermoregulatory Sweat Test (TST) and the Quantitative Sudomotor Axon Reflex Test (QSART). Autonomic nerve injury, a common consequence of diabetic neuropathy, is indicated by abnormal sudomotor function. The pace at which food exits the stomach is measured by Gastric Emptying Studies, which indicate autonomic regulation of gastrointestinal motility; delayed gastric emptying, or gastroparesis, is an additional sign of autonomic neuropathy in diabetes patients. To guide therapy options and prevent future consequences in diabetic neuropathy, these tests are essential for early identification of autonomic dysfunction (Tesfaye et al. 2010).

3.3.2.4 Laboratory Tests

Blood Tests Standard blood tests are conducted to evaluate the metabolic state and rule out further possible causes of neuropathy. These tests include fasting glucose, HbA1c, vitamin B12, thyroid function tests, and renal function tests. For instance, peripheral neuropathy is linked to both excessively high and low thyroid hormone levels. Hypothyroidism manifests as dry skin, weakness, sadness, and hair loss, constipation, weight gain, and neuropathies. Neuropathy, anxiety, insomnia, and weight loss are all signs of hyperthyroidism. Likewise, neuropathy, anaemia, weakness, spinal cord damage, and mental problems can result from low vitamin B12 levels, or deficiency. It is possible to test the amounts of methylmalonic acid. Neuropathy may result from excessive vitamin B6 intake. Mostly, motor neuropathy (weakness) is caused by thiamine insufficiency. Additionally, low amounts of copper have been linked to peripheral neuropathy, as have high levels of lead, mercury, thallium, zinc, and arsenic. Furthermore, to rule out autoimmune neuropathies, autoantibody screening may be necessary in some situations (Chowdhury et al. 2024).

3.3.2.5 Imaging Studies

Magnetic Resonance Imaging (MRI) When entrapment neuropathies or other structural reasons are suspected, MRI can be utilised to evaluate soft tissue and nerve structures. The early 1990s saw Filler et al. (1993). defined sequences with greater structural screen resolution and higher neuropathy contrast called magnetic resonance neurography (MRN). The primary MRI sequences utilised in DPN research are high-resolution MRN and quantitative MRI, which expands the importance of MRN in DPN. Examples of quantitative MRI techniques include quantitative relaxometry, diffusion-weighted imaging (DWI), diffusion tensor imaging (DTI), etc. They are not indicative of suspected mass lesions that compress nerves and are useful for DPN follow-up, severity assessment, and early screening. Peripheral nerve MRI signals alter in DPN patients, and these alterations are typically accompanied by morphological alterations. The cross-sectional area (CSA) of DPN patients was typically greater than that of healthy controls in both type 1 and type 2 diabetes. Patients with DPN often have this expression during the acute or subacute stages of the condition, and in the chronic state, atrophic fascicles with intraepidermal fat deposits are the hallmark (Zhang and Zhang 2023).

Ultrasonography Nerve ultrasonography is useful in identifying entrapment syndromes and discriminating between different kinds of neuropathy. It can also give real-time imaging of peripheral nerves. As an addition to NCS, high-frequency ultrasonography was initially employed in the clinical diagnosis of DPN. It may significantly increase the diagnostic efficacy of DPN and lower the rates of missed and misdiagnosed diagnoses by measuring the size, blood vessels, movement, echo, and damage to the nerve tissue. There are special diagnostic benefits to high-resolution ultrasonography for early or subclinical neuropathy. High-frequency ultrasonography can identify anomalies and subclinical peripheral nerve involvement in individuals with normal electrical diagnoses. Individuals diagnosed with DPN have increased nerve cross-sectional area (CSA) and longitudinal section, as well as notable increases in nerve echogenicity, border ambiguity, and nerve blood flow. Compared to mild DPN, the mean CSA in the nerves under examination was greater in moderate to severe DPN (Yu 2021).

3.4 Novel and Emerging Diagnostic Tools

3.4.1 Skin Biopsy

Skin biopsy enables intra-epidermal nerve fibre (IENF) morphometric measurement, which is reported as the IENF/mm, or intra-epidermal nerve fibres per section length. This method is highly reproducible and allows for the assessment of IENF pathology along with normative age-matched ranges using an accurate quantification methodology. Although IENFD declines with age and sex, it is not impacted by height or weight like neurophysiology. The European Federation of Neurological Societies has published instructions for its use in the diagnosis of peripheral neuropathy. However, the confirmation of small fibre neuropathy and the reference and cut-off values used determine the diagnostic yield of the skin biopsy (Murthy et al. 2013).

3.4.2 Corneal Confocal Microscopy

It is a nonsurgical in vivo ophthalmic examination approach that permits a detailed, high magnification, cellular investigation of the cornea. CCM imaging provides a higher magnification than traditional slit-lamp biomicroscopy by allowing for the delineation of the corneal layers at varying depths through the capture of several two-dimensional pictures. These characteristics enable CCM to obtain excellent pictures of the sub-basal nerve plexus's corneal C-fibres. Given the established correlation between harm to these fibres and diabetic peripheral neuropathy, there exists a possibility for their utilisation as a substitute biomarker for DPN (Carmichael et al. 2021).

Corneal disease detection and treatment constituted the main uses of CCM. But in the early 2000s, some studies indicated that patients with DPN that was developing worse had a steady loss of ocular sub-basal nerve endings. Following the confirmation of these results in various populations by several centres, the application of CCM in the research of DPN and neurodegeneration in general has increased dramatically.

3.5 Screening Recommendations for Diabetic Neuropathy

To diagnose diabetic peripheral neuropathy (DPN) and identify those susceptible for diabetic foot ulcers, the American Diabetes Association and the International Working Group on the Diabetic Foot advise frequently evaluating people with diabetes. For T2DM, monitoring should start upon diagnosis, and for T1DM, it should start five years later. After the initial screening, there should be annual examinations. According to the Toronto consensus, symptoms or indicators in conjunction with abnormal nerve conduction tests are the most accurate way to diagnose large fibre neuropathy. DPN is usually screened for using a screening test, foot examination, and neuropathic symptoms (Carmichael et al. 2021).

3.5.1 Scored Clinical Assessment Tools

Accomplished Standardised, quantitative, and objective measures are available in clinical evaluations to evaluate the degree of neuropathy impairments as well as the intensity of symptoms. These instruments, which have undergone rigorous validation research, are reasonably repeatable but only need a little amount of training. In many clinical studies, the key outcome measure is the neuropathy physical examination scale. Sensation, strength, and reflexes are evaluated semi-quantitatively using a range of approved measures. The Neuropathy Impairment Score (NIS) for neuropathic deficits (impairments), the University of Utah Early Neuropathy Scale (UENS), the Michigan Diabetes Neuropathy Scale (MDNS), the Toronto Clinical Neuropathy Score (TCNS, composite score of neuropathy symptoms sensory exam and reflexes), and the United Kingdom Screening Test (UKST) are popular exam scales. These measures balance the contribution of motor and sensory findings to the test score and are designed to be relevant to a broad range of neuropathy kinds and severity.

A diagnostic tool for evaluating distal symmetrical peripheral neuropathy in diabetics is the Michigan Neuropathy Screening Instrument (MNSI). It includes a lower extremity examination and a 15-item self-administered questionnaire. Unusual findings are valuable points. Fifteen questions about foot feeling, discomfort, numbness, and temperature sensitivity make up the first half of the MNSI. More neuropathic symptoms are indicated by a higher score. A quick physical examination, including testing for monofilament, ankle reflexes, vibration feeling at the dorsum of the great toe, and foot anomalies, is part of the second section. A person is deemed neuropathic and referred for further testing if they earn a score of more than two on a 10-point scale on the clinical portion of the MNSI (Herman et al. 2012).

The Utah Early Neuropathy Scale (UENS) was developed specifically to detect and quantify early small-fibre sensory neuropathy and to detect subtle changes in sensory distribution and severity. Pin sensation, large fibre and deep tendon reflex tests, allodynia/hyperesthesia, and motor deficits are all included, and the highest score is 42. Twenty-four of the 42 items on the UENS are devoted to a brief morphological characterisation of sensory loss to pin in the foot and lower leg (Singleton et al. 2008).

A quick, subjective test called the United Kingdom Screening Test (UKST) can be used to check for neuropathy. Five questions about the kind, intensity, and location of symptoms make up this symptom-based assessment. The questionnaire has a cut-off point of two or above, and a maximum score of nine (Fateh et al. 2015).

3.6 Challenges in Diagnosing Diabetic Neuropathy

There are significant differences in the diagnostic criteria for DN and painful DN in terms of questionnaires, electrophysiological methods, hierarchical classification systems, and diagnostic imaging and pathology tools. These factors all have advantages and disadvantages. It is necessary to determine categories and criteria that consider the needs of both general practice and research. For DN or painful DN, there isn't yet a single gold standard diagnostic test. Which clinical questionnaires and neurological evaluation metrics best separate those with DN and painful DN from those without these diseases is still a mystery. Not all cases of neuropathy in DM patients are related to or caused by diabetes, it should be highlighted. Thyroid, autoimmune, viral, vitamin deficiencies (particularly vitamin B12 deficiency), and intoxications are a few other conditions that might change this idea. It needs to be necessary to exclude these conditions from consideration in research settings or diagnosis in order to confirm "definite diabetic neuropathy," in addition to utilising appropriate laboratory evaluation instruments and clinical presentation (Ansari et al. 2021).

In addition, there are a number of symptoms, such movement abnormalities, that are similar to those of other neurological conditions that mimic diabetic neuropathy. Movement disorders are a broad category of neurological illnesses that are typified by abnormally elevated (hyperkinesia) or decreased (hypokinesia) sluggish motions. Peripheral neuropathy (PN) and movement abnormalities might share comparable pathophysiological pathways in a variety of neurological diseases, although the association between the two ailments is not extensively documented in the literature. Thus, they can exist side by side or serve as a continuum inside the same therapeutic range. Such illnesses include, for example, Parkinson's disease (PD), the second most prevalent neurodegenerative ailment in the elderly after Alzheimer's disease; asymmetric bradykinesia, stiffness, and rest tremor are prominent presentations of PD. About 16% of people with Parkinson's disease have peripheral neuropathy, which is a rather frequent condition in this group. It often manifests as an axonal neuropathy that is mostly sensory and is characterised by neuropathic pain, tingling, numbness, and occasionally ataxia. Furthermore, neurological symptoms associated with paraneoplastic illnesses can occur in cancer patients; these symptoms are frequently severe and go undiagnosed. When neuropathy and a certain mobility issue coexist, it may be an indication of a paraneoplastic syndrome, in which case neurological symptoms might be resolved by treating the underlying cancer. Parkinsonism and neuropathy have been linked to exposure to heavy metals, including carbon disulphide, manganese (Mn), and mercury. These metals have the potential to seriously harm the nervous system, which might result in both nerve damage and mobility problems (Gentile et al. 2022).

Painful DPN can be confused with inflammatory arthritis, which is confined to the joints and exacerbated by movement or activity. It is accompanied with morning stiffness that goes away as you move. Localised discomfort and soreness in the intertarsal region are symptoms of Morton's neuroma. Pain and numbness that radiate from under the medial malleolus to the sole are symptoms of plantar fasciitis and tarsal tunnel syndrome, which are exclusive to the inner side of the foot. DPN pain, on the other hand, is bilateral, symmetrical, and affects the entire foot, particularly the dorsum. It also interferes with sleep and gets worse at night (Anandhanarayanan et al. 2022).

3.7 Differential Diagnosis of Diabetic Neuropathy

When someone with diabetes or prediabetes has developed symptoms and indications and other possible causes of neuropathy have been ruled out, the existence of DPN is mostly established by clinical diagnosis. Since many other types of peripheral neuropathy can resemble or coexist with DPN and may be curable, a thorough differential diagnosis must be made in all cases before a definitive diagnosis of DPN can be made. Diabetic neuropathy can present with symptoms that mirror a number of other illnesses, thus it is important to take them into account when making a differential diagnosis:

Alcoholic Neuropathy Alcohol neuropathy is among the most common adverse consequences of chronic alcohol consumption. Alcohol damages the neurones through its direct toxicity as well as the hunger it produces. The aetiology is complicated and includes both dietary deficiencies and the direct harmful effects of alcohol metabolism on neurones. A history and physical check-up may be used to differentiate this condition from other forms of neuropathy. No specific laboratory test is available for diagnosis. Restoring vital nutrients and achieving alcohol abstinence should be the main objectives of therapy (Julian et al. 2019).

Nutritional Deficiencies Vitamin B12, B1 (thiamine), and E deficiencies might result in neuropathy. Vitamin B12 blood tests can be used to distinguish between diabetic neuropathy and these disorders. Vitamin B12 deficiency is one of the most prevalent DPN mimics in type 2 diabetics, and it can be completely resolved with therapy (Alam et al. 2020).

The primary cause of vitamin B12 deficiency is subacute combined degeneration of the spinal cord, which results in the loss of posterior columns and corticospinal pathways together with concurrent axonal sensory peripheral neuropathy. Likewise, as acquired copper insufficiency and vitamin B12 deficiency can present with very similar symptoms, those who come with myeloneuropathy should also have their acquired copper insufficiency checked. Furthermore, even though spinocerebellar disease is the primary neurologic deficit, there is often a concomitant large fibre sensory-predominant axonal peripheral neuropathy with vitamin E deficiency. Vitamin E insufficiency is linked to severe fat malabsorption (such biliary dysfunction or cystic fibrosis) or hereditary diseases (like abetalipoproteinemia or ataxia with vitamin E deficiency). Unusual for vitamin B6, it is linked to peripheral neuropathy when excesses or deficiencies occur. When using isoniazid for TB, vitamin B6 deficiency-related peripheral neuropathy primarily develops. This condition can be avoided by taking vitamin B6 supplements concurrently. Beriberi syndrome includes a progressive axonal sensory peripheral neuropathy brought on by a vitamin B1 (thiamine) deficiency. Atrophic skin alterations are also often observed. Thiamine deficiency can show in a variety of neuropathic ways, some of which may appear before systemic and cognitive symptoms (Staff and Windebank 2014).

Chronic Inflammatory Demyelinating Polyneuropathy (CIDP) A common symptom of this immune-driven neuropathy is progressive weakness and loss of feeling. It is an immune-mediated polyneuropathy with inflammation of peripheral nerves and nerve roots, as well as segmental demyelination and remyelination. More vibration and position senses are often affected than pain and temperature senses, and areflexia is frequently present as well. While idiopathic instances are frequent, CIPD can also be brought on by other illnesses. Its more pronounced motor involvement and response to immunotherapy set it apart from diabetic neuropathy. When symptoms and indicators appear suddenly and progress quickly, there should be serious suspicions (Gogia et al. 2026).

Mononeuropathies and Radiculopathies The loss or disruption of nerve function resulting from compression of one of the nerve roots near a vertebra is referred to as a radiculopathy. Lumbar radiculopathy, often known as sciatica, is a kind of radiculopathy that affects the lower back because the sciatic nerve is implicated in the nerve roots. A compressed nerve root in the neck causes cervical radiculopathy. Since the nerve roots in this area control sensations in your hands and arms, this is the area where symptoms are most likely to appear. Either a spinal injury or a gradual degeneration of the spine can cause radiculopathy. Furthermore, the damage of a single peripheral nerve, as can occur in a situation involving trauma or physical injury, is referred to as mononeuropathy. One kind of mononeuropathy is carpal tunnel syndrome, which occurs when the wrist nerve is compressed repeatedly. Peroneal nerve palsy, radial nerve palsy, and ulnar nerve palsy are further types of mononeuropathy. Research using imaging and electrophysiology can differentiate them from diabetic polyneuropathy (Robblee and Katzberg 2016).

Amyloidosis A diverse range of illnesses known as amyloidosis may be identified by the deposition of insoluble, incorrectly put together fibril proteins in the tissues, which eventually results in the disruption of normal tissue structure and function. Amyloid build-up in the nerves can result in peripheral neuropathy from this illness. This might result in neuropathies affecting sensory, motor, and/or autonomic fibres that are focal, multifocal, or diffuse. Specific procedures like as tissue biopsy and serum and urine protein electrophoresis can be used to differentiate it (Shin and Robinson-Papp 2012).

Vasculitic Neuropathy Systemic or non-systemic vasculitis, which only affects the peripheral nerve, is associated with a variety of peripheral nerve disorders known as vascular neuropathies. Inflammation of the vasa nervorum, mainly the nerve's epineural arteries, results in thrombosis and eventually ischaemic damage and is a feature common to all vasculitic neuropathies. Vasculitis-related neuropathy is frequently painful and might show up as systemic signs including fever, weight loss, or involvement of other organs. A biopsy of the afflicted blood vessels and nerves can confirm the diagnosis (Blaes 2015).

HIV-Associated Neuropathy A number of neuropathies can be brought on by an HIV infection. HIV-distal symmetrical polyneuropathy (HIV-DSP), also known as HIV-associated neuropathy, affects around 40% of these HIV-positive people. The causes of individuals' human immunodeficiency virus-distal symmetrical polyneuropathy may include the illness's debilitating nature and chronicity, as well as the impacts of post-antiretroviral treatment (ART). Pain, weakness, hypoesthesia, or no sensations at all are among the many symptoms that define HIV-peripheral neuropathy (HIV-PN). For distinction, a thorough medical history, a risk factor analysis, and particular HIV serological testing are necessary (Motwani et al. 2022).

Hereditary Neuropathies A number of hereditary peripheral neuropathies exist, including Charcot-Marie-Tooth disease (CMT, also called hereditary motor sensory neuropathy, or HMSN), small fibre neuropathies (SFN), and hereditary sensory and autonomic neuropathies (HSAN, also called hereditary sensory neuropathy, or HSN). A persistent course of gradually progressing weakness and sensory loss is the initial symptom of conditions like Charcot-Marie-Tooth illness. Loss of feeling in pain may be the primary clinical sign in HSN and HSAN, accompanied by further traumas and painless fractures. Autonomic symptoms, such as dyspepsia, anhidrosis, or cardiac arrhythmias, can exacerbate the condition depending on the subtype. To make a conclusive diagnosis, genetic testing might be necessary (Eggermann et al. 2018).

Drug-Induced Neuropathy A number of drugs, such as antibiotics, antiretrovirals, and chemotherapeutic medicines, can induce neuropathy. A complete medical history can be very helpful in identifying possible offenders. One of the main toxicities that limits dosage is peripheral neuropathy, which often reduces the quantity of chemotherapy that may be used to treat the underlying malignancies. Although toxicity is mostly determined by dosage, there is growing evidence that a patient's genetic composition and the type of cancer they have may also play a part. Even with more modern chemotherapy medications approved by the US Food and Drug Administration in recent years, peripheral neuropathy is still a frequent side effect.

The proteasome inhibitor bortezomib is the main therapy for multiple myeloma; nonetheless, it frequently causes a dose-depending sensory-predominant axonal neuropathy. It has been demonstrated that carfilzomib, a more modern proteasome inhibitor, causes less peripheral neuropathy than bortezomib (Staff and Windebank 2014).

3.8 Conclusion

Neuropathy symptoms might differ significantly throughout patients. Several symptom screening questionnaires with comparable scoring schemes have been created as a result. These are valuable tools for monitoring patients and evaluating their response to therapy. Diagnosing DPN requires a thorough clinical evaluation. Neurological examinations, laboratory studies, and clinical assessment are all essential for an accurate assessment of diabetic neuropathy. The need of a thorough evaluation cannot be overstated, as diabetic neuropathy can present with a wide variety of symptoms that may overlap with other neurological disorders. In order to customise successful treatment plans, avoid problems, and enhance patient outcomes, early and accurate diagnosis is essential. Healthcare professionals may treat glycaemic control as well as other risk factors including metabolic syndrome and cardiovascular disease by using a holistic strategy that helps them differentiate diabetic neuropathy from other forms of neuropathies. In order to ensure proper care and therapy, it is important to understand the differential diagnosis in order to rule out other illnesses that may resemble diabetic neuropathy. In the end, patients with diabetic neuropathy benefit from better care, an enhanced quality of life, and a decreased morbidity rate due to a thorough diagnosis procedure.

References

Alam U, Sloan G, Tesfaye S. Treating pain in diabetic neuropathy: current and developmental drugs. Drugs. 2020;80(4):363–84.

Anandhanarayanan A, Teh K, Goonoo M, Tesfaye S, Selvarajah D. Diabetic neuropathies. In: Endotext [Internet]. MDText.com, Inc; 2022.

Ansari P, Hannan JMA, Azam S, Jakaria M. Challenges in diabetic micro-complication management: focus on diabetic neuropathy. Int J Transl Med. 2021;1(3):175–86.

Blaes F. Diagnosis and therapeutic options for peripheral vasculitic neuropathy. Ther Adv Musculoskelet Dis. 2015;7(2):45–55.

Boulton AJ, Vinik AI, Arezzo JC, Bril V, Feldman EL, Freeman R, et al. Diabetic neuropathies: a statement by the American Diabetes Association. Diabetes Care. 2005;28(4):956–62.

Carmichael J, Fadavi H, Ishibashi F, Shore AC, Tavakoli M. Advances in screening, early diagnosis and accurate staging of diabetic neuropathy. Front Endocrinol. 2021;12:671257.

Chowdhury SR, Dey A, Mondal S, Gautam MK. Environmental microplastics and nanoplastics: effects on cardiovascular system. Toxicol Anal Clin. 2024;36(2):145–57.

Cruccu G, Sommer C, Anand P, Attal N, Baron R, Garcia-Larrea L, et al. EFNS guidelines on neuropathic pain assessment: revised 2009. Eur J Neurol. 2010;17(8):1010–8.

Deli G, Bosnyak E, Pusch G, Komoly S, Feher G. Diabetic neuropathies: diagnosis and management. Neuroendocrinology. 2014;98(4):267–80.

Eggermann K, Gess B, Häusler M, Weis J, Hahn A, Kurth I. Hereditary neuropathies: clinical presentation and genetic panel diagnosis. Dtsch Arztebl Int. 2018;115(6):91.

England JD, Gronseth GS, Franklin G, Miller RG, Asbury AK, Carter GT, et al. Distal symmetric polyneuropathy: a definition for clinical research: report of the American Academy of Neurology, the American Association of Electrodiagnostic Medicine, and the American Academy of Physical Medicine and rehabilitation. Neurology. 2005;64(2):199–207.

Fateh HR, Madani SP, Heshmat R, Larijani B. Correlation of Michigan neuropathy screening instrument, United Kingdom screening test and electrodiagnosis for early detection of diabetic peripheral neuropathy. J Diabetes Metab Disord. 2015;15:1–5.

Filler AG, Kliot M, Winn HR, Tsuruda JS, Hayes CE, Howe FA, et al. Magnetic resonance neurography. Lancet. 1993;341(8846):659–61.

Gentile F, Bertini A, Priori A, Bocci T. Movement disorders and neuropathies: overlaps and mimics in clinical practice. J Neurol. 2022;269(9):4646–62.

Gogia B, Rocha Cabrero F, Khan Suheb MZ, Lui F, Rai PK. Chronic Inflammatory Demyelinating Polyradiculoneuropathy. 2024 Mar 4. In: StatPearls [Internet]. Treasure Island (FL): StatPearls Publishing; 2026,

Hartemann A, Attal N, Bouhassira D, Dumont I, Gin H, Jeanne S, et al. Painful diabetic neuropathy: diagnosis and management. Diabetes Metab. 2011;37(5):377–88.

Herman WH, Pop-Busui R, Braffett BH, Martin CL, Cleary PA, Albers JW, et al. Use of the Michigan Neuropathy Screening Instrument as a measure of distal symmetrical peripheral neuropathy in Type 1 diabetes: results from the Diabetes Control and Complications Trial/Epidemiology of Diabetes Interventions and Complications. Diabet Med. 2012;29(7):937–44.

Hicks CW, Selvin E. Epidemiology of peripheral neuropathy and lower extremity disease in diabetes. Curr Diab Rep. 2019;19:1–8.

Islam A, Jena D, Mondal NS, Mondal S, Gautam MK. In-silico approaches for drug designing technology: bridging discovery and development. Curr Drug Discov Technol. 2025;22:E15701638326869. https://doi.org/10.2174/0115701638326869250207060616.

Julian T, Glascow N, Syeed R, Zis P. Alcohol-related peripheral neuropathy: a systematic review and meta-analysis. J Neurol. 2019;266(12):2907–19.

Khan KS, Pop-Busui R, Devantier L, Kristensen AG, Tankisi H, Dalgas U, et al. Falls in individuals with type 2 diabetes; a cross-sectional study on the impact of motor dysfunction, postural instability and diabetic polyneuropathy. Diabet Med. 2021;38(9):e14470.

Mondal NS, Islam A, Gautam MK, Mondal S, Ahmed M, Jamir SI, Khongwir LD, Mahajon B. Unraveling the mechanisms of diabetic wounds: insights into pathogenesis and advanced treatment strategies. Curr Diabetes Rev. 2025;2025:e15733998368222.

Motwani L, Asif N, Patel A, Vedantam D, Poman DS. Neuropathy in human immunodeficiency virus: a review of the underlying pathogenesis and treatment. Cureus. 2022;14(6):e25905.

Murthy S, Gautam MK, Goel S, Purohit V, Sharma H, Goel RK. Evaluation of in-vivo wound healing activity of Bacopa monniera on different wound models in rats. BioMed Res Int. 2013;2013:972028.

Pop-Busui R, Ang L, Boulton AJ, Feldman EL, Marcus RL, Mizokami-Stout K, et al. Diagnosis and treatment of painful diabetic peripheral neuropathy. Arlington: American Diabetes Association; 2022.

Robblee J, Katzberg H. Distinguishing radiculopathies from mononeuropathies. Front Neurol. 2016;7:111.

Rosenberger DC, Blechschmidt V, Timmerman H, Wolff A, Treede RD. Challenges of neuropathic pain: focus on diabetic neuropathy. J Neural Transm. 2020;127(4):589–624.

Saeedi P, Petersohn I, Salpea P, Malanda B, Karuranga S, Unwin N, et al. Global and regional diabetes prevalence estimates for 2019 and projections for 2030 and 2045: results from the International Diabetes Federation Diabetes Atlas. Diabetes Res Clin Pract. 2019;157:107843.

Shin SC, Robinson-Papp J. Amyloid neuropathies. Mt Sinai J Med J Transl Pers Med. 2012;79(6):733–48.

Singleton JR, Bixby B, Russell JW, Feldman EL, Peltier A, Goldstein J, et al. The Utah Early Neuropathy Scale: a sensitive clinical scale for early sensory predominant neuropathy. J Peripher Nerv Syst. 2008;13(3):218–27.
Sloan G, Selvarajah D, Tesfaye S. Pathogenesis, diagnosis and clinical management of diabetic sensorimotor peripheral neuropathy. Nat Rev Endocrinol. 2021;17(7):400–20.
Smith S, Normahani P, Lane T, Hohenschurz-Schmidt D, Oliver N, Davies AH. Prevention and management strategies for diabetic neuropathy. Life. 2022;12(8):1185.
Staff NP, Windebank AJ. Peripheral neuropathy due to vitamin deficiency, toxins, and medications. Contin Lifelong Learn Neurol. 2014;20(5):1293–306.
Tesfaye S, Boulton AJ, Dyck PJ, Freeman R, Horowitz M, Kempler P, et al. Diabetic neuropathies: update on definitions, diagnostic criteria, estimation of severity, and treatments. Diabetes Care. 2010;33(10):2285.
Yang Z, Zhang Y, Chen R, Huang Y, Ji L, Sun F, et al. Simple tests to screen for diabetic peripheral neuropathy. Cochrane Database Syst Rev. 2018;2018(7):CD010975.
Yasuda H, Terada M, Maeda K, Kogawa S, Sanada M, Haneda M, et al. Diabetic neuropathy and nerve regeneration. Prog Neurobiol. 2003;69(4):229–85.
Yu Y. Gold standard for diagnosis of DPN. Front Endocrinol. 2021;12:719356.
Zhang X, Zhang F. Peripheral neuropathy in diabetes: what can MRI do? Diabetes. 2023;72(8):1060–9.
Zimmet P, Alberti KGMM, Shaw J. Global and societal implications of the diabetes epidemic. Nature. 2001;414(6865):782–7.

Part II

Role of Antioxidants in Diabetic Neuropathy

4 Molecular Mechanism of Oxidative Damage in Diabetes

Kirtimaya Mishra, Diptimayee Jena, Arup Dey, Arghadeep Bhattachrjee, Manish Kumar Gautam, and Raja Chakraborty

4.1 Introduction

Diabetes and its consequences are strongly linked to oxidative stress. Oxidative stress plays a vital role in the onset and progression of diabetes, as well as its related consequences. This condition arises from a disruption in the equilibrium among generation of reactive oxygen species (ROS) and efficacy of body's antioxidant defense system, leads to cellular injury and dysfunction. Recent research has shown the enormous influence of oxidative stress upon human health, including its link to diabetes. Oxidative stress occurs when the body produces more reactive oxygen species (ROS) than its antioxidant defense system can neutralize. Reactive oxygen species (ROS), including free radicals and various other reactive compounds, are initiated as a byproduct of regular cellular metabolic activities. Oxidative stress arises when the body's antioxidants cannot prevent the excess generation of free radicals (Forman and Zhang 2021; Badawi et al. 2010). While internal environment of a cell compromising by factors such as infections, illness, toxins, or nutritional deficiencies, mitochondria altered the flow of electrons, resulting to the initiation of ROS & RNS; reactive oxygen species &reactive nitrogen species respectively. Such

K. Mishra
School of Pharmacy and Life Science, Centurion University of Technology & Management, Bhubaneswar, Odisha, India

D. Jena (✉)
School of Pharmacy and Life Science, Centurion University of Technology & Management, Bhubaneswar, Odisha, India

Faculty of Pharmaceutical Science, Assam down town University, Guwahati, Assam, India

A. Dey · A. Bhattachrjee · M. K. Gautam
Faculty of Pharmaceutical Science, Assam down town University, Guwahati, Assam, India

R. Chakraborty
Institute of Pharmacy, Assam Don Bosco University, Tepesia, Assam, India

M. K. Gautam et al. (eds.), *Antioxidants and Functional Foods for Diabetic Neuropathy*, https://doi.org/10.1007/978-981-95-9326-2_4

alteration results in a decrease in consuming oxygen. Such phenomenon, termed "oxidative shielding," functions like a protective mechanism of action by either limiting the absorption of harmful pathogens or toxins from the external environment inside the cell or inducing apoptosis in affected cells, thereby preventing the dissemination of these threats to adjacent cells. Consequently, the production of ROS is recognized as a physiological response to stress (Naviaux 2012). The concept regarding "oxidative stress" describes such condition characterized by elevated levels of ROS along with RNS, which may arise from either excessive production or inadequate removal. ROS and RNS are extremely reactive chemicals, and their excess cause impairment of proteins, lipids, as well as DNA (Johansen et al. 2005). The aggregation of reactive oxygen species (ROS) alongside reactive nitrogen species (RNS) is crucial to the occurrence of oxidative stress, which may outcome in physiological dysfunction, cells dying, and the beginning of various diseases, which comprises diabetic disorder along with cancer, and also contributes towards ageing process of organisms (Ceriello 2006). In aerobic environments, biological systems encounter oxidants, which may arise either as intentional agents or as metabolic byproducts. These oxidants are categorized into two types: reactive oxygen species (ROS) and reactive nitrogen species (RNS), both of which are paramagnetic free radicals. Reactive oxygen species (ROS) are chemical entities created during partial oxygen reduction, including superoxide anion (O2−), hydrogen peroxide (H2O2), and the hydroxyl radical. Various oxidation states and reactive forms of nitric oxide synthase (NOS) products, notably nitric oxide (NO), nitroxyl (NO-), S-nitrosothiol (RSNO), particularly peroxynitrite, are termed reactive nitrogen species (D'Autreaux and Toledano 2007). Historically these are regarded as detrimental, both ROS & RNS are conceded regarding essential functions in normal physiological processes. These reactive species are capable of initiating and facilitating enzyme along with gene-dependent reaction that occurs in both physiological along with pathological contexts. Homeostasis may be disrupted and sickness may occur from an imbalance marked by either an excess or a shortage of ROS and/or RNS. Disequilibrium in redox balance is thought to be a common pathway for several pathogenic processes (Palmieri and Sblendorio 2007).

4.2 Reactive Oxygen Species (ROS) & Their Significance Towards Diabetes

Research has attained that in diabetic conditions, reactive oxygen species (ROS) are generated in a number of tissues. Numerous processes, including membrane-bound NADPH oxidase, the mitochondrial electron transport chain, and nonenzymatic glycosylation, can generate ROS in cells. In diabetic animals, the glycation process can be observed throughout multiple organs & tissues, resulting in the formation of diverse glycated proteins comprising glycosylated hemoglobin, albumin, as well as lens crystalline, all created nonenzymatically by glycation processes. This process produces Amadori products, Schiff bases, including, ROS alongside AGEs; advanced glycosylation end products (Dandona et al. 1996; Sakurai and Tsuchiya

1988; Mohazzab et al. 1994; Brownlee 2001a; Baynes and Thorpe 1999; Harrison et al. 2003; Mondal et al. 2025). The electron transport chain found throughout mitochondria is an important mechanism for the formation of reactive oxygen species (ROS) (Fig. 4.1). In the scenario of diabetes, this chain is triggered and results in enhanced ROS production. NADPH oxidase, a membrane-associated enzyme, has been found to possess a major part in ROS production. This enzyme is made up of enclosed by the membrane subunits such as Nox1, Nox4,gp91 phox (Nox2) and p22 phox, as well as the oxidase's catalytic site as well as cytosolic components which include p47phox along with p67phox. NADPH oxidase can be triggered by many stimuli, notably advanced glycation end products (AGEs), insulin, and angiotensin II, which can all be raised in diabetic circumstances (Kaneto et al. 2010).

Acute exposure to elevated glucose levels enhances the expression of the insulin gene, whereas sustained exposure adversely affects β-cell functionality. Nevertheless, chronic hyperglycemia can hinder both the production and secretion of insulin. Diabetics typically experience β-cell glucose toxicity. In the framework of diabetes, hyperglycemia as a whole, together with the subsequent formation of reactive oxygen species (ROS), inhibits insulin gene expression and secretion, eventually leading to cell death (Weir et al. 2001; Chandra et al. 2025; Poitout et al. 1996; Poitout and Robertson 2002; Sharma et al. 1995; Moran et al. 1997; Evans et al. 2003; Gorogawa et al. 2002; Harmon et al. 2005; Ihara et al. 1999; Kajimoto et al. 1999; Maechler et al. 1999; Chowdhury et al. 2024; Kaneto et al. 1996, 2001, 2005; Matsuoka et al. 1997; Teli et al. 2025; Robertson 2004a; Tanaka et al. 1999, 2002; Gautam et al. 2015). Research indicates that a decrease in the expression of the insulin gene is linked to a decrease in the expression and/or DNA binding activity of the transcription factors MafA and pancreatic and duodenal homeobox-1 (PDX-1). Long-term exposure to elevated glucose levels decreases these transcription factors' DNA binding abilities and/or frequency of expression (Ohlsson et al.

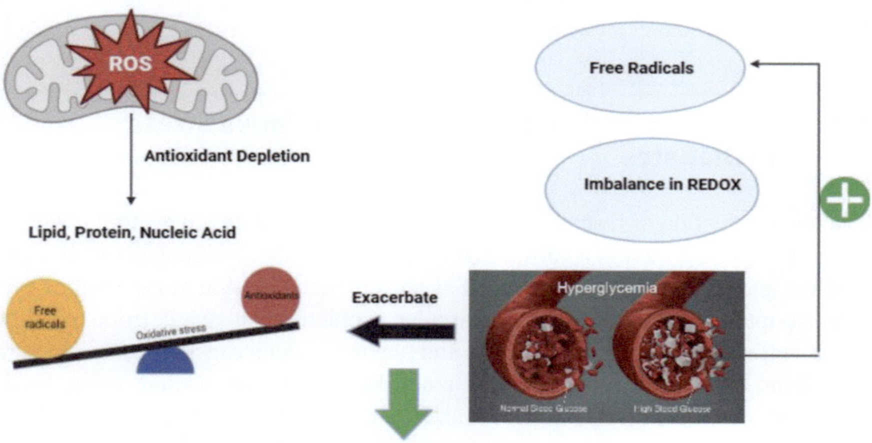

Fig. 4.1 Role of oxidative stress and redox imbalance in hyperglycemia-induced diabetic complications

1993; Gautam et al. 2012, 2013a, b; Miller et al. 1994; Jonsson et al. 1994; Stoffers et al. 1997; Ahlgren et al. 1998; Ferber et al. 2000; Holland et al. 2002; Miyazaki et al. 2004; Kaneto et al. 2007, 2008). PDX 1 is critical for pancreatic growth, β-cell distinctions, surrogate β cell induction, and maintaining mature β cell activity. MafA, a transcription factor exclusive to β cells, enhances insulin gene transcription (Olbrot et al. 2002; Kataoka et al. 2002; Matsuoka et al. 2003, 2004, 2007; Gautam et al. 2014).

4.3 Antioxidant Defense System in Diabetes

Diabetes mellitus induces cellular and organ damage primarily across the mechanisms of ROS as well as RNS. Types 1 and type 2 diabetes significantly increase oxidative and nitrosative stress, which impacts the activity of antioxidant enzyme systems, including glutathione peroxidase (GPx), catalase (CAT), superoxide dismutase (SOD), glutathione (GSH), and total antioxidant capacity (TAC). Diabetes can be detected by the ensuing deterioration of carbs, lipids, proteins, and nucleic acids. Diabetes mellitus lowers antioxidant defenses and damages mitochondrial function via promoting the NFkB and p38-MAPK signaling pathways through Rac, TNF-α, and TLR4. Hyperglycemia impairs the functional activity of pancreatic islet beta cells, leading to harmful oxidative stress (Puchau et al. 2010). Beta cells operate normally, but plasma TAC levels remain elevated (Miyazaki et al. 2007). Prediabetes begins with insulin resistance and hyperinsulinemia, progressing to glycemic control failure and diabetic hyperglycemia. As diabetes progresses, plasma TAC levels decrease gradually. Individuals experiencing acute pancreatitis due to alcohol consumption, biliary complications, trauma, or unknown causes exhibited a 40% reduction in serum TAC (Total Antioxidant Capacity). The availability of diabetes may impair antioxidant defenses, resulting in cumulative DNA damage over time (Song et al. 2007; Ferrari 2007; Rysz et al. 2007; Wang et al. 2011; Gautam and Goel 2014; Jain et al. 2009).

4.4 Mitochondrial Impairment and Oxidative Stress in Diabetes

The overproduction of reactive oxygen species (ROS) is a distinguishing feature of oxidative stress, which is a substantial risk factor for the development of diabetes mellitus and its consequences, as well as causing deterioration at the level of each cell. Apoptosis and diabetic microvascular problems can result from oxidative stress's effects on mitochondrial shape and function, which can also cause structural alterations and anomalies in the way macromolecules like; nucleic acids, lipids, alongside proteins, activity (Gorrini et al. 2013; Alfarhan et al. 2020; Jha et al. 2016; Pop-Busui et al. 2006; Burgos-Morón et al. 2019). Mitochondria are commonly accredited as the primary producer of reactive oxygen species (ROS); however, it is crucial to note that they do not constitute the sole organelle implicated in the oxidative stress linked to diabetes (Oguntibeju 2019; Vitale et al. 2013). There is emerging evidence that stress inside the endoplasmic reticulum (ER) plays a significant role in the development of diabetes (Rovira-Llopis et al. 2014; Sun et al. 2015; Pang et al. 2020; Rocha et al. 2016). Endoplasmic reticulum (ER), in possession of folding proteins into functional three-dimensional configurations, with disulfide bonds playing an important role in this process. Endoplasmic reticulum oxidoreductin 1 (ERO1), glutathione (GSH), and native protein disulfide isomerases (PDI) collaborate in a chaperone-like manner to address and rectify the improper disulfide bond formation. A redox imbalance is caused by this oxidative folding system, which produces significant quantities of ROS and reduces the accessibility of the GSH pool (Papa 2012; Pasupuleti et al. 2020). Changes in the oxidative conditions within the endoplasmic reticulum (ER) lumen can result in the unauthorized production of disulfide bonds, and the subsequent buildup of misfolded polypeptides, a phenomenon known as ER stress (Zeeshan et al. 2016; Murphy 2013). Most of the ROSs produced as a result of the oxidative folding machinery being activated to fix the abnormal bonds. The hyperoxidative environment in the endoplasmic reticulum (ER) lumen promotes the formation of misfolded proteins by interfering with PDI's correct activity. A prolonged state of oxidative and ER stress is one of the effects of this harmful loop, which ultimately impairs ER function (Malhotra and Kaufman 2007). All of the aforementioned organelles play a part in the production of reactive oxygen species (ROS). The endoplasmic reticulum (ER) mainly creates ROS during the formation of disulfide bonds in the protein folding process, whereas mitochondria make ROS through ETC; Electron Transport Chain (ETC). These pathways are involved in the pathophysiology of diabetes, particularly type 2 diabetes (T2D), which is linked with excessive quantities of glucose in the bloodstream. Mitochondria use glucose as an energy source through the ETC, which naturally creates ROS as a consequence. Elevated glucose levels can surpass the capability of antioxidant defenses, causing oxidative stress inside mitochondria and simultaneously boosting insulin synthesis. Furthermore, excessive ROS can overwhelm the antioxidant systems in the ER, contributing to oxidative stress in that organelle as well.

4.5 Oxidative Stress & Inflammatory Pathways in Diabetes

With respected to diabetes mellitus, primary source of inflammatory factors is predominantly visceral white adipose tissue. Hyperglycemia triggers cytokine and chemokine release through many pathways, including hypoxia and macrophage infiltration. Inflammatory factors, alongside OS; oxidative stress, act as an crucial factor in onset of diabetes by insulin resistant enhancement via alterations in beta-cell functionality & disruption of insulin signaling pathways. The significance of interleukins in inflammatory processes, encompassing each pro-inflammatory & anti-inflammatory function, has resulted to the examination of various interleukins in patients with T2DM. It is anticipated that the etiology of type 2 diabetes would involve significant roles for interleukin-1 beta (IL-1b), interleukin-6 (IL-6), and interleukin-10 (IL-10) (Tsalamandris et al. 2019; Oguntibeju 2019; Akbari and

Hassan-Zadeh 2018; Alfadul et al. 2022; Dror et al. 2017; Burke et al. 2018; Barry et al. 2016). Diabetes as well as oxidative stress (OS) promote vascular remodeling, resulting in changes in arterial structure that begin with endothelial dysfunction. Both inflammation and oxidative stress can cause endothelial dysfunction by reducing nitric oxide availability, raising the risk of hypertension via the renin-angiotensin-aldosterone pathway (Kartha et al. 2017; Shi and Vanhoutte 2017).

4.6 Oxidative Damage to Biomolecules in Diabetes

Diabetes causes hyperglycemia that affects all cells in the body. However, most cells can regulate glucose transfer to maintain glucose tolerance. Hyperglycemia has the potential to impact various cell types, including neuronal and Schwann cells in peripheral nerves, mesangial cells in the kidney glomerulus, and retinal capillary endothelial cells (Chowdhury et al. 2019). These particular cells do not possess this capability. Diabetes-induced tissue damage harms cells that cannot maintain a consistent glucose concentration during hyperglycemia, contributing to diabetes complications. To better understand diabetes problems, it's important to investigate the internal mechanics of these cells.

The mechanisms are as follows:

- *Polyol pathway flux:*
  Excessive oxidized DNA, protein, and lipid levels are indicators of elevated oxidative stress in different organs caused via diabetes. In addition to impairing these biomolecules' activity, this oxidative stress triggers a number of cellular reactions, which comprises as the activation of protein kinase C (PKC), the transcription factor NF-κB, and JNK stress-related kinases (Konishi et al. 1997; Mohamed et al. 1999; Ho et al. 2000). The pathophysiology of several diabetes problems is believed to be influenced by the improper activation of such important regulatory elements, which negatively impact cellular functioning (Rosen et al. 2001). The precise mechanism through which hyperglycemia results to enhanced oxidative stress remains ambiguous. It is a reduced capacity of cellular antioxidant defense system, coupled with heightened ranges of ROS, contributes to this phenomenon. The oxidative stress caused by hyperglycemia may be caused by a number of factors, including auto-oxidation of glucose, nonenzymatic glycation processes, the interaction of glycated products with their respective receptors, excessive ROS generation by mitochondria, and activation of the polyol pathway (Fig. 4.2) (Wolff and Dean 1987; Mullarkey et al. 1990; Schmidt et al. 1994; Nishikawa et al. 2000; Cheng and Gonzalez 1986).
- *Intracellular production of AGE precursors*
  Glycative stressors, arising from routine metabolic and glycolytic activities both within cells and in the bloodstream, exert an influence on nearly all cells throughout the body to varying degrees. Both hyperglycemia and ageing exacerbate these glycative stressors. Cells regulate the synthesis and degradation of AGE

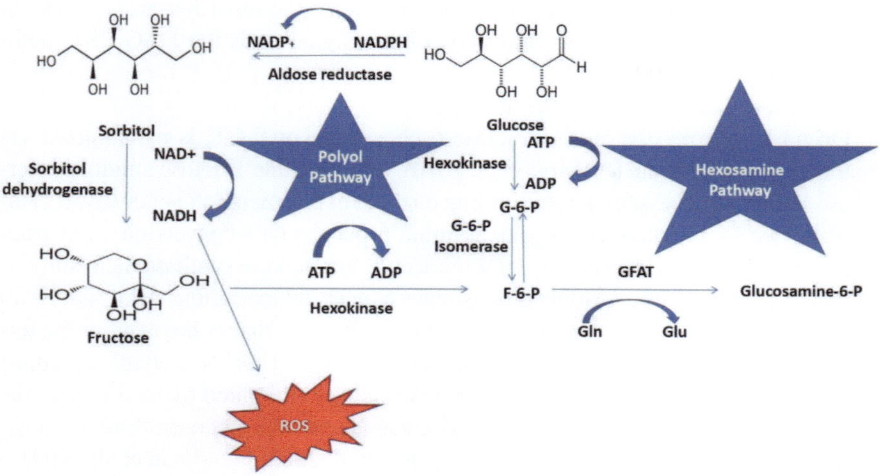

Fig. 4.2 Hyperglycemia-induced activation of the polyol and hexosamine pathways leading to ROS generation

precursors and AGEs through a range of repair mechanisms, as previously discussed. Numerous cell reactions may result from an aberrant buildup of AGEs in a cell. These consist of apoptotic cell death and/or damage healing or elimination. Any individual protein's glycation can have a variety of effects. AGE modification makes a lot of proteins functionally inactive. For instance, pentosidine or CML may glycate the αA-crystallin protein on lysines 11 and 78 (Abraham et al. 1994; Nagaraj et al. 2012). Numerous AGEs emanating from food sources, metabolic processes, or cellular degradation by products are present in the extracellular milieu. Free glycated amino acids, apolipoproteins, and serum albumins are examples of common circulating AGEs. Glycation frequently targets collagen proteins, with the crosslinking glucosepane being the most prevalent collagen AGE in skin. Extracellular matrix proteins serve as a significant reservoir for circulating AGEs and are commonly subjected with glycation processes. Another common characteristic of diabetes and ageing is glycation of the extracellular matrix proteins, which acts as significantly a major factor in the basement membrane enlargement associated with problems from diabetes (Dyer et al. 1993; Genuth et al. 2015).

- *Activation of PKC; Protein kinase-C*

  Examples pertaining to cellular processes include increased polyol pathway activation, nonenzymatic glycation and advanced glycation end product pathways, increased reactive oxygen species production, and the diacylglycerol (DAG)-protein kinase C (PKC) signaling pathway activation (Obrosova et al. 2003; Wendt et al. 2006; Brownlee 2001b; Koya and King 1998; Nishizuka 1992). The intracellular and extracellular abnormalities brought on by hyperglycemia most likely result in modifications to signal transduction pathways, which is having impact on both gene expression and protein function and ultimately originate cellular malfunction as well asdysruption. One of the most researched mechanisms in relation to diabetes-induced cellular signaling is the DAG-PKC pathway (Robertson 2004b).

- *Increased hexosamine pathway activity*
  Originating from glycolysis, fructose-6-phosphate, or F-6-P, is metabolized via the hexosamine route (Acharya et al. 2010). Glucosamine-fructose amidotransferase (GFAT) is an essential rate-limiting enzyme in this metabolic pathway. It catalyzes necessary intermediate glucosamine 6-phosphate's conversion from fructose-6-phosphate. The enzyme UDP-N-acetylglucosamine synthase then converts this intermediate into Uridine diphosphate-N-acetylglucosamine (UDP-GlcNAc) (Buse 2006) (Fig. 4.2). One vital metabolic substance that is involved in the formation of glycosyl chains in proteins and lipids is UDP-N-acetylglucosamine. Another important significance of this material is highlighted by the fact that the enzyme O-glucosamine-N-acetyl transferase is essential in controlling the post-translational modification of proteins (Fülöp et al. 2007; Freedman et al. 1999).

4.7 Contribution of AGEs to Oxidative Injury in Case of Diabetes

Maillard process creates AGEs. Protein amino groups undergo non-enzymatic reactions with reducing sugars such as glucose, resulting in the formation of Schiff bases. These Schiff bases subsequently rearrange to produce Amadori compounds. Research has shown that various substances, including pyridoxamine and aminoguanidine, can influence this process& limit the development of AGE by either trapping reactive carbonyl species or blocking the Maillard reaction, which stops them from interacting with proteins (Yamagishi and Imaizumi 2005). Following a very slow series of irreversible dehydration and condensation processes, these Amadori adducts finally give rise to AGEs, which are yellowish-brown substances with distinct fluorescence (Lo et al. 1994). Glyoxal, methylglyoxal, & 3-deoxyglucosone represent such degradation products of glucose. Additionally, they include dicarbonyl compounds produced through autoxidation, along with α-hydroxy aldehydes alike glyceraldehyde alongside glycolaldehyde (Ahmed et al. 2003; Uribarri et al. 2005). AGEs are continuously originated & accumulate within circulatory system along with other tissues in the event of chronic hyperglycemia, which leads to vascular problems in diabetes. AGEs can be ingested exogenously by food, even though they are endogenously formed by oxidative stress or hyperglycemia. It is known that the body retains between 6% and 7% of AGEs from meals for a specific amount of time (Yamagishi et al. 2007; Bierhaus et al. 1998). The upregulation regarding receptors for Advanced Glycation End-products (RAGEs) is stimulated by presence of AGEs. It is believed that the persistent activation of the AGE-RAGE pathway contributes to the legacy effect, also referred to as long-term metabolic memory (Naka et al. 2004). Because AGEs are recognized via RAGE; Receptor for Advanced Glycation End products (RAGE) in endothelial cells, oxidative stress is produced more rapidly in these cells. This process then results in the synthesis of several cytokines and growth factors by activating nuclear factor κB. Transcription factors are getting activated by triggering an inflammatory response and starts the production of adhesion molecules (Schmidt and Stern 2000). People with diabetes have far higher levels of advanced glycation end-products (AGE) in their atherosclerotic lesions, which is correlated with their level of blood glucose control (Magenta et al. 2014).

4.8 Diabetic Complications & Oxidative Stress

The progression of microvascular and cardiovascular complications associated with diabetes is profoundly affected by oxidative stress. Metabolic disturbances linked to diabetes result in an accumulation of mitochondrial superoxide within the cardiac tissue and the endothelial cells of both large and small blood vessels. Increased superoxide production triggers five main pathways that are involved in the pathophysiology of complications: the polyol pathway's flux, increased activity of the hexosamine pathway, AGEs origination, AGE receptor expression and its ligands activation, as well as the activating several isoforms related to protein kinase C. Additionally, two essential antiatherosclerotic enzymes, prostacyclin synthase and endothelial nitric oxide synthase, are directly inhibited by superoxide. These pathways produce high ranges of intracellular ROS, which in turns trigger many proinflammatory processes, prevent angiogenesis in response to ischemia, and result in long-lasting epigenetic modifications that cause pro inflammatory genes to continue to be expressed even after blood glucose levels return to normal—a phenomenon known as "hyperglycaemic memory". Two important factors that contribute to the development of atherosclerosis and cardiomyopathy in people with type 2 diabetes are pathway-selective insulin resistance, which is defined by an increase in mitochondrial reactive oxygen species (ROS) production from free fatty acids, and ROS-mediated inactivation of enzymes that counteract atherosclerosis. Superoxide dismutase over expression has been demonstrated to offer protection against diabetic cardiomyopathy, retinopathy, and nephropathy in investigations involving transgenic diabetic mice (Giacco and Brownlee 2010).

4.9 Therapeutic Approaches Aimed at Mitigating Oxidative Stress in Regards to Diabetes

It's a challenge for determining the time sequence of the interaction between elevated oxidative stress along with elevated inflammation in association with illnesses like type-2 diabetes because of their intricate and close relationship. A decrease in cellular antioxidant defenses is linked to an elevation in oxidative stress in diabetes patients. Numerous attempts have been made to use antioxidant supplements to lessen the cellular alterations in diabetic patients that are dependent on oxidative stress. Experimental and epidemiologic research provide the majority of the justification for the use of specific vitamins along with other substances in preventing and treating diabetes mellitus. Long-chain omega-3 fatty acids, antioxidants, and some medications represent some of the dietary components that can affect a person's susceptibility to inflammatory illnesses.

In this brief overview, we discuss the potential effects of nonenzymatic antioxidants on oxidative stress, including vitamins A, C, and E, folate, glutathione, coenzyme Q10, α-lipoic acid, carotenoids, flavonoids, and trace minerals like copper, zinc, magnesium, and selenium. Furthermore, we draw attention to the part that enzymatic processes—more especially, glutathione peroxidase, thioredoxin, catalase, and superoxide dismutase—play in regulating oxidative stress Furthermore, we will focus on antioxidant substances that have the potential to enhance vascular function in diabetic environments. As briefly described, endothelial dysfunction acts a crucial role regarding development of micro as well as macrovascular diabetic problems and also intimately correlated with oxidative stress, inflammation, & mitochondrial dysfunction. Therefore, treatment approaches that focus on these biological processes may be essential for controlling diabetes and its severe side effects. Here, we go over the primary anti-inflammatory and antioxidant benefits linked to some of the most effective pharmacological and non-pharmacological antidiabetic treatments.

4.10 Physical Exercise (Training)

Regular physical activity (training) has been shown to enhance metabolic health by improving insulin sensitivity and glycemic management, as well as by having a beneficial effect on visceral fat and belly circumference (de Lemos et al. 2007). Frequent exercise also positively influences blood pressure as well as lipid profiles, which are having other cardiovascular risk factors emerged to type 2 diabetes. This is made obvious by a decrease in total cholesterol, LDL cholesterol, and triglycerides and an increase in HDL, or high-density lipoprotein cholesterol (Bassuk and Manson 2005; de Lemos et al. 2009). Additionally, exercise enhances NO bioavailability, which in turn improves vascular function (Green et al. 2004; Kwon et al. 2011). Because active tissues use more oxygen during exercise, there is an increase in the production of free radicals (Teixeira-Lemos et al. 2011; Vierck et al. 2012; Radak et al. 2013). Only a tiny portion (about 2–5%) of the oxygen is transformed into ROS, even though the majority is used in the mitochondria for oxidative phosphorylation, where it is broken down into water (Cooper et al. 2002). By increasing cells' resilience to oxidative stress and lowering baseline oxidative damage, long-term, moderate-intensity physical activity (training) has a favorable impact on the oxidative homeostasis of cells and tissues. This is corroborated by observed increases in glutathione peroxidase and superoxide dismutase activities, which show that regular exercise efficiently boosts antioxidant capacity and shields cells from oxidative stress-induced damage.

4.11 Lifestyle Interventions

Nutritional counselling, the adoption of low-calorie, balanced meals, and the encouragement of physical activity are all examples of non-pharmacological anti-diabetic tactics. Significant alterations in lifestyle may be crucial for averting the development and/or advancement of diabetes, as well as for mitigating vascular complications. These changes are linked to several beneficiary impacts, including anti-inflammatory along with antioxidant properties.

4.12 Dietary Intervention

The primary goals of lifestyle intervention programs for people having type-2 diabetes, particularly those who are obese, have been calorie restriction and weight loss, as these strategies appear to have a significant positive influence on cardiovascular risk and glycemic control (Buse et al. 2006). A significant content of research has been undertaken to identify the optimal macronutrient composition that may postpone the development of disease such as CVD, cardiovascular disease. Nevertheless, ideal proportions of proteins, carbohydrates, as well as fats differ among individuals, complicating the formulation of a universal dietary guideline for these patients (Franz et al. 2004; Grundy et al. 2005).

4.13 Future Directions and Research Perspectives in the Field of Oxidative Damage in Diabetes

The development and evolution of diabetes, as well as its consequences, such as retinopathy, nephropathy, and cardiovascular illnesses, are significantly influenced by oxidative stress. New approaches to early detection, targeted medicines, and individualized treatment are opening up as our knowledge of the underlying mechanisms advances. The development and evolution of diabetes, as well as its consequences, such as retinopathy, nephropathy, and cardiovascular illnesses, are significantly influenced by oxidative stress. New approaches to early detection, targeted medicines, and individualized treatment are opening up as our knowledge of the underlying mechanisms advances. The absence of sensitive, targeted biomarkers that can accurately detect early oxidative stress is one of the primary obstacles to controlling oxidative damage in diabetes. Future studies will probably concentrate on finding and confirming new biomarkers that are unique to diabetes and that can indicate oxidative damage early on, enabling timely treatment.

4.14 Conclusion

The intricate interactions between metabolic dysfunctions, mitochondrial impairment, &overproduction of ROS are revealed via molecular mechanisms underpinning oxidative damage in diabetes. As a primary trigger, hyperglycemia starts processes that lead to excessive ROS generation, including the polyol pathway, protein kinase C activation, advanced glycation end-products (AGEs) synthesis, and oxidative phosphorylation. These processes overpower the body's antioxidant defenses, causing oxidative disruption to proteins, lipids, and also DNA at the molecular & cellular levels. In addition to affecting insulin signaling and aggravating glucose imbalance, this oxidative stress has a role in the development of diabetic sequelae, such as cardiovascular diseases, nephropathy, retinopathy, and neuropathy. A better comprehension of these processes has made it possible to develop therapeutic strategies that target particular oxidative pathways, improve mitochondrial function, and strengthen endogenous antioxidant defenses. Along with improvements in genetic and pharmaceutical therapies, future research aiming at pinpointing specific biomarkers holds promise for lowering oxidative damage and lessening the difficulties associated with diabetes. The field becomes closer to more individualized, efficient methods of managing oxidative stress in diabetes patients with further investigation of these molecular pathways, having ultimate goal of enhancing patient outcomes and quality of life.

References

Abraham EC, Cherian M, Smith JB. Site selectivity in the glycation of alpha A- and alpha B-crystallins by glucose. Biochem Biophys Res Commun. 1994;201:1451–6.

Acharya A, Das I, Chandhok D, Saha T. Redox regulation in cancerA double-edged sword with therapeutic potential. Oxidative Med Cell Longev. 2010;3(1):23–34.

Ahlgren U, Jonsson J, Jonsson L, Simu K, Edlund H. β-Cell-specific inactivation of the mouse Ipf1/Pdx1 generesults in loss of the β-cell phenotype and maturity onsetdiabetes. Genes Dev. 1998;12(12):1763–8.

Ahmed N, Thornalley PJ, Dawczynski J, Franke S, Strobel J, Stein G, Haik GM. Methylglyoxal-derived hydroimidazolone advanced glycation end-products of human lens proteins. Invest Ophthalmol Vis Sci. 2003;44:5287–92.

Akbari M, Hassan-Zadeh V. IL-6 signalling pathways and the development of type 2 diabetes. Inflammopharmacology. 2018;26(3):685–98.

Alfadul H, Sabico S, Al-Daghri NM. The role of interleukin-1b in type 2 diabetes mellitus: a systematic review and meta-analysis. Front Endocrinol. 2022;13:1–13.

Alfarhan M, Jafari E, Narayanan SP. Acrolein: a potential mediator of oxidative damage in diabetic retinopathy. Biomolecules. 2020;10(11):1579.

Badawi A, Klip A, Haddad P, Cole DE, Bailo BG, El-Sohemy A, Karmali M. Type 2 diabetes and inflammation: prospects for biomarkers of risk and nutritional intervention. Diabetes Metab Syndr Obes. 2010;3:173–86.

Barry JC, Shakibakho S, Durrer C, Simtchouk S, Jawanda KK, Cheung ST, et al. Hyporesponsiveness to the anti-inflammatory action of interleukin-10 in type 2 diabetes. Sci Rep. 2016;6:1–9.

Bassuk SS, Manson JE. Epidemiological evidence for the role of physical activity in reducing risk of type 2 diabetes and cardiovascular disease. J Appl Physiol. 2005;99:1193–204.

Baynes JW, Thorpe SR. Role of oxidative stressin diabetic complications: a new perspective on an oldparadigm. Diabetes. 1999;48(1):1–9.

Bierhaus A, Hofmann MA, Ziegler R, Nawroth PP. AGEs and their interaction with AGE-receptors in vascular disease and diabetes mellitus. I. The AGE concept. Cardiovasc Res. 1998;37:586–600.

Brownlee M. Biochemistry and molecular cell biology ofdiabetic complications. Nature. 2001a;414(6865):813–20.

Brownlee M. Biochemistry and molecular cell biology of diabetic complications. Nature. 2001b;414:813–20.

Burgos-Morón E, Abad-Jiménez Z, Marañón AM, Iannantuoni F, Escribano-López I, López-Domènech S, Salom C, Jover A, Mora V, Roldan I, Solá E, Rocha M, Víctor VM. Relationship between oxidative stress, ER stress, and inflammation in type 2 diabetes: the battle continues. J Clin Med. 2019;8(9):1385.

Burke SJ, Batdorf HM, Burk DH, Martin TM, Mendoza T, Stadler K, et al. Pancreatic deletion of the interleukin-1 receptor disrupts whole body glucose homeostasis and promotes islet b-cell de-differentiation. Mol Metab. 2018;14:95–107.

Buse MG. Hexosamines, insulin resistance, and the complications of diabetes: current status. Am J Physiol Endocrinol Metab. 2006;290:E1–8.

Buse JB, Ginsberg HN, Bakris GL, Clark NG, Costa F, Eckel R, et al. Primary prevention of cardiovascular diseases in people with diabetes mellitus: a scientific statement from the american heart association and the american diabetes association. Circulation. 2006;115:114–26.

Ceriello A. Oxidative stress and diabetes-associatedcomplications. Endocr Pract. 2006;12(1):60–2.

Chandra P, Shahjad, Porwal M, Rastogi V, Mittal P, Verma A, Sachan N, Prabahar EA, Gautam MK. Progresses in wound healing: integrating nutrition, physical therapy, traditional and alternative medicine, and novel technologies. Curr Diabetes Rev. 2025;22:e15733998350164.

Cheng HM, Gonzalez RG. The effect of high glucose and oxidative stress on lens metabolism, aldose reductase, and senile cataractogenesis. Metabolism. 1986;35:10–4.

Chowdhury S, Ghosh S, Das AK, Sil PC. Ferulic acid protects hyperglycemia-induced kidney damage by regulating oxidative insult, inflammation and autophagy. Front Pharmacol. 2019;10(27):1–24.

Chowdhury SR, Dey A, Gautam MK, Mondal S, Pawar SD, Ranade A. Immune-mediated bowel disease: role of intestinal parasites and gut microbiome. Curr Pharm Des. 2024;30(40):3164–74.

Cooper CE, Vollaard NB, Choueiri T, Wilson MT. Exercise, free radicals and oxidative stress. Biochem Soc Trans. 2002;30:280–5.

D'Autreaux B, Toledano MB. ROS as signalling molecules: mechanisms that generate specificity in ROS homeostasis. Nat Rev Mol Cell Biol. 2007;8:813–24.

Dandona P, Thusu K, Cook S, et al. Oxidative damage to DNA in diabetes mellitus. Lancet. 1996;347(8999):444–5.

de Lemos ET, Reis F, Baptista S, Pinto R, Sepodes B, Vala H, et al. Exercise training is associated with improved levels of C-reactive protein and adiponectin in ZDF (type 2) diabetic rats. Med Sci Monit. 2007;13:BR168–74.

de Lemos TE, Reis F, Baptista S, Pinto R, Sepodes B, Vala H, et al. Exercise training decreases proinflammatory profile in Zucker diabetic (type 2) fatty rats. Nutrition. 2009;25:330–9.

Dror E, Dalmas E, Meier DT, Wueest S, Thévenet J, Thienel C, et al. Postprandial macrophage-derived IL-1b stimulates insulin, and both synergistically promote glucose disposal and inflammation. Nat Immunol. 2017;18(3):283–92.

Dyer DG, Dunn JA, Thorpe SR, Bailie KE, Lyons TJ, McCance DR, Baynes JW. Accumulation of Maillard reaction products in skin collagen in diabetes and aging. J Clin Invest. 1993;91:2463–9.

Evans JL, Goldfine ID, Maddux BA, Grodsky GM. Are oxidative stress-activated signaling pathways mediatorsof insulin resistance and β-cell dysfunction? Diabetes. 2003;52:1–8.

Ferber S, Halkin A, Cohen H, et al. Pancreatic andduodenal homeobox gene 1 induces expression of insulingenes in liver and ameliorates streptozotocin-induced hyper-glycemia. Nat Med. 2000;6(5):568–72.

Ferrari CKB. Pathophysiology and clinics of metabolic syndrome. Arq Catarin Med. 2007;36:90–5.

Forman HJ, Zhang H. Targeting oxidative stress in disease: promise and limitations of antioxidant therapy. Nat Rev Drug Discov. 2021;20:689–709.
Franz MJ, Bantle JP, Beebe CA, Brunzell JD, Chiasson J-L, Garg A, et al. Nutrition principles and recommendations in diabetes. Diabetes Care. 2004;27(Suppl. 1):S36–46.
Freedman BI, et al. Design and baseline characteristics for the aminoguanidine clinical trial in overt type 2 diabetic nephropathy (ACTION II). Control Clin Trials. 1999;20(5):493–510.
Fülöp N, Marchase RB, Chatham JC. Role of protein *O*-linked *N*-acetyl-glucosamine in mediating cell function and survival in the cardiovascular system. Cardiovasc Res. 2007;73(2):288–97.
Gautam MK, Goel RK. Toxicological study of *Ocimum sanctum* Linn. leaves: hematological, biochemical, and histopathological studies. J Toxicol. 2014;2014:135654. https://doi.org/10.1155/2014/135654.
Gautam MK, Ghatule RR, Singh A, Purohit V, Gangwar M, Kumar M, Goel RK. Study of healing effects of extract of *Aegle marmelos* fruit on experimental colitis. Indian J Exp Biol. 2012;51:157–64.
Gautam MK, Goel S, Ghatule RR, Singh A, Nath G, Goel RK. Curative effect of *Terminalia chebula* extract on AA-induced experimental colitis: role of antioxidants, free radicals, and acute inflammatory markers. Inflammopharmacology. 2013a;21(5):377–83. https://doi.org/10.1007/s10787-012-0147-3.
Gautam MK, Goel S, Ghatule RR, Singh A, Joshi VK, Goel RK. *Azadirachta indica* attenuates colonic mucosal damage in experimental colitis induced by trinitrobenzene sulfonic acid. Indian J Pharm Sci. 2013b;75(5):602–6.
Gautam MK, Purohit V, Agarwal M, Singh A, Goel RK. In-vivo healing potential of *Aegle marmelos* in excision, incision, and dead space wound models. Sci World J. 2014;2014:740107. https://doi.org/10.1155/2014/740107.
Gautam MK, Gangwar M, Singh SK, Goel RK. Effects of *Azadirachta indica* on vascular endothelial growth factor and cytokines in diabetic deep wound. Planta Med. 2015;81:713–21.
Genuth S, Sun W, Cleary P, Gao X, Sell DR, Lachin J, Monnier VM. Skin advanced glycation end products glucosepane and methylglyoxal hydro-imidazolone are independently associated with long-term microvascular complication progression of type 1 diabetes. Diabetes. 2015;64:266–78.
Giacco F, Brownlee M. Oxidative stress and diabetic complications. Circ Res. 2010;107(9):1058–70.
Gorogawa S-I, Kajimoto Y, Umayahara Y, et al. Probucolpreserves pancreatic β-cell function through reduction ofoxidative stress in type 2 diabetes. Diabetes Res Clin Pract. 2002;57(1):1–10.
Gorrini C, Harris IS, Mak TW. Modulation of oxidative stress as an anticancer strategy. Nat Rev Drug Discov. 2013;12(12):931–47.
Green DJ, Maiorana A, O'Driscoll G, Taylor R. Effect of exercise training on endothelium-derived nitric oxide function in humans. J Physiol. 2004;561:1–25.
Grundy SM, Cleeman JI, Daniels SR, Donato KA, Eckel RH, Franklin BA, et al. Diagnosis and management of the metabolic syndrome: an American Heart Association/National Heart, Lung, and Blood Institute scientific statement: executive summary. Crit Pathw Cardiol. 2005;4:198–203.
Harmon JS, Stein R, Robertson RP. Oxidativestress-mediated, post-translational loss of MafA protein as acontributing mechanism to loss of insulin gene expression inglucotoxic beta cells. J Biol Chem. 2005;280(12):11107–13.
Harrison D, Griendling KK, Landmesser U, Hornig B, Drexler H. Role of oxidative stress in atherosclerosis. Am J Cardiol. 2003;91(3):7A–11A.
Ho FM, Liu SH, Liau CS, Huang PJ, Lin-Shiau SY. High glucose-induced apoptosis in human endothelial cells is mediated by sequential activations of c-Jun NH2-terminal kinase and caspase-3. Circulation. 2000;101:2618–24.
Holland AM, Hale MA, Kagami H, Hammer RE, MacDonald RJ. Experimental control of pancreaticdevelopment and maintenance. Proc Natl Acad Sci USA. 2002;99(19):12236–41.
Ihara Y, Toyokuni S, Uchida K, et al. Hyperglycemiacauses oxidative stress in pancreatic β-cells of GK rats, amodel of type 2 diabetes. Diabetes. 1999;48(4):927–32.

Jain N, Naseem I, Ahmad J. Evaluation of DNA damage and metabolic syndrome parameters in diabetic rabbits supplemented with antioxidants. Fund Clin Pharmacol. 2009;23:197–205.

Jha JC, Banal C, Chow BS, Cooper ME, Jandeleit-Dahm K. Diabetes and kidney disease: role of oxidative stress. Antioxid Redox Signal. 2016;25(12):657–84.

Johansen JS, Harris AK, Rychly DJ, Ergul A. Oxidative stress and the use of antioxidants in diabetes: linking basic science to clinical practice. Cardiovasc Diabetol. 2005;4:1–11.

Jonsson J, Carlsson L, Edlund T, Edlund H. Insulin-promoter-factor 1 is required for pancreas development inmice. Nature. 1994;371(6498):606–9.

Kajimoto Y, Matsuoka T, Kaneto H, et al. Induction ofglycation suppresses glucokinase gene expression in HIT-T15cells. Diabetologia. 1999;42(12):1417–24.

Kaneto H, Fujii J, Myint T, et al. Reducing sugars triggeroxidative modification and apoptosis in pancreatic β-cells byprovoking oxidative stress through the glycation reaction. Biochem J. 1996;320(3):855–63.

Kaneto H, Xu G, Song K-H, et al. Activation of thehexosamine pathway leads to deterioration of pancreatic β-cell function through the induction of oxidative stress. J Biol Chem. 2001;276(33):31099–104.

Kaneto H, Matsuoka T-A, Nakatani Y, et al. Oxidativestress, ER stress, and the JNK pathway in type 2 diabetes. J Mol Med. 2005;83(6):429–39.

Kaneto H, Miyatsuka T, Shiraiwa T, et al. Crucial role ofPDX-1 in pancreas development, β-cell differentiation, andinduction of surrogate β-cells. Curr Med Chem. 2007;14(16):1745–52.

Kaneto H, Miyatsuka T, Kawamori D, et al. PDX-1 andMafA play a crucial role in pancreatic β-cell differentiationand maintenance of mature β-cell function. Endocr J. 2008;55:235–52.

Kaneto H, Katakami N, Matsuhisa M, Matsuoka T-A. Role of reactive oxygen species in the progression of Type 2 diabetes and atherosclerosis. Mediat Inflamm. 2010;2010:453892. 11 pages.

Kartha CC, Ramachandran S, Pillai RM. Mechanisms of vascular defects in diabetes mellitus. Cham: Springer International Publishing; 2017. p. 574.

Kataoka K, Han S-I, Shioda S, Hirai M, Nishizawa M, Handa H. MafA is a glucose-regulated and pancreaticβ-cell-specific transcriptional activator for the insulin gene. J Biol Chem. 2002;277(51):49903–10.

Konishi H, Tanaka M, Takemura Y, Matsuzaki H, Ono Y, Kikkawa U, Nishizuka Y. Activation of protein kinase C by tyrosine phosphorylation in response to H2O2. Proc Natl Acad Sci USA. 1997;94:11233–7.

Koya D, King GL. Protein kinase C activation and the development of diabetic complications. Diabetes. 1998;47:859–66.

Kwon HR, Min KW, Ahn HJ, Seok HG, Lee JH, Park GS, et al. Effects of aerobic exercise vs. resistance training on endothelial function in women with type 2 diabetes mellitus. Diabetes Metab J. 2011;35:364.

Lo TW, Westwood ME, McLellan AC, Selwood T, Thornalley PJ. Binding and modification of proteins by methylglyoxal under physiological conditions. A kinetic and mechanistic study with N alpha-acetylarginine, N alpha-acetylcysteine, and N alpha-acetyllysine, and bovine serum albumin. J Biol Chem. 1994;269:32299–305.

Maechler P, Jornot L, Wollheim CB. Hydrogen per-oxide alters mitochondrial activation and insulin secretion inpancreatic beta cells. J Biol Chem. 1999;274(39):27905–13.

Magenta A, Greco S, Capogrossi MC, Gaetano C, Martelli F. Nitric oxide, oxidative stress, and $p66^{Shc}$ interplay in diabetic endothelial dysfunction. Biomed Res Int. 2014;2014:193095. 16 pages.

Malhotra JD, Kaufman RJ. Endoplasmic reticulum stress and oxidative stress: a vicious cycle or a double edged sword? Antioxid Redox Signal. 2007;9:227–2293.

Matsuoka T-A, Kajimoto Y, Watada H, et al. Glycation-dependent, reactive oxygen species-mediated suppression ofthe insulin gene promoter activity in HIT cells. J Clin Invest. 1997;99(1):144–50.

Matsuoka T-A, Zhao L, Artner I, et al. Members ofthe large Maf transcription family regulate insulin genetranscription in islet β cells. Mol Cell Biol. 2003;23(17):6049–62.

Matsuoka T-A, Artner I, Henderson E, Means A, Sander M, Stein R. The MafA transcription factor appears to be responsible for tissue-specific expression of insulin. Proc Natl Acad Sci USA. 2004;101(9):2930–3.

Matsuoka T-A, Kaneto H, Stein R, et al. MafA regulates expression of genes important to islet β-cell function. Mol Endocrinol. 2007;21(11):2764–74.

Miller CP, McGehee RE Jr, Habener JF. IDX-1: a new homeodomain transcription factor expressed inrat pancreatic islets and duodenum that transactivates thesomatostatin gene. EMBO J. 1994;13(5):1145–56.

Miyazaki S, Yamato E, Miyazaki J-I. Regulatedexpression of PDX-1 promotes in vitro differentiation ofinsulin-producing cells from embryonic stem cells. Diabetes. 2004;53(4):1030–7.

Miyazaki Y, Kawano H, Yoshida T, Miyamoto S, Hokamaki J, et al. Pancreatic B-cell function is altered by oxidative stress induced by acute hyperglycemia. Diabet Med. 2007;24:154–60.

Mohamed AK, Bierhaus A, Schiekofer S, Tritschler H, Ziegler R, Nawroth PP. The role of oxidative stress and NF-κB activation in late diabetic complications. Biofactors. 1999;10:157–67.

Mohazzab HKM, Kaminski PM, Wolin MS. NADH oxidoreductase is a major source of superoxideanion in bovine coronary artery endothelium. Am J Physiol. 1994;266(6):H2568–72.

Mondal NS, Islam A, Gautam MK, Mondal S, Ahmed M, Jamir SI, Khongwir LD, Mahajon B. Unraveling the mechanisms of diabetic wounds: insights into pathogenesis and advanced treatment strategies. Curr Diabetes Rev. 2025;22:e15733998368222.

Moran A, Zhang H-J, Olson LK, Harmon JS, Poitout V, Robertson RP. Differentiation of glucose toxicity frombeta cell exhaustion during the evolution of defective insulingene expression in the pancreatic islet cell line, HIT-T15. J Clin Invest. 1997;99(3):534–9.

Mullarkey CJ, Edelstein D, Brownlee M. Free radical generation by early glycation products: a mechanism for accelerated atherogenesis in diabetes. Biochem Biophys Res Commun. 1990;173:932–9.

Murphy MP. Mitochondrial dysfunction indirectly elevates ROS production by the endoplasmic reticulum. Cell Metab. 2013;18:145–6.

Nagaraj RH, Linetsky M, Stitt AW. The pathogenic role of Maillard reaction in the aging eye. Amino Acids. 2012;42:1205–20.

Naka Y, Bucciarelli LG, Wendt T, Lee LK, Rong LL, Ramasamy R, Yan SF, Schmidt AM. RAGE axis: animal models and novel insights into the vascular complications of diabetes. Arterioscler Thromb Vasc Biol. 2004;24:1342–9.

Naviaux RK. Oxidative shielding or oxidative stress? J Pharmacol Exp Ther. 2012;342:608.

Nishikawa T, Edelstein D, Du XL, Yamagishi S, Matsumura T, Kaneda Y, Yorek MA, Beebe D, Oates PJ, Hammes HP, Giardino I, Brownlee M. Normalizing mitochondrial superoxide production blocks three pathways of hyperglycaemic damage. Nature. 2000;404:787–90.

Nishizuka Y. Intracellular signaling by hydrolysis of phospholipids and activation of protein kinase C. Science. 1992;258:607–14.

Obrosova IG, Minchenko AG, Vasupuram R, White L, Abatan OI, Kumagai AK, Frank RN, Stevens MJ. Aldose reductase inhibitor fidarestat prevents retinal oxidative stress and vascular endothelialgrowth factor overexpression in streptozotocin-diabetic rats. Diabetes. 2003;52:864–71.

Oguntibeju OO. Type 2 diabetes mellitus, oxidative stress and inflammation: examining the links. Int J Physiol Pathophysiol Pharmacol. 2019;11(3):45–63.

Ohlsson H, Karlsson K, Edlund T. IPF1, ahomeodomain-containing transactivator of the insulingene. EMBO J. 1993;12(11):4251–9.

Olbrot M, Rud J, Moss LG, Sharma A. Identificationof β-cell-specific insulin gene transcription factor RIPE3b1as mammalian MafA. Proc Natl Acad Sci USA. 2002;99(10):6737–42.

Palmieri B, Sblendorio V. Oxidative stress tests: overview on reliability and use. Eur Rev Med Pharmacol Sci. 2007;11:309–42.

Pang L, Lian X, Liu H, Zhang Y, Li Q, Cai Y, Ma H, Yu X. Understanding diabetic neuropathy: focus on oxidative stress. Oxid Med Cell Longev. 2020;2020:952463513 pages.

Papa FR. Endoplasmic reticulum stress, pancreatic beta-cell degeneration, and diabetes. Cold Spring Harb Perspect Med. 2012;2:a007666.

Pasupuleti VR, Arigela CS, Gan SH, Salam SKN, Krishnan KT, Rahman NA, Jeffree MS. A review on oxidative stress, diabetic complications, and the roles of honey polyphenols. Oxid Med Cell Longev. 2020;2020:887817216 pages.

Poitout V, Robertson RP. Minireview: secondary β-cell failure in type 2 diabetes—a convergence of glucotoxicityandlipotoxicity. Endocrinology. 2002;143(2):339–42.

Poitout V, Olson LK, Robertson RP. Chronicexposure of βTC-6 cells to supraphysiologic concentrationsof glucose decreases binding of the RIPE3b1 insulin genetranscription activator. J Clin Invest. 1996;97(4):1041–6.

Pop-Busui R, Sima A, Stevens M. Diabetic neuropathy and oxidative stress. Diabetes Metab Res Rev. 2006;22(4):257–73.

Puchau B, Zulet MA, de Echávarri AG, Hermsdorff HHM, Martínez JA. Dietary total antioxidant capacity is negatively associated with some metabolic syndrome features in healthy young adults. Nutrition. 2010;26:534–41.

Radak Z, Zhao Z, Koltai E, Ohno H, Atalay M. Oxygen consumption and usage during physical exercise: the balance between oxidative stress and ros-dependent adaptive signaling. Antioxid Redox Signal. 2013;18:1208–46.

Robertson AP. Chronic oxidative stress as a centralmechanism for glucose toxicity in pancreatic islet beta cellsin diabetes. J Biol Chem. 2004a;279(41):42351–4.

Robertson RP. Chronic oxidative stress as a central mechanism for glucose toxicity in pancreatic islet Beta cells in diabetes. J Biol Chem. 2004b;279(41):42351–4.

Rocha M, Diaz-Morales N, Rovira-Llopis S, Escribano-Lopez I, Bañuls C, Hernandez-Mijares A, Diamanti-Kandarakis E, Victor VM. Mitochondrial dysfunction and endoplasmic reticulum Stress in diabetes. Curr Pharm Des. 2016;22(18):2640–9.

Rosen P, Nawroth PP, King G, Moller W, Tritschler HJ, Packer L. The role of oxidative stress in the onset and progression of diabetes and its complications: a summary of a Congress Series sponsored by UNESCO-MCBN, the American Diabetes Association and the German Diabetes Society. Diabetes Metab Res Rev. 2001;17:189–212.

Rovira-Llopis S, Banuls C, Apostolova N, Morillas C, Hernandez-Mijares A, Rocha M, Victor VM. Is glycemic control modulating endoplasmic reticulum stress in leukocytes of type 2 diabetic patients? Antioxid Redox Signal. 2014;21:1759–65.

Rysz J, Baszczak R, Banach M, Kedziora-Kornatowska K, Kornatowski T, et al. Evaluation of selected parameters of the antioxidative system in patients with type 2 diabetes in different periods of metabolic compensation. Arch Immunol Ther Exp. 2007;55:335–40.

Sakurai T, Tsuchiya S. Superoxide production fromnonenzymatically glycated protein. FEBS Lett. 1988;236(2):406–10.

Schmidt AM, Stern D. Atherosclerosis and diabetes: the RAGE connection. Curr Atheroscler Rep. 2000;2:430–6.

Schmidt AM, Hori O, Brett J, Yan SD, Wautier JL, Stern D. Cellular receptors for advanced glycation end products. Implications for induction of oxidant stress and cellular dysfunction in the pathogenesis of vascular lesions. Arterioscler Thromb. 1994;14:1521–8.

Sharma A, Fusco-DeMane D, Henderson E, Efrat S, Stein R. The role of the insulin control element and RIPE3b1activators in glucose-stimulated transcription of the insulingene. Mol Endocrinol. 1995;9(11):1468–76.

Shi Y, Vanhoutte PM. Macro- and microvascular endothelial dysfunction in diabetes. J Diabetes. 2017;9(5):434–49.

Song F, Jia W, Yao Y, Hu Y, Lei L, Lin J, et al. Oxidative stress, antioxidant status, and DNA damage in patients with impaired glucose regulation and newly diagnosed type 2 diabetes mellitus. Clin Sci. 2007;112:599–606.

Stoffers DA, Zinkin NT, Stanojevic V, Clarke WL, Habener JF. Pancreatic agenesis attributable to asingle nucleotide deletion in the human IPF1 gene codingsequence. Nat Genet. 1997;15(1):106–10.

Sun J, Cui J, He Q, Chen Z, Arvan P, Liu M. Proinsulin misfolding and endoplasmic reticulum stress during the development and progression of diabetes. Mol Asp Med. 2015;42:105–18.

Tanaka Y, Gleason CE, Tran POT, Harmon JS, Robertson RP. Prevention of glucose toxicity in HIT-T15 cellsand Zucker diabetic fatty rats by antioxidants. Proc Natl Acad Sci USA. 1999;96:10857–62.

Tanaka Y, Tran POT, Harmon J, Robertson RP. A role of glutathione peroxidase in protecting pancreatic βcells against oxidative stress in a model of glucose toxicity. Proc Natl Acad Sci USA. 2002;99:12363–8.

Teixeira-Lemos E, Nunes S, Teixeira F, Reis F. Regular physical exercise training assists in preventing type 2 diabetes development: focus on its antioxidant and anti-inflammatory properties. Cardiovasc Diabetol. 2011;10:12.

Teli A, Jena D, Saikia L, Mondal NS, Sen S, Mishra K, Gautam MK. Ethnobotanical use, phytochemistry, and pharmacological activities of *Clerodendrum colebrookianum* (Lamiaceae). Curr Nutr Food Sci. 2025;21(2):148–55.

Tsalamandris S, Antonopoulos AS, Oikonomou E, Papamikroulis GA, Vogiatzi G, Papaioannou S, et al. The role of inflammation in diabetes: current concepts and future perspectives. Eur Cardiol Rev. 2019;14(1):50–9.

Uribarri J, Cai W, Sandu O, Peppa M, Goldberg T, Vlassara H. Diet-derived advanced glycation end products are major contributors to the body's AGE pool and induce inflammation in healthy subjects. Ann N Y Acad Sci. 2005;1043:461–6.

Vierck HB, Darvin ME, Lademann J, Reißhauer A, Baack A, Sterry W, et al. The influence of endurance exercise on the antioxidative status of human skin. Eur J Appl Physiol. 2012;112:3361–7.

Vitale G, Salvioli S, Franceschi C. Oxidative stress and the ageing endocrine system. Nat Rev Endocrinol. 2013;9(4):228–40.

Wang Z, Li L, Zheng FP, Jia CF, Ruan Y, et al. Correlation between the amplitude of glucose excursion and the oxidative/antioxidative system in subjects with different types of glucose regulation. Biomed Environ Sci. 2011;24:68–73.

Weir GC, Laybutt DR, Kaneto H, Bonner-Weir S, Sharma A. β-Cell adaptation and decompensation during theprogression of diabetes. Diabetes. 2001;50(supplement 1):S154–9.

Wendt T, Harja E, Bucciarelli L, Qu W, Lu Y, Rong LL, Jenkins DG, Stein G, Schmidt AM, Yan SF. Rage modulates vascular inflammation and atherosclerosis in a murine model of type 2 diabetes. Atherosclerosis. 2006;185:70–7.

Wolff SP, Dean RT. Glucose autoxidation and protein modification. The potential role of 'autoxidative glycosylation' in diabetes. Biochem J. 1987;245:243–50.

Yamagishi S, Imaizumi T. Diabetic vascular complications: pathophysiology, biochemical basis and potential therapeutic strategy. Curr Pharm Des. 2005;11:2279–99.

Yamagishi S, Ueda S, Okuda S. Food-derived advanced glycation end products (AGEs): a novel therapeutic target for various disorders. Curr Pharm Des. 2007;13:283.

Zeeshan HM, Lee GH, Kim HR, Chae HJ. Endoplasmic reticulum stress and associated ROS. Int J Mol Sci. 2016;17:327.

Antioxidants: An Introduction to Their Role in Prevention of Diabetic Neuropathy

5

Swati Barche, Prashant Chandra, Pratibha Rathee, Tanuj Hooda, Amit Lather, and Ramchander Khatri

5.1 Introduction

Diabetes is defined as impairment in metabolism indicated by insulin insufficiency or resistance. The condition affects 382 million individuals worldwide, with 316 million with insufficient glucose tolerance, making it a significant health concern. This population may quadruple by 2030 (Zatalia and Sanusi 2013). Diabetes is main root of mortality as well as morbidity globally (Can et al. 2004). The condition causes indications like discomfort, weakness, dyssynergia, impotence, and afferent nerve loss, negatively impacting individuals' lifestyles. The condition is complex and progressive, including complications such as retinal vascular disease, renal disorder, cardiac muscle disease, liver disease, and neuronal disorder (Aljabri et al. 2010). Severe hyperglycemia, caused by either Type 1 or Type 2 DM, is mainly due to diabetic complications

S. Barche
RVSKVV – College of Agriculture Indore, Indore, Madhya Pradesh, India

P. Chandra
Department of Pharmacy Practice, Amrita School of Pharmacy, Amrita Vishwa Vidyapeetham, Kochi, Kerala, India

P. Rathee
Department of Pharmacology, MM College of Pharmacy Maharishi Markandeshwar (Deemed to be University), Mullana Haryana, India

T. Hooda
Department of Pharmaceutical Chemistry, Delhi Pharmaceutical Science and Research University, Delhi, New Delhi, India

A. Lather (✉)
Department of Pharmacology, Hindu College of Pharmacy, Sonipat, Haryana, India

R. Khatri
ACOE, Delhi Pharmaceutical Sciences and Research University, New Delhi, India

M. K. Gautam et al. (eds.), *Antioxidants and Functional Foods for Diabetic Neuropathy*, https://doi.org/10.1007/978-981-95-9326-2_5

in several organs (Wang et al. 2012). An anti-oxidant is a chemical which stops radical chain reaction by contributing an H_2 atom to a free radical (Vincent and Feldman 2004). Hyperglycemia causes free radical damage and nitrosative stress, which is linked to diabetes and its consequences (Negi et al. 2011). Hyperglycemia causes free radicals by oxidizing glucose and glycosylating proteins (Al-Faris et al. 2010). The raised ROS and RNS, when combined decreased Antioxidant activity, cause free radical damage and increased reactive nitrogen species. This will cause endothelium malfunction, insulin resistance, and changes pancreatic β cell function, ultimately leading to diabetic small and large blood vessel complexity (Bandeira et al. 2013). Enhanced productivity of ROS and RNS damages super molecule like carbs, lipids, proteins, and deoxyribonucleic acid, resulting in uncertainty and impairment (Murthy et al. 2013). The ROS and RNS may trigger cellular signaling cascades, leading to transcribe stress associated gene that increase risk of complications of diabetes (Ayepola et al. 2013). The NF-κB is a transcription component, impelled by elevated reactive oxygen species, leading to proinflammatory transcribed protein that worsen illness state. Proinflammatory chemotactic factors and cytokines, such as MCP 1, TNF-α, and Interleukin-1β and 6, linked to diabetes progression and complications (Ayepola et al. 2013). Elevated ROS levels caused by hyperglycemia can lead to tissue apoptosis. ROS can initiate Bax caspase pathway of programmed cell death by decreasing electrochemical potential across sarcosome membrane. This draining of cytochrome c oxidase into the cytoplasm impel caspases and causes programmed cell death (Graves et al. 2007). DN affects more than half of chronic diabetics and remains a poorly understood condition (Callaghan et al. 2012; Kasznicki et al. 2012). Diabetes is main reason of hospitalizations and iatrogenic amputations in US (Boulton et al. 2005). It occurs tardily in insulin dependent diabetes but initially in adult onset diabetes, and reason for its occurrence is unknown (Shaikh and Somani 2010). Diabetic neuropathy can be due to elevated free radical formation or lack of antioxidant defenses, leading to free radical damage (Ziegler et al. 2004). Diabetic neuropathies affect many sections of CNS and have different clinical symptoms. Antioxidants can be found naturally in the body or gained through diet. Enzymatic Antioxidants like SOD, catalase, GSTs, and GPx, also nonenzymatic antioxidants including GSH, lipoic acid, xanthophylls, flavonoid, retinoids, L-ascorbate, and tocopherols, are some examples. Superoxide dismutase transforms superoxide anions (O2−) into hydrogen peroxide, which is then broken down into water by the actions of catalase and GPx. Glutathione-s-transferase transforms electrophilic species to water loving species, easily eliminated compounds through amalgamation with Glutathione. Vitamins C, E, and lipoic acid help stop lipid peroxidation (Gautam et al. 2014). Flavonoids shows antioxidant effects. Peroxiredoxins, thioredoxins, and glutaredoxins are specialized proteins that operate as antioxidants (Rahman et al. 2006). This book chapter will explain different methods by which extraneous antioxidants indicate their antidiabetic property on diabetic complications specifically DN.

5.2 Diabetic Neuropathy Classification

Diabetic neuropathy can be categorized both pathophysiologically (Table 5.1) and clinically (Table 5.2). It is separated into symmetrical and asymmetrical neuropathies based on clinical manifestation. Diabetic peripheral polyneuropathy, dysautonomia, insulin neuritis, hypoglycemia neuropathy, and polyneuropathy linked to deteriorating glycemic control are the most common symmetrical neuropathies. Cranial neuropathy, diabetic amyotrophy, and localized mononeuropathies such oculomotor or abducens palsy, as well as neuropathies affecting the lumbosacral, thoracic, cervical, fibular, ulnar (elbow), and median (wrist) nerves, are examples of asymmetrical neuropathies. According to pathophysiology, diabetic metabolic problems (such as ketosis-prone diabetes neuropathy), damage from hypoxemia, and treatment-related factors like hypoglycemia or insulin neuritis can all result in neuropathy. Furthermore, diabetic lumbosacral radiculoplexus neuropathy, thoracic radiculopathy, and cranial neuropathy are linked to immune-inflammatory pathways, and compression/repetitive damage is a factor in localized entrapment neuropathies (median, ulnar, and peroneal) (Sinnreich et al. 2005).

Table 5.1 Classification based on pathophysiology

Presumed underlying pathophysiology	Subtype of neuropathy
Diabetes complications	Neuropathy of ketosis-prone diabetes
Hypoxemia	Diabetic polyneuropathy
Treatment related	Hypoglycemic neuropathy Insulin neuritis
Compression and repetitive injury	Median neuropathy Ulnar neuropathy Peroneal neuropathy
Inflammatory immune	Diabetic lumbosacral radiculoplexus neuropathy Diabetic thoracic radiculopathy Cranial neuropathy

Table 5.2 Classification on basis of clinical use

Symmetry	Type of neuropathy	Subtype
Symmetrical	Diabetic peripheral neuropathy Dysautonomia Insulin neuritis Hypoglycemic neuropathy Polyneuropathy having sugar deterioration	
Asymmetrical	Cranial neuropathy	Oculomotor palsy Abducens palsy
	Diabetic amyotrophy neuropathy	
	Mononeuropathy	Lumbosacral Thoracic Cervical Fibular neuropathy Ulnar neuropathies at elbow Median neuropathy at wrist

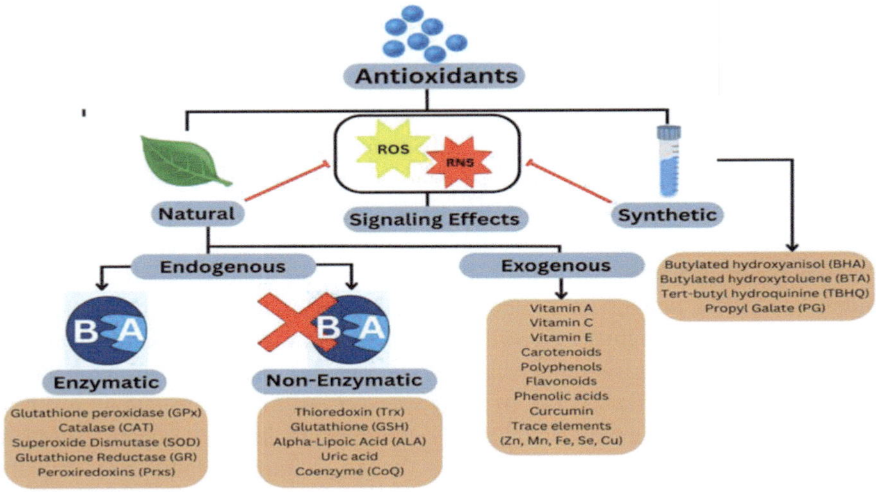

Fig. 5.1 Classification of Antioxidants

5.3 Classification of Antioxidant

The broad function of antioxidants in shielding the organism from oxidative stress by neutralizing dangerous reactive species like ROS and RNS. It emphasizes how antioxidants may come from both natural and artificial sources and work by lessening cellular damage caused by free radicals. Antioxidants assist protective signaling processes, prevent damage to lipids, proteins, and DNA, and preserve normal physiological activities by regulating oxidative reactions. Overall, the figure highlights how crucial antioxidants are for preserving physiological equilibrium and lowering problems brought on by oxidative stress (Fig. 5.1) (Fatima et al. 2023).

5.4 Diabetic Neuropathy Molecular Mechanism

Nerve is highly reactive to blood sugar changes due to their high glucose requirement and susceptibility to glycolytic and anaerobiotic metabolism. Nerve cell does not govern insulin-induced episodic glucose uptake. Neurons contain glucose transporter 3, allowing for uninterrupted glucose purporting because of its elevated affinity (KM less than normal fasting blood sugar levels) (Simpson et al. 2007; Meldgaard et al. 2018). Nerves rely on ATP, which is produced mostly through glucose metabolism, to function properly. Peripheral neuropathies are caused by disruption in glucose metabolism, and diabetes hyperglycemia can result in till quadruple rises in glucose levels. Extra glucose intake can alter intracellular metabolism and cause neuronal damage, known as glucose neurotoxicity (Gautam and Goel 2014). Excess intracellular glucose can cause a move in a lane of polyol pathway, converting glucose into sorbitol. Rise in sorbitol negatively impacts neuronal homeostasis in two ways. Firstly, it disrupts the osmoregulation, causing satisfactory outflow of myo-inositol, required for Nerve activity (Feldman et al. 2017). When glucose to sorbitol converts then it rises and leads to free radical damage, a key factor in DSPN pathophysiology (Maritim et al. 2003; Román-Pintos et al. 2016). The enzyme al-dose reductase and its cocatalyst NADPH enable the synthesis of sorbitol. NADPH normally utilized to regenerate the antioxidant glutathione. Excessive sorbitol synthesis depletes cellular NADPH resources, impairing the cell's defense against free radical damage and accumulating reactive oxygen species (Fig. 5.2) (Feldman et al. 2017). Hyperglycemia increases pyruvate production, increasing potential grade across endogenous mitochondria layer. This leads to superoxide formation, oxidative stress, and mitochondrial damage (Maritim et al. 2003). This alters bioenergetics and reduces ATP synthesis, potentially leading to apoptosis and neuronal degeneration (Román-Pintos et al. 2016). Accumulation of glucose leads to auto-oxidation, converting glucose to glyoxal. Glyoxal, along with methylglyoxal and 3-deoxyglucasone, is a key AGE precursor that forms irreversible interactions with proteins and lipids both inside and outside the cell. AGEs interact with RAGEs to activate protein kinase C and NF-κB, resulting in tending to cause inflammation cytokine formation. AGE-RAGE correlation depletes glutathione, leading to ROS production and vascular dysfunction by lowering nitric oxide levels needed for vasodilation (Shakeel 2015). AGE production can lead to endoneurial hypoxia and neuronal injury (Van Dam et al. 2013). Alterations in lipid metabolism can contribute to the development of DSPN (Athyros et al. 2018). Oxidation of LDL activates the LOX-1, leading to

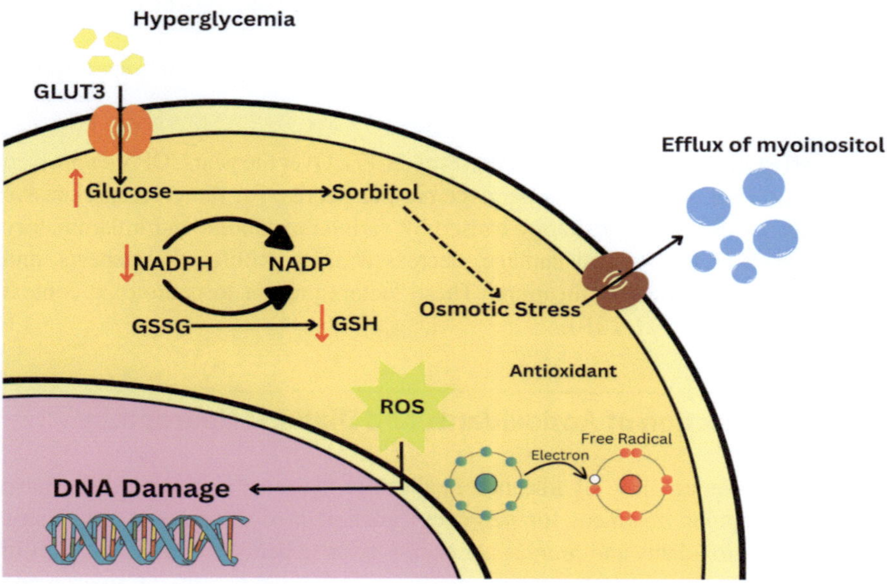

Fig. 5.2 Summary of the polyol mechanistic ways. High blood sugar induces increased glucose absorption in neurons via Glucose transporter-3. As a result, higher levels of sorbitol cause free radical damage and reactive oxygen species formation because of impaired rejuvenation of the inner antioxidant glutathione (Gautam and Goel 2014)

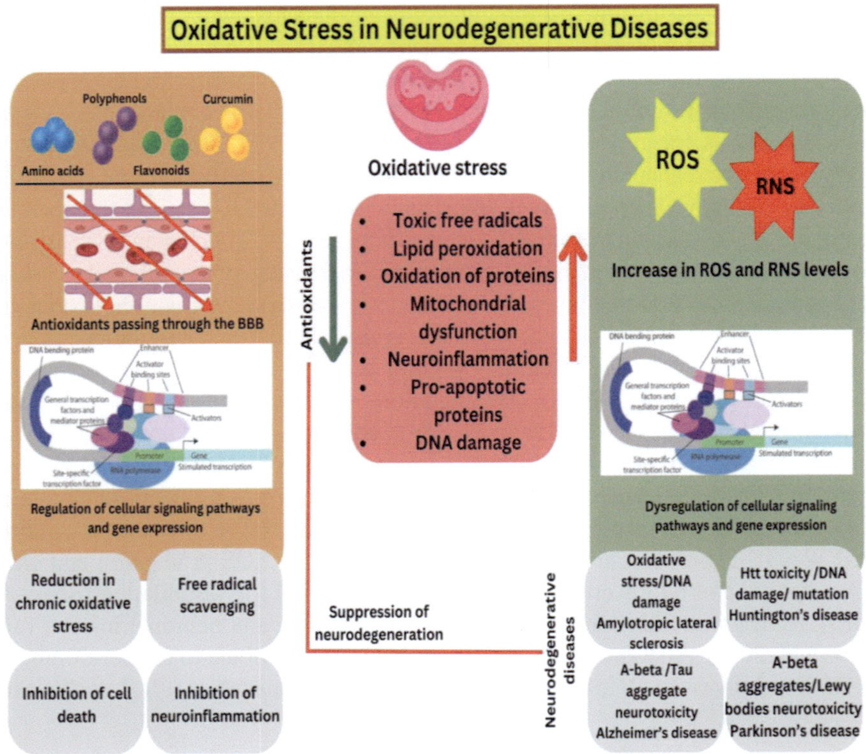

Fig. 5.3 Outline of the effect of free radical damage on brain disorders and involvement of antioxidants (Fatima et al. 2023)

increased ROS generation (Perez-Matos et al. 2017). Over the years, DPN-engrossed investigation has shifted from a perglycemia perspective for more comprehensive perception of the pathophysiology caused by various metabolic and inflammatory remark, such as free radical damage, decreased nerve proliferation aspects, and morphological vessels modifications. These factors present to neurotoxic context (Fig. 5.3) (Tesfaye et al. 2010).

5.5 Interaction of Antioxidants and Diabetic Neuropathy

Free radical damage has significance in diabetic neuropathy. Antioxidants have become a popular treatment for nerve dysfunction in diabetes over decennary. Numerous antioxidants and mimics are examined in experimental models; both In vitro as well as In vivo, with growing success (Cameron et al. 1994; Cameron and Cotter 1999; Mondal et al. 2025; Coppey et al. 2003; Sayyed et al. 2006; Kumar et al. 2007). Antioxidants include retinoids, ascorbic acid, and tocopherols, flavonoids, melatonin, and curcumin. Some Antioxidant in human clinical trials are still

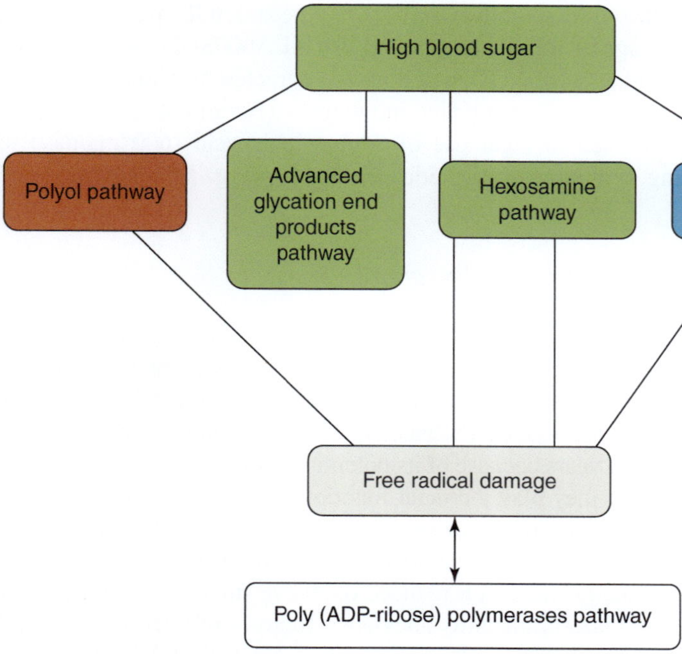

Fig. 5.4 Relation between free radical damage and various physiological pathways in diabetic neuropathy (Hosseini and Abdollahi 2013)

being evaluated, while others have been dropped due to safety or effectiveness concerns. Currently, the US FDA has not validated any antioxidant cure for DN. However, some European countries have approved alpha lipoic acid, appears to be a very efficacious antioxidant in clinical studies (Fig. 5.4) (SYDNEY Trial Authors, for the SYDNEY Trial Study Group, et al. 2003; Ziegler et al. 2006, 2011).

5.6 Antioxidants Used in Treating Diabetic Neuropathy

(a) *α-lipoic acid*

Alpha lipoic acid is considered highly effective antioxidant in medical research. It is single antioxidant that dissolves in water as well as lipids (Vallianou et al. 2009). Plants and animals can manufacture ALA, which is then converted to dihydrolipoic acid when absorbed into cells. Alpha lipoic acid and dihydro-lipoic acid are powerful free radical scavengers which help regenerate vitamins C, E, and oxiglutathione inside cells (Vallianou et al. 2009; Packer et al. 2001). Alpha lipoic acid is a compound for many mitochondrial processes (Packer et al. 2001). Experiments show that ALA reduces lipid peroxidation, free radical damage, and improves cerebral blood flow and flow of energy in diabetic mice (Nagamatsu et al. 1995; Gangwar et al. 2024). Clinical research has shown that ALA can help alleviate symptoms of

DN (Vallianou et al. 2009). ALA inhibits the hexosamine and AGEs pathway, which reduces oxidative damage. A recent study found that ALA600SOD, oral combination of alpha lipoic acid as well as SD, ameliorated characteristics and electroneurography markers in diabetic neuropathy individuals (Bertolotto and Massone 2012). This affirmation led to legal use of Alpha lipoic acid (600 mg/day) in Germany for treating symptomatic diabetic neuropathy (Fig. 5.5) (Shakher and Stevens 2011).

(b) *Retinoic acid, Ascorbic acid, and Tocopherol*

Antioxidant like retinol, ascorbic acid, and tocopherol help purify free radicals to coordinate through reuse tasks (Maritim et al. 2003). Antioxidant vitamins can boost immunity and prevent hereditary changes so as to prevent DNA impairment which occurs due to ROS metabolites (Salah et al. 2010). Vitamin C and E provides much attention in last few years because of its potentiality to scavenge free radicals. Several studies show that they play a crucial function in shielding cells from free radical damage. Vitamin E improves diabetic symptoms and complexity in creatures by reducing free radical damage indicators (Kunisaki et al. 1995; Çınar et al. 2001; Chang et al. 2003). Although vitamin E reduced oxidative stress molecular marker in clinical research, it do not significantly alleviate symptoms of microvascular and macrovascular consequences (GISSI-Prevenzione Investigators 1999; Yusuf et al.

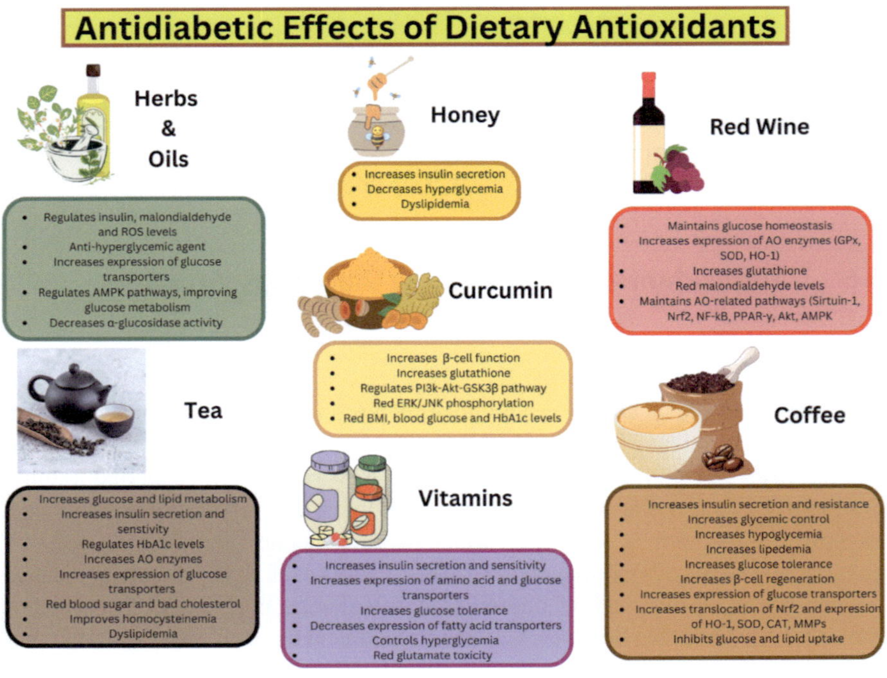

Fig. 5.5 Effects and actions of dietary antioxidants in type 2 diabetes (Fatima et al. 2023)

2000; Pruthi et al. 2001). The underperformance of tocopherol may be linked to the fact that representation of every research was indirectly pointed at diabetic endpoints like A1C test, b.p and abstains low blood sugar level events (Einarson et al. 2014). To fully reap the benefits of DN, it is crucial to prioritize it. Vitamin E modulates oxidative stress in the dorsal root ganglia, reducing neuropathic pain in diabetic rats produced with streptozotocin (Babaei et al. 2010). Ascorbic acid manifest to standardize sorbitol levels in blood, scavenge lipid oxidation, and repair decreased GSH in diabetes, but its effect in DN remains unclear (Cunningham et al. 1994; Eriksson and Kohvakka 1995).

(c) *Flavonoids*

Flavonoids, present in plentiful amount belongs to polyphenolic complex in herbs, present in stems, flowers, tea, cereals, fruit, vegetable, bark, wine, and roots (Lukačínová et al. 2008; Nijveldt et al. 2001). Flavonoid consist of many subdivisions that can explore free radicals and chelate metals (Arts and Hollman 2005; Nettleton et al. 2006). Flavonoid including hesperidin, fisetin, proanthocyanidin, luteolin, rutin, and quercetin, have antioxidant properties that protect against diabetic nephropathy (Cui et al. 2008; Wang et al. 2011; Ibrahim 2008).

(d) Tea and its components

Green tea includes free radical scavengers including flavan-3-ols and theanine, such as EGCG (Gangwar et al (2024)), which can protect against diabetes, its complications, and neurodegenerative disorders. Tea ingredients boost insulin activity and initialize insulin signal transduction system. They also neutralize free radicals and reduce swelling. In laboratory tests involving lipoprotein oxidation, the catechins and phenolic acids in tea have been shown to be more effective antioxidants than ascorbic acid, tocopherol, and β-carotene (Fatima et al. 2023).

(e) Caffeine

A cup of coffee contains approx. 250 mg of caffeine. Caffeine has a deleterious impact on insulin sensitivity and blood sugar management in both diabetics and in good health people. Several researches show that caffeine raises glycemia and reduces insulin sensitivity. Other research suggest that caffeine therapy can lower sorbitol levels in diabetic serum samples, potentially reducing the development of DN. The study found that sorbitol levels are linked to glucose levels, impaired insulin sensitivity, and insulin secretion inadequacy. In laboratory studies, caffeine has been shown to lower sorbitol levels, which might help slow the development of neurological issues related to diabetes and enhance the wellbeing for individuals with diabetes. Further study should investigate if caffeine's ability to reduce sorbitol reduces insulin resistance and indirectly improves insulin secretion. Improved glycemic control can further reduce the risk of neuropathy (Fatima et al. 2023; Srivastava et al. 2022).

(f) Resveratrol

Resveratrol is a phytoalexin that occurs naturally in a variety of vegetation, fruit, and seeds, as well as in grapes and wine. Resveratrol has diverse variety of natural actions, including antioxidant, reducing inflammation, and chemoprotectant properties. Resveratrol has been widely studied in diabetes and its associated complexities, indicating that it has anti-diabetic properties and protects against a variety of diabetic problems. Neurons are highly susceptible to damage caused by oxidants, likely because of their high oxygen consumption and limited antioxidant defense enzymes. Previously it was thought that the potential benefits of resveratrol in DN were mainly attributable to its ability to neutralize free radicals. However, numerous other linked or independent mechanisms for its positive effect on nerve dysfunction have lately been hypothesized, including activation of Nrf2, SIRT1, and suppression of NF-kB and AP-1 (Kumar et al. 2013).

(g) Common herbs and oils

Polyphenols found in seasonings like mint, parsley, basil, oregano, thyme, rosemary, as well as oils like olive oil and Persea Gratissima Oil, have antioxidant characteristics and may help prevent type-2 diabetes and brain-related illnesses. Rosmarinic acid is a phenolic acid present in many botanicals (Tiwari et al. 2019).

5.7 Conclusion

Diabetic neuropathy remains one of the unmet medical concerns. This book chapter covers current improvements in antioxidant methods, including those with dual antihyperglycemic and antioxidant end-points. Research suggests that free radical damage is a significant provider to diabetes-related deficiencies and complications. Two antioxidant techniques could hold the key. Traditional antioxidants like retinol, ascorbic acid, and tocopherol, as well as thioctic acid, were revealed effectively to scavenge free radicals in experimental models as well as clinical research for DN. Combinations of retinol, ascorbic acid, and tocopherol, alpha lipoic acid, and medicinal herbs having antihyperglycemic and antioxidant characteristics should be investigated. Advancements in technology may lead to more targeted findings than in previous trials. Novel medicines that modulate specific targets, such as mitochondrial stress and innate antioxidant defense, have potential for use as preventive therapy for diabetic neuropathy. Targeted antioxidant therapy has a promising future in treating diabetes and related complications, including neuropathy. However, there is still much work to be done.

Conflict of Interest The study reported in this publication, according to the authors, could not have been impacted by any of their relationships or known financial conflicts.

References

Al-Faris NA, Al-Sawadi AD, Alokail MS. Effect of samh seeds supplementation (Mesembryanthemum forsskalei Hochst) on liver enzymes and lipid profiles of streptozotocin (STZ)-induced diabetic Wistar rats. Saudi J Biol Sci. 2010;17(1):23–8. https://doi.org/10.1016/j.sjbs.2009.12.004.

Aljabri KS, Bokhari SA, Khan MJ. Glycemic changes after vitamin D supplementation in patients with type 1 diabetes mellitus and vitamin D deficiency. Ann Saudi Med. 2010;30(6):454–8. https://doi.org/10.4103/0256-4947.72265.

Arts IC, Hollman PC. Polyphenols and disease risk in epidemiologic studies. Am J Clin Nutr. 2005;81(1):317S–25S. https://doi.org/10.1093/ajcn/81.1.317S.

Athyros VG, Doumas M, Imprialos KP, Stavropoulos K, Georgianou E, Katsimardou A, Karagiannis A. Diabetes and lipid metabolism. Hormones. 2018;17:61–7. https://doi.org/10.1007/s42000-018-0014-8.

Ayepola OR, Chegou NN, Brooks NL, Oguntibeju OO. Kolaviron, a Garcinia biflavonoid complex ameliorates hyperglycemia-mediated hepatic injury in rats via suppression of inflammatory responses. BMC Complement Altern Med. 2013;13:1–9. https://doi.org/10.1186/1472-6882-13-363.

Babaei BF, Zare S, Heydari R, Farokhi F. Effects of melatonin and vitamin E on peripheral neuropathic pain in streptozotocin-induced diabetic rats. Iran J Basic Med Sci. 2010;13(2):1.

Bandeira SDM, Da Fonseca LJS, Guedes GDS, Rabelo LA, Goulart MO, Vasconcelos SML. Oxidative stress as an underlying contributor in the development of chronic complications in diabetes mellitus. Int J Mol Sci. 2013;14(2):3265–84. https://doi.org/10.3390/ijms14023265.

Bertolotto F, Massone A. Combination of alpha lipoic acid and superoxide dismutase leads to physiological and symptomatic improvements in diabetic neuropathy. Drugs R&D. 2012;12:29–34. https://doi.org/10.2165/11599200-000000000-00000.

Boulton AJ, Vinik AI, Arezzo JC, Bril V, Feldman EL, Freeman R, et al. Diabetic neuropathies: a statement by the American Diabetes Association. Diabetes Care. 2005;28(4):956–62. https://doi.org/10.2337/diacare.28.4.956.

Callaghan BC, Cheng HT, Stables CL, Smith AL, Feldman EL. Diabetic neuropathy: clinical manifestations and current treatments. Lancet Neurol. 2012;11(6):521–34. https://doi.org/10.1016/S1474-4422(12)70065-0.

Cameron NE, Cotter MA. Effects of antioxidants on nerve and vascular dysfunction in experimental diabetes. Diabetes Res Clin Pract. 1999;45(2–3):137–46. https://doi.org/10.1016/S0168-8227(99)00043-1.

Cameron NE, Cotter MA, Archibald V, Dines KC, Maxfield EK. Anti-oxidant and pro-oxidant effects on nerve conduction velocity, endoneurial blood flow and oxygen tension in non-diabetic and streptozotocin-diabetic rats. Diabetologia. 1994;37:449–59. https://doi.org/10.1007/s001250050131.

Can A, Akev N, Ozsoy N, Bolkent S, Arda BP, Yanardag R, Okyar A. Effect of Aloe vera leaf gel and pulp extracts on the liver in type-II diabetic rat models. Biol Pharm Bull. 2004;27(5):694–8. https://doi.org/10.1248/bpb.27.694.

Chang TI, Horal M, Jain SK, Wang F, Patel R, Loeken MR. Oxidant regulation of gene expression and neural tube development: insights gained from diabetic pregnancy on molecular causes of neural tube defects. Diabetologia. 2003;46:538–45. https://doi.org/10.1007/s00125-003-1063-2.

Çınar MG, Ülker S, Alper G, Evinç A. Effect of dietary vitamin E supplementation on vascular reactivity of thoracic aorta in Streptozotocin-diabetic rats. Pharmacology. 2001;62(1):56–64. https://doi.org/10.1159/000056072.

Coppey LJ, Gellett JS, Davidson EP, Yorek MA. Preventing superoxide formation in epineurial arterioles of the sciatic nerve from diabetic rats restores endothelium-dependent vasodilation. Free Radic Res. 2003;37(1):33–40. https://doi.org/10.1080/1071576021000028442.

Cui XP, Li BY, Gao HQ, Wei N, Wang WL, Lu M. Effects of grape seed proanthocyanidin extracts on peripheral nerves in streptozocin-induced diabetic rats. J Nutr Sci Vitaminol. 2008;54(4):321–8. https://doi.org/10.3177/jnsv.54.321.

Cunningham JJ, Mearkle PL, Brown RG. Vitamin C: an aldose reductase inhibitor that normalizes erythrocyte sorbitol in insulin-dependent diabetes mellitus. J Am Coll Nutr. 1994;13(4):344–50. https://doi.org/10.1080/07315724.1994.10718420.

Einarson TR, Garg M, Kaur V, Hemels MEH. Composite endpoints in trials of type-2 diabetes. Diabetes Obes Metab. 2014;16(6):492–9. https://doi.org/10.1111/dom.12226.

Eriksson J, Kohvakka A. Magnesium and ascorbic acid supplementation in diabetes mellitus. Ann Nutr Metab. 1995;39(4):217–23. https://doi.org/10.1159/000177865.

Fatima MT, Bhat AA, Nisar S, Fakhro VA, Akil ASAS. The role of dietary antioxidants in type 2 diabetes and neurodegenerative disorders: an assessment of the benefit profile. Heliyon. 2023;9(1). https://doi.org/10.1016/j.heliyon.2022.e12698.

Feldman EL, Nave KA, Jensen TS, Bennett DL. New horizons in diabetic neuropathy: mechanisms, bioenergetics, and pain. Neuron. 2017;93(6):1296–313. https://doi.org/10.1016/j.neuron.2017.02.005.

Gangwar M, Dey A, Chowdhury SR, Gautam MK, Prakash P, Nath G. Botanical warriors: Harnessing nature's antiviral arsenal–A comprehensive study of medicinal plants combatting pathogenic viral infections. Pharmacological Research-Natural Products. 2024 Jun 1;3:100043.

Gautam MK, Goel RK. Toxicological study of Ocimum sanctum Linn. leaves: hematological, biochemical and histopathological studies. J Toxicol. 2014:135654. https://doi.org/10.1155/2014/135654.

Gautam MK, Purohit V, Agarwal M, Singh A, Goel RK. In-vivo healing potential of Aegle marmelos in excision, incision and dead space wound models. Sci World J. 2014:740107. https://doi.org/10.1155/2014/740107.

GISSI-Prevenzione Investigators. Dietary supplementation with n-3 polyunsaturated fatty acids and vitamin E after myocardial infarction: results of the GISSI-Prevenzione trial. Lancet. 1999;354(9177):447–55. https://doi.org/10.1016/S0140-6736(99)07072-5.

Graves DT, Liu R, Oates TW. Diabetes-enhanced inflammation and apoptosis–impact on periodontal pathosis. Periodontology 2000. 2007;45(1). https://doi.org/10.1111/j.1600-0757.2007.00219.x.

Hosseini A, Abdollahi M. Diabetic neuropathy and oxidative stress: therapeutic perspectives. Oxidative Med Cell Longev. 2013;2013(1):168039. https://doi.org/10.1155/2013/168039.

Ibrahim SS. Protective effect of hesperidin, a citrus bioflavonoid, on diabetes-induced brain damage in rats. J Appl Sci Res. 2008;4(1):84–95.

Kasznicki J, Kosmalski M, Sliwinska A, Mrowicka M, Stanczyk M, Majsterek I, Drzewoski J. Evaluation of oxidative stress markers in pathogenesis of diabetic neuropathy. Mol Biol Rep. 2012;39:8669–78. https://doi.org/10.1007/s11033-012-1722-9.

Kumar A, Kaundal RK, Iyer S, Sharma SS. Effects of resveratrol on nerve functions, oxidative stress and DNA fragmentation in experimental diabetic neuropathy. Life Sci. 2007;80(13):1236–44. https://doi.org/10.1016/j.lfs.2006.12.036.

Kumar A, Negi G, Sharma SS. Neuroprotection by resveratrol in diabetic neuropathy: concepts & mechanisms. Curr Med Chem. 2013;20(36):4640–5.

Kunisaki MAKOTO, Bursell SE, Clermont AC, Ishii HIDEHIRO, Ballas LM, Jirousek MR, et al. Vitamin E prevents diabetes-induced abnormal retinal blood flow via the diacylglycerol-protein kinase C pathway. Am J Physiol Endocrinol Metab. 1995;269(2):E239–46. https://doi.org/10.1152/ajpendo.1995.269.2.E239.

Lukačínová A, Mojžiš J, Beňačka R, Rácz O, Ništiar F. Structure-activity relationships of preventive effects of flavonoids in alloxan-induced diabetes mellitus in rats. J Anim Feed Sci. 2008;17(3):411–21. https://doi.org/10.22358/jafs/66635/2008.

Maritim AC, Sanders A, Watkins Iii JB. Diabetes, oxidative stress, and antioxidants: a review. J Biochem Mol Toxicol. 2003;17(1):24–38. https://doi.org/10.1002/jbt.10058.

Meldgaard T, Olesen SS, Farmer AD, Krogh K, Wendel AA, Brock B, et al. Diabetic enteropathy: from molecule to mechanism-based treatment. J Diabetes Res. 2018;2018(1):3827301. https://doi.org/10.1155/2018/3827301.

Mondal NS, Islam A, Gautam MK, Mondal S, Ahmed M, Jamir SI, Khongwir LD, Mahajon B. Unraveling the mechanisms of diabetic wounds: insights into pathogenesis and advanced treatment strategies. Curr Diabetes Rev. 2025;2025:e15733998368222. https://doi.org/10.2174/0115733998368222250509185511.

Murthy S, Gautam MK, Goel S, Purohit V, Sharma H, Goel RK. Evaluation of in-vivo wound healing activity of Bacopa monniera on different wound models in rats. Biomed Res Int. 2013:972028. https://doi.org/10.1155/2013/972028.

Nagamatsu M, Nickander KK, Schmelzer JD, Raya A, Wittrock DA, Tritschler H, Low PA. Lipoic acid improves nerve blood flow, reduces oxidative stress, and improves distal nerve conduction in experimental diabetic neuropathy. Diabetes Care. 1995;18(8):1160–7. https://doi.org/10.2337/diacare.18.8.1160.

Negi G, Kumar A, Joshi RP, Ruby PK, Sharma SS. Oxidative stress and diabetic neuropathy: current status of antioxidants. IIOAB J. 2011;2(6):71–8.

Nettleton JA, Harnack LJ, Scrafford CG, Mink PJ, Barraj LM, Jacobs DR Jr. Dietary flavonoids and flavonoid-rich foods are not associated with risk of type 2 diabetes in postmenopausal women. J Nutr. 2006;136(12):3039–45. https://doi.org/10.1093/jn/136.12.3039.

Nijveldt RJ, Van Nood ELS, Van Hoorn DE, Boelens PG, Van Norren K, Van Leeuwen PA. Flavonoids: a review of probable mechanisms of action and potential applications. Am J Clin Nutr. 2001;74(4):418–25. https://doi.org/10.1093/ajcn/74.4.418.

Packer L, Kraemer K, Rimbach G. Molecular aspects of lipoic acid in the prevention of diabetes complications. Nutrition. 2001;17(10):888–95. https://doi.org/10.1016/S0899-9007(01)00658-X.

Perez-Matos MC, Morales-Alvarez MC, Mendivil CO. Lipids: a suitable therapeutic target in diabetic neuropathy? J Diabetes Res. 2017;2017(1):6943851. https://doi.org/10.1155/2017/6943851.

Pruthi S, Allison TG, Hensrud DD. Vitamin E supplementation in the prevention of coronary heart disease. Mayo Clin Proc. 2001;76(11):1131–6. Elsevier. https://doi.org/10.4065/76.11.1131.

Rahman I, Biswas SK, Kode A. Oxidant and antioxidant balance in the airways and airway diseases. Eur J Pharmacol. 2006;533(1–3):222–39. https://doi.org/10.1016/j.ejphar.2005.12.087.

Román-Pintos LM, Villegas-Rivera G, Rodríguez-Carrizalez AD, Miranda-Díaz AG, Cardona-Muñoz EG. Diabetic polyneuropathy in type 2 diabetes mellitus: inflammation, oxidative stress, and mitochondrial function. J Diabetes Res. 2016;2016(1):3425617. https://doi.org/10.1155/2016/3425617.

Salah SH, Abdou HS, Rahim EA. Modulatory effect of vitamins A, C and E mixtures against tefluthrin pesticide genotoxicity in rats. Pestic Biochem Physiol. 2010;98(2):191–7. https://doi.org/10.1016/j.pestbp.2010.06.006.

Sayyed SG, Kumar A, Sharma SS. Effects of U83836E on nerve functions, hyperalgesia and oxidative stress in experimental diabetic neuropathy. Life Sci. 2006;79(8):777–83. https://doi.org/10.1016/j.lfs.2006.02.033.

Shaikh AS, Somani RS. Animal models and biomarkers of neuropathy in diabetic rodents. Indian J Pharmacol. 2010;42(3):129–34. https://doi.org/10.4103/0253-7613.66833.

Shakeel M. Recent advances in understanding the role of oxidative stress in diabetic neuropathy. Diabetes Metab Syndr Clin Res Rev. 2015;9(4):373–8. https://doi.org/10.1016/j.dsx.2014.04.029.

Shakher J, Stevens MJ. Update on the management of diabetic polyneuropathies. Diabetes Metab Syndr Obes. 2011:289–305. https://doi.org/10.2147/DMSO.S11324.

Simpson IA, Carruthers A, Vannucci SJ. Supply and demand in cerebral energy metabolism: the role of nutrient transporters. J Cereb Blood Flow Metab. 2007;27(11):1766–91. https://doi.org/10.1038/sj.jcbfm.9600521.

Sinnreich M, Taylor BV, Dyck PJB. Diabetic neuropathies: classification, clinical features, and pathophysiological basis. Neurologist. 2005;11(2):63–79. https://doi.org/10.1097/01.nrl.0000156314.24508.ed.

Srivastava B, Sen S, Bhakta S, Sen K. Effect of caffeine on the possible amelioration of diabetic neuropathy: a spectroscopic study. Spectrochim Acta A Mol Biomol Spectrosc. 2022;264:120322. https://doi.org/10.1016/j.saa.2021.120322.

SYDNEY Trial Authors, for the SYDNEY Trial Study Group, Ametov AS, Barinov A, Dyck PJ, Hermann R, Kozlova N, et al. The sensory symptoms of diabetic polyneuropathy are improved with α-lipoic acid: the SYDNEY trial. Diabetes Care. 2003;26(3):770–6. https://doi.org/10.2337/diacare.26.3.770.

Tesfaye S, Boulton AJ, Dyck PJ, Freeman R, Horowitz M, Kempler P, et al. Diabetic neuropathies: update on definitions, diagnostic criteria, estimation of severity, and treatments. Diabetes Care. 2010;33(10):2285. https://doi.org/10.2337/dc10-1303.

Tiwari R, Siddiqui MH, Mahmood T, Bagga P, Ahsan F, Shamim A. Herbal remedies: a boon for diabetic neuropathy. J Diet Suppl. 2019;16(4):470–90. https://doi.org/10.1080/19390211.2018.1441203.

Vallianou N, Evangelopoulos A, Koutalas P. Alpha-lipoic acid and diabetic neuropathy. Rev Diabet Stud. 2009;6(4):230. https://doi.org/10.1900/RDS.2009.6.230.

Van Dam PS, Cotter MA, Bravenboer B, Cameron NE. Pathogenesis of diabetic neuropathy: focus on neurovascular mechanisms. Eur J Pharmacol. 2013;719(1–3):180–6. https://doi.org/10.1016/j.ejphar.2013.07.017.

Vincent AM, Feldman EL. New insights into the mechanisms of diabetic neuropathy. Rev Endocr Metab Disord. 2004;5:227–36. https://doi.org/10.1002/dmrr.287.

Wang GG, Lu XH, Li W, Zhao X, Zhang C. Protective effects of luteolin on diabetic nephropathy in STZ-induced diabetic rats. Evid Based Complement Alternat Med. 2011;2011(1):323171. https://doi.org/10.1155/2011/323171.

Wang WT, Lee P, Yeh HW, Smirnova IV, Choi IY. Effects of acute and chronic hyperglycemia on the neurochemical profiles in the rat brain with streptozotocin-induced diabetes detected using in vivo1H MR spectroscopy at 9.4 T. J Neurochem. 2012;121(3):407–17. https://doi.org/10.1111/j.1471-4159.2012.07698.x.

Yusuf S, Dagenais G, Pogue J, Bosch J, Sleight P. Vitamin E supplementation and cardiovascular events in high-risk patients. N Engl J Med. 2000;342(3):154–60. https://doi.org/10.1056/nejm200001203420302.

Zatalia SR, Sanusi H. The role of antioxidants in the pathophysiology, complications, and management of diabetes mellitus. Acta Med Indones. 2013;45(2):141–7.

Ziegler D, Sohr CG, Nourooz-Zadeh J. Oxidative stress and antioxidant defense in relation to the severity of diabetic polyneuropathy and cardiovascular autonomic neuropathy. Diabetes Care. 2004;27(9):2178–83. https://doi.org/10.2337/diacare.27.9.2178.

Ziegler D, Ametov A, Barinov A, Dyck PJ, Gurieva I, Low PA, et al. Oral treatment with α-lipoic acid improves symptomatic diabetic polyneuropathy: the SYDNEY 2 trial. Diabetes Care. 2006;29(11):2365–70. https://doi.org/10.2337/dc06-1216.

Ziegler D, Low PA, Litchy WJ, Boulton AJ, Vinik AI, Freeman R, et al. Efficacy and safety of antioxidant treatment with α-lipoic acid over 4 years in diabetic polyneuropathy: the NATHAN 1 trial. Diabetes Care. 2011;34(9):2054–60. https://doi.org/10.2337/dc11-0503.

Exploring Antioxidants: Types and Sources

6

Priyam Jyoti Das, Joyjeet Dey, Farid Jahan, and Saikat Sen

Abbreviation

AMR	Antimicrobial resistance
CAR-T	Chimeric antigen receptor T cells
CD4	Clusters of differentiation 4
CD8	Clusters of differentiation 8
CRISPR	Clustered regularly interspaced short palindromic repeats
DNA	Deoxyribonucleic acid
ESCs	Embryonic stem cell
EVs	Extracellular vehicles
IL-10	Interleukin 10
IL-35	Interleukin 35
LMICs	Low- and middle-income countries
MSCs	Mesenchymal stem cells
NMs	Nanomaterials
NPs	nanoparticles
NSCs	Neural stem cells
RNA	Ribonucleic acid
RPE	Retinal pigment epithelium
TGF-β	Transforming growth factor-beta
UTI	Urinary tract infection

P. J. Das · J. Dey · F. Jahan · S. Sen (✉)
Faculty of Pharmaceutical Science, Assam down town University, Guwahati, Assam, India
e-mail: saikat.sen@adtu.in

M. K. Gautam et al. (eds.), *Antioxidants and Functional Foods for Diabetic Neuropathy*, https://doi.org/10.1007/978-981-95-9326-2_6

6.1 Introduction

Antioxidants are compounds that in small amount can slow down or prevent the oxidative process, typically by undergoing oxidation themselves (Lotito and Frei 2006). Antioxidants are essential for maintaining tissue and cellular structure. Antioxidants act by neutralizing free radicals and stabilizing reactive species, and help in maintaining cellular and systemic health at optimal levels (Aziz et al. 2016). The oxidative properties of oxygen are fundamental to our biological processes. Although oxygen is necessary for life to exist, it can also contribute to cellular damage through oxidative mechanisms. Reactive oxygen species (ROS) and reactive nitrogen species (RNS), which arise as by-products of cellular redox processes, exhibit dual functionality. ROS and RNS serve vital roles at low or moderate concentrations, such as acting as signaling molecules and supporting immune responses. However, when present in excessive amounts, they can cause significant molecular damage (Azat Aziz et al. 2019; Sies 1993), and antioxidants are playing an important defensive role by scavenging/neutralizing ROS/RNS, thereby protecting from oxidative damage.

Antioxidants have a long history that dates back thousands of years. The Egyptians were able to display impressive technical proficiency by embalming dead corpses with plants whose extracts contain a high concentration of phenolic compounds. Free radical autoxidation reactions had been explained by the 1940s and a number of chain-breaking antioxidants had been found. By the late 1950s, it was demonstrated that oxidative reactions were involved in ageing and the progress of several diseases, and it was suggested that antioxidant molecules could decelerate the ageing process as well as disease progression and extend the life span (Panchatcharam et al. 2006). A stressful atmosphere is usually associated with ageing. The free radical hypothesis is one of several ideas that has drawn particular attention in relation to the ageing process (Birch-Machin and Bowman 2016). Antioxidant defense mechanisms balance out the body's production and use of free radicals. The defense system's antioxidant capability is hampered by many physiopathological conditions, which can lead to lipid peroxidation and the formation of reactive free radicals in the body. Eventually, this results in damage from oxidative stress (Chandra et al. 2020). A dietary antioxidant can scavenge ROS/RNS to prevent chain radical reactions or suppress the reactive oxidants from their initial formation (preventive). Antioxidants find extensive applications in dietary supplements to enhance well-being and avoid diseases, such as cancer and cardiovascular disease. Moreover, they are also employed as food preservatives. In food science, antioxidants consist of constituents that inhibit rancidity of fat. This compound greatly reduces the harmful effects of reactive oxygen species on the normal physiological activity of human being (Fukai 2009). By adding antioxidants to meals high in unsaturated fatty acids, scientists have been able to prolong food shelf life and help avoid rancidity, a nasty process in which unsaturated fats break down and release a rancid-like flavor and odor. In addition, the importance of antioxidants has been highlighted across different industries such as food, packaging, and cosmetic (Mehla et al. 2021).

6.2 Oxidative Stress Vs Antioxidants

Cellular redox homeostasis refers to the physiological state in which the levels of free radicals and reactive species are maintained within a normal range through genetic regulation and various enzymatic systems. Under typical conditions, cells manage redox balance by carefully regulating the generation and removal of ROS and RNS (Azat Aziz et al. 2019; MacDonald-Wicks et al. 2006). Antioxidants are compounds capable of neutralizing free radicals, thus decreasing the risk of the oxidative stress-induced damage. Antioxidants are indispensable in counteracting oxidative stress by neutralizing free radicals, ROS and RNS. In the body, the most prevalent oxidants are free radicals. Atoms, molecules, or ions with an unpaired electron, very unstable and highly reactive with other substances, are referred to as free radicals. The free radicals' unpaired electron makes them attack and steal electrons from other compounds in a bid to stabilize. Even when the odd electron has been paired upon the first assault, a chain reaction is occasioned by producing another free radical (Kumar 2011). This is required for reducing the harm that oxidative stress causes (Azat Aziz et al. 2019; MacDonald-Wicks et al. 2006). Oxidative stress is the disproportion between oxidants and antioxidants in the body because of antioxidant insufficiency or enhanced production of ROS and RNS, which causes harm to molecular structures like lipids, proteins, and nucleic acids. ROS is a general term that includes all highly reactive oxygen species, which include free radicals. Comprises hydroxyl radical (OḤ), perhydroxyl radical (HO2·), hypochlorous acid (HoCl), superoxide anion radical (O2·$^{-}$), hydrogen peroxide (H2O2), singlet oxygen (1O2), nitric oxide radical (NO·), hypochlorite radical (OCl·), peroxynitrite (ONOO), and various lipid peroxides. RNS is formed from nitric oxide by reaction with O2· to give ONOO, while RSS is readily generated from thiols by a reaction with ROS (Lotito and Frei 2006).

The tiny size and strong reactivity of free radicals allow them to harm a variety of substances, such as proteins, lipids, DNA, and RNA. Protein side chains of amino acids, bases in nucleic acids, and double bonds in unsaturated fatty acids—where ·OH is the greatest oxidant—can all be reacted with by ROS. Oxidative stress is the common phrase used to describe ROS damaging macromolecules. Normally, intracellular enzymes that maintain the levels of ROS at a minimal level of balance help cells defend themselves to oppose ROS damage.

Environmental stress and cellular malfunction significantly enhance ROS levels to a large extent and cause extensive damage to cells within the body. Therefore, the pathophysiology of inflammatory diseases, heart illness, diabetes, malignancies, Alzheimer's illness, eye diseases, autism, and ageing is significantly influenced by oxidative stress (Lotito and Frei 2006; Shih et al. 2007; Williamson and Manach 2005; Shahidi and Zhong 2010). To prevent or lessen the oxidative damage caused by ROS, the human body and other species have developed an antioxidant defense system that includes enzymatic, metallic chelating, as well as free radical neutralization capabilities to neutralize these radicals after they have been created. The ingestion of dietary antioxidants can also provide adequate antioxidants in the body. Oxidative reactions in humans are liable for generating different types of activated

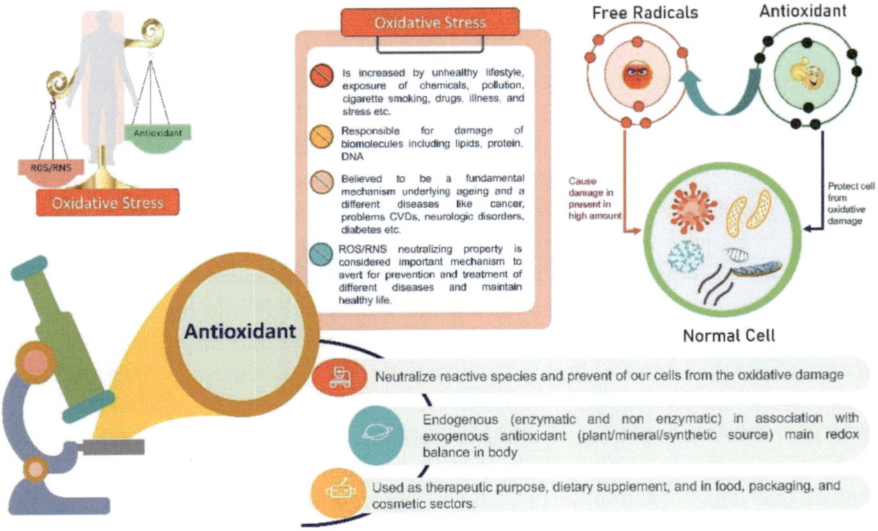

Fig. 6.1 Role of antioxidants

oxygen, the ROS. ROS are necessary for energy preservation, detoxification, and immune response constantly generated in the body (Panchatcharam et al. 2006). Aerobic metabolism involves the generation of reactive oxygen species even during basal conditions. Therefore, there is a constant need for the neutralization of these reactive oxygen species (Williamson and Manach 2005). The human system possesses several defense systems, including physical, antioxidant, repair, and preventative defenses, to defend against injury brought about by ROS/RNS. Endogenous antioxidants and exogenous antioxidants are significant in the body's defense mechanisms. The major role of antioxidants includes: (a) antioxidants are the initial step in defense because they stop free radicals from interacting and their derivatives with biological molecules in the body, thus play their preventive role; (b) play repairing effect by stopping radical oxidation reaction; (c) repair the structural damage through inactivation of reaction of free radical and their derivatives (Williamson and Manach 2005). Figure 6.1 includes the important role of antioxidants.

6.3 Types and Source of Antioxidants

Antioxidant substances can be classified based on their nature, source, properties, mechanism of action etc.

A. *Classification—Based on Source*

Antioxidants may be broadly divided into two groups, such as natural and synthetic, depending on the source.

(i) *Natural antioxidant*

Natural antioxidants can be obtained from outside natural sources or created by metabolic processes within the human body. Their chemical and physical properties, as well as the ways in which they work, greatly influence how effective they are. Endogenous and exogenous antioxidants are the two main groups into which the antioxidants may be divided (Azat Aziz et al. 2019).

Endogenous antioxidants: Endogenous antioxidants, which are synthesized by the metabolic processes in the body, can be subclassified as enzymatic or non-enzymatic.

Enzymatic antioxidants, superoxide dismutase (SOD), catalase (CAT), glutathione peroxidase (GPx), glutathione reductase (GR), and peroxiredoxins (Prxs) are crucial in the first line of defense against reactive species (Azat Aziz et al. 2019).

Non-enzymatic antioxidants involved also involved as a first line of defense against oxidative damage and are categorized as preventive antioxidants. Glutathione is a tripeptide and a powerful non-enzymatic antioxidant. The non-enzymatic antioxidants ceruloplasmin, ferritin, transferrin, and albumin found in the bloodstream chelate transition metal ions such as Fe and Cu to stop the production of new reactive species. Melatonin, myoglobin, bilirubin, cysteine, methionine, urate, lactoferrin and albumin are the example of other non-enzyme endogenous species of antioxidants 3, 15. Table 6.1 includes the characteristics of some of the endogenous antioxidants.

(a) *Exogenous antioxidants*: Exogenous antioxidant refer to antioxidant compounds provided to living system from external sources. These molecules can come in various forms, including dietary supplements or dietary elements such as vitamins, minerals, and phytochemicals. Exogenous antioxidants are essential for maintaining the health of a biological system, as they help to reduce and prevent oxidative damage to cells caused by free radicals. Exogenous antioxidants act synergistically with endogenous antioxidant to maintain or re-establish redox homeostasis (Azat Aziz et al. 2019). Table 6.2 includes the characteristics of some of the exogenous antioxidants.

(ii) *Synthetic antioxidants*

(iii) Synthetic antioxidants are chemical-derived compounds used in food products as preservatives to avoid the oxidation of lipids. A few synthetic antioxidants have stabilized fat and oil since natural antioxidants themselves possess inherent instability. Butylated hydroxytoluene (BHT) and butylated hydroxyanisole (BHA) are two popular synthetic antioxidants that originally were used for preventing petroleum to gum when oxidized (Simic and Karel 1980). However, these agents have been incorporated in food products, particularly in processed food, to enhance the shelf-life of the products (Sherwin 1976). In addition, BHT and BHA are also mixed in fats and oils because their chemical structures and antioxidant behavior is similar. Although both BHT and BHA are included on the list 'generally accepted as safe', but high doses of BHT have been implicated in several chronic toxicity studies as inducing tumors [18, 19]. Other synthetic antioxidants include tert-butylhydroxyquinone (TBHQ), propyl gal-

Table 6.1 Characteristics of endogenous antioxidants (A1)

Endogenous Antioxidants	Characteristics
Superoxide dismutase (SOD)	Enzyme that catalyzes the reduction of superoxide radicals to H_2O_2 and molecular oxygen, which saves cellular constituents from oxidative destruction. It exists in cytosol, mitochondria, and extracellular space.
Catalase (CAT)	A heme-containing enzyme, localized in peroxisomes, that catalyzes the decomposition of H_2O_2 into water and oxygen, alleviating oxidative stress.
Glutathione peroxidase (GPx)	Selenium-dependent enzyme that reduces H_2O_2 and lipid peroxides to water and alcohol, respectively. Requires glutathione as a cofactor and is present in cytosol and mitochondria.
Glutathione reductase (GR)	Enzyme that converts oxidized glutathione (GSSG) back into reduced glutathione (GSH) utilizing NADPH for maintaining cellular redox homeostasis. Localized in mitochondria and cytosol.
Peroxiredoxins (Prxs)	A thiol-specific peroxidase family that detoxifies hydrogen peroxide, peroxynitrite, and lipid peroxides, which act synergistically with thioredoxin and thioredoxin reductase
Glutathione (GSH)	A tripeptide and the most common intracellular non-protein thiol acting as a cofactor for GPx, directly scavenges reactive oxygen species (ROS), and is involved in redox signaling.
Melatonin	An indoleamine with powerful free radical scavenging activity, reducing hydroxyl and superoxide radicals, and decreasing oxidative damage by increasing the activity of antioxidant enzymes.
Ceruloplasmin	A copper-binding glycoprotein that catalyzes the oxidation of Fe^{2+} to Fe^{3+}, limiting the availability of free iron for Fenton-type reactions, thereby preventing ROS generation.
Transferrin	An iron-binding glycoprotein responsible for transporting iron in the blood. It sequesters free iron, minimizing its participation in ROS-generating reactions.
Bilirubin	A lipophilic end-product of heme degradation with antioxidant properties, efficiently scavenging peroxyl radicals and preventing lipid peroxidation at physiological concentrations.
Cysteine	A sulfur-containing amino acid that serves as a precursor for glutathione synthesis and donates electrons via its thiol group, directly reducing organic compounds and scavenging ROS.

late (PG), dodecyl gallate (DG), octyl gallate (OG), and ethylenediaminetetraacetic acid (EDTA).

B. *Classification—Based on Mechanism*

Antioxidant molecules modulate the levels of harmful ROS/RNS, and therefore contribute to regulating different pathways involved in redox signaling. Based on the mode of actions antioxidants can be classified into [3, 20]:

(i) *Catalytic system for the rerouting or neutralization of ROS*: SOD, CAT, GPx
(ii) *Metal ion binding or inactivation prevents Haber–Weiss processes from generating ROS:* Ferritin, caeruloplasmin, catechins

Table 6.2 Characteristics of exogenous antioxidants

Exogenous antioxidants	Characteristics
Vitamins	Essential micronutrients with significant antioxidant properties. Vitamin E acts as a lipid-soluble antioxidant, breaking lipid peroxidation chain reactions in membranes, while Vitamin C scavenges free radicals, regenerates Vitamin E, and neutralizes oxidants from activated neutrophils.
Trace elements	Inorganic micronutrients such as selenium, zinc, and manganese that serve as cofactors for antioxidant enzymes (e.g., GPx, SOD). They play a critical role in maintaining enzymatic activity and redox balance within cells.
Phytomolecules	Plant-derived bioactive compounds like polyphenols, flavonoids, and carotenoids with robust antioxidant activity. They quench singlet oxygen, scavenge ROS, and exhibit additional protective effects against oxidative stress-related diseases.

(iii) *Self-suicidal and chain-breaking antioxidants scavenge and destroy ROS*: Vitamin C, vitamin E, uric acid, glutathione, flavonoids
(iv) *Reduction of ROS and utilization of chemical sinks and traps to trap energy*: Carotenoids, anthocyanidins
(v) *Antioxidant action by selective oxidation*: Sodium Bisulphate, Thio urea, Vitamin C, Sodium Meta bisulphite.
(vi) *Oxidative chain reaction by blocking*: Alpha tocopherol, BHA, BHT,
(vii) *Antioxidant effect through co-operative action*: H3PO4, Citric acid.
(viii) *Act by binding the heavy metal traces*: EDTA salts

C. *Classification—Based on Solubility*

Based on solubility, antioxidants are divided into two categories (Maity et al. 2016):

(i) *Water soluble antioxidants:* Glutathione, Uric acid, Vitamin C, Lipoic acid
(ii) *Lipid soluble antioxidants:* Vitamin E, Carotenes, Ubiquinol

D. *Classification—Based on Size*

Antioxidants have been broadly divided into small-molecule and large-molecule antioxidants, each crucial in abating oxidative stress and promoting cellular wellness. Small-molecule antioxidants generally work by neutralizing ROS through a process termed radical scavenging. They transfer electrons or hydrogen ions to ROS and make them ineffective for removal by the body. The best examples for this class are vitamin C, vitamin E, carotenoids, and glutathione (GSH), which play a basic role in the prevention of oxidative damages to the cell components that contain lipids, proteins, and DNA.

There are also antioxidants which are known as large molecule antioxidants. Some of these comprise enzymatic systems and sacrificial proteins. It would make sense for enzymes to engage in ROS conversions to less-damaging compounds such as water and oxygen and thus reduce damage through superoxide dismutase (SOD), catalase (CAT), glutathione peroxidase (GSHPx), etc. sacrificed proteins or merely albumins buffer by being available to accommodate excess ROS with its ensuing preservation of other prominent protein and structural moieties from the oxidative assaults. Collectively, these antioxidant systems form an integrated defense system, ensuring the integrity of biological systems in opposition to oxidative damage (Table 6.3) (Nimse and Pal 2015).

Table 6.3 Different antioxidants and their sources

S. no	Antioxidant	Source	Importance
1.	Alpha lipoic acid	Beets, spinach, broccoli, potatoes, red meat, carrots.	Alpha lipoic acid reacts with radical and nonradical reactive oxygen species to convert to a less reactive radical cation. Vitamins C and E, can then convert this cation back.
2.	Alpha tocopherol	Sunflower seeds, almonds, hazelnuts, pine nuts, wheat germ, sunflower, corn, soybean oils, spinach, broccoli, vegetable oils, green leafy vegetables, fortified cereals, nuts, seeds	Alpha tocopherol protects cell membranes from oxidative damage by terminating free radicals by donating a hydrogen from its chromane ring.
3.	Apigenin	Parsley, chamomile, celery seeds, vine spinach, Chinese celery	Apigenin can prevent oxidative stress-induced cell death.
4.	Ascorbic acid	Citrus fruits, oranges, grapefruit, kiwifruit, strawberries, pineapple, mango, papaya, watermelon, broccoli, Brussels sprouts, cauliflower, cabbage, spinach, tomatoes, peppers, kale, guava	Ascorbic acid is an antioxidant that donates hydrogen atoms to lipid radicals, removing molecular oxygen, and quenching singlet oxygen. It also stabilizes vitamin E and folic acid.
5.	Astaxanthin	Algae, yeast, seafood, krill, copepods, marine algae, red yeast *Xanthophyllomyces dendrorhous,* Salmon, trout, shrimp, lobster	Astaxanthin is a carotenoid acts by neutralizing of reactive species, prevent chain reactions, inhibit of lipid peroxidation
6.	Baicalein	Fruit, root bark, leaves	These antioxidants protect cardiomyocytes against contractile failure and severe ischemia injury produced by mitochondrial ROS.
7.	Beta carotene	Vegetables, fruits, carrots, sweet potatoes, pumpkins, winter squash, broccoli, spinach, leafy greens, asparagus, beets, Swiss chard, mustard greens, mango, papaya, apricots, cantaloupe, nectarines, peaches, grapefruit	Beta carotene chelates oxygen-free radicals, scavenges reactive oxygen species, quenches singlet molecular oxygen, and also boost the immune system.

(continued)

Table 6.3 (continued)

S. no	Antioxidant	Source	Importance
8.	Beta cryptoxanthin	Tangerines, red peppers, pumpkin, oranges, papaya, peaches, watermelons	They inhibits osteoclast formation, reduces oxidative stress, improves insulin resistance, reduce risk of cancer and degenerative diseases, delay osteoporosis process, stimulates DNA repair.
9.	Catechin	Apples, blackberries, cherries, grapes, pears, raspberries, strawberries, blueberries, cocoa, coffee, red wine, tea, wine	These antioxidants scavenge lipid peroxidation products, regulate oxidative stress, and improve lipid metabolism disorders.
10.	Chlorogenic acid	Coffee beans, seeds and leaves of dicotyledonous plants, tomatoes, prunes, blueberries, fruits, vegetables, pears, plums,	They scavenge radicals, activates endogenous antioxidants
11.	Chrysin	Bee propolis, passionflower, Damiana, honey.	These antioxidants act by chelating metal, Glutathione induction, NF-κB inhibition, help in SOD, CAT, and GPx enzyme activity.
12.	Citric acid	Citrus fruits, lemons, oranges, grapefruits, mandarins, broccoli, carrots, berries, tomatoes, raspberries, pineapple.	They scavenging hydrogen peroxide, improve redox homeostasis, inhibit lipid peroxidation, improving immune system, increase bioavailability of minerals
13.	Coenzyme Q10	Reindeer meat, animal organ meats, Salmon, trout, mackerel, sardines, almonds, peanuts, sesame seeds, tofu, soy milk and yogurt, soybean, corn, olive, canola oils, fatty fish, nuts and seeds, broccoli, spinach	These antioxidants neutralize free radicals, reducing lipid peroxidation, supporting endogenous antioxidant mechanisms, improving energy production, and preventing blood clots are other roles.
14.	Copper	Organ meats, shellfish, nuts and seeds, legumes, dark chocolate, whole grains, liver, kidneys, heart, oysters, crabs, lobsters, almonds, cashews, sunflower seeds, sesame seeds, lentils, beans, and peas, brown rice and quinoa.	Copper is a component of antioxidant enzymes like ceruloplasmin and Cu/zinc superoxide dismutase (SOD)
15.	Curcuminoid	Turmeric, beverages, snacks, and supplements.	These antioxidants scavenge reactive oxygen species, inhibiting enzymes, chelating metals, regulating enzymes

(continued)

Table 6.3 (continued)

S. no	Antioxidant	Source	Importance
16.	Daidzein	Soybeans and soy products, red clove, alfalfa, kudzu, red lentils, tofu and textured vegetable protein	They act through oxidation. Oxidation of the 4-OH group on the B-ring and the 7-OH group on the A-ring. Antioxidants chelate metals, scavenge free radicals, and regulate signaling pathways.
17.	Dalphinidin	Berries, fruits, flowers, wine, bilberries, blackcurrants, cranberries, Concord grapes, maqui berries, and red raspberries, Jamun, pomegranates, and tomatoes	They donate electrons, chelate metals, and capture lipid free radicals.
18.	Tocopherol	Corn, palm, sunflower, barley, wheat germ, sesame, macadamia	These antioxidants trap radicals, regenerate, and inhibit tumor growth.
19.	Ellagalic acid	Fruits and nuts, strawberries, raspberries, blackberries, cherries, walnuts, pecans, pomegranates, grapes, black currants, and wolfberry.	They exhibit DPPH free radical scavenging activity and inhibit lipid peroxidation.
20.	Epicatechin gallate	Fruits, vegetables, cocoa, tea, blackberries, cherries, grapes, pears, raspberries, apples, cranberries, strawberries, kiwis, broad beans, dark chocolate, green tea, wine, grains, roots, stems, leaves, and flowers.	They neutralize reactive oxygen species (ROS), chelate metals, modulate signaling proteins, induce endogenous antioxidant proteins, and inhibit FOXO1.
21.	Epigalocatechin & Epigalocatechin gallate [EGCG]	Green tea, fruits, nuts, carob powder, white, oolong, and black teas, cranberries, strawberry, blackberry, kiwi, cherry, pear, peach, apple, avocado, pecans, pistachios, and hazelnuts.	These antioxidants chelate metal ions, inhibit pro-oxidant enzymes, scavenge free radicals, and inhibit angiogenesis and platelet aggregation.
22.	Gamma carotene	Dark leafy greens, carrots, sweet potatoes, palm oil, apricots, peaches, and bell peppers, sweet potatoes, particularly orange kinds, are high in gamma-carotene. Sweet potatoes, particularly the orange kind, are high in gamma-carotene.	They quench singlet oxygen, scavenge reactive species, and perform electron transfer, hydrogen abstraction, and metal chelation.

(continued)

Table 6.3 (continued)

S. no	Antioxidant	Source	Importance
23.	Gamma tocopherol	Nuts and seeds, vegetable oils, black walnuts, sesame seeds, pecans, pistachios, English walnuts, flaxseed, canola, corn, camelina, linseed, soybean, walnut oils, wheat germ oil, peanut butter, beets, collard greens, spinach, sunflower seeds, almonds, peanuts, and pumpkin.	These antioxidants trap reactive nitrogen species (NOx), block COX-1 and COX-2, inhibit 5-LOX, and activate protein phosphatase 2A.
24.	Genistein	Soybeans, fava beans, lupine, coffee beans, kudzu, pea, alfalfa, green lentils.	They inhibit lipid peroxidation, prevent NF-kB activation, increase antioxidant enzyme expression, inhibit hemolysis, and trap advanced glycation end products (AGEs).
25.	Hesperidin	Citrus fruits, orange juice, oranges, grapefruit, tangerines, limes, and lemons.	They counteract oxidative stress, prevent lipid peroxidation, inhibit inflammatory processes, restore dopamine levels, and promote neuronal survival.
26.	Isothiocynates	Broccoli, Brussels sprout, cabbage, cauliflower, kale, mustard greens, radishes, rutabagas, watercress.	These antioxidants decompose peroxides and hydroperoxides, reducing oxidative stress, and reducing cardiovascular problems
27.	Kaempferol	Vegetables, herbal medicines, tea, fruits, broccoli, spinach, cabbage, onions, Brussels sprouts, apples, grapes, berries (strawberries, blackberries, raspberries), green tea, Lovage, barbarum	They increase antioxidant enzymes, inhibiting lipid peroxidation, regulating malondialdehyde and glutathione, reducing reactive oxygen species (ROS) metabolism, disrupting anti-inflammatory membranes, restoring cell redox hemostasis.
28.	L tartaric acid	Grapes, bananas, avocados, tamarinds, and citrus fruits	They act as a sequestrant.
29.	Lipolic acid	Organ meats, nuts and seeds, fruits and vegetables, red meat, liver, kidney, and heart, walnuts, pecans, sunflower seeds, spinach, broccoli, tomatoes, Brussels sprouts, beef, lamb, pork.	These antioxidants perform free radical scavenging, metal chelation, recycle other antioxidants, and modulate signaling pathways.
30.	Lutein	Leafy greens, yellow and Orange fruits and vegetables, eggs, egg yolks, spinach, kale, collard greens, and broccoli, carrots, corn, bell peppers, and oranges.	They engage in electron transfer, chain reaction termination, and regulation of antioxidant enzymes.

(continued)

Table 6.3 (continued)

S. no	Antioxidant	Source	Importance
31.	Luteolin	Vegetables, herbs and spices, fruits, celery, parsley, broccoli, scallions, carrots, peppers, cabbages, peppermint, oregano, thyme, chamomile, rosemary, apples, olive oil.	They participate in hydrogen atom transfer (HAT), single electron transfer (SET), regulation of antioxidant enzymes, metal-catalyzed oxidation, reduction of oxidative stress, activation of the nuclear factor erythroid 2-related factor 2 (Nrf2) increased antioxidant gene expression. Inhibition of xanthine oxidase, resulting in reduced ROS generation.
32.	Lycopene	Tomatoes, watermelon, grapefruit, papaya, mango, red bell peppers, guava, asparagus, rose hips.	These antioxidants perform free radical scavenging, chain reaction inhibition, indirect antioxidant effects.
33.	Melatonin	Cherries, tomatoes, bananas, eggs, nuts and seeds, tart cherries, ripe tomato, yolk, almonds, walnuts, and flaxseeds	They engage in electron donation, chain reaction termination, stimulation of antioxidant enzymes, regulation of mitochondrial function, metal chelation.
34.	Myricetin	Fruits, vegetables. Berries (cranberries, blueberries, goji berries), oranges, tomatoes, fennel leaves, parsley, chard, tea, red wine, nuts	They participate in free radical scavenging, metal ion chelation, induction of antioxidant enzymes, prevention of lipid peroxidation.
35.	N acetyl cysteine	Supplements, foods, meat (chicken, beef, Turkey, fish, dairy products, eggs, legumes, seeds (sunflower, pumpkin), nuts (almonds, walnuts)	They participate in glutathione synthesis, free radical neutralization and regeneration.
36.	Naringenin	Oranges, certain herbs, cocoa, tomatoes, lemons	They participate in hydrogen atom transfer (HAT), electron transfer (ET), chelation of metal ions, inhibition of oxidative enzymes, induction of antioxidant enzymes, modulation of signaling pathways.

(continued)

Table 6.3 (continued)

S. no	Antioxidant	Source	Importance
37.	Octyl gallate	Peanut butter, processed nuts, packaged foods, cosmetics and skincare products, pharmaceutical product	Octyl gallate donates hydrogen atoms to free radicals, stabilizing them and preventing them from reacting with and destroying essential lipids, proteins, and DNA are components of cells. It prevents lipid peroxidation, which is the process by which lipids in food items or cell membranes are destroyed by free radicals. Cellular components include lipids, proteins, and DNA. It inhibits lipid peroxidation, a process in which free radicals destroy lipids in cell membranes or food products. This inhibition slows down the rancidity of fats, by reducing oxidation, octyl gallate extends the shelf life of foods, pharmaceuticals, and cosmetics by preventing degradation caused by oxygen exposure.
38.	Oestrogen	Fruits, grains, nuts and seeds, apples, berries, grapes, peaches, pears, plums, barley, oats, wheat germ, flaxseeds, peanuts, sesame seeds	When estrogen levels are low (e.g., during menopause), phytoestrogens can partially activate estrogen receptors and mimic the effect of estrogen. When estrogen levels are high, phytoestrogens can block receptors, reducing the effects of excess estrogen.
39.	Piperine	Black pepper, long pepper	Piperine is known to increase the bioavailability of various nutrients and drugs by inhibiting enzymes involved in their metabolism, such as cytochrome P450 3A4 (CYP3A4) and P-glycoprotein. Piperine stimulates thermogenesis (the production of heat in the body) by interacting with the transient receptor potential vanilloid 1 (TRPV1) receptor, enhancing fat metabolism and potentially aiding in weight loss.

(continued)

Table 6.3 (continued)

S. no	Antioxidant	Source	Importance
40.	Potassium L ascorbate	Bananas, avocados, spinach, potatoes, citrus fruits, orange, lemons, limes, ground potatoes, broccoli	As a powerful antioxidant, vitamin C reduces oxidative stress and neutralizes free radicals, which can help shield cells from harm and boost immunity in general.
41.	Proanthocyandins	Grape seeds, cranberries, grape seed extract, cocoa, cranberry extract	By neutralizing free radicals, Proanthocyanidins shield cells from oxidative harm. Chronic illnesses including cancer, neurological problems, and heart disease are less likely to occur as a result of this activity.
42.	Procysteine	Meat, fish, dairy products, eggs, legumes, nuts and seeds, cruciferous vegetable, chicken, Turkey, pork, beef, tuna, salmon, milk, cheese, yogurt, lentils, chickpeas, beans, sunflower seeds, walnuts, broccoli, Brussels sprouts, cauliflower	Glutathione helps neutralize free radicals and reduces oxidative stress. It aids in the detoxification of harmful substances in the liver.
43.	Proline	Grains, legumes, nuts and seed, meat, fish, dairy products, chicken, Turkey, pork, beef, tuna, salmon, milk, cheese, yogurt, lentils, chickpeas, beans, sunflower seeds, walnuts, broccoli, Brussels sprouts, cauliflower, whole grains like oats, quinoa, and barley	Proline is a key component of collagen, the most abundant protein in the body, which provides structural support to connective tissues, skin, and bones.
44.	Proplyl gallate	Synthesized through the reaction of gallic acid with propanol	Propyl gallate functions as an antioxidant by scavenging free radicals and preventing the oxidation of lipids and other sensitive compounds. It inhibits the oxidation of fats and oils in foods, thereby extending their shelf life and maintaining freshness.
45.	Pyrogallol	Plants, plant extracts, gallnuts, green tea leaves	Pyrogallol serves as a free radical scavenger, hence reducing oxidative stress in cells. Pyrogallol protects cells from oxidative damage and promotes general cellular health by neutralizing free radicals.

(continued)

Table 6.3 (continued)

S. no	Antioxidant	Source	Importance
46.	Quercetin	Fruits, vegetables, herbs and spices, beverages, apples, berries (e.g., blueberries, blackberries, strawberries), grapes citrus fruits (e.g., oranges, lemons), onions (especially red onions), kale, broccoli, tomatoes, Bell peppers	Quercetin neutralizes free radicals, reducing oxidative stress and protecting cells from damage.
47.	Rutin	Grains, herbs and spices, vegetables, fruits, apples citrus fruits (e.g., oranges, lemons) berries (e.g., blueberries, blackberries), onions Tomatoes, spinach, buckwheat (particularly high in rutin), Tea (green and black tea)	Rutin acts as a potent antioxidant by neutralizing free radicals and reducing oxidative stress. This helps protect cells and tissues from damage.
48.	Selenium	Sea foods, meats, organ meats, whole grains, grains and cereal, tuna, shrimp, crab, and salmon, whole grains like wheat, barley, and brown rice, beef, chicken, Turkey, and pork	Selenium is a crucial component of several antioxidant enzymes, including glutathione peroxidases and thioredoxin reductases. These enzymes help protect cells from oxidative stress by neutralizing harmful free radicals.
49.	Sulforaphane	Broccoli, cauliflower, sprouts and florets	Sulforaphane promotes the Nrf2 (nuclear factor erythroid 2-related factor 2) signaling pathway. Nrf2 is a transcription factor that controls the expression of antioxidant proteins and detoxification enzymes.
50.	Thiosulfides	Onion, garlic, leek, chive, broccoli	Thiosulfides, notably those present in garlic, have been proven to possess antioxidant effects. They can neutralize free radicals and lower oxidative stress.
51.	Thymol	Thyme	Thymol scavenges free radicals and reactive oxygen species (ROS), which reduces oxidative stress and protects cells.
52.	Tocopherols	Unprocessed vegetable oils, whole grains, leafy vegetables, legumes, cotton seed oil, peanut oil, sunflower oil, spinach, kale Swiss chard	Tocopherols act primarily as antioxidants. They protect cells from oxidative stress by neutralizing free radicals. They transfer electrons to free radicals, reducing oxidative damage to lipids, proteins, and DNA.

(continued)

Table 6.3 (continued)

S. no	Antioxidant	Source	Importance
53.	Ubiquinol-10	Meat, fish, whole grains, nuts and seeds, vegetables, legumes, organ meat, tuna, salmon, spinach, peas	Ubiquinol has antioxidant properties, meaning it helps neutralize free radicals and reduces oxidative stress. This protection extends to cell membranes and other cellular components.
54.	Uric acid	Meat, seafood, organ meats, alcohol, certain fish, beef, lamb, pork, anchovies, sardines, mussels, scallop, beer and other alcoholic beverages.	Uric acid has antioxidant properties and can help neutralize free radicals. This function can be beneficial, but excessive levels may lead to health issues.
55.	Vanillic acid	Wheat, rye, vegetables and herbs, fruits, vanilla beans, apples, oregano, cloves	Vanillic acid has strong antioxidant activity, which helps neutralize free radicals and reduce oxidative stress in the body. This can protect cells from damage and inflammation.
56.	Zeaxanthin	Egg yolk, fruits, green leafy vegetables, corn, green peas, eggs, kiwifruit, grapes, spinach, kale, and collard greens	Zeaxanthin is an antioxidant that protects cells from oxidative damage produced by free radicals.. This is crucial for maintaining cellular health and function.

6.4 Future Prospective

Antioxidants are crucial for avoiding oxidative stress, which plays a significant role in the development of ageing and a number of chronic illnesses. By neutralizing ROS and free radicals, antioxidants protect biological elements including proteins, lipids, and DNA from oxidative stress. Antioxidants are diverse, ranging from natural sources such as vitamins and phytochemicals to manufactured substances and enzymatic systems, demonstrating their varied character and vast range of effects. Antioxidants are diverse, ranging from natural sources such as vitamins and phytochemicals to manufactured substances and enzymatic systems, demonstrating their varied character and vast range of effects. These antioxidant defenses are integrated into body systems, playing the roles of direct radical scavengers, metal chelators, and enzyme modulators. The physiological roles are important; so, too, is their potential as therapy in cardiovascular diseases, diabetes mellitus, neurodegenerative diseases, and cancer. It stands to reason that increasing understanding of the mechanisms involved can deepen the scope for innovation of therapeutic strategy and diet that can make the full best of their potential. This would further emphasize their critical role in the maintenance of health and food preservation due to their presence in various food sources and their utilization in preventing lipid oxidation in food products.

References

Azat Aziz M, Shehab Diab A, Abdulrazak Mohammed A. Antioxidant categories and mode of action. Antioxidants. 2019;5:3–20. IntechOpen. https://doi.org/10.5772/intechopen.83544.

Aziz MA, Majeed GH, Diab KS, Al-Tamimi RJ. The association of oxidant–antioxidant status in patients with chronic renal failure. Ren Fail. 2016;38:20–6. https://doi.org/10.3109/0886022X.2015.1103654.

Birch-Machin MA, Bowman A. Oxidative stress and ageing. Br J Dermatol. 2016;175:26–9. https://doi.org/10.1111/bjd.14906.

Chandra P, Sharma RK, Arora DS. Antioxidant compounds from microbial sources: a review. Food Res Int. 2020;129:108849. https://doi.org/10.1016/j.foodres.2019.108849.

Fukai T. Extracellular SOD and aged blood vessels. Am J Physiol Heart Circ Physiol. 2009;297:H10–2. https://doi.org/10.1152/ajpheart.00425.2009.

Kumar S. Free radicals and antioxidants: human and food system. Adv Appl Sci Res. 2011;2:129–35.

Lotito S, Frei B. Consumption of flavonoid-rich foods and increased plasma antioxidant capacity in humans: cause, consequence, or epiphenomenon? Free Radic Biol Med. 2006;41:1727–46. https://doi.org/10.1016/j.freeradbiomed.2006.04.033.

MacDonald-Wicks LK, Wood LG, Garg ML. Methodology for the determination of biological antioxidant capacity in vitro: a review. J Sci Food Agric. 2006;86:2046–56. https://doi.org/10.1002/jsfa.2603.

Maity T, Banerjee S, Bishal A, Bhattacharjee R, Das SP, Maji M. An overview of antioxidant: classifications, therapeutic effect & its applications. ASIO J Exp Pharmacol Clin Res. 2016;1:27–34.

Mehla N, Kothari Chhajer A, Kumar K, Dahiya S, Mohindroo V. Applications of antioxidants: a review. 2021:1–29. https://doi.org/10.1007/978-3-030-45299-5_28-1.

Nimse SB, Pal D. Free radicals, natural antioxidants, and their reaction mechanisms. RSC Adv. 2015;5:27986–8006. https://doi.org/10.1039/C4RA13315C.

Panchatcharam M, Miriyala S, Gayathri VS, Suguna L. Curcumin improves wound healing by modulating collagen and decreasing reactive oxygen species. Mol Cell Biochem. 2006;290:87–96. https://doi.org/10.1007/s11010-006-9170-2.

Shahidi F, Zhong Y. Novel antioxidants in food quality preservation and health promotion. Eur J Lipid Sci Technol. 2010;112:930–40. https://doi.org/10.1002/ejlt.201000044.

Sherwin ER. Antioxidants for vegetable oils. J Am Oil Chem Soc. 1976;53:430–6. https://doi.org/10.1007/BF02605739.

Shih P-H, Yeh C-T, Yen G-C. Anthocyanins induce the activation of phase II enzymes through the antioxidant response element pathway against oxidative stress-induced apoptosis. J Agric Food Chem. 2007;55:9427–35. https://doi.org/10.1021/jf071933i.

Sies H. Strategies of antioxidant defense. Eur J Biochem. 1993;215:213–9. https://doi.org/10.1111/j.1432-1033.1993.tb18025.x.

Simic MG, Karel M. Autoxidation in food and biological systems. Boston: Springer US; 1980. https://doi.org/10.1007/978-1-4757-9351-2.

Williamson G, Manach C. Bioavailability and bioefficacy of polyphenols in humans. II. Review of 93 intervention studies. Am J Clin Nutr. 2005;81:243S–55S. https://doi.org/10.1093/ajcn/81.1.243S.

Role of Dietary Antioxidants in Mitigating Oxidative Stress and Nerve Dysfunction in Diabetic Neuropathy

7

Sejuti Ray Chowdhury, Tamal Mondal, Anamika Saha, Gourab Biswas, Biswajit Dash, and Partha Pratim Dutta

Abbreviations

4-Hne	4-Hydroxynonenal
Ages	Advanced glycation end products
Ala	Alpha-lipoic acid
Aris	Aldose reductase inhibitors
Cat	Catalase
Dhla	Dihydrolipoic acid
Dn	Diabetic neuropathy
Dpn	Diabetic peripheral neuropathy
Gpx	Glutathione peroxidase
Gsh	Glutathione
Gst	Glutathione S-transferase

S. R. Chowdhury
Eminent College of Pharmaceutical Technology, Barasat, Jagannathpur, West Bengal, India

T. Mondal
Department of Botany, Hiralal Mazumdar Memorial College for Women, Dakshineswar, Kolkata, India

A. Saha
The Neotia University, School of Pharmacy, Diamond Harbour, West Bengal, India

G. Biswas
Department of Pharmaceutical Technology, Brainware University, Kolkata, West Bengal, India

B. Dash (✉)
Amity Institute of Pharmacy, Amity University Kolkata, Kolkata, West Bengal, India

P. Pratim Dutta
Faculty of Pharmaceutical Science, Assam down town University, Panikhaiti, Guwahati, Assam, India

M. K. Gautam et al. (eds.), *Antioxidants and Functional Foods for Diabetic Neuropathy*, https://doi.org/10.1007/978-981-95-9326-2_7

Hfd	High fat diet
Lpo	Lipid peroxidation
Mapk	Mitogen activated protein kinases
Mda	Malondialdehyde
Nad	Nicotinamide adenine dinucleotide
Nadp	Nicotinamide adenine dinucleotide phosphate
Nadph	Nicotinamide adenine dinucleotide phosphate hydrogen
Ncv	Nerve conduction velocities
Ngf	Nerve growth factor
Nrf2	Nuclear factor erythroid 2-related factor 2
Pkc	Protein kinase C
Pufa	Poly unsaturated fatty acids
Rns	Reactive nitrogen species
Ros	Reactive oxygen species
Sod	Superoxide dismutase
Stz	Steptozotocin
Tpmp	Triphenylmethylphosphonium

7.1 Introduction

Diabetes is a chronic, complicated illness that must be managed with a combination of behavioral and pharmaceutical strategies to avoid or postpone complications and preserve the quality of life. These covers controlling blood sugar, weight, comorbidities, problems, and cardiovascular risk factors. This calls for the chronic care model's organised and structured approach to care, which incorporates a person-centred approach to upsurge involvement in self-care practices (Rodriguez-Gutierrez et al. 2016). Individualising treatment objectives and tactics requires cautious contemplation of socioeconomic causes of health, with the predilections of individuals with the disease diabetes (American Diabetes Association 2022). The category and interval of this disorder might determine how severe the signs and symptoms will be. People who have high blood sugar levels, especially youngsters who don't have any insulin at all, may have symptoms including weight loss, increased hunger, polydipsia, dysuria, and visual issues. Some diabetics, particularly those with type 2 diabetes in its early stages, may not exhibit any symptoms (Gautam et al. 2015). Diabetes that is not properly managed can result in many consequences, including unconsciousness, disorientation, and in rare instances, death from untreated nonketotic hyperosmolar syndrome or ketoacidosis (Poznyak et al. 2020). An estimated 537 million persons aged between 20 and 79 globally have diabetes (10.5% of all adults in this age bracket). Globally, almost near about 643 million people will have diabetes by the next 5 years, and by the year of 2045, that figure will upsurge to 783 million. The tenth edition of the IDF states that the pervasiveness of this disorder has been snowballing in countries in South-East Asia (SEA) for at least 20 years, and that current numbers have surpassed all earlier projections. Using data from the

IDF Diabetes Atlas (10th edition,) 2021, this analysis attempts to present current estimates and future forecasts of diabetes prevalence at the national and international levels. Thirty-five research studies were collected from over 60 previously published related publications that we examined for this review using a variety of sources, including PubMed and Google Scholar. But only 34 papers that specifically addressed diabetes and its prevalence in India, Southeast Asia, and the world were used (Kumar et al. 2024).

As more than 50% of people are suffering from diabetic neuropathy (DN), persistent diabetes appears as a mostly prevalent and poorly stated consequence (Callaghan et al. 2012; Gautam et al. 2014). DN is the primary cause of hospitalisations for diabetes with other severe conditions, such as nontraumatic subtractions in the US (Boulton et al. 2005). It is discovered early in the case of diabetes mellitus, whereas it is later discovered in type 1 diabetes for some idiopathic phenomenon (Shaikh and Somani 2010). The aetiology of diabetic neuropathy correlates with oxidative stress instigated by amplified free radical creation and/or a breakdown in antioxidant defences (Ziegler et al. 2004). Numerous nervous system components are impacted by diabetic neuropathies, which can present with a range of clinical symptoms (Boulton et al. 2005).

The body produces antioxidants naturally or through nutrition, and they are vital for cell defence. Enzymatic antioxidants such as superoxide dismutase (SOD), glutathione peroxidase (GPx), glutathione S-transferase (GST), and catalase (CAT) are among them. Abridged uric acid, carotenoids, glutathione (GSH), flavonoids, VIT-A, C, E, and lipoic acid are examples of nonenzymatic antioxidants. While GST transforms reactive electrophilic species into hydrophilic forms, SOD dismutase's superoxide which is generally is the anion to produce hydrogen peroxide (Kumar et al. 2013). There have also been reports of flavonoids' capacity to forage free radicals (Ayeleso et al. 2012). Additionally, certain specialised proteins, such as glutaredoxins, thioredoxins, and peroxiredoxins, have antioxidant properties (Rahman et al. 2006).

A deeper and more complete knowledge of the intricate relationship between diabetes mellitus and neuropathy is essential for creating focused therapies to lower the prevalence of diabetic neuropathy among diabetic patients. One of the outcomes of uncontrolled and ongoing hyperglycaemia is an increase in oxidative stress, which is also one of the primary risk factors for the expansion for the neuropathy in patients suffering from diabetes (Bana et al. 2023). In diabetic individuals, hyperglycaemia encourages the oxidation of glucose molecules, directing to an overabundance of advanced glycation end products (AGEs), which trigger the polyol pathway (Gugliucci 2017; Dewanjee et al. 2018). The peripheral nerve has high expression of aldose reductase, one of the primary catalysts engaged in the mechanism through the polyol pathways, indicating that these pathways have a role in the pathophysiology of DPN.

DPN development may also be impacted by a diminution in the quantity of antioxidant defence system-related catalysts in diabetic patients, such as catalase (CAT), glutathione (GSH), glutathione peroxidase (GPx), and superoxide dismutase (SOD) (Bandeira et al. 2012). and Free radicals and Reactive oxygen species (ROS) rise as a result of this decrease, which also speeds up the lipid peroxidation (LPO) process (Dworzański et al. 2020). Neurones have larger levels of polyunsaturated fatty acids (PUFA), which are more susceptible to excessive LPO when under circumstances the oxidative stress and free radicals (Shabalala et al. 2022). Diabetic individuals have considerably higher levels of by-products of LPO, including 4-hydroxynonenal (4-HNE)and malondialdehyde (MDA), which are linked to neuronal death in neurodegenerative diseases (Ghildiyal et al. 2012). The development of insulin resistance, dysfunction of β-cell, as well as diminished glucose tolerance are all linked to raised oxidative stress and increased LPO, which are harmful aspects that ultimately contribute to the pathophysiology of TIIDM's micro- and macrovascular difficulties, with diabetes associated neuropathy.

In recent times, there is a trend in growing interest in employing in vivo research to examine how antioxidants, particularly those found in natural products, affect DPN. Because of their vulnerability to diabetes-induced neuropathy, rats have been utilised in particular to explore the pathophysiology of DPN in a variety of animal models. Rat animal models may be used to induce DPN in a number of different ways, such as by administering a high-fat diet (HFD), a chemical such as streptozotocin (STZ), or a combination of the two (Mondal et al. 2025). In order to treat DPN and learn more about the pathophysiology and course of the illness, these in vivo investigations provide a useful screening platform. However, until recently, no evidence about these particular aspects besides trustworthy DPN in vivo models was recognised. Despite the fact that oxidative stress is recognised to have a part in the pathophysiology of DPN, little is known about how different natural products and substances with antioxidant qualities affect the disease. Complete information about the possible candidates and their impact on the DPN-associated metrics and oxidative stress parameters is crucial during the drug development process.

7.2 Diabetic Neuropathy and Oxidative Stress

Diabetes develops and progresses, leading to a number of microvascular and macrovascular problems (Gheibi et al. 2017). These consequences are mostly to blame due to the rise in illness and mortality among DM globally (Kiritoshi et al. 2003). They can damage the nerves (neuropathy), eyes (retinopathy), kidneys (nephropathy), or heart (cardiovascular disorders). Many alternative categories of DN are based on peripheral nerve dysfunctions involving various somatic and autonomic nervous system components (Fernyhough et al. 2010). The term diabetic peripheral neuropathy (DPN) mainly refers to polyneuropathies and a few uncommon types that can be further classified according to variations in pathogenesis, onset, duration, and clinical symptoms (Chandra et al. 2025).

Numerous animals, human, and experimental models of diabetes have reported nerve injury caused by oxidative stress (Fig. 7.1) (Hosseini and Abdollahi 2013; Coppey et al. 2001; Obrosova et al. 2002; Russell et al. 2002; Sahu et al. 2025). The processes that lead to neurological dysfunctions caused by DNA damage, lipid peroxidation associated with the free radical formations (Feldman and Vincent 2004;

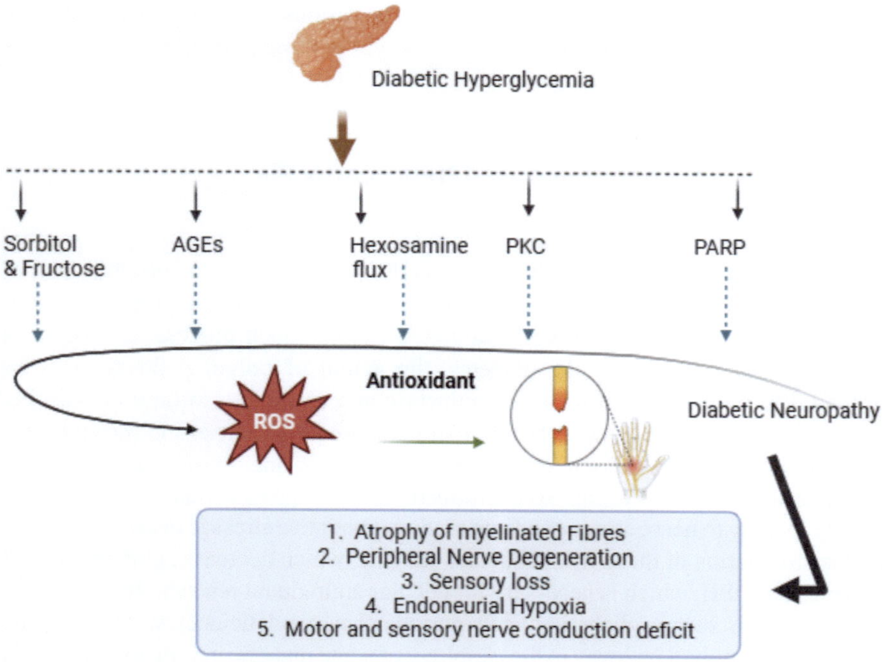

Fig. 7.1 A simple diagram that illustrates how antioxidants and reactive species play an important role in the development of diabetic neuropathy

Drel et al. 2006), production of reactive oxygen species, elevated reactive nitrogen species, and decreased cellular antioxidants (Obrosova et al. 2007). Axon loss and disturbance of the peripheral nervous system's microvasculature can occur from increased reactive oxygen and nitrogen species' ability to harm lipids found in the myelinated assemblies of nerve cells. The neuropathic pain linked to diabetic neuropathy is exacerbated by oxidative impairment to peripheral neurons, which results in hyperexcitability in the central neurons and afferent nociceptors. This leads to the production of impulsive impulses inside the nerves' dorsal root ganglions and in the axons (Ho et al. 2006; Lee et al. 2012). Moreover, degradation of the antioxidant defence system in individuals with diabetes mellitus, recent research links free radicals to the progression of neuropathy (Gugliucci 2017).

Studies demonstrated that elevated glucose levels raise peroxynitrite ions and superoxide anion, harming neurons in diabetes associated neuropathy (Ziegler et al. 2004). According to experimental research, excessive hyperglycaemia causes apoptosis in embryonic sensory neurones through a mitochondria-dependent pathway (Vincent et al. 2004). Some investigations hypothesised that an excess amount of glucose causes oxidative stress via many well-researched, interrelated pathways, which principally activate downstream signalling pathways involving NF-κB, mitogen-activated protein kinases (MAPK), proinflammatory cytokines, and gene transcriptions, ultimately resulting in nerve dysfunction (Negre-Salvayre et al. 2008). Increased hexosamine flow, glucose autoxidation, the overproduction of advanced glycation end-products (AGEs), the activation of various molecules such as protein kinase C and diacylglycerol, as well as the polyol pathway, are a few of the routes of oxidative stress carried on by diabetes (Jakus 2000).

7.2.1 Hyperglycaemia-Induced Neuropathy Via the Polyol Pathway

Approximately 3% of glucose that enters cells is transformed to sorbitol, demonstrating the critical function the polyol pathway plays a vital part in glucose metabolism. When the glycolytic route is saturated by too much glucose, as occurs in hyperglycaemic situations like diabetes, the action of polyol pathway rises by around 30%. Some enzymes, aldose reductase and as well as sorbitol dehydrogenase, catalyse this process, which turns extra glucose into fructose and sorbitol. This build-up inside cells results in hyperosmolarity and taurine, myoinositol, and adenosine efflux, which prevents ATP production and impairs axonal transport, ultimately leading to nerve structural deterioration. Oxidative stress is instigated mainly by the stimulation of the aldose reductase enzyme, which decreases glutathione and depletes NADPH, which is necessary for cellular antioxidant renewal. Reduced oxidative stress is seen in diabetic mice's peripheral nerve deficiencies, underscoring the role of the polyol pathway in the aetiology of acute diabetic neuropathy (Fig. 7.2) (Edwards et al. 2008; Ho et al. 2006; Mahmood et al. 2009).

7.2.2 Mechanism of the Antioxidant in DN

The two main objectives of antioxidants are generally to either increase the conventional defence systems through stimulating the actions of antioxidants and redeveloping additional proteins encompassed in different antioxidant pathways, or to

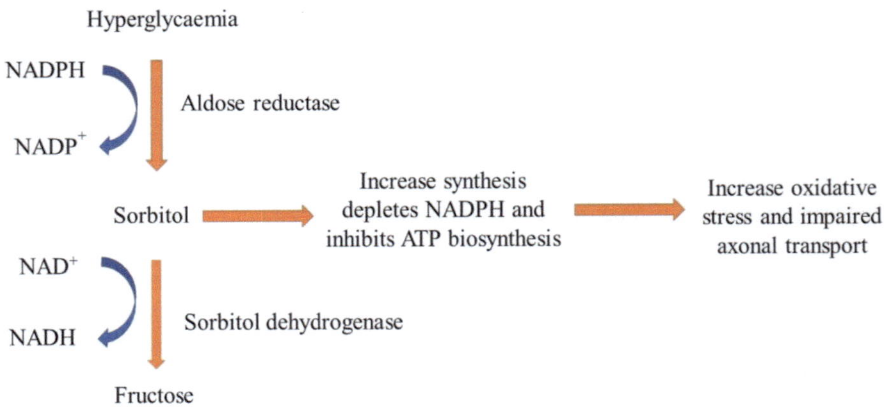

Fig. 7.2 The polyol pathway as a conceptual mechanism for hyperglycemia-induced neuropathy (Oyenihi et al. 2015)

lessen the negative outcomes of free radicals by additionally averting their creation or by scavenging and deactivating them. However, there are several methods used to prevent diabetic nerve damage when using various antioxidants. The antioxidants' kind, composition, and concentration all influence the approach that is used. It is also crucial to consider the disease's stage, severity, prevalence, and fundamental causes. A list of some of the tactics is provided below (Oyenihi et al. 2015).

7.2.2.1 Techniques Directly Targeting ROS and RNS

It is known that an upsurge in ROS, as well as similarly with RNS overproduction, is one of the causes of diabetes-induced nerve dysfunction. Consequently, it was postulated that mediators which directly scavenge free radicals, like as antioxidants, might lessen the occurrence or advancement of ROS processes, therefore reducing oxidative stress and enhancing DN conditions. Acetyl L-carnitine, taurine, melatonin, vitamins (A, C, E), and alpha-lipoic acid are a few of the most significant antioxidants. Only alpha-lipoic acid (ALA) is an antioxidant that dissolves in lipids and water. In both plants and animals, ALA may be biosynthesised and then absorbed into cells, where it is converted to dihydrolipoic acid (DHLA). The cell's ability to regenerate oxidised glutathione, ascorbic Acid, and Tocoferol is aided by the powerful free radical scavengers DHLA and ALA (Vallianou et al. 2010). Via the inhibition of the AGEs pathways and hexosamine, ALA is recognised to lower oxidative stress (Du et al. 2008).

In addition to directly detoxifying free radicals, vitamins like A, C, and E also work with reprocessing mechanisms to produce abridged versions of the vitamins (Maritim et al. 2003). Numerous biological functions of antioxidant vitamins include immunological activation and the protection of gene alterations through preventing DNA impairment brought on by metabolites of reactive oxygen species (Salah et al. 2010). NAD(P) H-dependent reductase and Ascorbate enzymes convert the stabilised phenolic radical that is created when tocopherols combine with a hydroxyl radical back with the phenol. Ascorbic Acid and vitamin E have acknowledged a lot of consideration for their capacity to scavenge free radicals. Numerous publications highlight their critical functions via shielding cells from oxidative damage (Rahman et al. 2006).

7.2.2.2 Techniques to Prevent Hyperglycaemia

The best way to postpone the development of DN and reduce its progression may be through glucose management (Shakher and Stevens 2011). The harmful outcomes of oxidative stress on neurones are lessened after blood glucose levels return to normal, which stops the overproduction of ROS caused by hyperglycaemia. Glycated haemoglobin and blood glucose levels were considerably lowered by vitamin E treatment (Nazıroğlu et al. 2004). and, without influencing intestinal area, intestinal wall thickness, or muscular tunic, exerted a neuroprotective impact on the whole myenteric population (Roldi et al. 2009). It has been shown that a naturally occurring antioxidant and antihyperglycemic plant can both prevent and improve diabetic neuropathy (Malik et al. 2011).

7.2.2.3 Techniques to Combat Personal Oxidative Stress Pathways

Potential treatment targets in DN include the previously outlined mechanisms of oxidative damage caused by hyperglycaemia. Certain treatments, such as PKC inhibitors, anti-AGE drugs, and aldose reductase inhibitors, have been developed for the consequences of some of the interventions. how the catalyst aldose reductase contributes to the build-up of fructose and sorbitol. Accordingly, aldose reductase inhibitors (ARIs) are substances which decrease the glucose flow into the polyol pathway, stopping the negative outcomes of too much fructose and sorbitol in neurons. In vitro and in vivo animal research have demonstrated the beneficial effects of aldose reductase inhibition on DN (Hotta et al. 1996). In addition to improving wrist and ankle F-wave delay and reducing neuropathic pain (Hotta et al. 2006), ARIs also increase sural sensory as well as motor nerve conduction velocities (NCV) (Kawai et al. 2010) and stop the progression of DN.

PKC affects the activation of some very important regulatory proteins which govern neuron activity and neurotransmitter production. Neuropathic pain has been demonstrated to be suppressed by PKC inhibition (Norcini et al. 2009). In diabetic neuropathy (as described in Fig. 7.1), a particular PKC-1b inhibitor with antioxidant properties enhances nerve conduction velocity (NCV) and endoneurial blood flow (Nakamura et al. 1999). Anti-AGE agents stop AGEs from forming and building up. Additionally, they inhibit the AGE-RAGE relations, which may exacerbate the damage caused by oxidative stress in neuropathy (Edwards et al. 2008). It reduces hyperglycaemic damage by increasing the activity of the transketolase enzyme, that routes AGE substrates into the pentose phosphate pathway. Additionally, it prevents a rise in UDP-N-acetylglucosamine (UDP-GlcNAc), which is mainly triggers the hexosamine pathway and, in turn, lowers tissue AGEs (Balakumar et al. 2010).

7.2.2.4 Techniques Aimed at Mitochondria

The relevance of the mitochondria into the redox state of the humanoid system is highlighted via the circumstance that it has the largest amount of antioxidants present in cells. Antioxidants are created to aim the mitochondria by conjugating with the hydrophobic cations and taking advantage of the organelle's negative membrane potential, which is around –140 mV, in order to get around this problem (Dikalov 2011). Lipophilic cations such as triphenylmethylphosphonium (TPMP) coupled with coenzyme Q10 as MitoQ10 or with tocoferol as MitovitE have been used successfully in this approach (Armstrong 2007). The efficient administration of healing concentrations of antioxidant medicines to the mitochondria, which are the primary location in cells where ROS are produced. This tactic may either significantly lower the levels of $*O_2$, *HO, and H_2O_2 that might cause oxidative cell damage or activate the antioxidants in the mitochondria to "mop up" the RNS and ROS generated.

7.2.3 Clinical Evidence for Antioxidant Therapy in Diabetic Neuropathy

Overall, this is the first clinical study to show that taking a supplement of tocotrienol-enrich vitamin E (Tocovid) for 2 months can enhance DPN as measured by upsurges in the conduction rate of nerves into the tibial motor, as well as in the median sensory, and sural sensory neurones. These enhancements are likewise consistent with higher serum NGF levels following 2 months of Tocovid intake as opposed to placebo. According to this clinical experiment, Tocovid most likely improves DPN by acting via the NGF pathway rather than oxidative stress or chronic inflammatory pathogeneses. To include Tocovid in the therapy regimen for illnesses including pathophysiological processes comparable to DPN, however, better research is essential to approve the specific pharmacodynamics of tocotrienol-rich vitamin E (Ng et al. 2020).

In one study, 187 individuals with diabetic symmetrical polyneuropathy saw improvements in neuropathic symptoms and impairments after receiving oral alpha-lipoic acid medication. 600 mg once a day was shown to have the best risk-benefit ratio, with 21% of participants experiencing side symptoms, mostly nausea. 509 participants in the ALADIN III experiment were given either a placebo or 0.6 g of alpha-lipoic acid. This treatment was linked to better nerve function; nevertheless, more research is required to evaluate its neurophysiological and therapeutic qualities. Deficits are a significant toxic effect for the progression of neuropathic foot ulcers, thus, this discovery is reassuring (Vallianou et al. 2010).

A case-control research discovered a nonlinear correlation between diabetic peripheral neuropathy (DPN) symptoms and blood 25-dihydroxyvitamin D. Compared to those with adequate levels, patients with less calciferol levels were more likely to experience symptoms of DPN. Patients were more likely to experience symptoms of DPN, nevertheless, if their 25-dihydroxy-vitamin D [25(OH)D] was more than 40 ng/mL. According to the study, controlling diabetic neuropathy requires regular monitoring and assessment of vitamin D levels (Esteghamati et al. 2016).

A clinical study that included male and female, both genders with type 2 diabetes and symptomatic DPN between the ages of 45 and 74 discovered that oral administration of nicotinamide, succinic acid, riboflavin (SINR) and inosine, medication, which is an antioxidant, effectively reduces the symptoms of neuropathy in individuals with type 2 diabetes. The SINR group's mean TSS change after 3 months was observed −2.65 (±1.46), whereas the placebo group was −1.73 (±1.51). Haemoglobin A1c levels did not affect the decrease of symptoms; however, individuals with initial TSS <7.5 showed improved outcomes. Among the study's drawbacks are its small impact size, moderate percentage of severe DPN symptoms, subjective evaluation of results, exclusion of people with uncontrolled and type 1 diabetes, and individuals who administered glucose-lowering drugs other than insulins (Kharitonova et al. 2022).

This is the first organised assessment and meta-analysis to assess the indication that vitamin supplements can enhance antioxidant status in T2DM patients in a variety of ways (e.g., by increasing GPx, SOD, and TAC levels and decreasing MDA and TBARS products). Glycaemic control, insulin resistance, and alterations in endothelial activities have been the subjects of earlier research (Xu et al. 2014). The findings showed that taking supplements of specific vitamins, particularly vitamin E, in people with diabetes mellitus can have a good result on parameters of glycaemic management and antioxidant status. There was less evidence for the impact of other vitamins (such as calciferol or B), but Ascorbic acid was additionally closely linked to deviations in antioxidant status. One possible explanation for vitamin E's positive benefits is that it lessens the harm that free radicals do to the structural and functional elements of blood vessels and cells (Ashor et al. 2015). One of the newest non-calcaemic functions proposed for vitamin D is its antioxidant impact (Mokhtari et al. 2016). Calciferol may have a momentous impact on reducing the incidence of diabetes, according to data from both human and animal models (Pittas et al. 2007).

Another meta-analysis research has shown that epalrestat might restrict the polyol pathway, lessen oxidative stress, and lessen the production of the antioxidant enzymes as well as aldose reductase. Inhibiting peroxidation, improving blood flow, raising nerve Na+–K+ ATPase activity, including safeguarding endothelial function are all possible benefits of ALA. It may also directly remove free radicals. Patients with diabetes are most commonly affected by DPN, the most prevalent neuropathic condition. Patients with DPN may have diabetic damage to their neurones (axons or myelin sheaths) as a result of hyperglycaemia and metabolic imbalance, mainly oxidative stress (Hosseini and Abdollahi 2013). Several related pathways, including hexosamine, polyol, protein kinase C, advanced glycation end products, and, poly-ADP-ribose polymerase were first triggered by oxidative stress (Tesfaye 2011). Furthermore, endothelial cell damage and vascular dysfunction brought on by oxidative stress lead to peripheral nerve ischaemia and hypoxia (Han et al. 2012).

Antioxidants should be used in conjunction with RAS blockers to treat DN in humans, as this medication is insufficient. In diabetic individuals, traditional antioxidants have not demonstrated any positive effects. Human DN may be improved by new multi-property antioxidants that exhibit strong anti-inflammatory effects, extended half-life, antioxidant capacity, and high permeability to mitochondria, and enhanced activity of antioxidant enzymes in the body. The potential benefits of rosmarinic acid, which act as a multi-property antioxidant, for DN patients must be demonstrated by clinical trials (Tavafi 2013).

7.2.4 Challenges and Limitations in Antioxidant Therapy for Diabetic Neuropathy

Since oxidative stress is a part of the underlying pathology of many illnesses and toxicities, the antioxidant defences and techniques discussed above provide some significant potential to avoid or lessen pathology. Our capacity to use antioxidant methods therapeutically is, however, limited by a number of factors.

Involvement of oxidative stress to pathogenesis: The oxidative stress plays a vital role in the pathology and also limits the efficacy of antioxidant defences. Preventing oxidative stress may not even significantly impact the course of a disease when it is a minor contributor, which is more common than when it is the primary cause. This is, in fact, one of the main reasons why antioxidants, despite their obvious improvement in antioxidant defence and reduction in oxidative stress indicators, have little to no impact on pathology (Forman and Zhang 2021).

Reaching efficient in vivo concentrations: An Alternative issue is that substances which trigger antioxidant defences may not be able to achieve efficient effects in case of in vivo concentrations; however, cyanoenones may be able to get around this. Supplemental NRF2 activators wouldn't be beneficial when healthy eating supplies sufficient amounts of NRF2 activators. Furthermore, if patients experience oxidative stress, NRF2 is often already somewhat engaged and there is little chance of further activation (Kostov et al. 2015).

Small molecule scavenging: Due to its concentration in membranes and comparatively quick rate of reactivity with lipid hydroperoxyl radicals, vitamin E is the only dietary antioxidant that is not limited in its ability to scavenge small molecules. However, in human trials, antioxidant medicines that appeared to be active in cell culture or non-human in vivo models have often fallen short of producing meaningful results. In vitro and in vivo exogenous agent ratios differ significantly, which is one of the main causes of this disparity (Forman and Zhang 2021).

Clinical translation is difficult because no other medications have received approval from FDA as of yet, partly because of clinical trial failures. An aldose reductase inhibitor, sorbinil, was demonstrated to prevent diabetes-induced neuron conduction debit in streptozotocin-induced diabetic rats, but less comparable outcomes in humans (Pang et al. 2020). This is one of the common reasons why clinical trials fail: the upgrading in the treated in vivo models is irreproducible compared to the human counterpart.

7.2.5 Future Aspects

Now would be a good moment to investigate antioxidants holistically as a possible solution to the persistent issue of diabetic neuropathy. The secret may lie in two antioxidant methods. The first is using conventional antioxidants, such as alpha lipoic acid and vitamins A, C, and E, which may quickly scavenge a range of free radicals in diabetic neuropathy clinical studies in humans and animal models. It is necessary to investigate combinational strategies that incorporate elements including alpha lipoic acid, Ascorbic acid, tocopherol, and vitamin A, as well as medicinal plant items with antioxidant and antihyperglycemic qualities. A natural incidence by certain elements being dietary ingredients, variety of actions pointing diverse phases in the development of neuropathy, and a usually low hazardous potential are the benefits of such combos. The ability to evaluate a wider variety of end-points is another significant benefit of this approach, which is why clinical trials should concentrate on it. The next is the efficient administration of beneficial concentrations of antioxidant medicines to the powerhouse mitochondria, the primary location in cells where ROS are produced (Oyenihi et al. 2015).

Nevertheless, using antioxidant nutrients presents difficulties because of their limited absorption, susceptibility to auto-oxidation, and potential for microbial metabolism. The effectiveness of antioxidant and medication combination therapy, as well as the appropriate dosages of each dietary antioxidant, will thus need to be further investigated. Additionally, this paper's analysis of studies shows that most of these natural antioxidants come from plants, highlighting new study avenues for plant-based diets (Pang et al. 2020).

7.3 Conclusion

New developments in antioxidant techniques have demonstrated encouraging outcomes in the therapy of neuropathy associated with diabetes. These methods have demonstrated promise in reducing oxidative stress through scavenging oxidants or blocking the processes which produce them. They have both antihyperglycemic and antioxidant end-points. These studies, however, have typically concentrated on certain end-points, which means they are not comprehensive. The adoption of techniques in combination is crucial since diabetes mellitus is a diverse illness with several aetiologies. Although nonclinical research has demonstrated positive outcomes, there has been little success in moving these tactics to clinical trials. The main feature of diabetes is hyperglycemia, which is connected with oxidative stress and diabetic neuropathy. The two primary processes underlying diabetes, decreased insulin production and insulin resistance, are both linked to oxidative stress. In order to prevent or cure diabetes and diabetic neuropathy, medicines that target oxidative stress and hyperglycaemia may be safe and affordable options.

References

American Diabetes Association Professional Practice Committee, & American Diabetes Association Professional Practice Committee. 6. Glycemic targets: standards of medical care in diabetes—2022. Diabetes care. 2022;45(Supplement_1):S83–96.

Armstrong JS. Mitochondrial medicine: pharmacological targeting of mitochondria in disease. Br J Pharmacol. 2007;151(8):1154–65.

Ashor AW, Siervo M, Lara J, Oggioni C, Afshar S, Mathers JC. Effect of vitamin C and vitamin E supplementation on endothelial function: a systematic review and meta-analysis of randomised controlled trials. Br J Nutr. 2015;113(8):1182–94.

Ayeleso AO, Oguntibeju OO, Brooks N. Bioactive phytochemicals: perspectives for modern medicine, vol. 1. New Delhi: Daya Publishing House; 2012.

Balakumar P, Rohilla A, Krishan P, Solairaj P, Thangathirupathi A. The multifaceted therapeutic potential of benfotiamine. Pharmacol Res. 2010;61(6):482–8.

Bana S, Kumar N, Sartaj A, Alhalmi A, Qurtam AA, Nasr FA, et al. Rubia cordifolia L. Attenuates diabetic neuropathy by inhibiting apoptosis and oxidative stress in rats. Pharmaceuticals. 2023;16(11):1586.

Bandeira SDM, Guedes GDS, Fonseca LJSD, Pires AS, Gelain DP, Moreira JCF, et al. Characterization of blood oxidative stress in type 2 diabetes mellitus patients: increase in lipid peroxidation and SOD activity. Oxidative Med Cell Longev. 2012;2012(1):819310.

Boulton AJ, Vinik AI, Arezzo JC, Bril V, Feldman EL, Freeman R, et al. Diabetic neuropathies: a statement by the American Diabetes Association. Diabetes Care. 2005;28(4):956–62.

Callaghan BC, Cheng HT, Stables CL, Smith AL, Feldman EL. Diabetic neuropathy: clinical manifestations and current treatments. Lancet Neurol. 2012;11(6):521–34.

Chandra P, Shahjad, Porwal M, Rastogi V, Mittal P, Verma A, Sachan N, Prabahar EA, Gautam MK. Progresses in wound healing: integrating nutrition, physical therapy, traditional and alternative medicine, and novel technologies. Curr Diabetes Rev. 2025;2025:e15733998350164.

Coppey LJ, Gellett JS, Davidson EP, Dunlap JA, Lund DD, Yorek MA. Effect of antioxidant treatment of streptozotocin-induced diabetic rats on endoneurial blood flow, motor nerve conduction velocity, and vascular reactivity of epineurial arterioles of the sciatic nerve. Diabetes. 2001;50(8):1927–37.

Dewanjee S, Das S, Das AK, Bhattacharjee N, Dihingia A, Dua TK, et al. Molecular mechanism of diabetic neuropathy and its pharmacotherapeutic targets. Eur J Pharmacol. 2018;833:472–523.

Dikalov S. Cross talk between mitochondria and NADPH oxidases. Free Radic Biol Med. 2011;51(7):1289–301.

Drel VR, Mashtalir N, Ilnytska O, Shin J, Li F, Lyzogubov VV, Obrosova IG. The leptin-deficient (Ob/Ob) mouse: a new animal model of peripheral neuropathy of type 2 diabetes and obesity. Diabetes. 2006;55(12):3335–43.

Du X, Edelstein D, Brownlee M. Oral benfotiamine plus α-lipoic acid normalises complication-causing pathways in type 1 diabetes. Diabetologia. 2008;51:1930–2.

Dworzański J, Strycharz-Dudziak M, Kliszczewska E, Kiełczykowska M, Dworzańska A, Drop B, Polz-Dacewicz M. Glutathione peroxidase (GPx) and superoxide dismutase (SOD) activity in patients with diabetes mellitus type 2 infected with Epstein-Barr virus. PLoS One. 2020;15(3):e0230374.

Edwards JL, Vincent AM, Cheng HT, Feldman EL. Diabetic neuropathy: mechanisms to management. Pharmacol Ther. 2008;120(1):1–34.

Esteghamati A, Fotouhi A, Faghihi-Kashani S, Hafezi-Nejad N, Heidari B, Sheikhbahaei S, et al. Non-linear contribution of serum vitamin D to symptomatic diabetic neuropathy: a case-control study. Diabetes Res Clin Pract. 2016;111:44–50.

Feldman EL, Vincent A. The prevalence, impact, and multifactorial pathogenesis of diabetic peripheral neuropathy. Adv Stud Med. 2004;4(8A):S642–9.

Fernyhough P, Roy Chowdhury SK, Schmidt RE. Mitochondrial stress and the pathogenesis of diabetic neuropathy. Expert Rev Endocrinol Metab. 2010;5(1):39–49.

Forman HJ, Zhang H. Targeting oxidative stress in disease: promise and limitations of antioxidant therapy. Nat Rev Drug Discov. 2021;20(9):689–709.

Gheibi S, Kashfi K, Ghasemi A. A practical guide for induction of type-2 diabetes in rat: incorporating a high-fat diet and streptozotocin. Biomed Pharmacother. 2017;95:605–13.

Gugliucci A. Formation of fructose-mediated advanced glycation end products and their roles in metabolic and inflammatory diseases. Adv Nutr. 2017;8(1):54–62.

Han T, Bai J, Liu W, Hu Y. Therapy of endocrine disease: a systematic review and meta-analysis of α-lipoic acid in the treatment of diabetic peripheral neuropathy. Eur J Endocrinol. 2012;167(4):465–71.

Ho EC, Lam KS, Chen YS, Yip JC, Arvindakshan M, Yamagishi SI, et al. Aldose reductase–deficient mice are protected from delayed motor nerve conduction velocity, increased c-Jun NH2-terminal kinase activation, depletion of reduced glutathione, increased superoxide accumulation, and DNA damage. Diabetes. 2006;55(7):1946–53.

Hosseini A, Abdollahi M. Diabetic neuropathy and oxidative stress: therapeutic perspectives. Oxidative Med Cell Longev. 2013;2013(1):168039.

Hotta N, Akanuma Y, Kawamori R, Matsuoka K, Oka Y, Shichiri M, et al. Long-term clinical effects of epalrestat, an aldose reductase inhibitor, on diabetic peripheral neuropathy: the 3-year, multicenter, comparative aldose reductase inhibitor-diabetes complications trial. Diabetes Care. 2006;29(7):1538–44.

Hotta N, Sakamoto N, Shigeta Y, Kikkawa R, Goto Y, Japan DNSG. Clinical investigation of epalrestat, an aldose reductase inhibitor, on diabetic neuropathy in Japan: multicenter study. J Diabetes Complicat. 1996;10(3):168–72.

Jakus V. The role of free radicals, oxidative stress and antioxidant systems in diabetic vascular disease. Bratisl Lek Listy. 2000;101(10):541–51.

Gautam MK, Purohit V, Agarwal M, Singh A, Goel RK. In-vivo healing potential of Aegle marmelos in excision, incision and dead space wound models. Sci World J. 2014:740107. https://doi.org/10.1155/2014/740107.

Kawai T, Takei I, Tokui M, Funae O, Miyamoto K, Tabata M, et al. Effects of epalrestat, an aldose reductase inhibitor, on diabetic peripheral neuropathy in patients with type 2 diabetes, in relation to suppression of Nε-carboxymethyl lysine. J Diabetes Complicat. 2010;24(6):424–32.

Kharitonova T, Shvarts YG, Verbovoy AF, Orlova NS, Puzyreva VP, Strokov IA. Efficacy and safety of the combined metabolic medication, containing inosine, nicotinamide, riboflavin and succinic acid, for the treatment of diabetic neuropathy: a multicenter randomized, double-blind, placebo-controlled parallel group clinical trial (CYLINDER). BMJ Open Diabetes Res Care. 2022;10(3):e002785.

Kiritoshi S, Nishikawa T, Sonoda K, Kukidome D, Senokuchi T, Matsuo T, et al. Reactive oxygen species from mitochondria induce cyclooxygenase-2 gene expression in human mesangial cells: potential role in diabetic nephropathy. Diabetes. 2003;52(10):2570–7.

Kostov RV, Knatko EV, McLaughlin LA, Henderson CJ, Zheng S, Huang JTJ, et al. Pharmacokinetics and pharmacodynamics of orally administered acetylenic tricyclic bis (cyanoenone), a highly potent Nrf2 activator with a reversible covalent mode of action. Biochem Biophys Res Commun. 2015;465(3):402–7.

Kumar A, Gangwar R, Ahmad Zargar A, Kumar R, Sharma A. Prevalence of diabetes in India: a review of IDF diabetes atlas 10th edition. Curr Diabetes Rev. 2024;20(1):105–14.

Lee KO, Nam JS, Ahn CW, Hong JM, Kim SM, Sunwoo IN, et al. Insulin resistance is independently associated with peripheral and autonomic neuropathy in Korean type 2 diabetic patients. Acta Diabetol. 2012;49:97–103.

Mahmood D, Singh BK, Akhtar M. Diabetic neuropathy: therapies on the horizon. J Pharm Pharmacol. 2009;61(9):1137–45.

Malik ZA, Singh M, Sharma PL. Neuroprotective effect of Momordica charantia in global cerebral ischemia and reperfusion induced neuronal damage in diabetic mice. J Ethnopharmacol. 2011;133(2):729–34.

Maritim AC, Sanders A, Watkins Iii JB. Diabetes, oxidative stress, and antioxidants: a review. J Biochem Mol Toxicol. 2003;17(1):24–38.

Mokhtari Z, Hekmatdoost A, Nourian M. Antioxidant efficacy of vitamin D. J Parathyr Dis. 2016;5(1):11–6.

Nakamura J, Kato K, Hamada Y, Nakayama M, Chaya S, Nakashima E, et al. A protein kinase C-beta-selective inhibitor ameliorates neural dysfunction in streptozotocin-induced diabetic rats. Diabetes. 1999;48(10):2090–5.

Nazıroğlu M, Şimşek M, Şimşek H, Aydilek N, Özcan Z, Atılgan R. The effects of hormone replacement therapy combined with vitamins C and E on antioxidants levels and lipid profiles in postmenopausal women with type 2 diabetes. Clin Chim Acta. 2004;344(1–2):63–71.

Negre-Salvayre A, Coatrieux C, Ingueneau C, Salvayre R. Advanced lipid peroxidation end products in oxidative damage to proteins. Potential role in diseases and therapeutic prospects for the inhibitors. Br J Pharmacol. 2008;153(1):6–20.

Ng YT, Phang SCW, Tan GCJ, Ng EY, Botross Henien NP, M. Palanisamy UD, et al. The effects of tocotrienol-rich vitamin E (Tocovid) on diabetic neuropathy: a phase II randomized controlled trial. Nutrients. 2020;12(5):1522.

Norcini M, Vivoli E, Galeotti N, Bianchi E, Bartolini A, Ghelardini C. Supraspinal role of protein kinase C in oxaliplatin-induced neuropathy in rat. Pain. 2009;146(1–2):141–7.

Obrosova IG, Huysen CV, Fathallah L, Cao X, Greene DA, Stevens MJ. An aldose reductase inhibitor reverses early diabetes-induced changes in peripheral nerve function, metabolism, and antioxidative defense. Federation of American Societies for Experimental Biology; 2002.

Obrosova IG, Ilnytska O, Lyzogubov VV, Pavlov IA, Mashtalir N, Nadler JL, Drel VR. High-fat diet–induced neuropathy of pre-diabetes and obesity: effects of "healthy" diet and aldose reductase inhibition. Diabetes. 2007;56(10):2598–608.

Oyenihi AB, Ayeleso AO, Mukwevho E, Masola B. Antioxidant strategies in the management of diabetic neuropathy. Biomed Res Int. 2015;2015(1):515042.

Pang L, Lian X, Liu H, Zhang Y, Li Q, Cai Y, et al. Understanding diabetic neuropathy: focus on oxidative stress. Oxidative Med Cell Longev. 2020;2020(1):9524635.

Pittas AG, Lau J, Hu FB, Dawson-Hughes B. The role of vitamin D and calcium in type 2 diabetes. A systematic review and meta-analysis. J Clin Endocrinol Metabol. 2007;92(6):2017–29.

Poznyak A, Grechko AV, Poggio P, Myasoedova VA, Alfieri V, Orekhov AN. The diabetes mellitus–atherosclerosis connection: the role of lipid and glucose metabolism and chronic inflammation. Int J Mol Sci. 2020;21(5):1835.

Rahman I, Biswas SK, Kode A. Oxidant and antioxidant balance in the airways and airway diseases. Eur J Pharmacol. 2006;533(1–3):222–39.

Rodriguez-Gutierrez R, Gionfriddo MR, Ospina NS, Maraka S, Tamhane S, Montori VM, Brito JP. Shared decision making in endocrinology: present and future directions. Lancet Diabetes Endocrinol. 2016;4(8):706–16.

Roldi LP, Pereira RV, Tronchini EA, Rizo GV, Scoaris CR, Zanoni JN, Natali MR. Vitamin E (α-tocopherol) supplementation in diabetic rats: effects on the proximal colon. BMC Gastroenterol. 2009;9:1–8.

Gautam MK, Gangwar M, Singh SK, Goel RK. Effects of Azadirachta indica on vascular endothelial growth factor and cytokines in diabetic deep wound. Planta Med. 2015;81(9):713–21.

Russell JW, Golovoy D, Vincent AM, Mahendru PIA, Olzmann JA, Mentzer A, Feldman EL. High glucose-induced oxidative stress and mitochondrial dysfunction in neurons. FASEB J. 2002;16(13):1738–48.

Salah SH, Abdou HS, Rahim EA. Modulatory effect of vitamins A, C and E mixtures against tefluthrin pesticide genotoxicity in rats. Pestic Biochem Physiol. 2010;98(2):191–7.

Shabalala SC, Johnson R, Basson AK, Ziqubu K, Hlengwa N, Mthembu SX, et al. Detrimental effects of lipid peroxidation in type 2 diabetes: exploring the neutralizing influence of antioxidants. Antioxidants. 2022;11(10):2071.

Shaikh AS, Somani RS. Animal models and biomarkers of neuropathy in diabetic rodents. Indian J Pharmacol. 2010;42(3):129–34.

Shakher J, Stevens MJ. Update on the management of diabetic polyneuropathies. Diabetes Metab Syndr Obes. 2011;4:289–305.

Tavafi M. Diabetic nephropathy and antioxidants. J Nephropathol. 2013;2(1):20.

Tesfaye S. Recent advances in the management of diabetic distal symmetrical polyneuropathy. J Diabetes Investig. 2011;2(1):33–42.

Kumar M, Gautam MK, Singh A, Goel RK. Healing effects of Musa sapientum var. paradisiaca in diabetic rats with co-occurring gastric ulcer: cytokines and growth factor by PCR amplification. BMC Complement Altern Med. 2013;13:305.

Vallianou N, Evangelopoulos A, Koutalas P. Alpha-lipoic acid and diabetic neuropathy. Rev Diabet Stud. 2010;6(4):230.

Mondal NS, Islam A, Gautam MK, Mondal S, Ahmed M, Jamir SI, Khongwir LD, Mahajon B. Unraveling the mechanisms of diabetic wounds: insights into pathogenesis and advanced treatment strategies. Curr Diabetes Rev. 2025;2025:e15733998368222.

Vincent AM, Russell JW, Low P, Feldman EL. Oxidative stress in the pathogenesis of diabetic neuropathy. Endocr Rev. 2004;25(4):612–28.

Xu R, Zhang S, Tao A, Chen G, Zhang M. Influence of vitamin E supplementation on glycaemic control: a meta-analysis of randomised controlled trials. PLoS One. 2014;9(4):e95008.

Ghildiyal S, Gautam MK, Gangwar M, Nath G, Goel RK, Joshi VK. Pharmacognostic evaluation and in-vitro antibacterial activity of Uraria picta Desv. root. Plant Arch. 2012;12(2):931–5.

Sahu BP, Gautam MK, Dey BK, Meher B. Assessment of hepatoprotective potential of Dendrobium bicameratum, a bryophyte from Eastern Ghats. Cuest Fisioter. 2025;54(4):5506–10.

Ziegler D, Sohr CG, Nourooz-Zadeh J. Oxidative stress and antioxidant defense in relation to the severity of diabetic polyneuropathy and cardiovascular autonomic neuropathy. Diabetes Care. 2004;27(9):2178–83.

Part III

Functional Foods for Diabetic Neuropathy

Functional Foods: Definition and Health Benefits

8

Pervej Alom Barbhuiya, Neha Ghimire, Gayatri Thapa, Punam Kumari, and Manash Pratim Pathak

8.1 Introduction

The term "food" refers to the elements necessary for several procedures that sustain life, including the generation of energy, nutrition supply, and support for diverse metabolic functions, and physiological development and well-being (Kaur and Das 2011). "Functional food" is essentially a marketing phrase; globally, it is not acknowledged by law (Henry 2010). Functional foods include biologically active components, such as minerals and vitamins, probiotics, anti-oxidants, and other chemical compounds that improve health or lower the risk of disease (Farr 1997). There are several definitions of functional foods exist. Health Canada defines functional foods as products that resemble traditional foods but have documented physiological advantages (Henry 2010; Kaur and Das 2011). Other definitions include that from the Food and Nutrition Board (FNB) of the National Academy of Sciences,

P. A. Barbhuiya · M. P. Pathak (✉)
Centre for Research on Ethnomedicine, Faculty of Pharmaceutical Science, Assam down town University, Guwahati, Assam, India

N. Ghimire
Regional College of Pharmaceutical Sciences, Kamrup, Assam, India

G. Thapa
Faculty of Pharmaceutical Science, Assam down town University, Guwahati, Assam, India

Himalayan Pharmacy Institute, Majitar, Sikkim, India

P. Kumari
Faculty of Pharmaceutical Science, Assam down town University, Guwahati, Assam, India

Centre for Research on Ethnomedicine, Faculty of Pharmaceutical Science, Assam down town University, Guwahati, Assam, India

Department of Pharmacy, School of Health Science, Central University of South Bihar, Gaya, Bihar, India

M. K. Gautam et al. (eds.), *Antioxidants and Functional Foods for Diabetic Neuropathy*, https://doi.org/10.1007/978-981-95-9326-2_8

Washington, USA/Thomas and Earl, Functional food refers to potentially beneficial foodstuffs, such as "any modified food or food ingredient that may provide a health benefit beyond that of the traditional nutrients in contains" (Kaur and Das 2011). Functional foods were developed in Japan in the 1980s, when they were officially recognized as "Foods for Specified Health Uses". The concept acquired global support as research linked nutrition to health and illness prevention (Weststrate et al. 2002). The phrase 'functional food' was first popularised in 1993 in the Nature news magazine with the title 'Japan explores the distinction among food and medicine' (Henry 2010). Functional foods provide health benefits above and beyond simple nutrition (Playne et al. 2003; Widyastuti and Rohmatussolihat 2014).

In today's world, modern lifestyles tend to be defined by the consumption of overly processed foods, decreased physical activity, and an increase in chronic health disorders. Obesity, diabetes, cardiovascular disease, and even some cancers have each become more prevalent as an outcome of these changes (Jew et al. 2009). Functional foods are important since they provide a distinctive response by linking nutrition and medicine. Regular consumption of functional foods may result in improved anti-inflammatory, insulin resistance, and antioxidant qualities—all of which are thought to be essential for avoiding and treating adult-onset diabetes and its consequences (Alkhatib et al. 2017). Functional foods have a comparable appearance to traditional foods, primarily offer energy and essential nutrients, functional foods are made with or naturally rich in substances that improve physiological functioning or disease-preventing characteristics that go beyond the basic role of providing nutrition. Supplemented cereals, probiotic yogurts, and naturally bioactive foods such as berries or nuts are some examples (Cencic and Chingwaru 2010; Kaur and Das 2011). The International Life Sciences Institute defines the basic criteria for categorising functional foods, the first role is to supply nutrition to the body for regular activities like growth and development. The second purpose is to deliver food satisfaction, such as tastes, sights, and fragrances, while also improving well-being and having the ability to regulate bodily physiological processes using substances other than the primary nutrients found in food (Jones 2002; Galanakis 2021).

Different types of functional foods include probiotic foods, omega-3-rich foods, whole grains, fruits and vegetables, and fortified foods. Probiotic foods like yogurt and kefir contains living microorganisms that, when consumed in proper quantities, enhance intestinal well-being and boost immunity (Tripathi and Giri 2014), omega-3-rich foods, such as fatty fish and the seeds of flax contain polyunsaturated fatty acids, primarily eicosapentaenoic acid (EPA) and docosahexaenoic acid (DHA) have triglyceride (TG) lowering effects (Rudkowska 2009), whole-grain products include oatmeal, grains such as rice and maize, millet, and quinoa intake help in digestion and lower cholesterol levels (Frølich et al. 2013), fruits and vegetables including berries, spinach, and green tea could be related to both antioxidant and anti-inflammatory mechanisms (Serafini and Peluso 2017), and fortified foods are the diet foods that can help to prevent or control deficiencies like zinc and iron, and vitamins A as well as D micronutrient deficiencies. Some of the examples of fortified foods are bread, flour, egg, almond milk, and breakfast cereals (Birch and Bonwick 2019; Olson et al. 2021).

The idea of functional foods is thoroughly examined in this chapter, covering their historical origins, characteristics, and role in modern dietary practices. By providing a thorough overview, it presents a comprehensive assessment of their revolutionary potential to change healthier food systems and encourage healthier lifestyles. This chapter emphasizes the important role functional foods can play in solving modern health concerns and developing a better, healthier future by analyzing their ability to improve public health outcomes.

8.2 Health Benefits of Functional Foods

Functional foods are crucial for providing protection against heart disorders, hyperglycemia, improving gut health, low bone density, boosting immunity, arthritis, mental disorders, etc. (Fig. 8.1) (Munekata et al. 2021). The primary source of functional foods is plants; however, other sources include probiotics obtained from good bacteria, polyunsaturated fatty acids derived from fish, and conjugated linoleic acid from dairy products and certain animal meats can also be sources of functional foods (Hasler 2002; Shaikh 2022). Common probiotics include the various strains of Lactobacillus sp., which when added into food may improve gut health and manage both urogenital infections and blood pressure (Sanders 2003). Breastfeeding also boosts the level of good bacteria in the babies, which is responsible for protecting against pathogens (Williamson 2009). Alongside probiotics, prebiotics are also

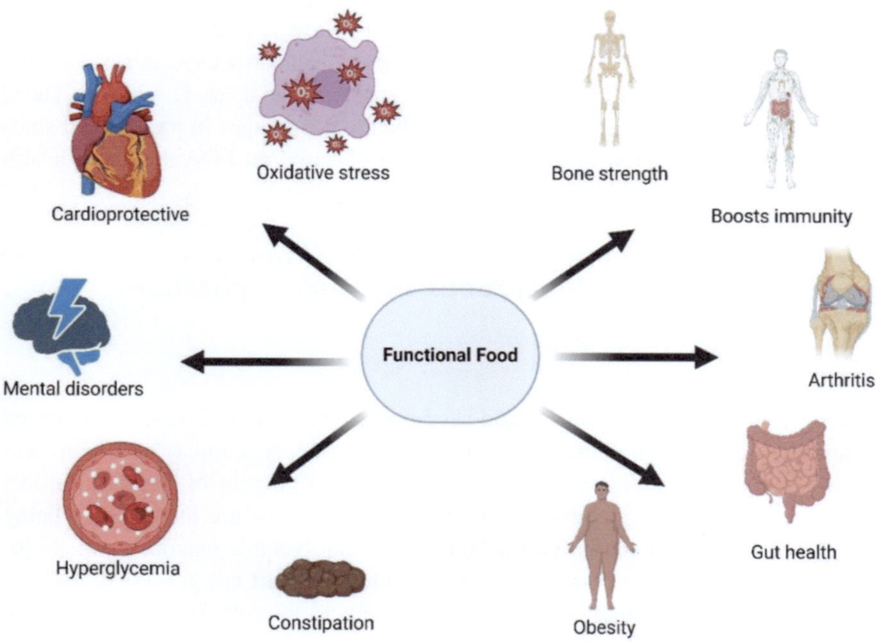

Fig. 8.1 Different health benefits of Functional foods

the source of functional foods, it promotes the development of beneficial gut flora in an ideal environment. Prebiotics are generally obtained from foods rich in dietary fibers as they contain the derivatives of carbohydrates such as fructans and galactans, which ultimately improve gut health (Gibson and Roberfroid 1995). It is possible that polyunsaturated fatty acids will help new-born's developing brains and eyes. (Crawford 2000). Furthermore, the addition of fatty acids in a CHD patient's diet will reduce the risk of sudden death (Bucher et al. 2002). However, excess consumption of fatty acid-rich foods may lead to cholesterol and glycemic imbalance. Therefore, regulatory bodies like the FDA recommend that the consumption of EPA and DHA supplements should not exceed 2 grams daily (Hasler 2002).

Data suggests that in animal models, conjugated linoleic acid is beneficial against breast cancer (Belury 1995), managing body weight by reducing the adipose tissues and building up the body mass (Park et al. 1999), and improves mineral mass of the bone (Ing and Belury 2011). Food from dairy farms, ruminants, and egg yolks contains a lot of conjugated linoleic acid (Hasler 2002). Out of many plant-derived functional foods, very few have been approved by the FDA, like beta-glucan derived from oat fibres, fibres obtained from Ispaghula seed husk, soy protein, and oleomargarine enriched with cholesterol-like substances called sterol and stanol esters, and curcuminoids enriched foods. These functional foods are beneficial in the management of constipation, regulating blood sugar level, production of satiety, the prime source of protein, lowering LDL levels, and many more (Hasler 2002). On the other hand, there are many unapproved health claim plant derived substances that can be potential functional foods such as proanthocyanidins from cranberry (treats UTI) (Howell et al. 1998), allicin from garlic contributes to elevated blood pressure (Silagy and Neil 1994) and high cholesterol levels (Stevinson et al. 2000), polyphenols from red grape acts as anti-oxidant (also beneficial for neurodegenerative disorders) (Adefegha 2018) and anti-platelet agent (Keevil et al. 2000). Those substances have been established to provide health advantages in pre-clinical studies and clinical trials, but they have still to be declared an FDA-approved health claim (Hasler 2002).

8.3 Common Functional Foods and Their Applications

8.3.1 Yogurt

A significant quantity of investigation has been undertaken on diverse dry-cultured materials, including yoghurt, acidophilus dairy products, kumiss, and kefir, frequently integrating additives to boost the nutritional content of the end product (Shiby and Mishra 2013). Several minor milk components are increasingly being recognized for their potential physiological benefits. Notable examples are conjugated linoleic acid (CLA), selenium, and β-ionone. Yoghurt is a gel-like dairy food produced via the fermentation of dairy by lactic acid microbial organisms, which naturally acidify the milk. Yoghurt, an essential in many dietary habits throughout the world, is well acknowledged for its potential as a functional food. It contains a lot of high-quality protein, which is important for muscle growth and repair. This is especially beneficial for people who engage in regular physical activity or strength training (Sumi et al. 2023). Building and keeping healthy bones and teeth need calcium. Appropriate intake of calcium prevents osteoporosis and promotes bone density, making yogurt a valuable dietary source for maintaining bone health. Calcium is vital for muscle contraction and function. It helps muscles contract and relax properly, which is important for overall muscle function and coordination (Fernandez and Marette 2017). Phosphorus, which is present in yogurt, plays several important roles related to bone health and energy metabolism. Phosphorus is a key component of hydroxyapatite, the mineral complex that makes up a significant portion of bone and tooth structure. The development and preservation of healthy teeth and bones depend on adequate phosphorus levels. Phosphorus functions together with calcium to develop and maintain healthy bones. It helps in the mineralization process, where calcium and phosphorus combine to form bone mineral. Phosphorus supports bone repair and regeneration. In the event of bone injuries or fractures, sufficient phosphorus is necessary for effective healing and recovery. Adenosine triphosphate (ATP), the main source of energy in cells, requires phosphorus as a necessary component. ATP supplies the energy required for different cellular functions, such as muscular contraction, neuronal activity, and metabolic activities (Cohen 1982; Zhang et al. 2011; Takeda et al. 2012; Serna and Bergwitz 2020; Peacock 2021). Yogurt is a valuable source of several B vitamins, each essential for maintaining overall health. Vitamin B2 (Riboflavin) supports energy production by converting nutrients into usable energy and helps protect cells from oxidative damage. Vitamin B5 (Pantothenic Acid) plays a key role in synthesizing coenzyme A for energy metabolism and hormone production, and also aids in wound healing. Vitamin B6 (Pyridoxine) is crucial for amino acid metabolism, brain function, and immune support. Vitamin B12 (Cobalamin) is essential for blood vessel development, neuron function, and the production of DNA. Folate (Vitamin B9) supports cell division, red blood cell production, and is particularly important for neural tube development during pregnancy. Incorporating yogurt into your diet helps provide these important B vitamins, contributing to various aspects of health and well-being (Marques de Brito et al. 2023; Kumar et al. 2024; Maurya et al. 2024). Yogurt's probiotic content is an important functional component. These are living microorganisms that offer health advantages when consumed in appropriate doses. Yogurt mainly features probiotics such as *Lactobacillus bulgaricus* and *Streptococcus thermophilus*, and it may also include additional strains like *Lactobacillus acidophilus* and *Bifidobacterium bifidum* in certain types. The probiotics and prebiotics in yogurt have been shown to offer various benefits, including aiding in weight loss, lowering cholesterol levels, and positively impacting overall heart illness, including managing high blood pressure (Hashemi Gahruie et al. 2015).

Vitamin D (in enriched types) and Ca, which are both critical for bone development, are abundant in yoghurt. Consuming it often may help reduce arthritis. Adequate calcium intake is crucial for bone density and strength. It also supports bone growth, particularly important for children, adolescents, and the elderly (Saleem 2023). It also has beneficial effects in regulating metabolic syndrome. It is an effective source of protein, and consuming it can assist preserve lean muscle while losing weight. Yogurt's protein contains the essential amino acids required for muscle repair and maintenance, which helps prevent muscle loss while you're in a calorie deficit. Additionally, yogurt often contains other nutrients that support overall health, making it a beneficial addition to a weight loss plan. Protein promotes feelings of fullness, which can lower the overall consumption of calories (Jacques and Wang 2014; Sumi et al. 2023). Low-fat and non-fat yogurts can help maintain healthy cholesterol levels because they contain less saturated fat compared to full-fat versions. Saturated fats can raise levels of LDL cholesterol, which is linked to heart disease. By choosing low-fat or non-fat yogurt, you're reducing your intake of saturated fats, which can help keep your cholesterol levels in a healthier range. According to studies, the structure of yoghurt effectively transports plant sterols and stanols, substances known to help decrease cholesterol levels. The matrix of yogurt allows these cholesterol-lowering substances to be delivered efficiently, enhancing their effectiveness in reducing LDL cholesterol and supporting heart health (Mensink et al. 2002; Clifton et al. 2004; Algorta Pineda et al. 2005; Hyun et al. 2005). Yoghurt and other whole-fat milk items may be linked to a decreased risk of hypertension, or high blood pressure. This is partly because these dairy products often contain potassium, a mineral which assists in maintaining blood pressure by neutralising the effects of sodium and reducing arterial pressure in the body. The potassium in yogurt contributes to promoting healthy blood pressure levels, which can reduce the risk of hypertension (Astrup 2014; Bhavadharini et al. 2020). Compared to enrich-glycemic meals, many yoghurts cause a lesser and more gradual rise in glucose levels in the body because they have a low glycaemic score. This slower impact helps maintain more stable blood sugar levels, which can be useful for regulating blood sugar and improving overall metabolic health (Astrup 2014). Protein and calcium may assist to regulate the level of blood sugar and improve general glucose metabolism. Carbohydrate digestion and absorption are slowed down by protein, which causes blood sugar to rise more gradually. In contrast, calcium influences insulin secretion and glucose metabolism. Together, these nutrients contribute to more stable blood sugar levels and improved control over glucose levels in the body (Yanni et al. 2020; Bhavadharini et al. 2020).

Yoghurt can have advantageous effects on the condition of the skin. It is commonly used as a facial mask and skincare treatment due to its moisturizing and soothing properties. Yogurt's lactic acid promotes a healthier complexion by acting as a mild exfoliator to exfoliate cellular debris. Additionally, yogurt's natural fats and proteins provide hydration, which can help keep the skin soft and supple. Its anti-inflammatory characteristics make it beneficial for relaxing inflamed or sensitive skin. As a result, yogurt can be a beneficial ingredient in homemade or commercial skincare products aimed at improving skin texture and comfort (Vaughn and Sivamani 2015). Yogurt's status as a functional food is well-supported by its rich nutrient profile, probiotic content, and various health benefits. Its role in improving digestive health, supporting bone density, aiding in weight management, and offering potential benefits for cardiovascular health makes it a valuable

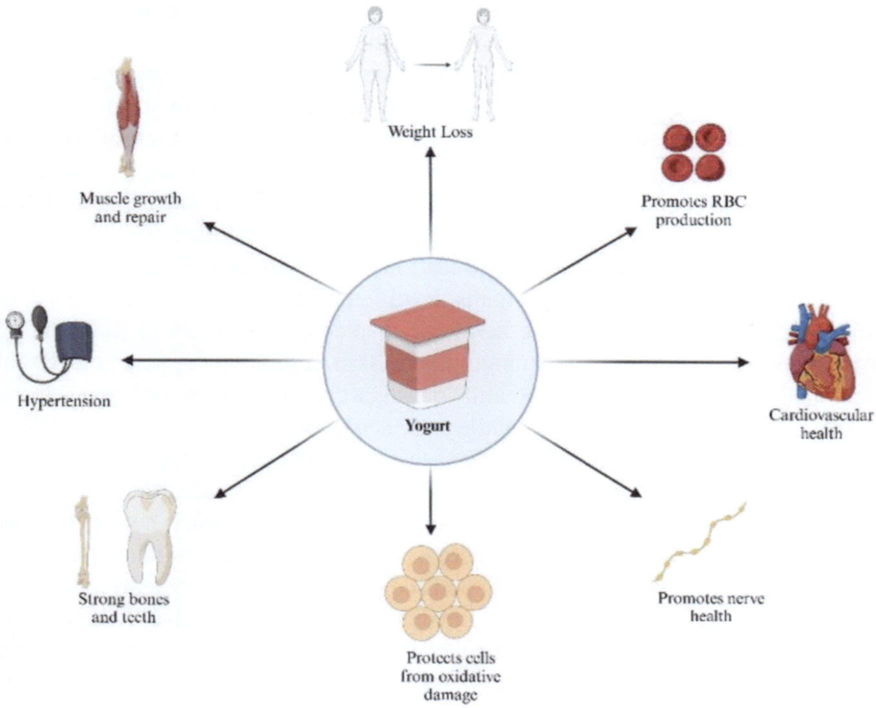

Fig. 8.2 Beneficial effects of yogurt

component of a balanced meals. Incorporating yoghurt into your daily habits provides favourably to in general wellness and good health (Fig. 8.2) (Chandan et al. 2017).

8.3.2 Fortified Milk

Fortified milk is regular cow's milk that has been enhanced with additional vitamins and minerals not typically present in substantial quantities in natural milk (Eichler et al. 2012). Fortified milk, enhanced with extra nutrients beyond its natural levels, is a key player in the category of functional foods. Fortified milk has further Ca and VitD, all of which are necessary for supporting good bone health. Calcium is a crucial element that makes up the framework of bones and teeth, making it indispensable for developing and sustaining bone density. Adequate calcium intake helps to strengthen bones and prevent conditions such as osteoporosis, a disease that produces fragile and weakened bones that are susceptible to fractures. Vitamin D, on the other hand, plays an essential role in facilitating the transport of calcium from the digestive system to the circulation. Without adequate amounts of VitD, the body is unable to adequately exploit the calcium it receives, leading to potential deficiencies and compromised bone health. By fortifying milk with these nutrients, it provides an easy and effective way to ensure that individuals, especially those who might not get enough calcium and vitamin D through their regular diet or sunlight exposure, can maintain strong, healthy bones and reduce their risk of osteoporosis. This fortification is particularly beneficial for populations at higher risk, including children, the elderly, and those with limited sun exposure (Yeh et al. 2017; Ilesanmi-Oyelere and Kruger 2020). Vitamin and mineral supplementation in fortified milk provides a crucial boost to overall health by addressing specific nutritional deficiencies and supporting well-being in populations with dietary gaps. Fortified milk often includes added vitamins such as A, B12, and folic acid, which play pivotal roles in maintaining various bodily functions. The strength of the skin, immunity support, and visual health all depend on vitamin A. Healthy neuronal function and bloodstream development depend on vitamin B12, while folic acid aids in the production of DNA, which is especially vital during pregnancy and for cell growth. Additionally, minerals such as zinc and magnesium are commonly added to fortified milk to further enhance its nutritional value. Zinc is essential for immunological health, healing of wounds, and cellular function, whereas magnesium promotes the functioning of muscles and nerves, bone health, and the generation of energy. By enriching milk with these essential nutrients, fortified milk helps to bridge dietary gaps, ensuring that individuals receive a more comprehensive range of vitamins and minerals necessary for maintaining good health. This supplementation is particularly beneficial for those who may have limited access to a balanced diet, thus lowering the possibility of deficiencies and improving general health (Vijayaraghavan 2002; Gracia-Marco et al. 2012; Sumi et al. 2023). Enriching fortified milk with vitamins like vitamin C and antioxidants significantly enhances its role in supporting immune function and aiding the body's ability to combat infections. By encouraging the development and activity of white blood cells, which are vital for warding off infections, vitamin C, which is well-known for its potent antioxidant qualities, plays a vital part in the body's defenses. Vitamin C also supports healthy skin and mucous layers, which serve as infection-resistant barriers. Vitamin C and other antioxidants shield cells from free radical and oxidative damage, which might otherwise compromise immune system performance and general health. By incorporating these nutrients into fortified milk, individuals benefit from an added layer of protection against illnesses, as well as support for maintaining a robust immune response. This fortified milk not only helps fill nutritional gaps but also contributes to enhanced overall wellness and resilience against infections (Hunt et al. 1994; Carr and Maggini 2017). The addition of plant-derived sterols and omega-3 fatty acids to fortified milk has major advantages for cardiac wellness, especially when it comes to controlling triglyceride concentrations. Omega-3 fatty acids, which are included in foods like seafood oil, are well-known for their capacity to minimize cholesterol levels and reduce inflammation, both of which can lessen the risk of heart disease. They help improve the overall lipid profile by raising levels of high-density lipoprotein (HDL), and lowering levels of low-density lipoprotein (LDL). Plant sterols, on the other hand, are naturally occurring substances present in meals derived from plants that can block the absorption of cholesterol in the intestines. Plant sterols reduce the amount of cholesterol that enters the bloodstream, thus contributing to reduce LDL cholesterol levels. Incorporating these components into fortified milk offers an accessible and practical method to integrate heart-healthy nutrients into the diet, hence promoting cardiovascular health and aiding in the maintenance of balanced cholesterol levels (Gonçalves et al. 2007; Fonollá et al. 2009; Jain et al. 2015). Incorporating probiotic strains and prebiotics into fortified milk offers substantial benefits for gut health and digestion. By boosting the quantity of beneficial bacteria in the digestive system, probiotics can aid in balancing the gut microbiota. This can improve digestion, boost immunity, and possibly lessen the symptoms of digestive diseases including constipation or irritable bowel syndrome (IBS). Prebiotics, on the other hand, are indigestible fibres that serve as food for these beneficial bacteria, encouraging their development and proliferation. By providing a dual approach-introducing beneficial bacterium through probiotics and fostering their growth with prebiotics-fortified milk can effectively promote a healthy gut environment. This combined strategy can improve overall digestive function, reduce bloating and discomfort, and contribute to a more balanced and resilient microbiome. Fortified milk combined with these ingredients is a convenient and pleasurable approach to improve digestive well-being for people with particular digestive problems or those who just want to maintain their gut health (Tuohy et al. 2003; Azad et al. 2018). Fortified milk can be tailored to support muscle maintenance and weight management by modifying its fat content and adding specific proteins. By offering a lower-fat version, fortified milk helps reduce overall calorie intake, which can be beneficial for weight management. Additionally, incorporating high-quality proteins, such as whey or casein, supports muscle repair and growth, making it an effective option for those looking to build or preserve muscle mass. This tailored approach not only contributes to a balanced diet but also assists in achieving fitness and health goals (Zemel 2004). Common nutritional deficits are successfully addressed with fortified milk, which provides essential nutrients for a child's well-being. When compared to ordinary cow's milk, it has been demonstrated that folic acid-fortified milk raises toddlers' iron levels. Additionally, especially in underdeveloped nations, milk and cereal products enhanced with several micronutrients can be a useful tactic to treat anaemia in children under three. In these areas, milk is frequently fortified with Fe, Zn, and B vitamins to help prevent iron deficiency anemia, which is a prevalent issue among young children (Eichler et al. 2012; Qudsia et al. 2015). Fortified milk, which is enhanced with vitamins and minerals including calcium, vitamin D, & A, promotes healthy bone formation, immune system performance, and general body growth. Incorporating these vital nutrients into children's diets helps fill in the gaps and guarantees that they get the building blocks needed for their physical and cognitive development, which is especially important during the critical early growth years (Eichler et al. 2012). Fortified milk is essential for bone health, supplying critical minerals like Ca and VitD, which are necessary for preserving bone density and strength. As people age, their ability to absorb these nutrients may decline, leading to an elevated risk of bone-related issues such as osteoporosis. By enriching milk with these critical nutrients, it helps counteract age-related nutritional declines and ensures that individuals receive adequate support for bone maintenance and overall health, promoting a healthier and more active lifestyle in later years (Gaucheron 2000; Park et al. 2008; Thorning et al. 2016). Including fortified milk in daily diets allows individuals to gain targeted health benefits, enhancing overall well-being and meeting specific nutritional requirements.

8.3.3 Whole Grains

Whole grains are regarded as a fundamental component of functional foods because of their extensive health benefits and nutritional value (Guo et al. 2022). Whole grains are a valuable source of essential nutrients crucial for overall health. They offer adequate dietary fiber, enhancing digestion, regulating blood sugar, and promoting cardiovascular health by reducing cholesterol levels. Vitamin B, which are vital for metabolic processes as well as neurological system function, are found in whole grains. Furthermore, they encompass essential minerals including iron, which facilitates oxygen transfer; magnesium, crucial for muscle and nerve function; and selenium, an antioxidant that safeguards cells from harm. Together, these nutrients promote better digestion, cardiovascular health, and increased energy levels (Slavin 2004). The high fiber content is advantageous for digestive health as it facilitates regular bowel movements by increasing stool size, hence aiding in the prevention of constipation. Additionally, fibre serves as a prebiotic, supporting a healthy and balanced gut microbiota by feeding beneficial microbes in the GI tract. For effective digestion and overall gut health, this balance is necessary (Slavin et al. 1999). Chronic Disease Management: Whole grains are associated with a reduced risk of cardiovascular diseases because they positively impact several key factors related to heart health. They help improve cholesterol levels by increasing the intake of soluble fiber, which can lower LDL cholesterol. Whole grains also contribute to lower blood pressure through their high content of potassium and other beneficial nutrients. Furthermore, they enhance endothelial function, which refers to the health of the lining of blood vessels, helping to maintain proper blood flow and reduce inflammation. Collectively, these factors diminish the risk of cardiovascular disease (Salmerón et al. 1997; Brown et al. 1999). Whole grains are particularly beneficial for both the avoidance and treatment of type 2 diabetes because they can help control the blood sugar level and improve the reaction to insulin. Their elevated fiber content decelerates the absorption of carbohydrates into the bloodstream, hence mitigating fast fluctuations in blood sugar levels. Additionally, the nutrients in whole grains, such as magnesium and chromium, support the body's ability to use insulin more effectively. This improved insulin sensitivity lowers the chance of insulin resistance, a major contributing factor to type II diabetes, and helps maintain normal glucose levels in body (Newby et al. 2003). The fiber and protein in whole grains enhance satiety, hence decreasing overall caloric consumption and aiding in weight maintenance or loss (Pereira et al. 1998; Slavin 2004). Whole grains provide prebiotics, which are indigestible fibers that nourish good gut microbes. Whole grains enhance digestion and promote a healthier gut by fostering the proliferation of beneficial microbes. This can subsequently improve immune function by fostering a healthy and efficient gut microbiome (Slavin et al. 1999). According to research, eating whole grains may lower your risk of developing numerous malignancies, including colon cancer, mostly due to their elevated fiber content and antioxidant characteristics. The fiber in whole grains facilitates regular bowel movements and assists in eliminating possible carcinogens from the digestive system. The antioxidants found in whole grains can safeguard cells from harm and diminish inflammation, so further aiding in the lowering of cancer risk (Jacobs et al. 1998; Nyanzi and Jooste 2012). Incorporating whole grains into one's diet allows individuals to harness their various health benefits, promoting overall wellness and significantly reducing the risk of chronic diseases.

8.3.4 Fortified Cereals

Fortified cereals exemplify functional foods, engineered to provide supplementary health advantages beyond fundamental nutrition (McKevith 2004). Fortified cereals are enhanced with supplementary vitamins and minerals, such as Fe, Ca, vit D, and B etc. This supplementation addresses nutritional inadequacies prevalent in various diets, assisting in their prevention. Fortified grains enhance general health and well-being by supplying necessary nutrients (Ruel et al. 2004). Increased Nutrient Density: By adding essential nutrients to cereals, these products can provide a significant portion of daily recommended intakes, particularly for populations at risk of nutrient deficiencies (McIntosh et al. 2003; Drewnowski and Fulgoni 2014). Increased fibre content in some fortified cereals promotes routine bowel motions and prevents bowel problems, both of which are beneficial to the intestinal tract. Fibre encourages sustained fullness, hence assisting in weight management by lowering total caloric consumption (Lee and Lip 2003). Moreover, including omega-3 fatty acids into fortified cereals can improve cardiovascular health and cognitive performance. Omega-3 fatty acids promote cardiovascular health by reducing inflammation and lowering cholesterol levels, while also enhancing cognitive function, potentially increasing memory and overall brain functioning (Businaro 2022). Because fortified cereals include vital nutrients that support overall wellness as well as health, regular consumption may help avoid chronic diseases like osteoporosis and some types of cancer (La Vecchia et al. 2003). By assisting in the development and upkeep of strong bones, cereals enhanced containing Ca and Vit. D improve bone health. This is especially helpful for children, adolescents, and older adults, who need extra support for growing or maintaining bone density. Adding B vitamins to cereals supports energy production and metabolism, which helps keep you energized and supports overall vitality. Additionally, these vitamins are essential for preserving cognitive function, which promotes focus and mental clarity (Newmark et al. 2004; O'Sullivan 2012; Kennedy 2016). Incorporating fortified cereals into daily diets can significantly enhance nutritional intake, support specific health needs, and contribute to overall well-being, making them a valuable component of a balanced and health-conscious diet.

8.3.5 Berries

Berries are recognised for their health benefits and are considered powerful functional foods. Essential vitamins like C and K, as well as minerals like manganese and folate, are abundant in berries. These nutrients support immunological response and general health (Nile and Park 2014). Additionally, berries such as blueberries, strawberries, raspberries, and blackberries are packed with antioxidants like anthocyanins, flavonoids, and polyphenols. These antioxidants protect your cells from oxidative stress and inflammation, which can lead to aging and other health problems (Bennett et al. 2010; Baby et al. 2018; Pereira et al. 2018). The elevated levels of antioxidants and fiber in berries can enhance vascular function, decrease blood pressure, and diminish the risk of cardiovascular disease (Zhu et al. 2017; Yousefi et al. 2021). Berries may aid in lowering LDL cholesterol levels while preserving HDL cholesterol, hence enhancing overall cardiovascular health (Millar et al. 2018; Sun et al. 2021). Berries include antioxidants that may help prevent cognitive loss due to age and are linked to enhanced cognitive performance. Frequent berry eating has been associated with improved learning and memory (Basu et al. 2010; Krikorian et al. 2010). The bioactive chemicals in berries mitigate inflammation, this supports the treatment of inflammation-related illnesses such as gout along with various chronic conditions (Shi et al. 2017; Basu et al. 2018). Due to their high dietary fiber content, berries help to improve gut flora and encourage frequent bowel movements (Farzaei et al. 2015; Makki et al. 2018). Berries are low in calories and high in fiber, making them a filling snack that can support weight management and reduce overall calorie intake (Tetens and Alinia 2009; Stubbs et al. 2010). The fiber and natural sugars in berries can assist in regulating blood glucose levels and enhancing insulin sensitivity, which is advantageous for those managing diabetes or prediabetes (Russell et al. 2016; Castro-Acosta et al. 2016). Vitamin C and other antioxidants found in berries help to repair and produce collagen, which keeps skin looking young and shields it from UV ray damage (Lee et al. 2014; Miracle Uwa 2017). Including a range of berries in one's diet can have several health advantages, supporting overall wellness and potentially reducing the risk of chronic diseases. Their rich nutrient profile and functional properties make them a valuable addition to a balanced and health-conscious diet.

8.3.6 Citrus Fruits

Citrus fruits, including grapefruits, lemons, tangerines, and oranges, are well known for their practical culinary applications and health benefits. Citrus fruits are celebrated for their elevated vitamin C levels, which bolster immune function, improve skin health via collagen formation, and serve as a powerful antioxidant. They also include vital nutrients such as dietary fiber, potassium, folate, and several B vitamins, enhancing overall health (Zou et al. 2016; Ahmed and Saeid 2021). Citrus fruits' high amount of Vit. C content is crucial for strengthening the body's defences system, aiding the body in combating infections and diseases. Citrus fruits include vitamin C and flavonoids that neutralize free radicals, thereby diminishing oxidative stress and inflammation (Farhan Aslam et al. 2017; Miles and Calder 2021; Saini et al. 2022; Akbari et al. 2022). The high dietary fibre content of citrus fruits promotes consistent stool elimination, aids in digestion, and supports a healthy gut microbiota. The elevated water content in citrus fruits aids in hydration and enhances overall digestive function (Musumeci et al. 2020; He et al. 2023). Citrus fruits contain fiber, potassium, and antioxidants that improve cardiovascular health by decreasing BP, cholesterol, and other cardiovascular parameters. Citrus chemicals, such as flavonoids, exhibit anti-inflammatory properties that may enhance cardiovascular health and diminish the likelihood of chronic diseases (Chau 2004; Parhiz et al. 2015; Mahmoud et al. 2019). Citrus fruits include Vitamin C, which facilitates collagen synthesis, crucial for preserving skin suppleness and diminishing the visibility of wrinkles. The antioxidant qualities of citrus fruits can enhance skin luminosity and diminish the effects of UV damage and hyperpigmentation (West et al. 2020; Michalak 2022). Citrus fruits include antioxidants that may help shield cells from harm and lower the chance of developing some types of cancer, including oral and esophageal cancers. Citrus fruits contain compounds that support the body's natural detoxification processes, potentially reducing cancer risk (Block 1993; Silalahi 2002). Integrating citrus fruits into a balanced diet can improve general health, facilitate numerous physiological activities, and aid in disease prevention. Their abundant nutrient composition and functional attributes render them an essential element of a health-oriented diet.

8.3.7 Leafy Greens

Leafy greens, including spinach, kale, collard greens, and Swiss chard, are esteemed for their health-enhancing attributes and functional food advantages. Leafy greens are abundant in vital vitamins, including A, C, K, and several B vitamins, along with minerals like calcium, magnesium, and potassium. These nutrients facilitate several physiological functions and enhance overall wellness (Randhawa et al. 2015). They have a variety of antioxidants, including lutein, carotenoids, and flavonoids, that effectively scavenge free radicals and shield cells from oxidative damage. Leafy greens are superior providers of vitamin K, essential for bone health since it facilitates calcium absorption and bone mineralization. This may mitigate the risk of fractures and osteoporosis (Office of the Surgeon General (US) 2004). The high fiber content in leafy greens helps lower cholesterol levels and improve heart health. Additionally, the potassium and nitrates found in these greens support healthy blood pressure levels (Blekkenhorst et al. 2018). Leafy greens include phytochemicals and antioxidants with anti-inflammatory qualities that may reduce the incidence of heart conditions (Griffiths et al. 2016). The high dietary fibre content of leafy greens promotes regular bowel motion and enhances gut health, both of which aid in a healthy digestive system. This may aid in the prevention of constipation and promote a healthy microbiota. Carotenoids such as lutein and zeaxanthin, present in leafy greens like kale and spinach, promote ocular health. They assist in safeguarding against age-related macular degeneration (AMD) and other ocular concerns (Abdel-Aal et al. 2013). The preventive effects of phytochemicals as well as antioxidants found in leafy greens on cells have been associated with a lower incidence of colon cancer and other malignancies (Ganesh et al. 2009; Alzate-Yepes et al. 2023). Incorporating leafy greens into a daily diet offers numerous health benefits, supporting various aspects of wellness from bone and cardiovascular health to digestive function and disease prevention. Their rich nutrient profile and functional properties make them a valuable addition to a health-conscious diet.

8.3.8 Green Tea

Green tea, obtained from the leaves of the *Camellia sinensis* plant, is well acknowledged for its health advantages and functional culinary uses. Green tea is an abundant source of catechins, especially epigallocatechin gallate (EGCG), which are potent antioxidants that safeguard cells against oxidative stress and inflammation. A systematic review on herbal tea reported that herbal tea has shown promising effects both clinically and pre-clinically with minimal side effects (Barbhuiya et al. 2024). Polyphenols have several health benefits, including anti-inflammatory and anti-cancer properties. By increasing HDL and decreasing LDL and triglycerides in the blood, green tea has been shown to improve cardiovascular health. Regular use of green tea may lower BP, which might reduce the risk of stroke and heart failure. (Zhao et al. 2022). The catechins in green tea, especially EGCG, can enhance metabolic rate and fat oxidation, aiding in weight management and potentially supporting fat loss. Green tea may assist in appetite regulation and diminish total caloric consumption, hence aiding weight management endeavours (Yang et al. 2016). The polyphenols in green tea may enhance a healthy gut microbiota by fostering the proliferation of good bacteria and suppressing harmful bacteria. The anti-inflammatory effects of green tea can enhance gastrointestinal health by diminishing inflammation and promoting digestive comfort (Truong and Jeong 2022). The antioxidants in green tea, especially EGCG, are linked to enhanced cognitive performance and may offer protection against neurodegenerative disorders such as Alzheimer's and Parkinson's (Zhao et al. 2022). By reducing free radical damage and inhibiting the growth of cancer cells, green tea's catechins and polyphenols have been studied for their potential to reduce the risk of several cancers, including as colorectal, prostate, and breast cancers (Zhao et al. 2022).

8.3.9 Functional Drinks

Functional beverages are liquids fortified with supplementary ingredients intended to deliver particular health advantages beyond fundamental nutrition. Functional beverages, including sports drinks and electrolyte solutions, assist in replenishing vital minerals such as salt, potassium, and magnesium, which are depleted through perspiration, hence facilitating hydration and recovery during and post-exercise. They are engineered to enhance fluid absorption and sustain hydration levels, which is essential for general health and optimal physiological function (Maughan 1998; Aoi et al. 2006). Some functional drinks contain caffeine, taurine, or adaptogens like ginseng and ashwagandha, which can enhance alertness, increase energy levels, and reduce fatigue. Energy drinks and functional beverages containing B vitamins and amino acids enhance physical and mental performance, rendering them favoured by athletes and busy professionals (Orrù et al. 2018). Functional beverages such as probiotic yogurts and kefirs encompass advantageous bacteria (probiotics) and fibers (prebiotics) that enhance gut health by fostering a balanced intestinal microbiome and facilitating digestion. Certain beverages contain digestive enzymes to facilitate the digestion and absorption of nutrients, hence improving overall digestive efficiency (Raman et al. 2019). Functional beverages enriched with vitamins (e.g., vitamin C & D) and antioxidants (e.g., elderberry, echinacea) bolster the immune system and augment the body's capacity to combat infections and diseases. Ingredients like ginger and turmeric offer anti-inflammatory and immune-enhancing advantages, promoting general health and well-being (Sahoo et al. 2019; Abokhwat et al. 2024). Functional drinks cater to diverse health needs and preferences, providing tailored solutions for hydration, energy, digestive health, immune support, and more. Their ability to offer specific benefits makes them a valuable addition to a health-conscious lifestyle.

8.3.10 Nuts and Seeds

Owing to their high nutritional profiles, nuts and seeds are essential components of functional meals and are praised for their many health advantages. Nuts and seeds include a lot of unsaturated fats, especially omega-3 as well as omega-6 fatty acids, which are vital for cardiac health, inflammation decreasing, and cognitive support. They supply essential minerals, including vitamin E, magnesium, zinc, and selenium, which enhance overall health, particularly immune function and dermal health. Consistent intake of nuts, including almonds, walnuts, and pistachios, may reduce LDL cholesterol levels and enhance HDL cholesterol levels, so promoting cardiovascular health. Nuts and seeds are excellent sources of potassium and magnesium, which aid in regulating blood pressure and supporting cardiovascular function (Sathe et al. 2008). Because nuts and seeds have an excellent fiber food, they help maintain a healthy gut microbiota, improve frequent bowel movements, and avoid constipation. Certain seeds, like flaxseeds and chia seeds, have prebiotic properties that nourish beneficial gut bacteria, enhancing gut health (Simopoulos 2002;

Praznik et al. 2015). Nuts and seeds, like almonds and chia seeds, are excellent suppliers of calcium and magnesium, vital for sustaining bone density and averting osteoporosis. Nuts such as cashews and seeds like pumpkin seeds are sources of vitamin K, which contributes to bone health by facilitating calcium metabolism (Karpouzos et al. 2017). Nuts, especially walnuts, are high in omega-3 fatty acids and antioxidants that support cognitive function, improve memory, and might aid in reducing age-associated cognitive deterioration. Nutrients like magnesium and vitamin E found in nuts and seeds can help support mood regulation and reduce the risk of mood disorders (Theodore et al. 2021). There are several health advantages to including a range of nuts and seeds in a nutritious meal plan, from supporting cardiovascular and digestive health to aiding in weight management and improving skin health. Their rich nutrient profiles and functional properties make them valuable additions to a health-conscious diet.

8.3.11 Legumes

The wide group of plants known as legumes includes peas, chickpeas, beans, and lentils. They have been gaining attention as functional foods due to their numerous health benefits and versatility. Legumes are rich in proteins, fiber, vitamins, and minerals. Their inclusion in the diet can enhance overall nutritional quality and help address deficiencies (Martín-Cabrejas 2019). The elevated fiber content in beans, especially soluble fiber, can impede the absorption of sugars and assist in regulating blood glucose levels. This renders them advantageous for individuals with diabetes or those predisposed to its development (Singhal et al. 2014). Legumes can help lower LDL cholesterol levels due to their soluble fiber and plant sterol content. Consistent use can lower the risk of heart conditions and improve the condition of the heart (Martins et al. 2004). The dietary fiber in legumes promotes healthy digestion and regular bowel movements. Additionally, it can promote the development of advantageous intestinal microbes, supporting a balanced microbiome (Clemente and Olias 2017). Research indicates that the antioxidants and phytonutrients present in beans may confer preventive benefits against specific cancer forms, including colorectal cancer (Zhu et al. 2015). Legumes provide a minimal environmental impact in comparison to animal-derived proteins. These are nitrogen-fixing plants that enhance soil health and diminish the necessity for synthetic fertilizers, rendering them a viable option for the future. Legumes are progressively utilized as components in diverse useful diets and supplements. Chickpea flour is employed in gluten-free baking, whereas lentil protein is utilized in plant-based meat substitutes (Foschia et al. 2017). They are a great resource of P, Ca, and Mg, among other elements crucial for healthy bones. They can contribute to maintaining bone density and reducing the risk of osteoporosis (Maphosa and Jideani 2017). By integrating legumes into diets, individuals can leverage these functional foods for better health outcomes while also supporting environmental sustainability.

8.4 Conclusion and Future Perspective

Beyond simple nourishment, dietary supplements provide additional health advantages, but they require careful consideration. While ω-3 fatty acids can protect against cardiovascular disease, overeating may disrupt cholesterol and blood sugar levels. Many functional foods, despite showing promise in research, lack FDA-approved health claims. Fortified foods, like rice and oils mandated in India, aim to combat malnutrition but introduce synthetic micronutrients with potential health risks. Overconsumption of fortified foods can lead to exceeding recommended nutrient limits, posing health concerns. Selecting functional foods, particularly those that are fortified, necessitates meticulous assessment. Evaluate the added nutrients, their quantities, and the total food consumption. Consulting a healthcare expert or certified dietitian can facilitate educated decisions and guarantee these items enhance a balanced diet without adverse health effects.

References

Abdel-Aal E-SM, Akhtar H, Zaheer K, Ali R. Dietary sources of lutein and zeaxanthin carotenoids and their role in eye health. Nutrients. 2013;5:1169–85. https://doi.org/10.3390/nu5041169.

Abokhwat BM, Mahmoud RM, Ashoush IS. Antioxidant status, physicochemical and immune-boosting characteristics of new formulated functional beverages. Egypt J Nutr. 2024;39:159–70. https://doi.org/10.21608/enj.2024.360751.

Adefegha SA. Functional foods and nutraceuticals as dietary intervention in chronic diseases; novel perspectives for health promotion and disease prevention functional foods and nutraceuticals as dietary intervention in chronic diseases ; novel perspectives for health promotion and disease prevention. 2018; https://doi.org/10.1080/19390211.2017.1401573.

Ahmed M, Saeid A. Citrus fruits: nutritive value and value-added products. In: Sarwar Khan M, Ahmad Khan I, editors. Citrus – research, development and biotechnology. IntechOpen; 2021.

Akbari B, Baghaei-Yazdi N, Bahmaie M, Mahdavi Abhari F. The role of plant-derived natural antioxidants in reduction of oxidative stress. Biofactors. 2022;48:611–33. https://doi.org/10.1002/biof.1831.

Algorta Pineda J, Chinchetru Ranedo MJ, Aguirre Anda J, Francisco Terreros S. Hypocholesteremic effectiveness of a yogurt containing plant stanol esters. Rev Clin Esp. 2005;205:63–6. https://doi.org/10.1157/13072497.

Alkhatib A, Tsang C, Tiss A, et al. Functional foods and lifestyle approaches for diabetes prevention and management. Nutrients. 2017;9:1310. https://doi.org/10.3390/nu9121310.

Alzate-Yepes T, Pérez-Palacio L, Martínez E, Osorio M. Mechanisms of action of fruit and vegetable phytochemicals in colorectal cancer prevention. Molecules. 2023;28:4322. https://doi.org/10.3390/molecules28114322.

Aoi W, Naito Y, Yoshikawa T. Exercise and functional foods. Nutr J. 2006;5:15. https://doi.org/10.1186/1475-2891-5-15.

Astrup A. Yogurt and dairy product consumption to prevent cardiometabolic diseases: epidemiologic and experimental studies. Am J Clin Nutr. 2014;99:1235S–42S. https://doi.org/10.3945/ajcn.113.073015.

Azad MAK, Sarker M, Li T, Yin J. Probiotic species in the modulation of gut microbiota: an overview. Biomed Res Int. 2018;2018:9478630. https://doi.org/10.1155/2018/9478630.

Baby B, Antony P, Vijayan R. Antioxidant and anticancer properties of berries. Crit Rev Food Sci Nutr. 2018;58:2491–507. https://doi.org/10.1080/10408398.2017.1329198.

Barbhuiya PA, Dey J, Saikia K, et al. Herbal tea used globally targeting metabolic syndrome: a systematic review. Int J Diabetes Dev Ctries. 2024;45:400–15. https://doi.org/10.1007/s13410-024-01361-1.
Basu A, Rhone M, Lyons TJ. Berries: emerging impact on cardiovascular health. Nutr Rev. 2010;68:168–77. https://doi.org/10.1111/j.1753-4887.2010.00273.x.
Basu A, Schell J, Scofield RH. Dietary fruits and arthritis. Food Funct. 2018;9:70–7. https://doi.org/10.1039/C7FO01435J.
Belury MA. Conjugated dienoic linoleate: a polyunsaturated fatty acid with unique chemoprotective properties. Nutr Rev. 1995;53:83–9. https://doi.org/10.1111/j.1753-4887.1995.tb01525.x.
Bennett LE, Singh DP, Clingeleffer PR. Micronutrient mineral and folate content of Australian and imported dried fruit products. Crit Rev Food Sci Nutr. 2010;51:38–49. https://doi.org/10.1080/10408390903044552.
Bhavadharini B, Dehghan M, Mente A, et al. Association of dairy consumption with metabolic syndrome, hypertension and diabetes in 147 812 individuals from 21 countries. BMJ Open Diabetes Res Care. 2020;8:e000826. https://doi.org/10.1136/bmjdrc-2019-000826.
Birch CS, Bonwick GA. Ensuring the future of functional foods. Int J Food Sci Technol. 2019;54:1467–85. https://doi.org/10.1111/ijfs.14060.
Blekkenhorst LC, Sim M, Bondonno CP, et al. Cardiovascular health benefits of specific vegetable types: a narrative review. Nutrients. 2018;10:595. https://doi.org/10.3390/nu10050595.
Block G. Vitamin C, cancer and aging. Age. 1993;16:55–8. https://doi.org/10.1007/BF02435037.
Brown L, Rosner B, Willett WW, Sacks FM. Cholesterol-lowering effects of dietary fiber: a meta-analysis. Am J Clin Nutr. 1999;69:30–42. https://doi.org/10.1093/ajcn/69.1.30.
Bucher HC, Hengstler P, Schindler C, Meier G. N-3 polyunsaturated fatty acids in coronary heart disease: a meta-analysis of randomized controlled trials. Am J Med. 2002;112:298–304. https://doi.org/10.1016/s0002-9343(01)01114-7.
Businaro R. Food supplements to complement brain functioning: the benefits of a combination of magnesium, folic acid, omega-3 fatty acids and vitamin E. F1000 Res. 2022;11:140. https://doi.org/10.12688/f1000research.75856.1.
Carr AC, Maggini S. Vitamin C and immune function. Nutrients. 2017;9:1211. https://doi.org/10.3390/nu9111211.
Castro-Acosta ML, Lenihan-Geels GN, Corpe CP, Hall WL. Berries and anthocyanins: promising functional food ingredients with postprandial glycaemia-lowering effects. Proc Nutr Soc. 2016;75:342–55. https://doi.org/10.1017/S0029665116000240.
Cencic A, Chingwaru W. The role of functional foods, nutraceuticals, and food supplements in intestinal health. Nutrients. 2010;2:611–25. https://doi.org/10.3390/nu2060611.
Chandan RC, Gandhi A, Shah NP. Yogurt. In: Yogurt in health and disease prevention. Elsevier; 2017. p. 3–29.
Chau C. Investigation of the cholesterol-lowering action of insoluble fibre derived from the peel of Citrus sinensis L. cv. Liucheng. Food Chem. 2004;87:361–6. https://doi.org/10.1016/j.foodchem.2003.12.006.
Clemente A, Olias R. Beneficial effects of legumes in gut health. Curr Opin Food Sci. 2017;14:32–6. https://doi.org/10.1016/j.cofs.2017.01.005.
Clifton PM, Noakes M, Sullivan D, et al. Cholesterol-lowering effects of plant sterol esters differ in milk, yoghurt, bread and cereal. Eur J Clin Nutr. 2004;58:503–9. https://doi.org/10.1038/sj.ejcn.1601837.
Cohen P. The role of protein phosphorylation in neural and hormonal control of cellular activity. Nature. 1982;296:613–20. https://doi.org/10.1038/296613a0.
Crawford M. Placental delivery of arachidonic and docosahexaenoic acids: implications for the lipid nutrition of preterm infants. Am J Clin Nutr. 2000;71:275S–84S. https://doi.org/10.1093/ajcn/71.1.275S.
Drewnowski A, Fulgoni VL. Nutrient density: principles and evaluation tools. Am J Clin Nutr. 2014;99:1223S–8S. https://doi.org/10.3945/ajcn.113.073395.

Eichler K, Wieser S, Rüthemann I, Brügger U. Effects of micronutrient fortified milk and cereal food for infants and children: a systematic review. BMC Public Health. 2012;12:506. https://doi.org/10.1186/1471-2458-12-506.

Farhan Aslam M, Majeed S, Aslam S, Irfan JA. Vitamins: key role players in boosting up immune response-a mini review. Vitam Miner. 2017;06:1. https://doi.org/10.4172/2376-1318.1000153.

Farr DR. Functional foods. Cancer Lett. 1997;114:59–63. https://doi.org/10.1016/S0304-3835(97)04626-0.

Farzaei M, Rahimi R, Abdollahi M. The role of dietary polyphenols in the management of inflammatory bowel disease. Curr Pharm Biotechnol. 2015;16:196–210. https://doi.org/10.2174/1389201016666150118131704.

Fernandez MA, Marette A. Potential health benefits of combining yogurt and fruits based on their probiotic and prebiotic properties. Adv Nutr. 2017;8:155S–64S. https://doi.org/10.3945/an.115.011114.

Fonollá J, López-Huertas E, Machado FJ, et al. Milk enriched with "healthy fatty acids" improves cardiovascular risk markers and nutritional status in human volunteers. Nutrition. 2009;25:408–14. https://doi.org/10.1016/j.nut.2008.10.008.

Foschia M, Horstmann SW, Arendt EK, Zannini E. Legumes as functional ingredients in gluten-free bakery and pasta products. Annu Rev Food Sci Technol. 2017;8:75–96. https://doi.org/10.1146/annurev-food-030216-030045.

Frølich W, Åman P, Tetens I. Whole grain foods and health—a Scandinavian perspective. Food Nutr Res. 2013;57:18503. https://doi.org/10.3402/fnr.v57i0.18503.

Galanakis CM. Functionality of food components and emerging technologies. Foods. 2021;10:128. https://doi.org/10.3390/foods10010128.

Ganesh B, Talole SD, Dikshit R. A case-control study on diet and colorectal cancer from Mumbai, India. Cancer Epidemiol. 2009;33:189–93. https://doi.org/10.1016/j.canep.2009.07.009.

Gaucheron F. Iron fortification in dairy industry. Trends Food Sci Technol. 2000;11:403–9. https://doi.org/10.1016/S0924-2244(01)00032-2.

Gibson GR, Roberfroid MB. Dietary modulation of the human colonic microbiota: introducing the concept of prebiotics. J Nutr. 1995;125:1401–12. https://doi.org/10.1093/jn/125.6.1401.

Gonçalves S, Maria AV, Silva-Herdade AS, et al. Milk enriched with phytosterols reduces plasma cholesterol levels in healthy and hypercholesterolemic subjects. Nutr Res. 2007;27:200–5. https://doi.org/10.1016/j.nutres.2007.03.003.

Gracia-Marco L, Valtueña J, Ortega FB, et al. Iron and vitamin status biomarkers and its association with physical fitness in adolescents: the HELENA study. J Appl Physiol Bethesda Md 1985. 2012;113:566–73. https://doi.org/10.1152/japplphysiol.01492.2011.

Griffiths K, Aggarwal BB, Singh RB, et al. Food antioxidants and their anti-inflammatory properties: a potential role in cardiovascular diseases and cancer prevention. Dis Basel Switz. 2016;4:28. https://doi.org/10.3390/diseases4030028.

Guo H, Wu H, Sajid A, Li Z. Whole grain cereals: the potential roles of functional components in human health. Crit Rev Food Sci Nutr. 2022;62:8388–402. https://doi.org/10.1080/10408398.2021.1928596.

Hashemi Gahruie H, Eskandari MH, Mesbahi G, Hanifpour MA. Scientific and technical aspects of yogurt fortification: a review. Food Sci Human Wellness. 2015;4:1–8. https://doi.org/10.1016/j.fshw.2015.03.002.

Hasler CM. Functional foods: benefits, concerns and challenges – a position paper from the American Council on Science and Health. J Nutr. 2002;132:3772–81. https://doi.org/10.1093/jn/132.12.3772.

He C, Qi J, Liao J, et al. Excellent hydration properties and oil holding capacity of citrus fiber: effects of component variation and microstructure. Food Hydrocoll. 2023;144:108988. https://doi.org/10.1016/j.foodhyd.2023.108988.

Henry CJ. Functional foods. Eur J Clin Nutr. 2010;64:657–9. https://doi.org/10.1038/ejcn.2010.101.

Howell AB, Vorsa N, Der Marderosian A, Foo LY. Inhibition of the adherence of P-fimbriated Escherichia coli to uroepithelial-cell surfaces by proanthocyanidin extracts from cranberries. N Engl J Med. 1998;339:1085–6.

Hunt C, Chakravorty NK, Annan G, et al. The clinical effects of vitamin C supplementation in elderly hospitalised patients with acute respiratory infections. Int J Vitam Nutr Res. 1994;64:212–9.

Hyun YJ, Kim OY, Kang JB, et al. Plant stanol esters in low-fat yogurt reduces total and low-density lipoprotein cholesterol and low-density lipoprotein oxidation in normocholesterolemic and mildly hypercholesterolemic subjects. Nutr Res. 2005;25:743–53. https://doi.org/10.1016/j.nutres.2005.08.004.

Ilesanmi-Oyelere BL, Kruger MC. The role of milk components, pro-, pre-, and synbiotic foods in calcium absorption and bone health maintenance. Front Nutr. 2020;7:578702. https://doi.org/10.3389/fnut.2020.578702.

Ing SW, Belury MA. Impact of conjugated linoleic acid on bone physiology: proposed mechanism involving inhibition of adipogenesis. Nutr Rev. 2011;69:123–31. https://doi.org/10.1111/j.1753-4887.2011.00376.x.

Jacobs DR, Marquart L, Slavin J, Kushi LH. Whole-grain intake and cancer: an expanded review and meta-analysis. Nutr Cancer. 1998;30:85–96. https://doi.org/10.1080/01635589809514647.

Jacques PF, Wang H. Yogurt and weight management. Am J Clin Nutr. 2014;99:1229S–34S. https://doi.org/10.3945/ajcn.113.073031.

Jain AP, Aggarwal KK, Zhang P-Y. Omega-3 fatty acids and cardiovascular disease. Eur Rev Med Pharmacol Sci. 2015;19:441–5.

Jew S, AbuMweis SS, Jones PJH. Evolution of the human diet: linking our ancestral diet to modern functional foods as a means of chronic disease prevention. J Med Food. 2009;12:925–34. https://doi.org/10.1089/jmf.2008.0268.

Jones PJ. Clinical nutrition: 7. Functional foods–more than just nutrition. CMAJ. 2002;166:1555–63.

Karpouzos A, Diamantis E, Farmaki P, et al. Nutritional aspects of bone health and fracture healing. J Osteoporos. 2017;2017:4218472. https://doi.org/10.1155/2017/4218472.

Kaur S, Das M. Functional foods: an overview. Food Sci Biotechnol. 2011;20:861–75. https://doi.org/10.1007/s10068-011-0121-7.

Keevil JG, Osman HE, Reed JD, Folts JD. Grape juice, but not orange juice or grapefruit juice, inhibits human platelet aggregation. J Nutr. 2000;130:53–6. https://doi.org/10.1093/jn/130.1.53.

Kennedy D. B vitamins and the brain: mechanisms, dose and efficacy—a review. Nutrients. 2016;8:68. https://doi.org/10.3390/nu8020068.

Krikorian R, Shidler MD, Nash TA, et al. Blueberry supplementation improves memory in older adults. J Agric Food Chem. 2010;58:3996–4000. https://doi.org/10.1021/jf9029332.

Kumar V, R A, Ahire JJ, Taneja NK. Techno-functional assessment of riboflavin-enriched yogurt-based fermented milk prepared by supplementing riboflavin-producing probiotic strains of Lactiplantibacillus plantarum. Probiotics Antimicrob Proteins. 2024;16:152–62. https://doi.org/10.1007/s12602-022-10026-6.

La Vecchia C, Chatenoud L, Negri E, Franceschi S. Session: whole cereal grains, fibre and human cancer wholegrain cereals and cancer in Italy. Proc Nutr Soc. 2003;62:45–9. https://doi.org/10.1079/PNS2002235.

Lee KW, Lip GYH. The role of omega-3 fatty acids in the secondary prevention of cardiovascular disease. QJM. 2003;96:465–80. https://doi.org/10.1093/qjmed/hcg092.

Lee SJ, Choi HR, Lee J-C, et al. The anti-aging effects of various berries in the human skin keratinocyte (HaCaT) cells. Korean J Food Sci Technol. 2014;46:198–204. https://doi.org/10.9721/KJFST.2014.46.2.198.

Mahmoud AM, Hernández Bautista RJ, Sandhu MA, Hussein OE. Beneficial effects of citrus flavonoids on cardiovascular and metabolic health. Oxidative Med Cell Longev. 2019;2019:5484138. https://doi.org/10.1155/2019/5484138.

Makki K, Deehan EC, Walter J, Bäckhed F. The impact of dietary fiber on gut microbiota in host health and disease. Cell Host Microbe. 2018;23:705–15. https://doi.org/10.1016/j.chom.2018.05.012.

Maphosa Y, Jideani VA. The role of legumes in human nutrition. In: Hueda MC, editor. Functional food – improve health through adequate food. InTech; 2017.

Marques de Brito B, de Menezes Campos V, Neves FJ, et al. Vitamin B12 sources in non-animal foods: a systematic review. Crit Rev Food Sci Nutr. 2023;63:7853–67. https://doi.org/10.1080/10408398.2022.2053057.

Martín-Cabrejas MA. Chapter 1. Legumes: an overview. In: Martín-Cabrejas MÁ, editor. Food chemistry, function and analysis. Cambridge: Royal Society of Chemistry; 2019. p. 1–18.

Martins JM, Riottot M, De Abreu MC, et al. Dietary raw peas (Pisum sativum L.) reduce plasma total and LDL cholesterol and hepatic esterified cholesterol in intact and Ileorectal anastomosed pigs fed cholesterol-rich diets. J Nutr. 2004;134:3305–12. https://doi.org/10.1093/jn/134.12.3305.

Maughan RJ. The sports drink as a functional food: formulations for successful performance. Proc Nutr Soc. 1998;57:15–23. https://doi.org/10.1079/pns19980005.

Maurya VK, Shakya A, Kolla V. B vitamins fortification. In: Food fortification. 1st ed. Boca Raton: CRC Press; 2024. p. 133–48.

McIntosh GH, Noakes M, Royle PJ, Foster PR. Whole-grain rye and wheat foods and markers of bowel health in overweight middle-aged men. Am J Clin Nutr. 2003;77:967–74. https://doi.org/10.1093/ajcn/77.4.967.

McKevith B. Nutritional aspects of cereals. Nutr Bull. 2004;29:111–42. https://doi.org/10.1111/j.1467-3010.2004.00418.x.

Mensink RP, Ebbing S, Lindhout M, et al. Effects of plant stanol esters supplied in low-fat yoghurt on serum lipids and lipoproteins, non-cholesterol sterols and fat soluble antioxidant concentrations. Atherosclerosis. 2002;160:205–13. https://doi.org/10.1016/s0021-9150(01)00562-7.

Michalak M. Plant-derived antioxidants: significance in skin health and the ageing process. Int J Mol Sci. 2022;23:585. https://doi.org/10.3390/ijms23020585.

Miles EA, Calder PC. Effects of citrus fruit juices and their bioactive components on inflammation and immunity: a narrative review. Front Immunol. 2021;12:712608. https://doi.org/10.3389/fimmu.2021.712608.

Millar CL, Norris GH, Jiang C, et al. Long-term supplementation of black elderberries promotes hyperlipidemia, but reduces liver inflammation and improves HDL function and atherosclerotic plaque stability in apolipoprotein E-knockout mice. Mol Nutr Food Res. 2018;62:e1800404. https://doi.org/10.1002/mnfr.201800404.

Miracle Uwa L. The anti-aging efficacy of antioxidants. Curr Trends Biomed Eng Biosci. 2017;7:555716. https://doi.org/10.19080/CTBEB.2017.07.555716.

Munekata PES, Pateiro M, Zhang W, et al. Health benefits, extraction and development of functional foods with curcuminoids. J Funct Foods. 2021;79:104392. https://doi.org/10.1016/j.jff.2021.104392.

Musumeci L, Maugeri A, Cirmi S, et al. Citrus fruits and their flavonoids in inflammatory bowel disease: an overview. Nat Prod Res. 2020;34:122–36. https://doi.org/10.1080/14786419.2019.1601196.

Newby P, Muller D, Hallfrisch J, et al. Dietary patterns and changes in body mass index and waist circumference in adults. Am J Clin Nutr. 2003;77:1417–25. https://doi.org/10.1093/ajcn/77.6.1417.

Newmark HL, Heaney RP, Lachance PA. Should calcium and vitamin D be added to the current enrichment program for cereal-grain products? Am J Clin Nutr. 2004;80:264–70. https://doi.org/10.1093/ajcn/80.2.264.

Nile SH, Park SW. Edible berries: bioactive components and their effect on human health. Nutr Burbank Los Angel Cty Calif. 2014;30:134–44. https://doi.org/10.1016/j.nut.2013.04.007.

Nyanzi R, Jooste PJ. Cereal-based functional foods. In: Rigobelo E, editor. Probiotics. InTech; 2012.

O'Sullivan T. Exploring B vitamins and their impact on cognitive function and mood from conception to early adulthood. In: Riby L, Smith M, Foster J, editors. Nutrition and mental performance. London: Macmillan Education; 2012. p. 53–82.

Office of the Surgeon General (US). Bone health and osteoporosis: a report of the surgeon general. Rockville: Office of the Surgeon General (US); 2004.

Olson R, Gavin-Smith B, Ferraboschi C, Kraemer K. Food fortification: the advantages, disadvantages and lessons from sight and life programs. Nutrients. 2021;13:1118. https://doi.org/10.3390/nu13041118.

Orrù S, Imperlini E, Nigro E, et al. Role of functional beverages on sport performance and recovery. Nutrients. 2018;10:1470. https://doi.org/10.3390/nu10101470.
Parhiz H, Roohbakhsh A, Soltani F, et al. Antioxidant and anti-inflammatory properties of the citrus flavonoids hesperidin and hesperetin: an updated review of their molecular mechanisms and experimental models. Phytother Res PTR. 2015;29:323–31. https://doi.org/10.1002/ptr.5256.
Park Y, Storkson JM, Albright KJ, et al. Evidence that the trans-10,cis-12 isomer of conjugated linoleic acid induces body composition changes in mice. Lipids. 1999;34:235–41. https://doi.org/10.1007/s11745-999-0358-8.
Park S, Johnson M, Fischer JG. Vitamin and mineral supplements: barriers and challenges for older adults. J Nutr Elder. 2008;27:297–317. https://doi.org/10.1080/01639360802265855.
Peacock M. Phosphate metabolism in health and disease. Calcif Tissue Int. 2021;108:3–15. https://doi.org/10.1007/s00223-020-00686-3.
Pereira MA, Jacobs DR, Slattery ML, et al. The Association of Whole Grain Intake and Fasting Insulin in a biracial cohort of young adults: the CARDIA study. CVD Prev J Int Soc Fed Cardiol. 1998;1:231–42.
Pereira CC, Do Nascimento Da Silva E, De Souza AO, et al. Evaluation of the bioaccessibility of minerals from blackberries, raspberries, blueberries and strawberries. J Food Compos Anal. 2018;68:73–8. https://doi.org/10.1016/j.jfca.2016.12.001.
Playne M, Bennett L, Smithers G. Functional dairy foods and ingredients. Aust J Dairy Technol. 2003;58:242–64.
Praznik W, Loeppert R, Viernstein H, et al. Dietary fiber and prebiotics. In: Ramawat KG, Mérillon J-M, editors. Polysaccharides. Cham: Springer; 2015. p. 891–925.
Qudsia F, Saboor M, Khosa SM, et al. Comparative analysis of serum iron, serum ferritin and red cell folate levels among breast fed, fortified milk and cow's milk fed infants. Pak J Med Sci. 2015;31:706–9. https://doi.org/10.12669/pjms.313.6937.
Raman M, Ambalam P, Doble M. Probiotics, prebiotics, and fibers in nutritive and functional beverages. In: Nutrients in beverages. Elsevier; 2019. p. 315–67.
Randhawa MA, Khan AA, Javed MS, Sajid MW. Green leafy vegetables: a health promoting source. In: Handbook of fertility. Elsevier; 2015. p. 205–20.
Rudkowska I. Functional foods for health: focus on diabetes. Maturitas. 2009;62:263–9. https://doi.org/10.1016/j.maturitas.2009.01.011.
Ruel MT, Menon P, Loechl C, Pelto G. Donated fortified cereal blends improve the nutrient density of traditional complementary foods in Haiti, but iron and zinc gaps remain for infants. Food Nutr Bull. 2004;25:361–76. https://doi.org/10.1177/156482650402500406.
Russell WR, Baka A, Björck I, et al. Impact of diet composition on blood glucose regulation. Crit Rev Food Sci Nutr. 2016;56:541–90. https://doi.org/10.1080/10408398.2013.792772.
Sahoo N, Mishra SK, Swain RK, et al. Effect of turmeric and ginger supplementation on immunity, antioxidant, liver enzyme activity, gut bacterial load and histopathology of broilers. Indian J Anim Sci. 2019;89(7):774–9. https://doi.org/10.56093/ijans.v89i7.92046.
Saini RK, Ranjit A, Sharma K, et al. Bioactive compounds of citrus fruits: a review of composition and health benefits of carotenoids, flavonoids, limonoids, and terpenes. Antioxid Basel Switz. 2022;11:239. https://doi.org/10.3390/antiox11020239.
Saleem A. Adult bone health and consumption of dairy-based products: a review. FSAMR. 2023;2:68–77.
Salmerón J, Ascherio A, Rimm EB, et al. Dietary fiber, glycemic load, and risk of NIDDM in men. Diabetes Care. 1997;20:545–50. https://doi.org/10.2337/diacare.20.4.545.
Sanders ME. Probiotics: considerations for human health. Nutr Rev. 2003;61:91–9. https://doi.org/10.1301/nr.2003.marr.91-99.
Sathe S, Monaghan E, Kshirsagar H, Venkatachalam M. Chemical composition of edible nut seeds and its implications in human health. In: Shahidi F, Alasalvar C, editors. Tree nuts. CRC Press; 2008.
Serafini M, Peluso I. Functional foods for health: the interrelated antioxidant and anti-inflammatory role of fruits, vegetables, herbs, spices and cocoa in humans. Curr Pharm Des. 2017;22:6701–15. https://doi.org/10.2174/1381612823666161123094235.

Serna J, Bergwitz C. Importance of dietary phosphorus for bone metabolism and healthy aging. Nutrients. 2020;12:3001. https://doi.org/10.3390/nu12103001.
Shaikh S. Sources and health benefits of functional food components. In: Shiomi N, Savitskaya A, editors. Current topics in functional food. Rijeka: IntechOpen; 2022.
Shi M, Loftus H, McAinch AJ, Su XQ. Blueberry as a source of bioactive compounds for the treatment of obesity, type 2 diabetes and chronic inflammation. J Funct Foods. 2017;30:16–29. https://doi.org/10.1016/j.jff.2016.12.036.
Shiby VK, Mishra HN. Fermented milks and milk products as functional foods—a review. Crit Rev Food Sci Nutr. 2013;53:482–96. https://doi.org/10.1080/10408398.2010.547398.
Silagy CA, Neil HA. A meta-analysis of the effect of garlic on blood pressure. J Hypertens. 1994;12:463–8.
Silalahi J. Anticancer and health protective properties of citrus fruit components. Asia Pac J Clin Nutr. 2002;11:79–84. https://doi.org/10.1046/j.1440-6047.2002.00271.x.
Simopoulos AP. Omega-3 fatty acids in wild plants, nuts and seeds. Asia Pac J Clin Nutr. 2002;11(s6):S163–73. https://doi.org/10.1046/j.1440-6047.11.s.6.5.x.
Singhal P, Kaushik G, Mathur P. Antidiabetic potential of commonly consumed legumes: a review. Crit Rev Food Sci Nutr. 2014;54:655–72. https://doi.org/10.1080/10408398.2011.604141.
Slavin J. Whole grains and human health. Nutr Res Rev. 2004;17:99–110. https://doi.org/10.1079/NRR200374.
Slavin JL, Martini MC, Jacobs DR, Marquart L. Plausible mechanisms for the protectiveness of whole grains. Am J Clin Nutr. 1999;70:459S–63S. https://doi.org/10.1093/ajcn/70.3.459s.
Stevinson C, Pittler MH, Ernst E. Garlic for treating hypercholesterolemia. A meta-analysis of randomized clinical trials. Ann Intern Med. 2000;133:420–9. https://doi.org/10.7326/0003-4819-133-6-200009190-00009.
Stubbs J, Whybrow S, Lavin J. Dietary and lifestyle measures to enhance satiety and weight control: practical weight-control solutions. Nutr Bull. 2010;35:113–25. https://doi.org/10.1111/j.1467-3010.2010.01827.x.
Sumi K, Tagawa R, Yamazaki K, et al. Nutritional value of yogurt as a protein source: digestibility/absorbability and effects on skeletal muscle. Nutrients. 2023;15:4366. https://doi.org/10.3390/nu15204366.
Sun P, Zhao L, Zhang N, et al. Bioactivity of dietary polyphenols: the role in LDL-C lowering. Foods. 2021;10:2666. https://doi.org/10.3390/foods10112666.
Takeda E, Yamamoto H, Yamanaka-Okumura H, Taketani Y. Dietary phosphorus in bone health and quality of life. Nutr Rev. 2012;70:311–21. https://doi.org/10.1111/j.1753-4887.2012.00473.x.
Tetens I, Alinia S. The role of fruit consumption in the prevention of obesity. J Hortic Sci Biotechnol. 2009;84:47–51. https://doi.org/10.1080/14620316.2009.11512594.
Theodore LE, Kellow NJ, McNeil EA, et al. Nut consumption for cognitive performance: a systematic review. Adv Nutr Bethesda Md. 2021;12:777–92. https://doi.org/10.1093/advances/nmaa153.
Thorning TK, Raben A, Tholstrup T, et al. Milk and dairy products: good or bad for human health? An assessment of the totality of scientific evidence. Food Nutr Res. 2016;60:32527. https://doi.org/10.3402/fnr.v60.32527.
Tripathi MK, Giri SK. Probiotic functional foods: survival of probiotics during processing and storage. J Funct Foods. 2014;9:225–41. https://doi.org/10.1016/j.jff.2014.04.030.
Truong V-L, Jeong W-S. Antioxidant and anti-inflammatory roles of tea polyphenols in inflammatory bowel diseases. Food Sci Human Wellness. 2022;11:502–11. https://doi.org/10.1016/j.fshw.2021.12.008.
Tuohy KM, Probert HM, Smejkal CW, Gibson GR. Using probiotics and prebiotics to improve gut health. Drug Discov Today. 2003;8:692–700. https://doi.org/10.1016/S1359-6446(03)02746-6.
Vaughn AR, Sivamani RK. Effects of fermented dairy products on skin: a systematic review. J Altern Complement Med. 2015;21:380–5. https://doi.org/10.1089/acm.2014.0261.
Vijayaraghavan K. Control of micronutrient deficiencies in India: obstacles and strategies. Nutr Rev. 2002;60:S73–6. https://doi.org/10.1301/00296640260130786.

West BJ, Deng S, Kehaati P'A. Vitamin C, grape seed extract and citrus bioflavonoids protect the skin against photoaging: a review. J Biosci Med. 2020;08:116–34. https://doi.org/10.4236/jbm.2020.812012.

Weststrate JA, Van Poppel G, Verschuren PM. Functional foods, trends and future. Br J Nutr. 2002;88:S233–5. https://doi.org/10.1079/BJN2002688.

Widyastuti Y, Rohmatussolihat FA. The role of lactic acid bacteria in milk fermentation. Food Nutr Sci. 2014;05:435–42. https://doi.org/10.4236/fns.2014.54051.

Williamson C. Functional foods: what are the benefits ? Br J Community Nurs. 2009;14:4–7.

Yang CS, Zhang J, Zhang L, et al. Mechanisms of body weight reduction and metabolic syndrome alleviation by tea. Mol Nutr Food Res. 2016;60:160–74. https://doi.org/10.1002/mnfr.201500428.

Yanni AE, Kartsioti K, Karathanos VT. The role of yoghurt consumption in the management of type II diabetes. Food Funct. 2020;11:10306–16. https://doi.org/10.1039/D0FO02297G.

Yeh EB, Barbano DM, Drake M. Vitamin fortification of fluid Milk. J Food Sci. 2017;82:856–64. https://doi.org/10.1111/1750-3841.13648.

Yousefi M, Shadnoush M, Khorshidian N, Mortazavian AM. Insights to potential antihypertensive activity of berry fruits. Phytother Res PTR. 2021;35:846–63. https://doi.org/10.1002/ptr.6877.

Zemel MB. Role of calcium and dairy products in energy partitioning and weight management. Am J Clin Nutr. 2004;79:907S–12S. https://doi.org/10.1093/ajcn/79.5.907S.

Zhang R, Lu Y, Ye L, et al. Unique roles of phosphorus in endochondral bone formation and osteocyte maturation. J Bone Miner Res Off J Am Soc Bone Miner Res. 2011;26:1047–56. https://doi.org/10.1002/jbmr.294.

Zhao T, Li C, Wang S, Song X. Green tea (Camellia sinensis): a review of its phytochemistry, pharmacology, and toxicology. Mol Basel Switz. 2022;27:3909. https://doi.org/10.3390/molecules27123909.

Zhu B, Sun Y, Qi L, et al. Dietary legume consumption reduces risk of colorectal cancer: evidence from a meta-analysis of cohort studies. Sci Rep. 2015;5:8797. https://doi.org/10.1038/srep08797.

Zhu Y, Sun J, Lu W, et al. Effects of blueberry supplementation on blood pressure: a systematic review and meta-analysis of randomized clinical trials. J Hum Hypertens. 2017;31:165–71. https://doi.org/10.1038/jhh.2016.70.

Zou Z, Xi W, Hu Y, et al. Antioxidant activity of citrus fruits. Food Chem. 2016;196:885–96. https://doi.org/10.1016/j.foodchem.2015.09.072.

The Global Regulatory Landscape and Future Trends in Functional Foods

9

S. Amatha, Brijesh Shivhare, Virendra S. Ligade, Mrunalini Harish Kulkarni, Sreejan Manna, Manish Kumar Gautam, and Vivek

9.1 Introduction

In the era of improving health quality, human beings are more focused on a nutrient-rich diet. Functional foods act as the best choice for a balanced diet. In addition to offering essential nutrients, functional foods provide qualities that help one to attain or preserve excellent health. Comprehensive dietary adjustments to improve health and to prevent disease have been the main focus of attention for managing diet-related diseases. Accordingly, the need for functional foods has grown significantly in developed countries (Verbeke 2005). While functional food has grown in popularity, many countries still need to fully understand its legal standing in food regulations (Duttaroy 2019). Functional foods are those present in daily intake that have some functional benefits on the human body, such as improvement in overall physiology, including skin, digestive

S. Amatha · V. S. Ligade (✉)
Department of Pharmaceutical Regulatory Affairs and Management, Manipal College of Pharmaceutical Sciences, Manipal, Udupi, Karnataka, India
e-mail: virendra.sl@manipal.edu

B. Shivhare
Department of Botany, Faculty of Sciences, Baba Mastnath University, Rohtak, Haryana, India

M. H. Kulkarni
School of Pharmacy, Vishwakarma University, Pune, Maharashtra, India

S. Manna
Department of Pharmaceutical Technology, Brainware University, Kolkata, West Bengal, India

M. K. Gautam
Faculty of Pharmaceutical Science, Assam down town University, Guwahati, Assam, India

Vivek
Metro College of Health Sciences and Research, Greater Noida, India

M. K. Gautam et al. (eds.), *Antioxidants and Functional Foods for Diabetic Neuropathy*, https://doi.org/10.1007/978-981-95-9326-2_9

system, dental system, and cardiovascular system. This food also controls several immune systems like blood pressure, bone strength, cholesterol level, glucose level, and so on (Shimizu 2019). The "Korean Health/Functional Food Act" of 2004 mandates that these products can available as measurable dosages like pills (Vettorazzi et al. 2020). Establishing and setting into strict practices to standardized manufacturing steps/methods, as well as quality assurance and control methods, is crucial. It is not easy to evaluate these items and guarantee their efficacy and safety via authorization and supervision (Bahorun et al. 2019). The current chapter aims to explore the global regulatory framework for the approval and market authorization of functional food and examine the challenges and future trends that influence the upcoming regulations.

Table 9.1 Definitions of functional foods by various international regulatory agencies and organizations

Sl. no	Agency	Definition	References
1	NCCAM (National Center for Complementary and Alternative Medicine)	"Components of the usual diet that may have biologically active components (e.g., polyphenols, phytoestrogens, fish oils, carotenoids) that may provide health benefits beyond basic nutrition"	FDA (2006)
2	Canada (Health Canada)	"A functional food is similar in appearance to, or may be, a conventional food, is consumed as part of a usual diet, and is demonstrated to have physiological benefits and/or reduce the risk of chronic disease beyond basic nutritional functions".	Health Canada
3	EFSA (European Food Safety Authority)	"A food, which beneficially affects one or more target functions in the body, beyond adequate nutritional effects, in a way that is relevant to either an improved state of health and Well-being and/or reduction of risk of disease. A functional food can be a natural food or a food to which a component has been added or removed by technological or biotechnological means, and it must demonstrate their effects in amounts that can normally be expected to be consumed in the diet."	Duttaroy (2019)
4	FOSHU	"Foods which are expected to have certain health benefits, and have been licensed to bear a label claiming that a person using them for specified health use may expect to obtain the health use through the consumption thereof"	Bigliardi and Galati (2013)
5	East–West Perspectives on Functional Foods	"Foods that improve or affect body functions over and above their normal nutritional values"	Kwak and Jukes (2001)
6.	ILSI	"As a food which by virtue of physiologically active food components, provides health benefits beyond basic nutrition"	Milner (2002)

9.2 Definitions

Although there are several definitions of functional food items, none are officially or universally accepted (Bahorun et al. 2019). Some of the definitions of functional foods are mentioned in Table 9.1.

9.3 Global Perspective of Functional Foods

Japan, United States, Europe, India, China, and Indonesia are the key market players having well established regulations for functional food. Each country and its contribution are discussed below.

Japan A 10-year long-term research on food functions was carried out in Japan, serving as the world's first commercialization policy in the field of functionally active food items. Such food materials were called "Food for Specified Health Uses" (FOSHU), later internationally recognized as functional food. As a regulatory framework for functional foods, the Japanese government Ministry of Health, Labour and Welfare (MHLW) established FOSHU in 1991. The mechanism of action and active components must be identified for the FOSHU system. The first step in obtaining FOSHU certification is to provide scientific evidence of the mode of action via in vitro and in vivo testing, followed by human clinical trials to demonstrate efficacy (Ohama et al. 2006). By 2009, MHLW power to regulate functional food came under the Japanese Consumer Commission. The Consumer Affairs Agency put forth a guideline for the experimental approach of human research but later in 2016 it has demanded the need to receive FoSHU clearance. The guideline specifies the types of experimental designs that should be used (such as randomized control trials), the total number of participants that should be used, the requirements of the subjects (such as healthy adults or subjects who are on the verge of becoming healthy), the foods to be tested, the duration of the study of treatment, safety study at different doses, clinical indicators that need to be measured, and the statistical approaches (Shimizu 2019). FOSHU demands the chemical properties, biological properties, physical properties, safety study reports and stability studies of the active ingredient, as well as the analytical techniques documents for the Food Safety Commission (FSC) review (Yamada et al. 2008). The flow of FOSHU product approval was depicted in Fig. 9.1. Despite FOSHU's excellent public image, no post-purchase research has been done on the consequences of its administration of functional foods in a situation without oversight.

US The "Food and Drug Administration" (FDA) is the regulatory agency upholding US food safety regulations under the "Federal Food, Drug, and Cosmetic Act" (FFDCA) (Storey 2004) and the "Dietary Supplement Health and Education Act" (DSHEA). Since there is no formal definition of functional foods in the FFDCA, there is no official regulatory category and guidance for functional food research/ functional food industry (Crawford 2019). The FDA controls dietary supplements

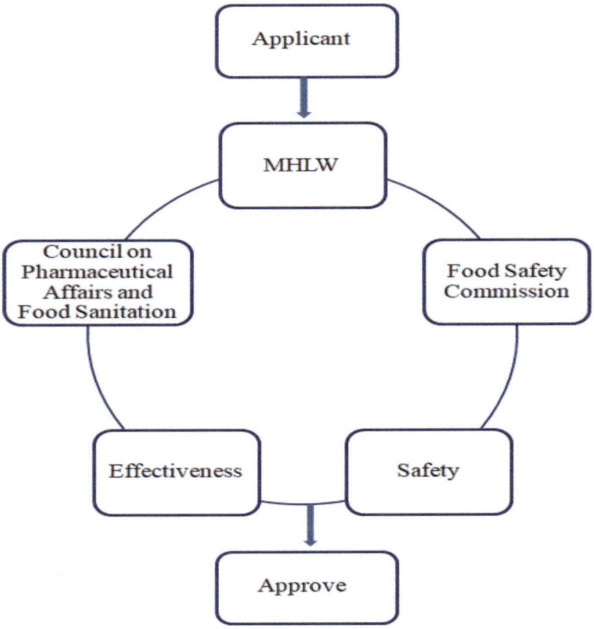

Fig. 9.1 Flow chart of FOSHU approval

and foods according to their intended use, established by medical claims and label details. The US established the "Functional Food Center" (FFC), an autonomous scientific and academic organization intending to develop and estimate regulatory requirements based on an assessment of the available scientific data. The FFC publishes textbooks, peer-reviewed publications, and online courses. It also hosts international conferences that link a diverse group of experts in functional food research, encouraging collaboration with government agencies (Martirosyan et al. 2022).

EU The food-related EU regulatory frameworks are of mainly five types: General food regulations (EU 178/2002), Regulations for food supplements/specific bioactive molecules (Directive 2002/46/EC), Regulations related to fortification of food materials(EC 1925/2006), Regulations on food for specific age group (EU 609/2013) and novel food regulations (EU 2015/2283 and EU 2017/2470). Most functional foods are naturally derived, so they must follow any of the above five regulations or a mixture of two or more regulations. An online registry of botanicals that may be used with food has been developed by the European Food and Safety Authority (EFSA) to provide some standardized data and help in risk assessment (Vettorazzi et al. 2020). To ensure customer safety with rising diversity of functional foods were promoted in EU with the approval of health claims from 2006 (the rule [EC] 1924/2006) (Martini et al. 2018). The main criteria for the approval of nutrition claim of functional food label are that it should be understandable and supported by scientific data (Duttaroy 2019).The approval in EU is depicted in Fig. 9.2.

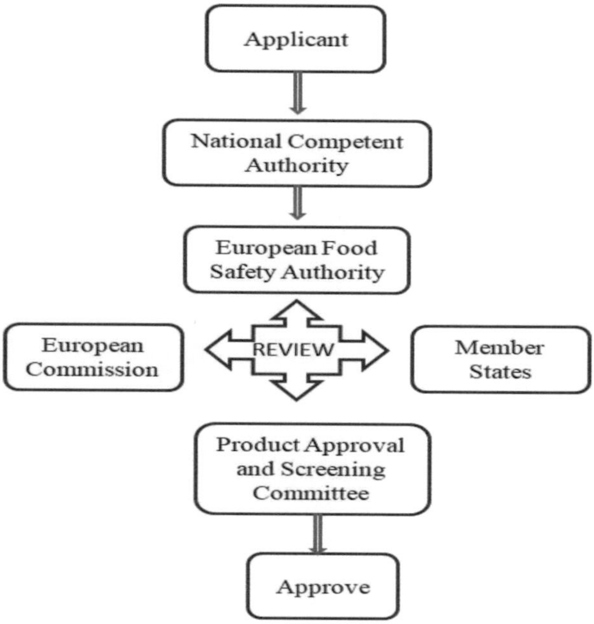

Fig. 9.2 Flow chart of EFSA approval

India The Food Safety and Standard Act was passed in 2006, which serves as a single reference rule to regulate and ensure the safety and efficacy of food products containing functional food and nutraceuticals (Bhupathiraju et al. 2019). In India, apart from the US, EU, and Japan, there is no further categorization or differentiation between functional food and nutraceuticals. Herbal derivatives, spices, fruits, and nutrient-wise enhanced foods with or without added functional components are examples of functional foods in India. According to the assessment conducted in 2003 by an expert group appointed by the Ministry of Health, the existing food regulations need to be amended to include new classifications for regulating functional foods. It is recommended that these items should undergo required safety testing before market approval. The Bureau of Indian Standards and National Standards with experts from several stakeholder groups in the food industry, sets voluntary norms in India. These standards include consumer affairs, system certification and testing, product authorization, and quality standards (FICCI 2006). On August 5, 2011, the "Food Safety and Standards Regulations" 2011, published in the Indian Gazette, went into effect. They governed the production, marketing, and distribution of functional foods in India. (Keservani et al. 2019). The registration of a functional food in India was depicted in Fig. 9.3.

China In 2016, China became the world's second-largest marketplace for functional foods. On 2003, the Ministry of Health gave all the power on functional food regulation to the China Food and Drug Administration (CFDA). "Administrative Measures of Health Functional Foods Registration and Notification" (Shimizu 2019) was issued by the CFDA by the Food Safety Law, which went into force from

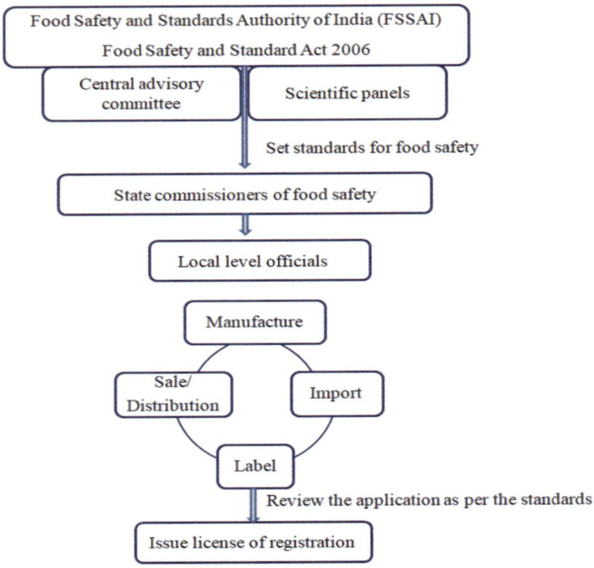

Fig. 9.3 Flow chart of functional food approval in India

July 1, 2016. The certification and notification processes for approving health-functional foods are described in this clause. The applicant should send a product dossier to CFDA as part of the notification process to show the effectiveness, safety, and quality of their product. The dossier will then be inspected and will give approval (Hu 2019). Later in early 2018, the China Drug Administration took over this responsibility.

Indonesia With its historical understanding of the health benefits of herbs and spices, Indonesia, the second-largest country with the highest levels of biodiversity, possesses some aspects of nature that could be turned into functional foods. Functional foods were not subject to the regulations of Indonesia's National Agency for Drug and Food Control. Functional foods have recently been developed using "design thinking" techniques, which center on troubleshooting around people rather than technologies or structure. A network model for the fabrication of functional foods can be modeled by a penta helix partnership involving academics, businesses, government agencies, and non-governmental organizations (NGOs) (Lestari 2021).

9.4 Codex Alimentarius Standards for Harmonizing Regulations of Functional Foods

A collection of globally recognized dietary guidelines, norms, rules, and recommendations pertaining to foods production, along with food safety are known as the Codex Alimentarius. The Codex Alimentarius is regarded by the World Trade Organisation as a universally accepted standard for settling issues pertaining to food safety and safeguarding customers. Thus, it is a global indicator for national food policymakers, regulatory agencies, the international food supply chain, producers, and customers. The Food and Agricultural Organisation (FAO) and the World Health Organisation (WHO) jointly established the Codex Alimentarius Commission (CAC) in 1963 to frame food standards, recommendations, and related literature, including codes of practice for the Joint FAO/WHO Food Standards Program. As of the end of 2020, the CAC comprises 128 individual countries, and the EU is a member organization (Heggum 2022). Codex records are particularly significant for countries with weak national laws or without laws, regulations, or recommendations.

9.5 Health-Related Claims

The FDA wants to safeguard consumers from false or deceptive health claims while still providing them with useful data regarding nutrition (Ziker). Health Canada issued a Consultation Document in June 2000 called "Standards of Evidence for Evaluating Foods with Health Claims: A Proposed Framework." A proposed strategy for monitoring nutritional claims unique to a product was released by Health Canada. A claim identification number (CIN), which appears on product labels, would indicate an authorized claim (Fitzpatrick 2004). Depending on personal taste, people start using different functional foods. Health-related claims, messages, or symbols attract attention toward a particular food product. The various promotional strategies like product labels, packaging, advertisements, and promotional campaigns have promoted innovation and competition among food companies. Even though the health claims in the US, Europe, and Japan seem similar, there are vast differences in the regulatory approval procedure. As we all know, the Japanese first coined the term, definition and classification. The regulatory framework for functional foods was also first framed by Japan because it is the birthplace of the functional food market. Nutrition Labeling and Education Act in 21 CFR functions as a regulatory guideline information relating to the manufacture, labeling, including claims, and distribution of functional foods in the US. According to US regulations, health claims are further classified into nutrient-content expressed or implied claims, functional or structural claims, and authorized and qualified health claims.

The Canadian Food Inspection Agency has established a list of acceptable function claims that speak directly of scientifically supported beneficial physiological effects resulting from ingesting a food product and provide good health or performance (Powers et al. 2019). Each of these claims should follow specific FDA standards; for example, in nutrient content claims, it is mandatory to specify the nutrient threshold; functional or structural claims indicate the effect of the nutrient in the human body, not mainly focusing on a disease, and finally, health claims should be based on available scientific literature or based on emerging scientific evidence-based studies. In Japan, the regulations of health claims are governed by Food for Specified Health Use (FOSHU) and Food with Nutrient Function Claims. These agencies are making rules for health-related claims with

(FOSHU needs approval from MHLW) or without pre-market approval. In 2001, criteria for FoSHU were integrated into the "food with health claim" (FHC) system. "Standardized FoSHU" (appropriate concentration or having scientific evidence) and "Qualified FoSHU" (no scientific evidence but specific efficacy data) systems were introduced to relax the frame of FHC. The coexistence of both FoSHU and FFC increases the credibility of the efficacy and safety of FFC products (Shimizu 2003)

Europe has no set definition or regulatory guideline for functional food products. However, the regulatory standards applied to these functional products vary depending on the product type because they may contain components controlled in the European market, such as nutrients or bioactive compounds. The EU has a specific register called the "EU Register of Nutrition and Health Claims", a reference for permissible claims because only a few claims related to functional food are approved in the EU market. EFSA evaluates the scientific evidence before granting regulatory approval for claims. From all these, it is clear that the regulatory approval of health-related claims is more stringent in the EU and Japan than in the US because of its complex approval process, including pre-market review of scientific evidence. Diaz et al., in the study of claims, highlight the need for

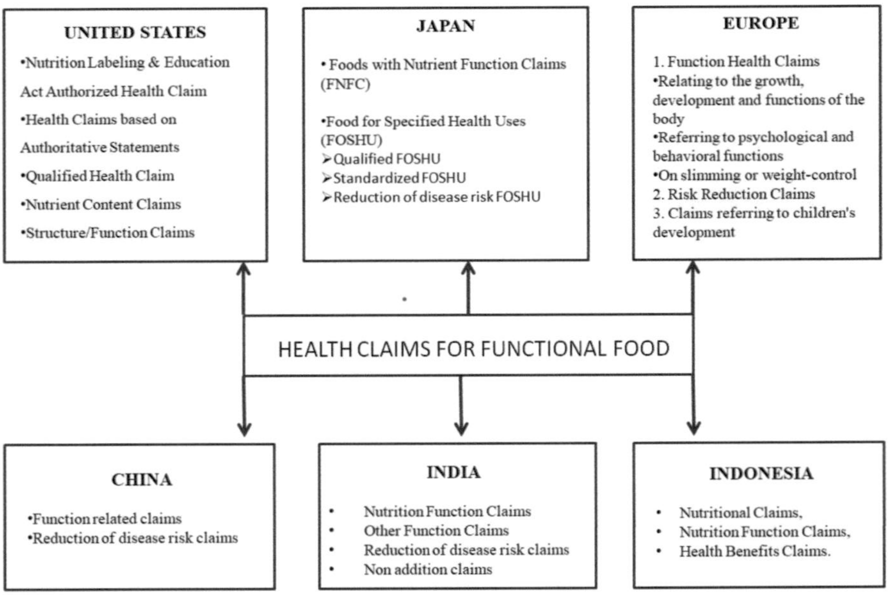

Fig. 9.4 An overview of health claims in different countries

internationally harmonized guidelines for health-related claims on functional food products. In the Spanish market, claims related to functional foods, highlighting the need for a stringent regulatory review before approval (Domínguez Díaz et al. 2020). Sato explores how the "Food with Function Claims" system has influenced the regulatory environment and the pace of innovation in the dietary supplements sector. The study highlights the balance between ensuring consumer safety through promoting transparency and consumer trust while encouraging innovation by reducing barriers and expanding market growth (Sato et al. 2023). Health claims on different countries are given below in Fig. 9.4.

9.6 Clinical Data Requirement

Clinical research should include well-designed patients who are either healthy or on the verge of disease. It is necessary to conduct double-masked, randomized, placebo-controlled trials with enough participants to demonstrate meaningful differences. In vivo experiments employing animals or cell-based assays are typically used to conduct mechanism research on the physiological effects of dietary components. Human clinical research is the most crucial step in confirming the efficacy and safety of FoSHU products. However, it had just been 20 years since clinical studies for FoSHU were standardized. A specification for designing experiments of the clinical human trial necessary for FoSHU approval was drafted and made public in 2016 based on the ongoing talks held by the CAA(Shimizu 2019).

9.7 Regulatory Challenges Associated with Functional Food

- Lack of definition and classification

A primary challenge encountered by functional food is the need for more well-defined and uniform rules in various countries and regions. Depending on the market, functional foods may be classified as food items, dietary supplements, or pharmaceuticals, and each category has various criteria for labeling, claims, safety, and efficacy. Furthermore, it can be challenging for manufacturers of functional foods to be up to date with the most recent standards and requirements due to the frequently changing and developing regulatory frameworks. Therefore, manufacturers of functional foods may encounter expenses, delays, or entrance hurdles in particular markets (Health Canada).

- Hurdles in the development of scientific evidence

An additional challenge for manufacturers of functional foods is presenting adequate and reliable scientific proof of the health advantages of their products. Functional foods must undergo extensive testing and evaluation to show that they are healthy and safe for the target population and improve particular health outcomes. These studies are often costly, time-consuming, and complex because many variables and factors can affect the outcome. FDA functional food approval took a minimum of 6–10 months, delaying the market entry (Martirosyan et al. 2022). Moreover, manufacturers of functional foods must clearly and concisely explain their research to regulators, who may have varying degrees of expertise and expectations.

- Health claim-related challenges

Getting FDA approval for health claims is a complex process. Although there are many studies on risk reduction, the FDA demands strong proof, particularly for claims about disease identification. This analysis highlights the strict requirements that functional foods and must fulfill to validate their health advantages and give customers accurate information. Consumers may need to be more cautious of food claims due to the weak restrictions found in EFSA regulations. However, this also poses problems, like the need for a defined category for functional foods and the possibility that patients must be aware of the complexities of health claims (Martirosyan et al. 2022).

- Lengthy approval process

There are currently no legal requirements for establishing therapeutic or health claims on the usage of the product or its adverse effects when it undergoes regulation as a food. The drug identification number (DIN) is required for any functional food or supplement with a health claim. After reviewing an application, the Therapeutic Products Directorate of the Health Protection Branch issues a DIN. The authorization would be granted product by product. As seen by each of the five recently authorized generic claims, this is a time-consuming process that can take up to 3 years to finish the regulatory process.

- Safety concerns

The Federal Food, Drug, and Cosmetic Act (FDCA) has regulated safety through pre-market regulation: they may file a petition for the new ingredient, or they may show that the ingredient is "generally recognized as safe" (GRAS) (Burdock et al. 2006). The manufacturers of some functional foods legitimately complain that following food additive procedures would significantly delay product marketing. Pearson standard and Bean Rule ensure functional food's post-market regulatory standards (Ziker).

( ) To get their individual opinions on how to help spread the word to consumers about the importance of healthy eating habits in accomplishing and maintaining good health, the US FDA has met with experts. The issue with health claims for functional food lies in the fact that they are not the same as those for drugs, which exhibit a dosage response. For instance, research has been conducted on garlic.

(a) It was said that the acceptable higher intake amounts of the essential nutrients or components to be employed in the health claim must be a worry for the FDA. Utilization data is examined when the FDA considers an application for an additive to food, and it was proposed that health claim requests be examined similarly.

There were concerns that instead of using traditional drugs, customers could instead eat. To reduce cholesterol, for instance, a consumer can eat muesli rather than take prescription drugs as directed by their doctor. These variables must be considered when deciding whether to approve a qualified health claim. It was mentioned that the FDA's decision-making process needs to consider food and dietary supplement interactions. It was mentioned that interactions with dietary items and drugs must be considered when making decisions. Concerns regarding infants and functional diets were also raised (Program 2024).

9.8 Recent Trends

The emergence of functional foods is an ongoing development with enormous potential for market share, and government laws, corporate expenditures, and consumption choices all depend on the information flows produced by research. Sociodemographic trends, advancement, and societal alterations prefer functional food. As a result, the food sector is seeing an ongoing sustainable trend with these items. The level of competition in this industry is increasing. Consumer investigation is essential to manufacturing functional foods and necessitates additional investigation and advancement (Bigliardi and Galati 2013).

Trends in Product Development

The use of traditional methods for customized nutrition has increased dramatically. Certain technologies used in the production of functional foods, which include vacuum infiltration, edible films, coatings, and microencapsulation, stop biologically active substances like the bioactive component or biologically active components and viable probiotic bacteria from degrading throughout processing, storage, and transportation. Nutritional genomics could lead to the development of specialized food products or personalized dietary recommendations for individuals or communities. Currently, it is still regarded as an emerging science (Betoret et al. 2011; Chua-Lim et al. 2023).

Nutrigenomic research studies found genes like insulin receptor genes, proteinkinase B genes glucokinase genes whose expression has been linked to diabetes. Recent advances in diabetes nutrigenomics studies have listed advanced functional fibers such as probiotic sweeteners and low-glycemic oligosaccharides to prevent and control type 2 diabetes by triggering adaptations in the human intestinal microbiota. Nutrigenetics and nutrigenomics studies shed light on particular genetic indicators in patients and lead to precision dietary intake (Berná et al. 2014).

Technological Trend

The naturally occurring chemicals, structurally diverse or those derived from various sources, are the primary ingredient used in manufacturing functional foods. Government grant sponsors contribute money to the industry's development and testing of innovative approaches in order to address this issue. One example is the Croatian Science Foundation-funded 3D-SustJuice project, which uses 3D printing and technological barriers to process and preserve fruit juice sustainably. The 3D-SustJuice project develops, implements, and promotes healthier and more ecologically friendly food processing through sustainable and innovative technology. Innovative techniques like hydrostatic pressure, ultrasounds, pulsed electric fields, and microwaves have become viable substitutes for traditional food processing methods during the last 20 years (Granato et al. 2020). These alternatives use less energy, prevent microbial growth, and improve the preservation of natural nutrients in fruits and vegetables. They are environmentally beneficial and can be used to use by-products (Putnik et al. 2017). As a result, they have a favorable effect on the production of new valuable items.

9.9 Conclusion

Further developments in producing food products that can promote optimal health may also be aided by the uniqueness of functional food technical innovation (cholesterol-lowering goods, chewing gum artificial sweeteners, fermenting, and genetically altered products) (Vettorazzi et al. 2020). Continued progress in nutritional food technologies are essential for advancing functional foods growth (Duttaroy 2019).

References

Bahorun T, Aruoma OI, Neergheen-Bhujun VS. Phytomedicines, nutraceuticals, and functional foods regulatory framework. In: Nutraceutical and functional food regulations in the United States and around the world [Internet]. Elsevier; 2019. p. 509–21. [cited 2024 Sep 25]. Available from: https://linkinghub.elsevier.com/retrieve/pii/B9780128164679000320.

Berná G, Oliveras-López MJ, Jurado-Ruíz E, Tejedo J, Bedoya F, Soria B, Martín F. Nutrigenetics and nutrigenomics insights into diabetes etiopathogenesis. Nutrients. 2014;6(11):5338–69.

Betoret E, Betoret N, Vidal D, Fito P. Functional foods development: trends and technologies. Trends Food Sci Technol. 2011;22(9):498–508.

Bhupathiraju K, Krishnaraju AV, Sengupta K, Golakoti T, Akolkar SK, Datla P. Regulations on nutraceuticals, functional foods, and dietary supplements in India. In: Nutraceutical and functional food regulations in the United States and around the world [Internet]. Elsevier; 2019. p. 445–64. [cited 2024 Sep 25]. Available from: https://linkinghub.elsevier.com/retrieve/pii/B9780128164679000290.

Bigliardi B, Galati F. Innovation trends in the food industry: the case of functional foods. Trends Food Sci Technol. 2013;31(2):118–29.

Burdock GA, Carabin IG, Griffiths JC. The importance of GRAS to the functional food and nutraceutical industries. Toxicology. 2006;221(1):17–27.

Chua-Lim LA, Vergara AS, Ulamarulama RM, Valencia EKA, Vergara ARN, Dable-Tupas G, et al. Role of nutrigenomics in diabetes care and prevention. In: Role of nutrigenomics in modern-day healthcare and drug discovery [Internet]. Elsevier; 2023. p. 115–33. [cited 2024 Sep 25]. Available from: https://linkinghub.elsevier.com/retrieve/pii/B9780128244128000011.

Crawford WM. The food safety modernization act (FSMA)'s role in the safety of functional foods and their ingredients. In: Nutraceutical and functional food regulations in the United States and around the world [Internet]. Elsevier; 2019. p. 61–74. [cited 2024 Sep 25]. Available from: https://linkinghub.elsevier.com/retrieve/pii/B9780128164679000058.

Domínguez Díaz L, Fernández-Ruiz V, Cámara M. An international regulatory review of food health-related claims in functional food products labeling. J Funct Foods. 2020;68:103896.

Duttaroy AK. Regulation of functional foods in European Union. In: Nutraceutical and functional food regulations in the United States and around the world [Internet]. Elsevier; 2019. p. 267–76. [cited 2024 Sep 25]. Available from: https://linkinghub.elsevier.com/retrieve/pii/B9780128164679000198.

FDA. Guidance for industry: complementary and alternative medicine products and their regulation by the Food and Drug Administration. FDA [Internet]. 2006. [Cited on: 2024 Sep]. Available from: https://www.fda.gov/media/76323/download.pdf

FICCI study on implementation of food safety and standard act 2006: an industry perspective at http://www.indiaenvironmentportal.org.in/Files/food_safety_study.pdf.

Fitzpatrick KC. Regulatory issues related to functional foods and natural health products in Canada: possible implications for manufacturers of conjugated linoleic acid. Am J Clin Nutr. 2004 Jun;79(6):1217S–20S.

Granato D, Barba FJ, Bursać Kovačević D, Lorenzo JM, Cruz AG, Putnik P. Functional foods: product development, technological trends, efficacy testing, and safety. Annu Rev Food Sci Technol. 2020;11(1):93–118.

Health Canada. Nutraceuticals/functional foods and health claims on foods. Government of Canada [Internet]. [Cited on: 2024 Sep]. Available online: https://www.canada.ca/en/health-canada/services/food-nutrition/food-labelling/health-claims/nutraceuticals-functional-foods-health-claims-foods-policy-paper.html

Heggum C. Codex Alimentarius. In: Encyclopedia of dairy sciences [Internet]. Elsevier; 2022. p. 726–36. [cited 2024 Sep 25]. Available from: https://linkinghub.elsevier.com/retrieve/pii/B9780128187661002415.

Hu C. Change in raw materials and claims of health food regulations in China 2013–17. In: Nutraceutical and functional food regulations in the United States and around the world [Internet]. Elsevier; 2019. p. 465–95. [cited 2024 Sep 25]. Available from: https://linkinghub.elsevier.com/retrieve/pii/B9780128164679000307.

Keservani RK, Jarouliya U, Sharma AK. Nutraceutical and functional food regulations in India. In: Nutraceutical and functional food regulations in the United States and around the world [Internet]. Elsevier; 2019. p. 431–44. [cited 2024 Sep 25]. Available from: https://linkinghub.elsevier.com/retrieve/pii/B9780128164679000289.

Kwak NS, Jukes DJ. Functional foods. Part 1: the development of a regulatory concept. Food Control. 2001;12(2):99–107.

Lestari LA. Opportunities and challenges of functional foods and nutraceuticals development during Covid-19 pandemic. In: Gunadi YT, Pramana AAC, Ophinni Y, Gusnanto A, Kusuma WA, et al., editors. , vol. 41. BIO Web Conference; 2021. p. 02009.

Martini D, Innocenti A, Cosentino C, Bedogni G, Zavaroni I, Ventura M, Galli D, Mirandola P, Vitale M, Dei Cas A, Bonadonna RC, Passeri G, Pruneti C, Del Rio D. Claimed effects, outcome variables and methods of measurement for health claims proposed under regulation (EC) 1924/2006 and related to cognitive function in adults. Arch Ital Biol. 2018;156:64–86.

Martirosyan D, Lampert T, Ekblad M. Classification and regulation of functional food proposed by the functional food center. Funct Food Sci. 2022;2(2):25.

Milner JA. Functional foods and health: a US perspective. Br J Nutr. 2002;88(Supplement 2):S1518.

Ohama H, Ikeda H, Moriyama H. Health foods and food with health claims in Japan. Toxicology. 2006;221:95–111.

Powers JP, Farrell M, McMullin C, Retik L, White J. Regulation of dietary supplements and functional foods in Canada. In: Nutraceutical and functional food regulations in the United States and around the world [Internet]. Elsevier; 2019. p. 235–52. [cited 2024 Sep 25]. Available from: https://linkinghub.elsevier.com/retrieve/pii/B9780128164679000174.

Program HF. Consumer health information for better nutrition initiative – Attachment H – Stakeholder meetings. FDA [Internet]. 2024. [cited 2024 Oct 7]; Available from: https://www.fda.gov/food/nutrition-food-labeling-and-critical-foods/consumer-health-information-better-nutrition-initiative-attachment-h-stakeholder-meetings

Putnik P, Kovačević DB, Režek Jambrak A, Barba F, Cravotto G, et al. Innovative "green" and novel strategies for the extraction of bioactive added value compounds from citrus wastes: a review. Molecules. 2017;22(5):E680.

Sato K, Kodama K, Sengoku S. Optimizing the relationship between regulation and innovation in dietary supplements: a case study of food with function claims in Japan. Nutrients. 2023;15(2):476.

Shimizu T. Health claims on functional foods: Japanese regulations and an international comparison. Nutr Res Rev. 2003;16:242–52.

Shimizu M. History and current status of functional food regulations in Japan. In: Nutraceutical and functional food regulations in the United States and around the world [Internet]. Elsevier; 2019. p. 337–44. [cited 2024 Sep 25]. Available from: https://linkinghub.elsevier.com/retrieve/pii/B9780128164679000228.

Storey ML. Regulatory issues of functional foods, feeds, and nutraceuticals. Vet Clin North Am Small Anim Pract. 2004;34(1):329–38.

Verbeke W. Consumer acceptance of functional foods: socio-demographic, cognitive and attitudinal determinants. Food Qual Prefer. 2005;16:45–57.

Vettorazzi A, López De Cerain A, Sanz-Serrano J, Gil AG, Azqueta A. European regulatory framework and safety assessment of food-related bioactive compounds. Nutrients. 2020;12(3):613.

Yamada K, Sato-Mito N, Nagara J, Umegaki K. Health claim evidence requirements in Japan. J Nutr. 2008;138:1192S–8S.

Ziker D. Regulating functional foods: pre- and post-market strategy.

10 Influence of Bioactives in Functional Foods for Management of Diabetic Neuropathy

Sangita Mazumder, Anagha Ranade, Rabinarayan Acharya, and Sharad D. Pawar

10.1 Introduction

In order to improve texture, flavour, and variety, a wide range of foods are referred to as "food adjuncts" and are eaten with staple foods. Pickles come in liquid and dehydrated forms and are commonly served with a range of chutneys, preserves, wadis, papads, salads, and a range of sauces. All these items combined together present a strong line of traditional side dishes that go to complement the diversity and flavour content of Indian food quite well. The large diversity and multi-dimensional composition of these diverse adjuncts, resulting from an incredibly large variety of raw materials including a large range of fruits, a large variety of vegetables, a range of legumes, a limited set of oilseeds, and a large set of spices, render it almost impossible to build end conclusions regarding the nutritional content of these adjuncts. Even if they are taken in modest amounts, adjuncts are essential for healthy eating and general well-being. They are abundant in functional compounds like vitamins, minerals, proteins, and phytochemicals. Bioactive compounds are found in most essential ingredients employed in culinary as well as medicinal applications. Examples of such include curry leaves, coriander leaves, mint leaves, green chilies, tomatoes, black gram, Bengal gram, sesame seeds, and flax seeds. The bioactive compounds present in these products have a range of valuable properties, including antibacterial, antifungal, anti-inflammatory, antiallergic, antidiabetic, and anticancer properties. In addition to these exceptional properties, they also exhibit significant antioxidant activity, thereby contributing to general well-being and health (Khedkar 2022).

S. Mazumder · S. D. Pawar (✉)
Central Ayurveda Research Institute (CARI), Kolkata, West Bengal, India

A. Ranade · R. Acharya
Central Council for Research in Ayurvedic Sciences (Ministry of Ayush, Government of India), New Delhi, India

M. K. Gautam et al. (eds.), *Antioxidants and Functional Foods for Diabetic Neuropathy*, https://doi.org/10.1007/978-981-95-9326-2_10

The potential health benefits attainable from such bioactive compounds have been emphasized and underscored by recent evidence in research. For instance, it has been established that phenolic compounds in coriander have remarkable antibacterial activities, and flavonoids in curry leaves have been highly promising in the ability to inhibit inflammation and modulate the level of oxidative stress in the body. Consuming green chilies, which are high in capsaicin, has been associated with a better metabolism and less pain perception. Additionally, by adding dietary adjuncts to regular meals, their prebiotic actions can improve gut health and increase the bioavailability of vital nutrients (Bahadori et al. 2018).

Understanding the therapeutic benefits of food additives in the age of functional foods might help integrate them as a single ingredient or as an adjuvant to a regular diet for chronic metabolic illnesses like diabetes and diabetic neuropathy. In addition to creating new chances for the creation of novel medications based on established knowledge, this might guarantee less adverse effects and lower expenses, making it more accessible to the general public. This also leads to a greater understanding of the role that dietary adjuncts play in traditional diets and how they can enhance health and wellness (Gokhale et al. 2021).

10.2 Functional Foods for Diabetic Neuropathy and the Significance of Bioactive Substances

The World Health Organization, also widely referred to as WHO, released a significant report indicating that diabetes was the leading cause of 1.6 million deaths in 2021. Of these, a staggering 47% were those patients under the age of 70, highlighting a frightening trend of diabetes-related mortality among young adults. Additionally, as reported by the American Diabetic Association, it is significant that almost half of all patients with diabetes are at increased risk of developing diabetic neuropathy, a significant complication of the disease (Cheng et al. 2022). Approximately 50% of individuals with diabetes diagnosed as Type 2 Mellitus (T2DM) do not follow the recommended dietary and lifestyle changes. The hunt for novel foods and nutritional therapies that can successfully improve general health and lower the chance of acquiring associated disorders is thus severely hampered by such patient non-compliance. Finding such dietary remedies may also lessen the financial strain of managing chronic illnesses over the long term. The food business is crucial in this regard because it is the main organization in charge of the creation, manufacturing, and distribution of food items to customers (Ahmad et al. 2022a, b).

What are today referred to as "functional foods" emerged as a cost-effective, practical, and efficient solution to the management and treatment of chronic illnesses that most people experience. The idea of functional food was first proposed in 1984, but since then, it has undergone tremendous change and adaptation to fit different national and cultural contexts worldwide (Surono et al. 2020). If a food product has been proven to have health benefits beyond basic nourishment, such as improving physiological processes or lowering the risk of disease, it is often considered to be "Functional Food". Food industry and nutrition science work together to ensure that new compounds and food items are generated and made available to the general population. This is why functional foods are important (Pimentel et al. 2021).

The functional food movement had its start in 1984 when the Japanese government funded studies on foods with medical uses. In Japan, the official definition of the term "functional foods" was initially adopted by means of the official usage of the term "fortified foods with special ingredients that have positive physiological effects." Foods licensed to be marked under the Foods for Specified Health Use (FOSHU) category are required to meet a number of regulatory requirements in order to qualify as functional (Fig. 10.1) (Martirosyan et al. 2021). Among the regulatory requirements, three main nutritional characteristics are required to be met by such food products in order to qualify for FOSHU status:

- Natural components must be used in the food.
- It needs to be ingested as a regular component of a diet.
- It needs to provide a certain physiological purpose to the human body, like: strengthening defence system of our body.
- Preventing some diseases or facilitating their recovery.
- Regulating physical and mental conditions.
- Delaying the aging process.

Following that, nations in Europe embraced this FOSHU nutritional requirements idea. The phrase "functional foods" was first used in 1994 by the Nutrition Board and the National Academy of Food Sciences, who defined it as "foods or ingredients that have been modified and can enhance health beyond their inherent nutritional content." The American Dietetic Association expanded the concept in

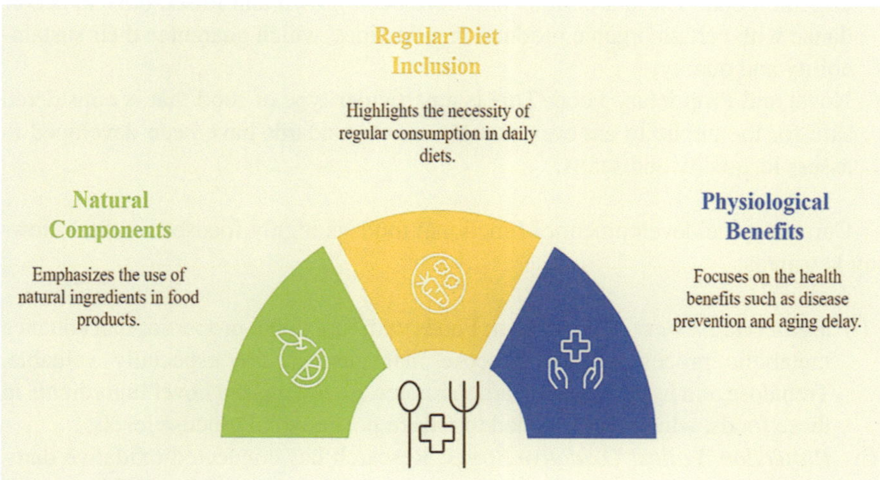

Fig. 10.1 FOSHU nutritional requirements

2004 when they claimed that functional foods improve health when taken in the right amounts and as part of a balanced diet. A wide variety of food items were covered under the concept, such as whole foods, fortified foods, enriched foods, and specially prepared foods.

The Nutrition Board and National Academy of Food Sciences established the term “functional foods” in 1994 to refer to “foods or ingredients that add to health beyond their nutritional value.” This was further elaborated by the American Dietetic Association in 2004, which claimed that functional foods improve health when taken correctly and as a component of a diet that is balanced overall. This idea covers a wide range of foods, including processed meals with extra components and nutrients as well as whole, fortified, enriched, and specifically prepared foods (Rashidinejad et al. 2021).

- Foods that include extra ingredients.
- Foods that have had parts removed.
- Foods that have had one or more constituents’ natures changed.
- Foods that have improved a certain component’s bioavailability.
- Mixtures of the previously listed traits.

As per FSSAI India Nutraceuticals, food supplements, special foods for dietary purposes, and functional foods are all functional foods. These are specially formulated foods for meeting dietary requirements due to diseases.

- Examples of non-normal foods include minerals, vitamins, proteins, animal foods, botanicals, and supplements.
- Genetically altered or engineered foods are those that contain genetically modified organisms (GMOs) from innovation or are made from GMOs but do not contain them.
- Organic Food: These are foods that have been grown and raised only in accordance with certain organic production guidelines, which guarantee their sustainability and quality.
- Novel and Proprietary Food: This is a particular type of food that is considered safe for the public to eat even though no set standards have been developed to assess its quality and safety.

Currently, the development of functional foods is highly focused on the following key areas

(i) *Metabolic Process Management*: Foods with functional properties that enhance metabolic processes, such glucose management, are especially valuable. Trehalose and hydrogenated carbohydrates are among the novel ingredients in these foods, which are designed to reduce postprandial glucose levels.
(ii) *Protection Against Oxidative Stress*: Research has connected oxidative damage, which is frequently brought on by reactive oxygen species (ROS), to age-

ing, cancer, heart disease, and neurological illnesses like Parkinson's and Alzheimer's. Antioxidant-rich functional foods such as carotenoids, polyphenols, vitamins E and C, may help prevent oxidative stress & it's health related hazards (Alongi and Anese 2021).

(iii) *Cardiovascular Health*: Consuming functional foods can help reduce a number of coronary heart disease risk factors, including hypertension, vascular integrity, dyslipidaemia, and elevated homocysteine levels. Some of the bioactive components of functional foods, including omega-3 fatty acids, dietary fibre, phytosterols, and flavonoids, have been shown to improve vascular health in general and reduce LDL cholesterol (LDL-C). Additionally, it was postulated that consuming more folate, vitamins B6, and B12 could aid in lowering homocysteine plasma levels, which could contribute significantly to improved cardiovascular protection (Fernández et al. 2019).

(iv) *Digestive Health*: The wide range and equilibrium of the gut microbiota are critical components of general health. Prebiotics, probiotics, and symbiotic foods are examples of functional foods that have been shown to enhance immunity and digestion by fostering a healthy gut flora (Bezirtzoglou et al. 2021).

The numerous bioactive substances found in functional food will be covered in this section along with their potential applications in the treatment and prevention of diabetes and diabetic neuropathy (Fig. 10.2).

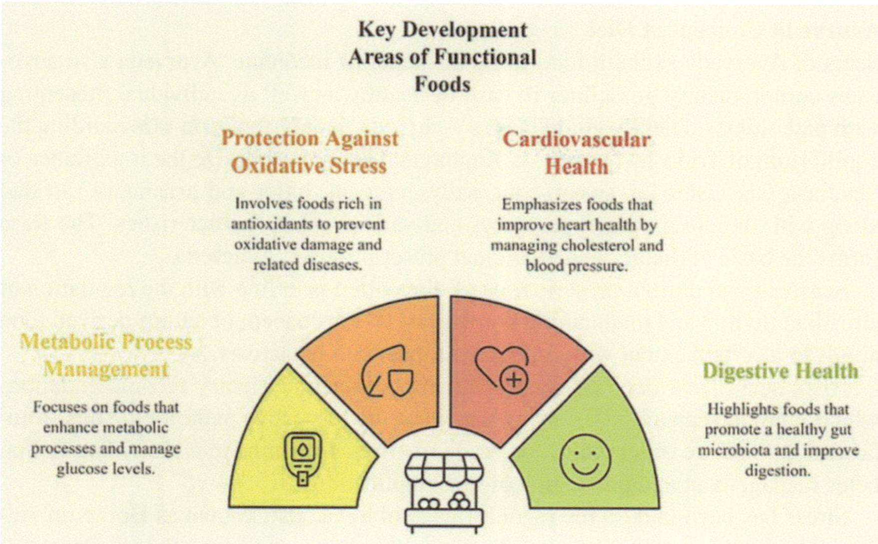

Fig. 10.2 Utility of function foods

10.3 Unlocking Nature's Remedies: The Role of Bioactive Compounds in Functional Foods in Diabetic Neuropathy

Cascades of diabetic complications as per Ayurveda:

In Ayurveda, Prameha is a group of pathologies associated with dysfunction of Urinary system which has different stages of involvement of tridosha and the manifestation of diabetes-like symptoms. As the chronicity increases, the stage becomes difficult to treat due to increasing predominance of Vata dosha. In Ayurveda, Madhumeha, one of the types of Vataja prameha is often regarded to share pathogenesis and clinical implication with Diabetes mellitus. The management of Diabetes in Ayurveda depends upon the type of patient too viz. Obese diabetic and lean diabetic. Dietary recommendations along with oral medication (Samshaman) is indicated in lean diabetics wherein they also need nutrition. Obese diabetic patient are often advised with detoxification therapies that include emesis/purgation/enema depending upon the involvement of dosha and tissue therein.

Further, complications of Diabetes in Ayurveda occur as a result of further vitiation of vata dosha. In case of Diabetic neuropathy, dysfunction of Vyana Vata is major which is reported to produce symptoms of pain (ruk), burning sensation (daha), tingling (harsha), numbness (supti), carbuncles (prameha pidaka), muscle wasting (shosha), bladder dysfunction, GIT symptoms coincide with the autonomic and peripheral neuropathy (Parveen et al. 2019).

The management of DPN in Ayurveda includes medication (shodhana/shamana), dietary and lifestyle recommendation. Here, an account of dietary recommendations and contra-indications are being provided:

Ayurveda Concept of Diet

Seers of Ayurveda exclaim food to be the superior medicine. Ayurveda also advocates certain dietary guidelines in case of healthy as well as individual presenting with pathologies. Dietary guidelines of Ayurveda mainly focus on safeguarding the equilibrium of Tridosha (Vata Pitta Kapha) and hence emphasize the importance of Shadrasa (six tastes i.e. sweet, sour, salty, pungent, bitter and astringent.) in diet along with the dose, nature, processes and various other characteristics. The Rasa forms the basis of dietary and medicinal prescriptions in Ayurveda.

Seasonal and diurnal prescriptions of diets often is in line with the regulation of circadian rhythm and maintaining homeostasis. This concept of nature-derived food aligns to a certain extent with functional foods with bioactives.

Also, seers have denoted disease/anomaly—specific dietary recommendations which if abided can aid in the management of the respective pathology. Thus, similar annotations are observed in Ayurveda treatise and compendia pertaining to diabetes management and prevention of its complications.

Stress has been laid on the regular intake of Yava, also known as Hordeum vulgare. The intake can be through different preparations, one being Saktu, a preparation of roasted and dehusked barley powder. Another one is Mantha, a cooling preparation derived by mixing 4 pala, or 192 ml, of water and 1 pala, or 48 g, of coarse-ground barley. Another unusual preparation is Apoopa, which is whole grain flour snacks, like wheat, barley, and other millets, and is primarily recommended for diabetic patients. Another preparation is roasted and popped cereals, also known as Laja, which is also recommended for regular intake. Boiling water with the pith of khadira, scientifically known as Acacia catechu, and Darbha grass has been universally emphasized in different ancient texts on therapy. Another specially prepared energy drink is Panak, also known as Sharbat, and sour preparations known as raga or shadava, which are made of ingredients like Kapittha, also known as Feronia elephantanum, pepper, rai, and other spices (Chikitsasthana, Chapter 14, verse 13).

Charak Samhita has enlisted use of oils viz. sesame, Flax oil, ingudi oil (*Balanites aegyptiaca L.* seed), mustard oil, etc. Further, it is also stated to triturate the cereals with the medicinal decoction and can used in various dietary preparations.

Apart from this, similar preparations from millets namely Shyamak, *Echinochloa frumentace* Linn; Kodrava (*Paspalum scrobiculatum* Linn.); wild Kodo, Chanaka (Bengal gram), Kulatta (horse gram) have been recommended which are stored for a year are recommended.

Apart from this, inclusion of acorus calamus, holarrhena bark, cardamomum, pepper, mustard, premna herbacea, asafoetida, embelia ribes, zingiber, Cyperus rhizomes in dietary preparations of diabetic people. All bitters vegetables (Tikta shaka) also form major part of dietary recommendation. Honey is also a major component in dietary recommendations of obesity and diabetes management in classics.

Bhaishyajya ratnavali has given a comprehensive list of recommended foods which are similar to above mentioned list. Apart from that, it has also added gruel, soup preparations using barley and other millets along with old cereals. Further, vegetables include drumstick (*Moringa oelifera*), *Momordica dioica* Roxb. (Karkotaki), *Costus speciosus* (Koenig) Sm. (Kebuka), Karavellaka (*M. charantia*), Lashuna (*Allium sativum*), Kathillka (*B. diffusa*), Guduchi (*Tinospora cordifolia*), etc. further description of vegetables in various lexicons is also available. Fruits include Jambu (*Eugenia jambolna*), Kasheru (*Scirpus kysoor*), watermelon (kalinga), Kharjura (*Phoenix sylvestris*), rhizome of nelumbo, etc.

Dietary contra-indications in Diabetes have been also denoted by Vangasen, Vrinda madhav and *Bhaishyajya ratnavali*. Curd, Ash gourd, Starchy foods, confectionaries, sugarcane juice, coconut water, nishpav (flat beans), alcoholic beverages like sauvira (fermented wheat), shukta (vinegar like fermented product), excessive consumption of sour items, jaggery, lipids, garlic, salt, alkali rich food, etc. are contra-indicated for consumption.

10.4 Unlocking Natures Remedies: The Role of Bioactive Compounds in Functional Foods for Diabetic Neuropathy

10.4.1 Fenugreek (*Trigonella foenum-graecum*)

Fenugreek has been recorded for centuries in Ayurveda's divine scriptures, particularly in primary texts like the Charaka Samhita and the Sushruta Samhita. Ancient scriptures verify that it was used to cure numerous health conditions, including arthritis, gastrointestinal diseases, and Madhumeha or diabetes. In the past, fenugreek leaves were used to enhance lactation in breastfeeding mothers and inhibit inflammation in the body. However, fenugreek seeds were powdered and used to stabilize blood sugar levels in patients with metabolic disorders. Being a cooling and bitter herb in nature, it is a central ingredient in curing metabolic and gastrointestinal disorders, especially in patients with Pitta-dominant constitutions (Table 10.1) (Prasad et al. 2022) (Singh et al. 2022).

10.4.1.1 Bioactive Compounds and Mechanisms of Action

Fenugreek seeds contain a rich array of bioactive compounds that are responsible for their therapeutic properties (Fig. 10.3). Below are the five key bioactive compounds, each with its detailed mechanism of action:

(a) *Diosgenin*

After 8 weeks of a high-fat diet, male C57 mice received a 2-day intraperitoneal injection of 100 mg/kg streptozotocin (STZ). Diosgenin reduces oxidative stress in DPN, improving behavioral and morphological alterations. The Nrf2/HO-1 signaling pathway was implicated in its neuroprotective properties. Following diosgenin treatment, symptoms were greatly reduced. Histopathological micrographs of HE-stained sciatic nerves demonstrated improvement with diosgenin treatment. Diosgenin increased SOD and GPx activity while decreasing MDA levels. Nrf2, HO-1, and NQO1 expression were all increased by diosgenin (Fig. 10.8) (Prasad et al. 2022).

(b) *Trigonelline*

Trigonelline (Fig. 10.8), the main active element of Fenugreek, has been shown to protect rats from type 2 diabetes and diabetic peripheral neuropathy. Streptozotocin, given at a dose of 160 mg/kg intraperitoneally, was used to induce experimental diabetes in murine models. Research studies have shown that trigonelline exhibits protective effects against the development of type 2 diabetes and streptozotocin-induced peripheral neuropathy in rat models. Trigonelline acts via various mechanisms, including lowering blood glucose and lipid levels, increasing insulin sensitivity, correcting the micropathology of the sciatic nerve, and modulating the glucagon-like peptide-1 receptor/p38 mitogen-activated protein kinase signaling pathway.

Table 10.1 List of bioactive compounds of the functional foods, their effect, in-vitro, in-vivo models along with their case studies

Functional food	Bioactive compounds	Effect on diabetes and diabetic neuropathy	In-vitro model	In-vivo model	Case studies	Reference
1. Cinnamon (*Cinnamomi cassia C. zeylanicum*)	Cinnamic aldehyde	Anti-diabetic, neuroprotective	Inhibition of α-amylase and α-glucosidase activities in cell cultures (Tolmie et al. 2021)	TDMM animal model (C57BIKsj db/db)	A study found significant reductions in HbA1c and fasting glucose levels in patients consuming cinnamon daily for 8 weeks (Zelicha et al. 2024)	Tolmie et al. (2021), Xu et al. (2021), Kim et al. (2006), Niknejad et al. (2024), Banerjee et al. (2022), Zelicha et al. (2024)
	Cinnamic acid					
	Methylhydroxychalcone polymer (MHCP)					
	Tannin		In vitro studies in human neuroblastoma cells line (SH-SY5Y) (Xu et al. 2021)	Diabetic mice treated with cinnamon showed decreased fasting glucose and improved insulin sensitivity (Kim et al. 2006)		
	Linallol					
2. Bitter melon (*Momordica charantia*)	Karaviloside XI	Anti-diabetic Neuroprotective	Bitter melon extracts increased glucose uptake in L6 myotubes	HFD/STZ-induced type 2 diabetic rats	A cohort study demonstrated a 30% reduction in fasting glucose levels among type 2 diabetes patients consuming bitter melon extract	
	Momordicoside S					
	Momordicoside T			Diabetic rats treated with bitter melon showed reduced blood glucose and improved insulin sensitivity		
	Charantoside C					
	Polypeptide-p					
	Vicine					
4. Green tea (*Camellia sinensis*)	Polyphenols		EGCG promoted GLUT4 translocation in 3T3-L1 adipocytes	Mice fed green tea extract showed lower fasting glucose and reduced inflammation		Yu et al. (2020)

(continued)

Table 10.1 (continued)

Functional food	Bioactive compounds	Effect on diabetes and diabetic neuropathy	In-vitro model	In-vivo model	Case studies	Reference
5. Curcumin (*Curcuma longa*)	Cathechin	Nerve regeneration	Curcumin inhibited NF-κB activation in cultured macrophages	STZ induced male SD rats	A clinical trial showed improved inflammatory markers and glucose metabolism in diabetic patients supplemented with curcumin	Orellana-Paucar (2024), Lin et al. (2024)
	Epigalllocatechin					
	Quercetin					
	3–6% essential oil β-turmerone 14.7–18 α-turmerone 30.1–32 Curcumol Curdione					
6. Grapes (*Vitis vinifera*)	Resveratrol	Anti-diabetic	Resveratrol enhanced glucose uptake in skeletal muscle cells.	Diabetic rats showed improved insulin sensitivity and reduced fasting blood glucose after grape polyphenol supplementation (Wang et al. 2021).		Wang et al. (2021)
	Quercetin					
	Anthocyanins					
7. Berries (Blueberry, raspberry)	Anthocyanins	Improved insulin sensitivity	Anthocyanin-rich extracts improved insulin sensitivity in adipocytes	Rats supplemented with berry extracts showed reduced oxidative stress and improved insulin sensitivity (Zhang et al. 2011).		Zhang et al. (2011)
	Polyphenols					
	Cyanidin					
	Delphinidin					

8. Neem	Diterpenek (sugiol and nimbiol)	Anti-diabetic Neuroprotective	Nimbin and salannin has Neuroprotective effect by preventing tau aggregation and increasing cell viabilityon HEK293T cells (Sandhir et al. 2021)	Neuropathic pain induced by chronic constriction injury to sciatic nerve of Wistar rat (Biswas et al. 2022).		El-Beltagy et al. (2020), Rekha et al. (2022), Sandhir et al. (2021), Biswas et al. (2022)
	Triterpenes (beta-sitosterol and stigmasterol)					
	Terta-terpenoid (azadirachtin-k)					
	Sulphur containing compounds (nimbin, nimbidin, nimbinin and nimbidol)					
	Limonoids (maliantriol and nimbendiol)					
	Flavonol glycosides (nimaton, quercetin, myricetin)					
	Kaempferol					
9. Amla (*Emblica officinalis*)	Isostrictiniin		Amla extracts reduced ROS levels in cultured β-cells (Majeed et al. 2022)	Arsenic Induced metabolic disorder in mice Streptozotocin-induced Diabetes mellitus in male rats Diabetic rats treated with Amla extract showed significant reductions in fasting glucose and HbA1c	A randomized trial found that 500 mg of Amla extract daily for 12 weeks significantly lowered fasting blood glucose and HbA1c in type 2 diabetes patients	Majeed et al. (2022)
	Apigenin,					
	Chebulinic acid					
	Quercetin					
	Gallic acid					
	Chebulagic acid					
	Emlicanin A, B					
	Phyllaemblicin					
	Emblicanin B					
	Punigluconine					
	Vit. C					
	Quercetin and b-sitosterol					

(continued)

Table 10.1 (continued)

Functional food	Bioactive compounds	Effect on diabetes and diabetic neuropathy	In-vitro model	In-vivo model	Case studies	Reference
10. Omega-3 fatty acids (fish oil, flaxseed)	EPA, DHA	Neuroprotective	EPA and DHA reduced inflammatory cytokines by inhibiting NLRP3 inflammasome (Lamantia et al. 2024).	Diabetic rats showed reduced neuropathic pain (Durán et al. 2024)	A meta-analysis demonstrated that omega-3 supplementation significantly improved metabolic parameters in diabetic patients (Khalili et al. 2021).	Lamantia et al. (2024), Yorek (2018)
11. Aloe vera (*Aloe barbadensis*)	Carbohydrates (galactogalacturan)	Antidiabetic	An endophytic isolate Sch1 (isolated from *Aloe vera*), identified to be *Schizophyllum commune* Fr. on molecular basis, exhibiting more than 90% α-glucosidase inhibitiory activity (Sharma et al. 2021a, b)	Intraperitoneal injection of streptozotocin (STZ) at a dose of 40 mg/kg body weight on wistar rats showed restoration of insulin levels (Babu et al. 2021).		Babu et al. (2021), Sharma et al. (2021a, b)
	Pedtides & polypeptides (lectins)					
	Vitamins & Minerals Calcium, copper					
	Antrhaquinones (aloe-emodin)					
	Saponins					
	Amino acids					

12. Garlic (*Allium sativum* L.)	Diallyl disulfide Diallyl trisulfide Allyl propyl sulfide Thiosulfinatealliin S-allyl-cysteine Diallyl polysulfides Vinyldithins Alliinase Manganese, vitamin B6, vitamin C, selenium, fiber	Improved hepatic glucose metabolism		Kunming mice were fed with GP (1.25, 2.5, 5.0 g/kg·bw) for 5 weeks after insulin resistance (IR) engendering by a high-oil-high-sugar diet and (STZ) showed improvement with treatment of Garlic (Xie et al. 2023a, b).	A cohort study involving 768 participants with type 2 diabetes revealed a decrease in fructosamine and glycosylated hemoglobin (Yoneda et al. 2022).	Xie et al. (2023a, b), Ajanaku et al. (2022), Yoneda et al. (2022)
13. Tulsi (*Ocimum sanctum*)	Vicenin 2 Luteolin-7-O-glucuronide Luteolin Caffeic acid	Antidiabetic	The in vitro α-glucosidase inhibitory	Without causing hypo-glycamic seizures, a single dose of O. sanctum extended the effects of glimepiride from the fourth to the sixth hour in STZ induced diabetic rats.		Chaudhary et al. (2020), Singh et al. (2023), Mousavi et al. (2020)
14. Sweet potato (*Ipomoea batatas* L.)	Carbohydrate Dietary fiber Protein Vitamins Lipids Minerals Flavonoids, carotenoids	Improves insulin sensitivity	In vitro using isolated hepatocytes decreases glucose production (Shyur et al. 2021)	Sweet potatoes ↑ skeletal muscle insulin sensitivity in normal rats, ↓ postprandial blood glucose levels.		Kinoshita et al. (2023), Shyur et al. (2021)

(continued)

Table 10.1 (continued)

Functional food	Bioactive compounds	Effect on diabetes and diabetic neuropathy	In-vitro model	In-vivo model	Case studies	Reference
15. Sugar beet (*Beta vulgaris* L.)	Vitamins	Anti-diabetic	Compared to α-amylase, beta vulgaris extract had a greater inhibitory impact on the α-glucosidase enzyme.	Hypoglycaemic and hypolipidemic effects of sugar beet extract on C57BL/KsJ db/db mice (Wu et al. 2022).		Wu et al. (2022), Üst et al. (2024)
	Minerals					
	Phenolics					
	Carotenoids					
	Nitrate					
	Ascorbic acids					
	Betalains					
	Epicatechin					
	Rutin					
16. Gudmar (*Gymnema sylvestre*)	Gymnemasaponins (Gymnemagenin and Gymnestrogenin)	Anti-diabetic	The methanol crude extract demonstrated encouraging inhibitory qualities against the enzymes α-amylase and α-glucosidase in vitro and in silico.	It has been demonstrated that Gymnema sylvestre ethanol and dichloromethane extracts significantly reduce blood sugar levels in STZ-indued type 1 diabetic rats.	The mean HbA1c values for 22 gymnema patients decreased from 9% to 8.2% after 16–18 months, and the average insulin dosage decreased from 60 units to 30 units.	Devi et al. (2024), Ditchou et al. (2024), Kumar et al. (2018)
	Triterpenoid saponins					
	Gymnemic acid A1					
	Steroids and saponins					
	Polypeptide (Gudmarin)					

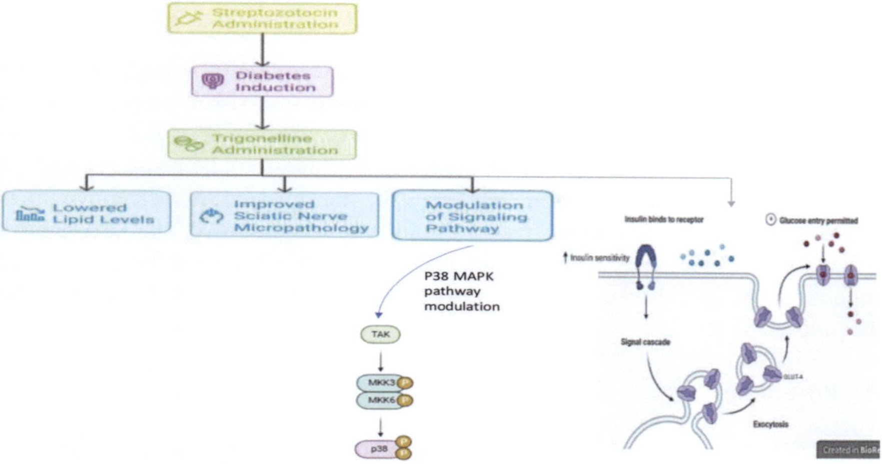

Fig. 10.3 Mechanistic pathways of *Trigonelline*

(c) *4-Hydroxyisoleucine*

This specific amino acid derivative plays a crucial role in the stimulation of insulin release from the pancreatic β-cells, particularly under hyperglycaemic conditions of elevated blood glucose levels. In addition, it potentiates the stimulation of the AMPK signalling pathway, resulting in enhanced glucose uptake in skeletal muscle cells and enhanced glycogen synthesis. The drug 4-hydroxyisoleucine (HI) has been reported to be a potent insulin mimetic and is characterized by its high degree of neuroprotective activity, which is part of the overall therapeutic action of the drug. Additionally, 4-HI has been reported to exhibit antioxidant activity, along with anti-apoptotic and anti-inflammatory activity, mainly by activating IGF-1 and GLP-1 signalling pathways.

(d) *Galactomannans*

This group's soluble dietary fibre is important for slowing gastric emptying and controlling carbohydrate absorption in the gastrointestinal tract. Such a mechanism aids in more effectively regulating postprandial blood glucose levels, a process known as postprandial blood sugar management. Galactomannans are also very interesting as prebiotics since they directly regulate and modulate the gut microbiota while also playing an important function in reducing systemic inflammation within the organism (Fig. 10.8).

(e) *Vitexin*

Vitexin, a flavonoid glycoside, increases insulin secretion by stimulating GLP-1 receptors and decreases oxidative stress via its strong antioxidant properties. It has been shown to inhibit α-glucosidase, delaying the breakdown of carbohydrates into glucose. Vixetin modulates glutamate transporters, cholesterol homeostasis, and ER stress. GABA synaptic inhibition suppresses the transmission of nociceptive information, including neuropathic pain. Vixetin extracts improve rats' explorative behaviour and reaction to thermal pain, restore oxidant/antioxidant balance, reduce inflammation, increase brain monoamine levels, and modulate MAO-A and Ache enzyme activity.

(f) *Protodioscin*

Another steroidal saponin included in fenugreek, protodioscin, stimulates the PI3K/Akt signaling cascade to improve glucose transport. It increases phosphorylation of the insulin receptor substrate, which increases insulin-mediated absorption of glucose in peripheral tissues. It also affects lipid metabolism by reducing serum levels of LDL cholesterol and triglycerides, both of which regulate metabolic syndrome (Zhao et al. 2021a, b). Protodioscin inhibited hyperglycemia-induced neurotoxicity in rat cortical primary neurons by reducing reactive oxygen species. Because it protects neurons isolated from the central nervous system, protodioscin can be used in the prevention and treatment of neurodegenerative diseases in diabetics.

(g) *Kaempferol*

By causing AMPK activation, this flavonoid promotes skeletal muscle's absorption of glucose and oxidation of fat. Additionally, kaempferol prevents the liver from producing gluconeogenesis, which lowers blood sugar. Kaempferol supplementation improves Streptozotocin-induced diabetic neuropathy by reducing oxidative stress and pro-inflammatory cytokine production, which may contribute to nerve damage caused by diabetes.

10.4.2 Bitter Melon (*Momordica charantia*)

Bitter melon, also known as Karela, is a basic medicinal plant that is very important in the management of diabetes in Ayurveda. This intriguing plant is a member of the Cucurbitaceae family, which is a fascinating and complex family that includes roughly 45 different species of the genus Momordica. Momordica charantia is the most widely utilized of these species for a variety of medicinal purposes, and it is extremely important in traditional medicine. Bitter melon is indigenous to tropical and subtropical areas; nonetheless, its most widespread cultivation is located in India, Africa, and Southeast Asia. Bitter melon is cultivated extensively in India, with particular frequency recorded in West Bengal, Karnataka, and Maharashtra states. Ayurvedic texts often describe bitter melon as a powerful Tikta (bitter) herb with Kapha and Pitta pacifying properties, primarily used for managing Madhumeha (diabetes) and promoting liver health (Borah et al. 2023). It is highly regarded for its ability to cleanse the blood and enhance digestion, often consumed as part of daily diets in India to maintain blood sugar levels and improve skin health (Table 10.1).

10.4.2.1 Bioactive Compounds and Mechanisms of Action

Bitter melon contains a variety of bioactive compounds known for their potent hypoglycemic activity. Below are the key bioactive compounds:

Charantin

Charantin is a steroidal saponin that activates the AMPK pathway, enhancing glucose uptake by muscle cells and improving insulin sensitivity. It reduces hepatic gluconeogenesis and promotes glycogen storage in the liver, leading to lower blood glucose levels (Saeed et al. 2021). MC's pain-relieving impact in neuropathic pain is thought to be due to its PPAR-gamma agonistic activity, anti-inflammatory capabilities, and antioxidant capacity (Biswas et al. 2022).

The neem tree's seeds, bark, and leaves are utilized in herbal medicine for daily use to treat a wide range of diseases, from digestive conditions to skin disorders, such as diabetic neuropathy.

10.4.3 Neem (*Azaradicta indica*)

10.4.3.1 Bioactive Compounds and their Mechanism of Action

(a) *Azadirachtin*

Azadirachtin lowers the amount of glucose produced in liver cells by suppressing the expression of glucose-6-phosphatase, an essential enzyme in gluconeogenesis (Rahmani et al. 2018). Azadirachtin has been shown in diabetic rats to significantly reduce blood glucose levels and boost insulin sensitivity by enhancing GLUT4 translocation to the cell membrane (Vidhya Rekha et al. 2022). Reduces hepatic glucose output by inhibiting gluconeogenic enzymes. The oral treatment of Azaradictin to STZ-induced diabetic rats provided neuroprotection against degenerative oxidative stress associated with metabolic and histological abnormalities (Biswas et al. 2022).

(b) *Nimbidin*

Nimbidin demonstrates its anti-inflammatory properties by inhibiting the production of pro-inflammatory cytokines such as IL-6 and tumour necrosis factor in macrophages exposed to high glucose levels (Sandhir et al. 2021). Nimbidin's neuroprotective properties have also been noted in animal models. Allodynia, hyperalgesia, motor coordination, and motor nerve conduction velocity were significantly reduced in animal models of peripheral neuropathy caused by partial sciatic nerve ligation, but these behavioral changes were significantly lessened after long-term administration of this extract (Rahmani et al. 2018).

A number of studies have demonstrated that nimbidin dramatically reduces oxidative stress and inflammation in the pancreatic tissue of diabetic mice, which in turn raises the animals' glucose tolerance level (Deshmukh et al. 2024). Nimbidin is responsible for the rise in insulin sensitivity, which in turn causes these diabetic patients to have higher glucose tolerance by reducing inflammation in the body (Fig. 10.4; Table 10.1).

(c) *Nimbin*

Nimbin improves glucose uptake in adipocytes via promoting insulin receptor expression, which leads to improved glucose metabolism (Ramteke et al. 2024).

Nimbin supplementation has been associated with better glycemic control and lower blood cholesterol levels in diabetic animals, both of which improve overall metabolic health (Ismail et al. 2024). Increases insulin receptor expression, facilitating glucose uptake in peripheral tissues.

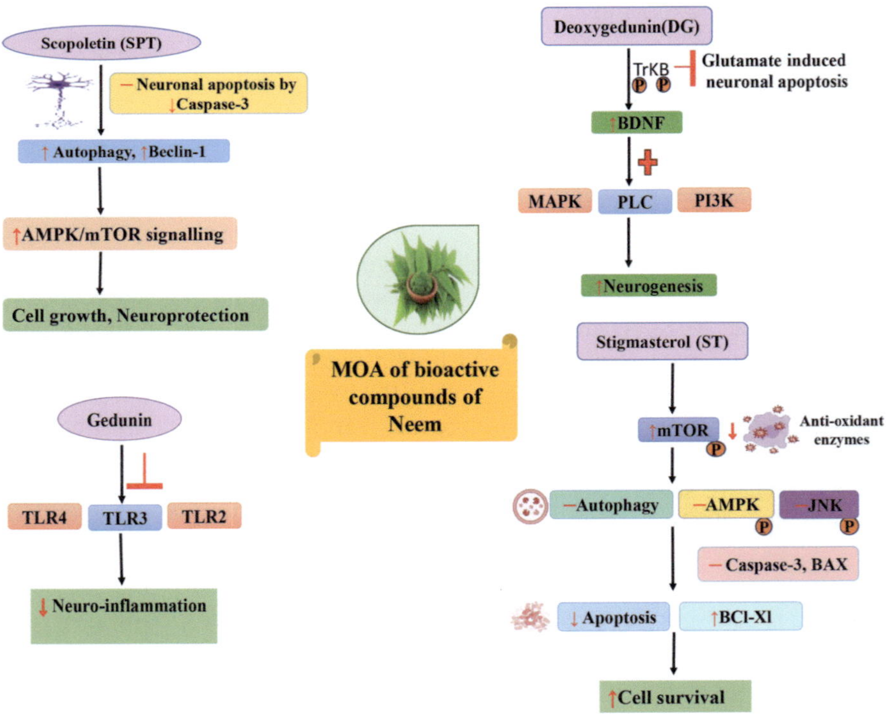

Fig. 10.4 Mechanism of chief bioactive compounds of fenugreek

(d) *Quercetin*

Quercetin, another flavonoid that has drawn a lot of interest from researchers, is widely known for having strong anti-inflammatory and antioxidant properties. This bioactive substance can lower pro-inflammatory indicators like inflammatory cytokines IL-6 and TNF- α and control the expression of genes regulated by oxidative stress, including the important Nrf2 gene. Quercetin helps people with Type 1 and Type 2 diabetes because, in addition to promoting the release of insulin, it has been demonstrated to lessen oxidative damage in pancreatic cells, which enhances pancreatic function. In a rat model of peripheral diabetic neuropathy, it has also been discovered that inflammation is reduced more precisely by controlling the intricate TLR4/MyD88/NF-κB signalling pathway (Zhao et al. 2021a, b), which is crucial during inflammatory reactions. It has been discovered that electrospun quercetin-loaded nanofibers can repair focal entrapment neuropathy in a streptozotocin-induced diabetic rat model. The investigations' findings showed that diabetic rats with sciatic nerve injury had significantly better oxidative stress state, especially those given low and intermediate dosages of zein-based nanofibers loaded with quercetin (Fig. 10.8) (Khursheed et al. 2021).

Quercetin improves glycaemic control by inhibiting SGLT1 and SGLT2 in the kidneys and intestines, respectively. This lowers glucose reabsorption (Sok Yen et al. 2021). By lowering oxidative damage in neurons, it also exhibits neuroprotective benefits. In order to prevent diabetic neuropathy, it also by reduces inflammation and protects against oxidative stress by modulating pathways including AMPK and NF-κB. By altering insulin signaling pathways, quercetin improves insulin activity and encourages muscle cells to absorb glucose (Fig. 10.5) (Shi et al. 2019).

10.4.4 Cinnamon (*Cinnamomum verum*)

Ayurvedic scriptures have often mentioned cinnamon, or Dalchini, and its many medical uses. It has historically been employed to harmonize the Vata and Kapha doshas. It is described as a treatment for cough and cold in Charaka Samhita and as a therapy for respiratory and digestive problems in Sushruta Samhita. It has been

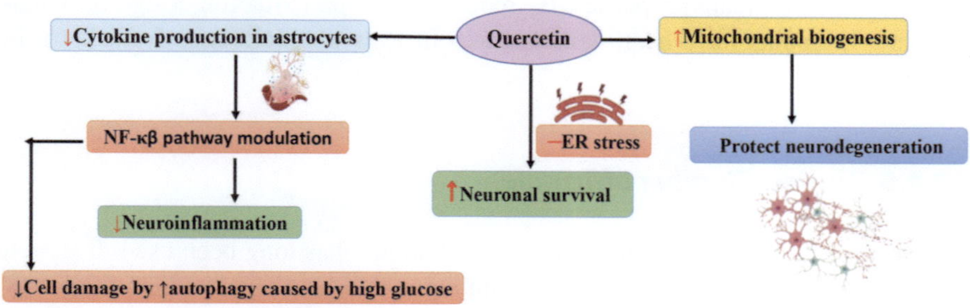

Fig. 10.5 Pathway of quercetin in the reduction in cell damage

used for ages to treat diabetes because of its hypoglycemics and antioxidant qualities (Table 10.1) (Semwal et al. 2023).

10.4.4.1 Bioactive Compounds and Mechanism of Action

(a) *Cinnamic Acid*

Inhibits the enzymes glucosidase and amylase to control postprandial glucose boosts (Niknejad et al. 2024). improves insulin sensitivity and decreases fasting blood sugar in diabetic mice. Controls hyperglycaemia by blocking the enzymes that break down carbs. In rats, oral administration of cinnamic acid (10, 20, and 40 mg/kg) reduced oxaliplatin (6 mg/kg)-induced cold (acetone test) and mechanical (von Frey test) allodynia. It significantly reduced oxaliplatin-induced hyperexcitation of wide dynamic range neurons in the spinal cord that respond to mechanical (brush, push, and pinch) and cold (acetone drop) stimuli on the hind paw (Lee et al. 2021).

(b) *Proanthocyanidins*

Proanthocyanidins activate the PI3K/Akt pathway, which lowers oxidative stress and improves insulin sensitivity. They decrease blood sugar levels of patients who takes this substance after meal while inhibiting pancreatic enzymes including α-amylase and α-glucosidase. Reduce free radical damage pancreatic-β cells. In diabetes models, it increases beta-cell function and reduces the oxidation of lipids (Rajasekhar et al. 2023). Proanthocyanidin B2 has been shown to cure high glucose-induced neuropathy in posterior spinal root ganglion neurons by activating the PI3K/Akt signalling pathway, which is necessary for cell survival and proliferation (Pan et al. 2018).

(c) *Tannins*

Tannins prevent diabetes problems by slowing the absorption of glucose, inhibiting digestive enzymes, and reducing inflammation (Kabir et al. 2021). Tannins decrease total glucose absorption by blocking glucose transporters. Promotes the breakdown of glucose by inhibiting DPP-IV (Kamiloglu et al. 2024). Tannins help reduce oxidative stress, a significant factor in the pathophysiology of DN, due to their strong antioxidant properties. By reducing lipid peroxidation, a sign of oxidative stress, tannins have been demonstrated in tests to shield cells from harm in diabetics.

10.4.5 Zinger (*Zingiber officinale*)

In Ayurveda, ginger—also called Adrak or Shunthi—has long been used. It is well recognized to support digestion, reduce inflammation, and regulate blood sugar by balancing Vata and Kapha. The Charaka Samhita mentions ginger's therapeutic qualities and suggests use for its anti-diabetic and digestive effects (Table 10.1).

10.4.5.1 Bioactive Compounds and Mechanism of Action

(a) *Gingerol*

Promotes glucose uptake by upregulating GLUT expression. Lowers blood glucose and promotes insulin sensitivity in diabetic rats. Promotes GLUT to increase glucose absorption. According to Shen et al. (2024), giving gingerol to rats with diabetic neuropathy reduced pain, anxiety, and depression as well as neuroimmune cells while increasing gut microbiota diversity and mitochondrial efficiency.

(b) *Shagol*

Boosts the release of insulin via lowering oxidative damage. Greatly enhances diabetic mice's glucose control (Fajrin et al. 2020). For 3 weeks, 6-shogaol 15 mg/kg BW daily dramatically decreased hyperalgesia and allodynia in a mouse model of peripheral diabetic neuropathy produced by STZ disease. Lower spinal cord expressions of TRPV1 and NMDAR2B mediate this activity, which has little bearing on pancreatic islet function. Hyperalgesia may result from damage to unmyelinated C fibres in diabetic mice.

(c) *Zingiberene*

Antioxidants protect beta cells from damage caused by oxidation. Insulin resistance is reduced in diabetic rat models (Borgonetti et al. 2023). In a mouse model of peripheral diabetic neuropathy (PDN) produced by STZ, 6-shogaol 15 mg/kg BW for 3 weeks each day significantly decreased allodynia and hyperalgesia. Reduced spinal cord expressions of TRPV1 and NMDAR2B mediate this action, which has no impact on pancreatic islet function. Damage to unmyelinated C fibres, which generate hyperalgesia in diabetic mice, may be the aetiology of PDN (Fajrin et al. 2020).

10.4.6 Grapes (*Vitis vinifera*)

The Rasayana (rejuvenating and nourishing) qualities of grapes (Draksha) are well-known in Ayurveda. Grapes, in addition to being high in nourishment and water, are used to balance the Pitta and Vata doshas. They have traditionally been recommended for longevity, better digestion, and detoxification. Grapes have long been used in Ayurvedic recipes to treat tiredness, excessive thirst, liver problems, and respiratory disorders.

Ayurvedic formulations frequently use the dried form, Munakka, highlighting its Snigdha (unctuous) and Madhura (sweet) qualities, which help manage Prameha (diabetes) by enhancing metabolism and controlling blood sugar levels (Gupta et al. 2020). Because of their antioxidant and restorative qualities, grapes are commonly used in preparations like Chyawanprash and Drakshasava. These remedies complement Ayurveda's all-encompassing approach to diabetes management by promoting immune system function, metabolic balance, and healthy aging (Table 10.1).

10.4.6.1 Bioactive Compounds and Mechanisms of Action

Grapes are rich in polyphenols and other bioactive compounds with promising antidiabetic and neuroprotective properties. Key bioactive compounds include:

(a) *Resveratrol*

Resveratrol increases insulin responsiveness and glucose metabolism by engaging the SIRT1 and AMPK pathways (Fig. 10.8). Up until the 12th week following the STZ injection, the resveratrol group received 10 ml/kg of 10% resveratrol intragastrially once daily. The study on the effects of resveratrol on diabetic peripheral neuropathy (DPN) produced the following results: (i) treatment with resveratrol led to a decrease in pain sensitivity and temperature response in diabetic mice; (ii) there was an increase in Nrf2 expression within the peripheral nerves of diabetic mice treated with resveratrol, and blocking the NF-kB pathway contributed to nerve protection during resveratrol treatment for neuropathy; and (iii) resveratrol altered the inflammatory state of peripheral nerves by enhancing Nrf2 activation and p-p65 levels (Zhang et al. 2021; Farkhondeh et al. 2020).

(b) *Anthocyanins*

Anthocyanins have been shown to improve GLUT4 expression and function, which in turn promotes the utilization of glucose in adipocytes. By triggering the PI3K/Akt pathway, anthocyanins promote the absorption of glucose by adipocytes (Les et al. 2021). Anthocyanin-rich extracts enhanced insulin sensitivity and decreased oxidative stress markers in diabetic rats, improving glucose tolerance and lowering diabetes consequences, such as neuropathy. By altering PPARγ and improving adipocyte glucose absorption, anthocyanins lower insulin resistance. Additionally, they prevent against inflammation and oxidative stress, two major causes of diabetic neuropathy. Studies conducted in vivo revealed that anthocyanin-treated diabetic rats had better glucose tolerance and fewer oxidative stress indicators (Zhu et al. 2023).

10.4.7 Berries (*Vaccinium* Spp)

Despite not being considered traditional Ayurvedic herbs, berries like blueberries and cranberries are linked to qualities like those of other Rasayana plants, which include promoting longevity, balancing digestion, and enhancing metabolic health. Because of their Madhura (sweet) and Kashaya (astringent) qualities, which balance Pitta and Kapha doshas, these fruits are in line with Ayurvedic principles (Tian et al. 2022). Berries and other fruits with antioxidant qualities are suggested by Ayurveda to lower inflammation, control blood sugar, and support heart health. Berries can be used in place of other fruits in Ayurvedic formulations to aid in detoxification and digestive health, which will help control diabetes and its related issues, such as neuropathy (Table 10.1).

10.4.7.1 Bioactive Compounds and Mechanisms of Action

Anthocyanins, flavonoids, polyphenols, and other bioactive substances found in berries have been demonstrated to have neuroprotective and anti-diabetic properties. These are the main bioactive compounds (Gowd et al. 2017).

(a) *Pterostillbene*

Due to its strong antioxidant properties, pterostilbene is essential for lowering oxidative stress, a primary cause of diabetic neuropathy. By activating the Nrf2 pathway, it increases the expression of antioxidant enzymes such as glutathione S-transferase (GST) and haem oxygenase-1 (HO-1), protecting neuronal cells from oxidative damage (Dodda et al. 2020). By suppressing NF-κB and activating Nrf2, it also lowers oxidative stress, which helps prevent diabetes complications, especially neuropathy (Tian et al. 2022). Pterostilbene has been shown to enhance insulin secretion in pancreatic beta cells and improve glucose uptake in muscle cells (Zhao et al. 2021a, b). Pterostilbene treatment improved glucose homeostasis, decreased oxidative stress, increased insulin sensitivity, and prevented diabetic neuropathy in diabetic rats (Patil et al. 2023). In sciatic nerves, proteostilbane was similarly successful in treating peripheral neuropathy by modifying neurotransmitters and oxidative stress (Fig. 10.6) (Amin et al. 2022).

(b) *Ellagic Acid*

By altering the Nrf2 and NF-κB pathways, ellagic acid, a strong antioxidant, lowers inflammation and oxidative stress. It is advantageous for controlling diabetes and preventing associated consequences, such as neuropathy, because it enhances insulin sensitivity and glucose metabolism (Ahmad et al. 2022a, b). Comparing EA-treated rats to untreated rats, the former displayed a damaged blood vessel, neuronal hydropic degenerative changes, and disordered fibrillary degenerative changes, along with a reduced microhaemorrhage focus. In alloxan-induced diabetic mice, a 2-week course of EA (10 mg/kg/day) improved peripheral nerve function (and the resulting antinociceptive activity) and reduced brain levels of myeloperoxidase and several caspases (Fig. 10.7) (Amor et al. 2020).

(c) *Kaempferol and Myricetin*

Myricetin and kaempferol are examples of flavanol's that lower blood glucose levels by blocking the enzymes α-amylase and α-glucosidase, which are essential

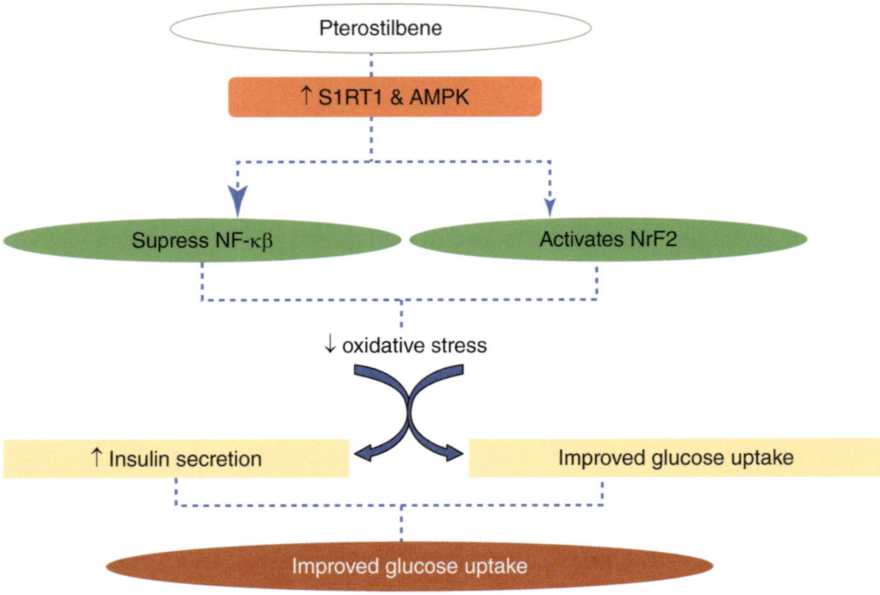

Fig. 10.6 Mechanism of pterostilbene

Fig. 10.7 Role of ellagic acid

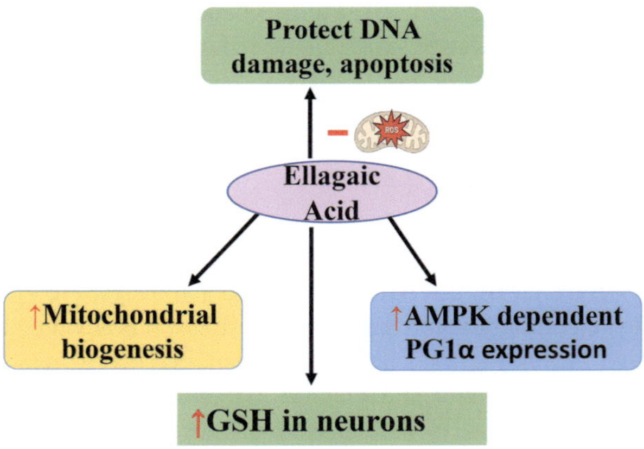

for the digestion of carbohydrates. They also protect neurons from damage caused by hyperglycaemia by improving insulin sensitivity and lowering oxidative stress (Altındağ et al. 2021; Bondonno et al. 2021).

Flavanol's prevented intestinal cells' α-glucosidase from absorbing glucose. Myricetin and kaempferol-treated diabetic rats demonstrated enhanced insulin sensitivity, decreased postprandial glucose levels, and neuroprotective benefits against diabetic neuropathy (Ma et al. 2022). Increases insulin receptor sensitivity by activating PI3K/AKT signaling. Lowers inflammation and insulin resistance in experiments on diabetic animals (Rao et al. 2021). Increase sensitivity by stimulating insulin signaling pathway.

10.4.8 Giloy (*Tinospora cardifolia*)

In Ayurveda, Giloy, also called guduchi, is valued for its potent Rasayana (revitalizing) qualities. In addition to improving vitality and immune function, it is frequently used to treat a variety of illnesses, including as fever, diabetes (Madhumeha), and inflammation. Because giloy is said to balance the three doshas—Pitta, Kapha, and Vata—it can be used to treat a variety of metabolic conditions, including diabetes. Because of its revitalizing and life-sustaining properties, it is referred to in Sanskrit writings as an Amrita (nectar of immortality) (Gupta et al. 2024). Guduchi Satva and other Ayurvedic preparations are frequently recommended for the treatment of diabetes and its complications (Table 10.1).

10.4.8.1 Bioactive Compounds and Mechanisms of Action

Alkaloids, diterpenoid lactones, glycosides, steroids, and polysaccharides are among the many bioactive compounds found in giloy. These substances exhibit excellent neuroprotective, antioxidant, and anti-diabetic properties (Balkrishna et al. 2023).

(a) *Tinosporadide*

By improving sensitivity to insulin, tinosporaside regulates glucose metabolism. It affects the PI3K/Akt pathway, which controls peripheral tissues' intake of glucose. Tinosporaside increases the insulin signaling pathway, by inhibiting α-Amylase & α-Glucosidase which improves muscle cell utilization of glucose. Tinosporaside enhanced insulin sensitivity and decreased blood glucose levels in diabetic rats (Sharma et al. 2021a, b). By blocking soluble epoxide hydrolase, T. cordifolia stem may mitigate neuropathic pain (Joshi et al. 2024). The substance increases insulin secretion via modifying the activity of pancreatic beta cells and their defence against oxidative stress. Cordifolioside A causes beta cells to secrete insulin in response to hyperglycaemia (Mandar et al. 2021). Cordifolioside A-treated diabetic rats showed better beta cell function and lower blood glucose levels (Table 10.1) (Nandan et al. 2023).

(b) *Berberine*

Berberine increases insulin sensitivity and decreases hepatic gluconeogenesis (Xu et al. 2020). via activating AMPK. Additionally, it protects against diabetic neuropathy through its anti-inflammatory properties. Berberine inhibits gluconeogenesis, which lowers glucose production in hepatocytes. Berberine reduced blood sugar levels and protected diabetic Zucker fatty rats by improving the gut microbiota and increasing glucagon-like peptide-2 secretion in the intestine (Wang et al. 2021). Higher berberine dosages (20 and 40 mg/kg) markedly increased paw withdraw latency and progressively enhanced thermal nociception. At 40 mg/kg dose, berberine treatment profoundly suppressed the formation of MDA and ROS and, at the same time, elevated catalase activity. Further, it decreased the STZ-induced release of TNF-α and IL-6 profoundly, indicating that berberine decreases the

STZ-induced neuroinflammation in rats. Both MOR protein and mRNA expression were markedly restored by berberine in the spinal cord and DRG of STZ-treated rats, and these levels were noticeably greater than those of the control rats (Table 10.1) (Dong et al. 2019)

(c) *Magnoflorine*

Magnoflorine reduces inflammation by inhibiting the NF-κB pathway. It also helps mitigate oxidative stress, a condition that contributes to diabetic neuropathy. In cultured neurons, magnesium reduces oxidative stress and inflammation, perhaps preventing diabetic neuropathy (Joshi et al. 2024).

Enhanced Akt phosphorylation and inhibited autophagy and proteolysis, which may be connected to magnoflorine ability to reduce blood sugar (Table 10.1) (Khanal et al. 2019).

10.4.9 Punarnava (*Boerhavia diffusa*)

It has been reported that *B. diffusa* rotenoids have neuroprotective properties. Boeravinone B inhibited neuronal damage by lowering cerebral water content, antioxidant enzymes (e.g., GSH, GPx, CAT, MDA, glutathione reductase), inflammatory cytokines (e.g., IL-4, IL-1, TNF-α, IL-10, IL-6, and IL-1β), and inflammatory mediators (e.g., PGE2, NF-κB, and COX-2). In obese type 2 diabetic mice, syringaresinol mono-β-D-glucoside (Eleuthroside E) enhanced insulin sensitivity. It enhanced hepatic glucose metabolism by increasing glycolysis and decreasing gluconeogenesis, while protecting pancreatic α and β cells from diabetic damage (Table 10.1) (Zhai and Wang 2018).

10.4.10 Curcumin (*Curcuma longa*)

One of the oldest medicinal plants in Indian medicine, curcumin is extracted from the rhizome of Curcuma longa L. and has been used for thousands of years to cure a wide range of ailments (Zhang et al. 2021). One of the many therapeutic benefits of curcumin is its antibacterial qualities (Park et al. 2012). anti-inflammatory, anticancer, and antioxidant. Curcumin and its derivatives possess potent antioxidant qualities due to the presence of O-methoxy phenolic groups. Curcumin alters the expression of enzymes, cytokines, and transcription factors that are necessary for cell survival. Curcumin contains catechin, Epigallocatechin, Quercetin, 3–6% essential oil, Quercetin, β-turmerone 14.7–18%, Curcumol, Curdione, α-turmerone. There is mounting evidence that the prevention of diabetic peripheral neuropathy (DPN) is largely dependent on boosting nerve growth factor (NGF) production in sciatic nerves (Lin et al. 2024). and reducing the apoptosis of Schwann cells (SCs). One of turmeric's main bioactive ingredients, curcumin, has several properties and has been demonstrated to have numerous therapeutic benefits for a range of illnesses. Curcumin decreased the apoptosis of SCs in sciatic nerves at doses of 50, 100, and 150 mg/kg. Furthermore, curcumin at 150 mg/kg was most effective at raising serum NGF levels and sciatic nerve NGF protein expression. In vitro, curcumin may prevent apoptosis and prevent Schwann cells damaged by elevated glucose levels. Curcumin may be used therapeutically to improve DPN by blocking Schwann cell death through Hmox1. Curcumin (CUR), one such supplement can help prevent underlying low-grade inflammation in diabetic neuropathy. The findings were further supported by the mechanistic investigations, which revealed that double-headed nano systems containing curcumin (nCUR) or nCUR+INS (long-acting subcutaneous insulin) preserved nestin and NF200 expression in the sciatic nerve while significantly reducing TUNEL positive cells, NLRP3 mRNA expression, IL-1β, and macrophage infiltration (Table 10.1).

10.4.11 Amla (*Emblica officinalis*)

Ayurveda, an indigenous medical system in India, holds *Emblica officinalis* (EO) in high regard. It is considered the first tree in the universe, according to folklore from ancient India. The Euphorbiaceae family includes it. Indian gooseberry, *Phyllanthus Emblica*, and amla are some additional names for it (Dawane et al. 2016). Gallic, ellagic, chebulinic, chebulagic, emblicanin A, emblicanin B, punigluconin, pedunculagin, citric acid, ellagotannin, trigallaylglucose, pectin, and isostrictiniin are also present in the fruit. Flavonoids including quercetin and kaempferol are also present. β-carotene, calcium, arginine, ascorbic acid, aspartic acid, boron, carbohydrates, chembulagic acid, chibulinic acid, chembulaginic acid, and chembic acid Chloride Cysteine, copper, corilagin, corilagic acid, d-fructose, d-glucose, glucogallin, glucose, glutamic acid, gallic acid, gallic acid ethyl ester, ethyl gallate (syn. Phyllemblin), emblicanin-A, B, emblicol, Isoleucine, Leucine, Glycine, Histidine, Iron, Lysine, and Malic Acid. Phyllemblinic acid, manganese, magnesium, methionine, polysaccharide, potassium, proline, protein, quercetin, ribofavin, rutin, selenium, serine, silica, sodium, starch, sucrose, sulfur, tannin, terchebin, thiamin, threonine, trigalloyl glucose, tryptophan, phyllantidine, and phyllantine (Baset et al. 2020). Since ancient times, amla fruits have been used to treat sickness. Research based on in vitro and in vivo methods have demonstrated the role of amla in preventing illness. It is a strong scavenger of free radicals on cytokine, oxidative stress, and nociceptive levels. By lowering the oxidative stress-mediated release of proinflammatory cytokines, which cause diabetes-induced nerve damage, the aqueous extract of *Emblica officinalis* improves diabetic neuropathy. Type 2 Streptozotocin (STZ)-induced diabetic Wistar rats are treated for diabetic neuropathy with Nishamalaki (NA), a mixture of Curcuma longa and *Emblica officinalis*. The medicine NA seemed to be effective. It has enhanced motor and sensory components. The most significant mode of action of NA, aside from other effects, seems to be its antioxidant function. Extracts of *Emblica officinalis* dramatically reduced neuron loss. Male Wistar rats were given streptozotocin to develop diabetes, and amla fruit extracts preserved their neurons. Amla protected against neurohistopathological alterations as vacuolation, gliosis, and neuronal degeneration (Table 10.1) (Koshikawa et al. 2021; Lagani et al. 2013).

10.4.12 Ashwagandha (*Withania somnifera*)

In a alloxan induced diabetes rat, treatments with ashwagandha root extract resulted in noticeably quicker reaction times to heat discomfort during HPT. Both diabetes groups that received ashwagandha extract treatment saw a considerable reduction in the total amount of time needed to cross the beam during NBT. In terms of glucose levels, CPK, oxidative stress biomarkers, and pro-inflammatory cytokines, the ashwagandha root extract-treated group mitigated the effects of diabetes. Histological analysis demonstrated that ashwagandha root extract restored the damage to the pancreas in diabetic rats as well as the altered infiltration of inflammatory cells in the sciatic nerve. Findings suggest that ashwagandha root extract, which is high in phenolic compounds, could be used in conjunction with conventional medicinal treatments to treat inflammation caused by problems from diabetes (Jayaraman et al. 2023). Numerous secondary metabolic classes, including phenols, flavonoids, tannins, saponins, alkaloids, steroids, terpenoids, glycosides, and reducing sugar, were identified through chemical testing as the cause of this plant's anti-inflammatory, anti-spasmodic, anti-analgesic, neuroprotective, and diuretic properties. It was discovered that WMT significantly decreased the neuropathic pain behaviors brought on by STZ. It prevented the genotoxicity and histological alterations linked to STZ in the spinal cord and sciatic nerve. Furthermore, WMT reduced STZ-induced changes in the biochemical components of the sciatic nerve's myelin sheath, according to results from Fourier transform infrared (FTIR) spectroscopy. It significantly decreased the immunoreactivity of mitogen-activated protein kinases (MAPKs) signaling components in the spinal cord, including activator-protein 1 (AP-1), p38-MAPK, c-Jun N-terminal kinase (JNK), and extracellular signal-regulated kinase (ERK). Additionally, it decreased the expression levels of inducible nitric oxide synthase (iNOS), cyclooxygenase-2 (COX-2), and nuclear factor-kappa-B (NF-κB). WMT significantly decreased the inflammatory cytokine production. The fundamental mechanism of *Withania somnifera* antidiabetic neuropathy may be the normalization of HbA1c levels, the enhancement of insulin secretagogue potential, and the reduction of oxidative stress (Table 10.1) (Dinata et al. 2023).

10.4.13 Haritaki (*Terminilia chebula*)

The medium- to large-sized *Terminalia chebula* Retzius (*T. chebula* Retz) tree is a member of the Combretaceae family and is found across Asia. Some common names for it include haritaki (Sanskrit and Bengali), dark myrobalan, ink tree, or chebulic myrobalan (Nigam et al. 2020). *T. chebula* has a wide range of bioactive substances. A comparative investigation by Bhatt et al. (2017) found that out of the ten wild edible fruit specimens examined, *T. chebula* fruit had the highest levels of gallic acid, catechin, total phenols, total flavonoids, chlorogenic acid, and coumaric acid (Sheng et al. 2018). Additionally, it is most likely the most plentiful source of ascorbic acid.

The fruit's most prevalent tannins are punicalagin, terflavin A, corilagin, galloyl-glucose, ellagic acid, gallic acid, chebulic acid, and chebulinic acid. Flavonoids such as quercetin, catechin, and kaempferol are also present at the same time. Additionally, fruits include saccharides such as D-glucose, D-fructose, quinic acid, and shikimic acid. Fruits have been used to isolate sterols such daucosterol and β-sitosterol (Suryavanshi et al. 2021). In a STZ induced diabetes model, Triphala churna treatment prevented neuronal injury by blocking the inflammatory cascade, TNF-α, IL1-β, and TGF-β, which in turn affected the extracellular matrix and caused neurons to demyelinate. NGF expression in peripheral nerves (Sciatic nerve) increased after triphala administration, suggesting that it has a protective effect against nerve injury (Suryavanshi et al. 2021; Haghani et al. 2021). After sciatic nerve damage in rats, the analgesic effects of TC extract are mediated by the down-regulation of brain NGF and NF-κB protein expression. Additionally, the analgesic effects of TC extract are mediated by the improvement of brain oxidative balance through the suppression of GSH and MDA content (Forouzanfar et al. 2023). Its anti-oxidative and anti-inflammatory properties make it potentially useful as an adjuvant to conventional treatment. In Wistar rats, chronic constriction injury (CCI) of the right sciatic nerve increased mechanical allodynia, cold allodynia, and thermal hyperalgesia, all of which were reduced by treatment with Ellagic acid (EA) (50 or 100 mg/kg). CCI also increased TNF-α, NO, and MDA levels and decreased thiol content in the spinal cord, all of which were reversed by administering EA (50 or 100 mg/kg) (Table 10.1).

10.4.14 Shatavari (*Asparagus racemosus*)

Shatavari is regarded as one of the key herbal remedies in Ayurveda. Six significant rasayana reference Shatavari. Rasayana are herbal remedies that boost immunity and cellular vitality to improve a person's overall health. Patients' pitta and vata doshas can be adjusted with shatavari (Alok et al. 2013). The plant *A. racemosus* is endowed with a variety of phytochemical components, such as carboxylic acid, polycyclic alkaloid (asparagamine A), steroidal saponins (shatavarin I and IV), Sarsasapogenin, asparagusic acid, racemofuran, and a few trace elements (Selvaraj et al. 2019; Pandey et al. 2023). Sarsasapogenin, often known as Sar, is a naturally occurring steroidal molecule that exhibits anti-inflammatory, cognitive, and neuroprotective properties. It reduced the memory loss linked to diabetes in rats with streptozotocin-induced diabetes. However, in high glucose-cultured SH-SY5Y cells, Sar reduced the effects of high glucose on neuronal damage, NLRP1 inflammasome activation, and PAR-1 up-regulation. Using the PAR-1 knockdown assay, it has been revealed that the effect of PAR-1 up-regulation in high glucose conditions was mediated by NLRP1 inflammasome and NF-κB activation. In conclusion, this study showed that Sar inhibited neuro-inflammation from activated NLRP1 inflammasome and NF-κB regulated by cerebral PAR-1, hence preventing memory impairment brought on by diabetes (Kong et al. 2021). The reparative effect of SA on SCI in rats was validated by in vivo tests, which further indicated that SA could be able to repair SCI by modifying the immunological milieu. In vitro studies provided insight into how SA inhibits the MAPK/NF-kB pathways to control the immunological microenvironment (Table 10.1).

10.4.15 Mandukaparni (*Centella asiatica* (Linn.) Urban)

Typically, naive stems and leaves of mandukaparni are sliced into little pieces and served as vegetables (Chauhan et al. 2010). At a dosage of 250 mg/kg, the ethanolic and methanolic extracts of *C. asiatica* leaves demonstrated strong anti-diabetic effects (Rahman et al. 2011). Triterpenes from *Centella asiatica* (CA) have been shown to improve microcirculation in previous clinical investigations and to have neurotropic effects in preclinical models, which may have a disease-modifying effect in DN. When comparing the CA ($n = 21$) and placebo ($n = 22$) groups, only the CA-treated groups showed significant decreases from baseline in TSS ($p < 0.01$) and paresthesia ($p < 0.01$). Numbness was substantially higher in the Placebo group ($p < 0.05$) than in the CA group ($p < 0.001$), and it only increased from baseline in the Placebo group. There was less burning in both groups ($p < 0.01$). Neurotropic concentrations in vitro were reflected in plasma triterpene levels in patients receiving CA (Table 10.1) (Lou et al. 2018).

10.4.16 *Allium sativum* Linn. (Lashuna/Garlic)

The cellular components of garlic (*Allium sativum* L.) create the defensive compound allicin (diallylthiosulfinate) (Gangwar et al. 2024). When tissue is damaged, the enzyme alliinase catalyzes the conversion of the non-proteinogenic amino acid alliin (S-allylcysteine sulfoxide) to allicin. Allicin's nonpolar nature enables it to easily cross cell membranes and operate as a reactive sulfur species (RSS) inside of cells. Redox-dependent pathways mediate most of the neuroprotective benefits of allicin. Allicin lowers neuroinflammation by inhibiting the production of ROS and the TLR4/MyD88/NF-κB, P38, and JNK pathways. In acute traumatic spinal cord injury (SCI), allicin protects neuron damage via regulating inflammation, apoptosis, and the synthesis of Nrf2 (nuclear factor erythroid 2-related factor 2). Additionally, patients' cognitive abilities can be enhanced and metal-induced neurotoxicity can be lessened (Table 10.1; Fig. 10.8) (Qian et al. 2023).

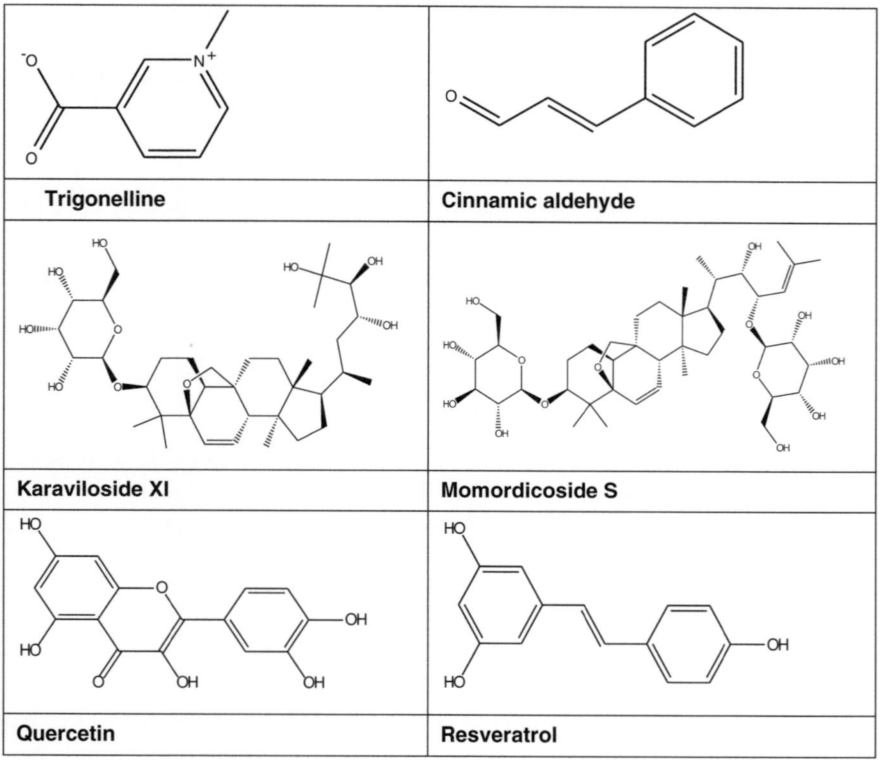

Fig. 10.8 List of active phytoconstituents from the functional foods

Future Perspective

The current scenario of diabetes and diabetic neuropathy (DN) research poses a significant difference in therapeutic approaches, particularly regarding bioactive compounds.

1. Therapeutic Crossover Gap
   - Several bioactive compounds (e.g., berberine, quercetin) demonstrate robust anti-diabetic properties but remain unexplored for DN
   - Conversely, compounds showing promise in DN management (e.g., ellagic acid, pterostilbene) have limited investigation in primary diabetes treatment
   - This bifurcation in research focus has potentially delayed the development of comprehensive treatments
2. Mechanistic Understanding Disparities
   - While the glucose-lowering mechanisms of many compounds are well-documented, their potential neuroprotective pathways remain poorly understood

- Some compounds showing neuroprotective effects in DN have unexplored potential for β-cell protection in diabetes
- Limited investigation into how compounds might simultaneously address both conditions through shared pathways

3. Clinical Translation Challenges

   - Despite extensive preclinical evidence, few bioactive compounds have progressed to clinical trials for either condition
   - Research often focuses on single compounds rather than investigating synergistic combinations
   - Lack of standardization in dosing and delivery methods hampers clinical application

4. Unexplored Traditional Knowledge

   - Many traditional medicinal plants used for diabetes have not been evaluated for DN protection
   - Ancient medical systems often treated these conditions holistically, yet modern research tends to separate them
   - Potential loss of valuable therapeutic insights due to fragmented research approaches

5. Future Research Priorities

   - Need for systematic screening of known anti-diabetic compounds for neuroprotective properties
   - Investigation of compounds showing promise in DN for potential anti-diabetic effects
   - Development of comprehensive therapeutic approaches targeting both conditions simultaneously
   - Standardization of bioactive compound formulations for clinical use

We may discover more effective therapeutic strategies that address both diabetes and its neuropathic complications concurrently, potentially leading to better patient outcomes.

References

Ahmad S, Alouffi S, Khan S, Khan M, Akasha R, Ashraf JM, et al. Physicochemical characterization of *in vitro* LDL glycation and its inhibition by ellagic acid (EA): an *in vivo* approach to inhibit diabetes in experimental animals. Biomed Res Int. 2022a;2022(1):5583298.

Ahmad E, Lim S, Lamptey R, Webb DR, Davies MJ. Type 2 diabetes. Lancet. 2022b;400(10365):1803–20.

Ajanaku CO, Ademosun OT, Atohengbe PO, Ajayi SO, Obafemi YD, Owolabi OA, et al. Functional bioactive compounds in ginger, turmeric, and garlic. Front Nutr. 2022;9:1012023.

Alok S, Jain SK, Verma A, Kumar M, Mahor A, Sabharwal M. Plant profile, phytochemistry, and pharmacology of *Asparagus racemosus* (Shatavari): a review. Asian Pac J Trop Dis. 2013;3(3):242–51.

Alongi M, Anese M. Re-thinking functional food development through a holistic approach. J Funct Foods. 2021;81:104466.

Altındağ F, Rağbetli MÇ, Özdek U, Koyun N, Alhalboosi JKI, Elasan S. Combined treatment of sinapic acid and ellagic acid attenuates hyperglycemia in streptozotocin-induced diabetic rats. Food Chem Toxicol. 2021;156:112443.

Amin A, Akhtar MF, Saleem A, Sharif A, Shah S, Khan MI, et al. Pterostilbene improves CFA-induced arthritis and peripheral neuropathy through modulation of oxidative stress, inflammatory cytokines and neurotransmitters in Wistar rats. Inflammopharmacology. 2022;30(6):2285–300.

Amor AJ, Gómez-Guerrero C, Ortega E, Sala-Vila A, Lázaro I. Ellagic acid as a tool to limit the diabetes burden: updated evidence. Antioxidants. 2020;9(12):1226.

Babu SN, Govindarajan S, Noor A. Aloe vera and its two bioactive constituents in alleviation of diabetes–proteomic & mechanistic insights. J Ethnopharmacol. 2021;280:114445.

Bahadori MB, Zengin G, Bahadori S, Dinparast L, Movahhedin N. Phenolic composition and functional properties of wild mint (*Mentha longifolia var. calliantha* (Stapf) Briq.). Int J Food Prop. 2018;21(1):183–93.

Balkrishna A, Haldar S, Varshney A. OECD-407 Driven 28-day-repeated-dose non-clinical safety evaluation of *Tinospora cordifolia* (Giloy) stem aqueous extract in Sprague-Dawley rats under GLP compliance. Front Pharmacol. 2023;14:1095083.

Banerjee A, Mohammed Breig SJ, Gómez A, Sánchez-Arévalo I, González-Faune P, Sarkar S, et al. Optimization and characterization of a novel exopolysaccharide from *Bacillus haynesii* CamB6 for food applications. Food Biosci. 2022;12(6):834.

Baset M, Ali T, Elshamy H, El Sadek A, Sami D, Tawfik Badawy M. Anti-diabetic effects of fenugreek (*Trigonella foenum-graecum*): a comparison between oral and intraperitoneal administration—an animal study. Int J Funct Nutr. 2020;1(1):1–9.

Bezirtzoglou E, Stavropoulou E, Kantartzi K, Tsigalou C, Voidarou C, Mitropoulou G, et al. Maintaining digestive health in diabetes: the role of the gut microbiome and the challenge of functional foods. Microorganisms. 2021;9(3):516.

Bhatt ID, Rawat S, Badhani A, Rawal RS. Nutraceutical potential of selected wild edible fruits of the Indian Himalayan region. Food Chem. 2017;215:84–91.

Biswas P, Shahnaz M, Akhter M, Ripa AP, Ali T, Rafiq K. Effects of *Azadirachta indica* on neuropathic pain induced by chronic constriction injury to sciatic nerve of Wistar rat. J Adv Vet Anim Res. 2022;9(3):359–68.

Bondonno NP, Dalgaard F, Murray K, Davey RJ, Bondonno CP, Cassidy A, et al. Higher habitual flavonoid intakes are associated with a lower incidence of diabetes. J Nutr. 2021;151(11):3533–42.

Borah N, Shrirnag JM, Chandra P. Probiotic fermented bitter melon juice as a potential adjuvant therapy for type 2 diabetes: evidence from animal model research. Rev Electron Vet. 2023;24(2):222–32.

Borgonetti V, Governa P, Manetti F, Galeotti N. Zingiberene, a non-zinc-binding class I HDAC inhibitor: a novel strategy for the management of neuropathic pain. Phytomedicine. 2023;111:154670.

Bulbul MR, Chowdhury MN, Naima TA, Sami SA, Imtiaj MS, Huda N, Uddin MG. A comprehensive review on the diverse pharmacological perspectives of *Terminalia chebula* Retz. Heliyon. 2022;8(8):e10220.

Chaudhary A, Sharma S, Mittal A, Gupta S, Dua A. Phytochemical and antioxidant profiling of *Ocimum sanctum*. J Food Sci Technol. 2020;57(10):3852–63.

Chauhan PK, Pandey IP, Dhatwalia VK. Evaluation of the anti-diabetic effect of ethanolic and methanolic extracts of *Centella asiatica* leaves extract on alloxan induced diabetic rats. Adv Biol Res. 2010;4(1):27–30.

Chen S, Zhao Z, Ke L, Li Z, Li W, Zhang Z, et al. Resveratrol improves glucose uptake in insulin-resistant adipocytes via Sirt1. J Nutr Biochem. 2018;55:209–18.

Cheng Y, Cao W, Zhang J, Wang J, Liu X, Wu Q, Lin Q. Determinants of diabetic peripheral neuropathy and their clinical significance: a retrospective cohort study. Front Endocrinol. 2022;13:934020.

Deng Z, Ren C, Tang C, Chen S, Li J, Wei J, et al. Syringin alleviates hepatic fibrosis by enhancing autophagic flux and attenuating ER stress-TRIB3/SMAD3 in diabetic mice. Tissue Cell. 2023;83:102159.

Deshmukh P, Phuse A, Nimkar S, Ruikar D, Raut A. A review: anti-diabetic plants and its recent application. J Ayur Herb Med. 2024;10(2):56–66.

Devi MB, Kelapure NN, Kadam ML, Salve RV, Wadatkar HV, Bisarya D. Phytochemical composition, traditional uses, and health benefits of *Gymnema sylvestre* (Gudmar) leaves powder: a review. Bioscene; 2024.

Dinata IG, Udrayana O, Suparna K. Diabetic foot ulcer healing potential mechanism from *Azadirachta indica* (neem) leaves. J Ayur Herb Med. 2023;9:169–73.

Ditchou YON, Leutcha PB, Miaffo D, Mamoudou H, Ali MS, Ngnoung GAAÀ, et al. In vitro and in silico assessment of antidiabetic and antioxidant potencies of secondary metabolites from *Gymnema sylvestre*. Biomed Pharmacother. 2024;177:117043.

Dodda D, Rao AR, Veeresham C. In vitro and in vivo evaluation of pterostilbene for the management of diabetic complications. J Ayurveda Integr Med. 2020;11(4):369–75.

Dong J, Zuo Z, Yan W, Liu W, Zheng Q, Liu X. Berberine ameliorates diabetic neuropathic pain in a rat model: involvement of oxidative stress, inflammation, and μ-opioid receptors. Naunyn Schmiedebergs Arch Pharmacol. 2019;392:1141–9.

Durán AM, Zamora F, De León M. Dietary docosahexaenoic acid-rich supplementation decreases neurotoxic lipid mediators in participants with type 2 diabetes and neuropathic pain. Nutrients. 2024;16(23):4025.

El-Beltagy AEFB, Abou El-Naga AM, El-Habibi ESM, Shams SESM. Ameliorative role of neem (Azadiracta indica) leaves ethanolic extract on testicular injury of neonatally diabetic rats induced by streptozotocin. Egypt J Basic Appl Sci. 2020;7(1):210–25.

Fajrin FA, Nugroho AE, Nurrochmad A, Susilowati R. Ginger extract and its compound, 6-shogaol, attenuates painful diabetic neuropathy in mice via reducing TRPV1 and NMDAR2B expressions in the spinal cord. J Ethnopharmacol. 2020;249:112396.

Farkhondeh T, Folgado SL, Pourbagher-Shahri AM, Ashrafizadeh M, Samarghandian S. The therapeutic effect of resveratrol: focusing on the Nrf2 signaling pathway. Biomed Pharmacother. 2020;127:110234.

Fernández J, Ledesma E, Monte J, Millán E, Costa P, de la Fuente VG, et al. Traditional processed meat products re-designed towards inulin-rich functional foods reduce polyps in two colorectal cancer animal models. Sci Rep. 2019;9(1):14783.

Forouzanfar F, Tanha NK, Pourbagher-Shahri AM, Mahdianpour S, Esmaeili M, Ghazavi H. Synergistic effect of ellagic acid and gabapentin in a rat model of neuropathic pain. Metab Brain Dis. 2023;38(4):1421–32.

Gokhale JS, Lele SS, Ananthanarayan L. Indian traditional foods and diets: combining traditional wisdom with modern science of nutraceuticals and functional foods. In: Nutrition, food and diet in ageing and longevity. Cham: Springer; 2021. p. 357–92.

Gowd V, Jia Z, Chen W. Anthocyanins as promising molecules and dietary bioactive components against diabetes – a review of recent advances. Trends Food Sci Technol. 2017;68:1–13.

Gul K, Singh AK, Jabeen R. Nutraceuticals and functional foods: the foods for the future world. Crit Rev Food Sci Nutr. 2016;56(16):2617–27.

Gupta M, Dey S, Marbaniang D, Pal P, Ray S, Mazumder B. Grape seed extract: having a potential health benefits. J Food Sci Technol. 2020;57:1205–15.

Gupta A, Gupta P, Bajpai G. *Tinospora cordifolia* (Giloy): an insight on the multifarious pharmacological paradigms of a most promising medicinal ayurvedic herb. Heliyon. 2024;10:e26125.

Haghani M, Jaafari M, Meftahi GH, Behzadnia MJ, Bahari Z, Jangravi Z. Analgesic effects of *Terminalia chebula* are mediated by suppression of brain protein expression of NGF and NF-κB and oxidative markers following neuropathic pain in rats. Metab Brain Dis. 2021;36(4):875–88.

Ismail RM, Salman HA, Ibrahim HJ. Study the phytochemical, biological and pharmacological aspects of *Azadirachta indica*: a review. GSC Biol Pharm Sci. 2024;29:33–41.

Jayaraman S, Krishnamoorthy K, Prasad M, Veeraraghavan VP, Krishnamoorthy R, Alshuniaber MA. Glyphosate potentiates insulin resistance in skeletal muscle through the modulation of IRS-1/PI3K/Akt mediated mechanisms: an *in vivo* and *in silico* analysis. Int J Biol Macromol. 2023;242:124917.

Jiang J, Zhang Z, Zhao J, Liu Y. The effect of non-covalent interaction of chlorogenic acid with whey protein and casein on physicochemical and radical-scavenging activity of *in vitro* protein digests. Food Chem. 2018;268:334–41.

Joshi P, Patel S, Paliwal A, Jain S, Verma K, Dwivedi J, Sharma S. *Tinospora cordifolia* ameliorates paclitaxel-induced neuropathic pain in albino rats. J Ethnopharmacol. 2024;321:117559.

Kabir MT, Tabassum N, Uddin MS, Aziz F, Behl T, Mathew B, et al. Therapeutic potential of polyphenols in the management of diabetic neuropathy. Evid Based Complement Alternat Med. 2021;2021(1):9940169.

Kamiloglu S, Günal-Köroğlu D, Ozdal T, Tomas M, Capanoglu E. Recent advances on anti-diabetic potential of pigmented phytochemicals in foods and medicinal plants. Phytochem Rev. 2024;24:1–31.

Khalili L, Valdes-Ramos R, Harbige LS. Effect of n-3 (omega-3) polyunsaturated fatty acid supplementation on metabolic and inflammatory biomarkers and body weight in patients with type 2 diabetes mellitus: a systematic review and meta-analysis of RCTs. Metabolites. 2021;11(11):742.

Khanal P, Mandar BK, Patil BM, Hullatti KK. In silico antidiabetic screening of borapetoside C, cordifolioside A and magnoflorine. Indian J Pharm Sci. 2019;81(3):550–5.

Khedkar R. Traditional food adjuncts: sustainable and healthy option for functional foods. In: Bioactive components: a sustainable system for good health and well-being. Singapore: Springer Nature; 2022. p. 503–15.

Khursheed R, Singh SK, Wadhwa S, Gulati M, Kapoor B, Awasthi A, et al. Opening eyes to therapeutic perspectives of bioactive polyphenols and their nano formulations against diabetic neuropathy and related complications. Expert Opin Drug Deliv. 2021;18(4):427–48.

Kim SH, Hyun SH, Choung SY. Anti-diabetic effect of cinnamon extract on blood glucose in db/db mice. J Ethnopharmacol. 2006;104(1–2):119–23.

Kinoshita A, Nagata T, Furuya F, Nishizawa M, Mukai E. White-skinned sweet potato (*Ipomoea batatas* L.) acutely suppresses postprandial blood glucose elevation by improving insulin sensitivity in normal rats. Heliyon. 2023;9(4):e14719.

Kong L, Liu Y, Zhang YM, Li Y, Gou LS, Ma TF, Liu YW. Sarsasapogenin ameliorates diabetes-associated memory impairment and neuroinflammation through down-regulation of PAR-1 receptor. Phytother Res. 2021;35(6):3167–80.

Koshikawa N, Kida Y, Yasui N, Shinozaki Y, Tsuji K, Watanabe T, et al. A linear five-ring pyrrole-imidazole polyamide-triphenylphosphonium conjugate targeting a mitochondrial DNA mutation efficiently induces apoptosis of HeLa cybrid cells carrying the mutation. Bioorg Med Chem. 2021;576:93–9.

Kumar S, Kumar D, Bhat A, Kumar A. Phytochemicals in clinical studies: current perspective. In: Functional food and human health. Singapore: Springer; 2018. p. 471–511.

Lagani V, Koumakis L, Chiarugi F, Lakasing E, Tsamardinos I. A systematic review of predictive risk models for diabetes complications based on large-scale clinical studies. J Diabetes Complicat. 2013;27(4):407–13.

Lal S, Van Gossum A, Joly F, Bozzetti F, Cuerda C, Lamprecht G, et al. Considerations for the management of home parenteral nutrition during the SARS-CoV-2 pandemic: a position paper from the Home Artificial Nutrition and Chronic Intestinal Failure Special Interest Group of ESPEN. Clin Nutr. 2020;39(7):1988–91.

Lamantia V, Bissonnette S, Beaudry M, Cyr Y, Rosiers CD, Baass A, Faraj M. EPA and DHA inhibit LDL-induced upregulation of human adipose tissue NLRP3 inflammasome/IL-1β pathway and its association with diabetes risk factors. Sci Rep. 2024;14(1):27146.

Lee JH, Kim N, Park S, Kim SK. Analgesic effects of medicinal plants and phytochemicals on chemotherapy-induced neuropathic pain through glial modulation. Pharmacol Res Perspect. 2021;9(6):e00819.

Les F, Cásedas G, Gómez C, Moliner C, Valero MS, López V. The role of anthocyanins as antidiabetic agents: from molecular mechanisms to in vivo and human studies. J Physiol Biochem. 2021;77(1):109–31.

Lin Z, Wang S, Cao Y, Lin J, Sun A, Huang W, et al. Bioinformatics and validation reveal the potential target of curcumin in the treatment of diabetic peripheral neuropathy. Neuropharmacology. 2024;260:110131.

Lou JS, Dimitrova DM, Murchison C, Arnold GC, Belding H, Seifer N, et al. *Centella asiatica* triterpenes for diabetic neuropathy: a randomized, double-blind, placebo-controlled, pilot clinical study. Esper Dermatol. 2018;20(2 Suppl 1):12.

Ma J, Liu J, Chen Y, Yu H, Xiang L. Myricetin improves impaired nerve functions in experimental diabetic rats. Front Endocrinol. 2022;13:915603.

Majeed M, Mundkur L, Paulose S, Nagabhushanam K. Novel Emblica officinalis extract containing β-glucogallin vs. metformin: a randomized, open-label, comparative efficacy study in newly diagnosed type 2 diabetes mellitus patients with dyslipidemia. Food Funct. 2022;13(18):9523–31.

Mandar BK, Khanal P, Patil BM, Dey YN, Pasha I. In silico analysis of phytoconstituents from *Tinospora cordifolia* with targets related to diabetes and obesity. In Silico Pharmacol. 2021;9(1):3.

Martirosyan D, Adany A, Kanya H. Japan's health food industry: an analysis of the efficacy of the FOSHU system. Bioact Compd Health Dis. 2021;4(4):63–78.

Mousavi L, Salleh RM, Murugaiyah V. Antidiabetic and in vitro enzyme inhibition studies of methanol extract of *Ocimum tenuiflorum* Linn leaves and its fractions. Trop Life Sci Res. 2020;31(1):141.

Nandan A, Sharma V, Banerjee P, Sadasivam K, Venkatesan S, Prasher B. Deciphering the mechanism of *Tinospora cordifolia* extract on Th17 cells through in-depth transcriptomic profiling and in silico analysis. Front Pharmacol. 2023;13:1056677.

Nigam M, Mishra AP, Adhikari-Devkota A, Dirar AI, Hassan MM, Adhikari A, et al. Fruits of *Terminalia chebula* Retz.: a review on traditional uses, bioactive chemical constituents and pharmacological activities. Phytother Res. 2020;34(10):2518–33.

Niknejad A, Razavi SM, Hosseini Y, Arab ZN, Abdolghaffari AH, Momtaz SJ. Cinnamon modulates toll-like receptors: a new therapeutic approach for diabetes. Res Biochem Funct Biol. 2024;34(2):223–35.

Omar N, Ismail CAN, Long I. Tannins in the treatment of diabetic neuropathic pain: research progress and future challenges. Front Pharmacol. 2022;12:805854.

Orellana-Paucar AM. Turmeric essential oil constituents as potential drug candidates: a comprehensive overview of their individual bioactivities. Molecules. 2024;29(17):4210.

Pan C, Wang C, Zhang L, Song L, Chen Y, Liu B, et al. Procyanidins attenuate neuropathic pain by suppressing matrix metalloproteinase-9/2. J Neuroinflammation. 2018;15:1–12.

Pandey V, Shri M, Dubey S, Saema S, Tiwari S. An insight into phytochemicals of Shatavari (*Asparagus racemosus*). In: Plants for immunity and conservation strategies. Singapore: Springer Nature; 2023. p. 169–205.

Patil R, Aswar U, Vyas N. Pterostilbene ameliorates type-2 diabetes mellitus–induced depressive-like behavior by mitigating insulin resistance, inflammation and ameliorating HPA axis dysfunction in rat brain. Brain Res. 2023;1817:148494.

Pimentel TC, Da Costa WKA, Barão CE, Rosset M, Magnani M. Vegan probiotic products: a modern tendency or the newest challenge in functional foods. Food Res Int. 2021;140:110033.

Prasad M, Jayaraman S, Rajagopal P, Veeraraghavan VP, Kumar PK, Piramanayagam S. Diosgenin inhibits ER stress-induced inflammation in aorta via iRhom2/TACE mediated signaling in experimental diabetic rats: an *in vivo* and *in silico* approach. Chem Biol Interact. 2022;358:109885.

Qian R, Chen H, Lin H, Jiang Y, He P, Ding Y, et al. The protective roles of allicin on type 1 diabetes mellitus through AMPK/mTOR mediated autophagy pathway. Front Pharmacol. 2023;14:1108730.

Rahman S, Jamal MM, Parvin A, Mahfuz-Al-Mamun M, Islam MR. Antidiabetic activity of *Centella asiatica* (L.) urbana in alloxan induced type 1 diabetic model rats. J Biosci. 2011;19:23–7.

Rahmani A, Almatroudi A, Alrumaihi F, Khan A. Pharmacological and therapeutic potential of neem (Azadirachta indica). Pharmacogn Rev. 2018;12(24):250–5.

Rajasekhar S, Subramanyam MV, Asha Devi S. Grape seed proanthocyanidin extract suppresses oxidative stress in the rat pancreas of type-1 diabetes. Arch Physiol Biochem. 2023;129(5):1045–57.

Ramteke PK, Jambhekar HA, Jambhekar VB, Kamthe KT, Ghule MR. Antidiabetic plants, phytoconstituents, and nano formulations for diabetes treatment. In: Antidiabetic medicinal plants. Academic Press; 2024. p. 167–91.

Rao PN, Mainkar O, Bansal N, Rakesh N, Haffey P, Urits I, et al. Flavonoids in the treatment of neuropathic pain. Curr Pain Headache Rep. 2021;25(7):43.

Rashidinejad A, Boostani S, Babazadeh A, Rehman A, Rezaei A, Akbari-Alavijeh S, et al. Opportunities and challenges for the nano delivery of green tea catechins in functional foods. Food Res Int. 2021;142:110186.

Rekha UV, Anita M, Bhuminathan S, Sadhana K. Known data on the therapeutic use of Azadiracta indica (neem) for type 2 diabetes mellitus. Bioinformation. 2022;18(2):82.

Saeed F, Sultan MT, Riaz A, Ahmed S, Bigiu N, Amarowicz R, Manea R. Bitter melon (*Momordica charantia* L.) fruit bioactives charantin and vicine potential for diabetes prophylaxis and treatment. Plants. 2021;10(4):730.

Sandhir R, Khurana M, Singhal NK. Potential benefits of phytochemicals from *Azadirachta indica* against neurological disorders. Neurochem Int. 2021;146:105023.

Selvaraj K, Sivakumar G, Veeraraghavan VP, Dandannavar VS, Veeraraghavan GR, Rengasamy G. *Asparagus racemosus*—a review. Syst Rev Pharm. 2019;10(1):1–6.

Semwal P, Painuli S, Jamloki A, Rauf A, Rahman MM, Olatunde A, et al. Himalayan wild fruits as a strong source of nutraceuticals, therapeutics, food and nutrition security. Food Rev Intl. 2023;39(9):6500–36.

Sharma R, Bolleddu R, Maji JK, Ruknuddin G, Prajapati PK. *In-vitro* α-amylase, α-glucosidase inhibitory activities and *in-vivo* anti-hyperglycemic potential of different dosage forms of guduchi (*Tinospora cordifolia* [Willd.] Miers) prepared with ayurvedic bhavana process. Front Pharmacol. 2021a;12:642300.

Sharma A, Kaur R, Kaur J, Garg S, Bhatti R, Kaur A. An endophytic *Schizophyllum commune* Fr. exhibits *in-vitro* and *in-vivo* antidiabetic activity in streptozotocin induced diabetic rats. AMB Express. 2021b;11:1–11.

Shen CL, Wang R, Santos JM, Elmassry MM, Stephens E, Kim N, Neugebauer V. Ginger alleviates mechanical hypersensitivity and anxio-depressive behavior in rats with diabetic neuropathy through beneficial actions on gut microbiome composition, mitochondria, and neuroimmune cells of colon and spinal cord. Nutr Res. 2024;124:73–84.

Sheng Z, Zhao J, Muhammad I, Zhang Y. Optimization of total phenolic content from *Terminalia chebula* Retz. fruits using response surface methodology and evaluation of their antioxidant activities. PLoS One. 2018;13(8):e0202368.

Shi GJ, Li Y, Cao QH, Wu HX, Tang XY, Gao XH, et al. In vitro and in vivo evidence that quercetin protects against diabetes and its complications: a systematic review of the literature. Biomed Pharmacother. 2019;109:1085–99.

Shyur LF, Varga V, Chen CM, Mu SC, Chang YC, Li SC. Extract of white sweet potato tuber against TNF-α-induced insulin resistance by activating the PI3K/Akt pathway in C2C12 myotubes. Bot Stud. 2021;62(1):7.

Singh N, Yadav SS, Kumar S, Narashiman B. Ethnopharmacological, phytochemical and clinical studies on fenugreek (*Trigonella foenum-graecum* L.). Food Biosci. 2022;46:101546.

Singh TM, Swati Y, Karunakar S. *In–vitro* drug interaction between Tulsi and Glimepride. Int J Pharm Sci Med. 2023;8:150–63.

Sok Yen F, Shu Qin C, Tan Shi Xuan S, Jia Ying P, Yi Le H, Darmarajan T, et al. Hypoglycemic effects of plant flavonoids: a review. Evid Based Complement Alternat Med. 2021;2021(1):2057333.

Surono IS, Wardana AA, Waspodo P, Saksono B, Verhoeven J, Venema K. Effect of functional food ingredients on gut microbiota in a rodent diabetes model. Nutr Metab. 2020;17(1):1–9.

Suryavanshi SV, Barve K, Addepalli V, Utpat SV, Kulkarni YA. *Triphala churna*—a traditional formulation in Ayurveda mitigates diabetic neuropathy in rats. Front Pharmacol. 2021;12:662000.

Tian JL, Si X, Shu C, Wang YH, Tan H, Zang ZH, et al. Synergistic effects of combined anthocyanin and metformin treatment for hyperglycemia in vitro and in vivo. J Agric Food Chem. 2022;70(4):1182–95.

Tolmie M, Bester MJ, Apostolides Z. Inhibition of α-glucosidase and α-amylase by herbal compounds for the treatment of type 2 diabetes: a validation of in silico reverse docking with in vitro enzyme assays. J Diabetes. 2021;13(10):779–91.

Üst Ö, Yalçin E, Çavuşoğlu K, Özkan B. LC–MS/MS, GC–MS and molecular docking analysis for phytochemical fingerprint and bioactivity of Beta vulgaris L. Sci Rep. 2024;14(1):7491.

Vidhya Rekha U, Anita M, Bhuminathan S, Sadhana K. Known data on the therapeutic use of *Azadirachta indica* (neem) for type 2 diabetes mellitus. Bioinformation. 2022;18(2):82–7.

Gangwar M, Dey A, Chowdhury SR, Gautam MK, Prakash P, Nath G. Botanical warriors: Harnessing nature's antiviral arsenal–A comprehensive study of medicinal plants combatting pathogenic viral infections. Pharmacological Research-Natural Products. 2024;3:100043.

Wang Y, Liu H, Zheng M, Yang Y, Ren H, Kong Y, et al. Berberine slows the progression of prediabetes to diabetes in Zucker diabetic fatty rats by enhancing intestinal secretion of glucagon-like peptide-2 and improving the gut microbiota. Front Endocrinol. 2021;12:609134.

Wu Z, Gong H, Shen Y, Han H, Zhu W, Tian J, et al. Hypoglycemic and hypolipidemic effects of sugar beet (Beta vulgaris L.) extract on C57BL/KsJ db/db mice. Rend Fis Acc Lincei. 2022;33(1):205–12.

Xie Y, Deng Q, Guo M, Li X, Xian D, Zhong J. Proanthocyanidins: A novel approach to Henoch-Schonlein purpura through balancing immunity and arresting oxidative stress via TLR4/MyD88/NF-κB signaling pathway. Exp Ther Med. 2023a;25(6):1–12.

Xie C, Gao W, Li X, Luo S, Wu D, Chye FY. Garlic (*Allium sativum* L.) polysaccharide ameliorates type 2 diabetes mellitus (T2DM) via the regulation of hepatic glycogen metabolism. NFS J. 2023b;31:19–27.

Xu XH, Hu Q, Zhou LS, Xu LJ, Zou X, Lu FE, Yi P. Berberine inhibits gluconeogenesis in skeletal muscles and adipose tissues in streptozotocin-induced diabetic rats via LKB1-AMPK-TORC2 signaling pathway. Curr Med Sci. 2020;40(3):530–8.

Xu Q, Chen Z, Zhu B, Li Y, Reddy MB, Liu H, et al. Neuroprotective effects of b-type cinnamon procyanidin oligomers on MPP+-induced apoptosis in a cell culture model of Parkinson's disease. Molecules. 2021;26(21):6422.

Yoneda T, Kojima N, Matsumoto T, Imahori D, Ohta T, Yoshida T, et al. Construction of sulfur-containing compounds with anti-cancer stem cell activity using thioacrolein derived from garlic based on nature-inspired scaffolds. Org Biomol Chem. 2022;20(1):196–207.

Yorek MA. The potential role of fatty acids in treating diabetic neuropathy. Curr Diab Rep. 2018;18:1–10.

Yu X, Li Y, He C, Zhou J, Chen Y, Yu Z. Nonvolatile metabolism in postharvest tea (*Camellia sinensis* L.) leaves: effects of different withering treatments on nonvolatile metabolites, gene expression levels, and enzyme activity. Food Chem. 2020;327:126992.

Zelicha H, Yang J, Henning SM, Huang J, Lee RP, Thames G, et al. Effect of cinnamon spice on continuously monitored glycemic response in adults with prediabetes: a 4-week randomized controlled crossover trial. Am J Clin Nutr. 2024;119(3):649–57.

Zhai L, Wang X. Syringaresinol-di-O-β-D-glucoside, a phenolic compound from Polygonatum sibiricum, exhibits an antidiabetic and antioxidative effect on a streptozotocin-induced mouse model of diabetes. Mol Med Rep. 2018;18(6):5511–9.

Zhang B, Kang M, Xie Q, Xu B, Sun C, Chen K, Wu Y. Anthocyanins from Chinese bayberry extract protect β cells from oxidative stress-mediated injury via HO-1 upregulation. J Agric Food Chem. 2011;59(2):537–45.

Zhang W, Yu H, Lin Q, Liu X, Cheng Y, Deng B. Anti-inflammatory effect of resveratrol attenuates the severity of diabetic neuropathy by activating the Nrf2 pathway. Aging (Albany NY). 2021;13(7):10659.

Zhao X, Shi A, Ma Q, Yan X, Bian L, Zhang P, Wu J. Nanoparticles prepared from pterostilbene reduce blood glucose and improve diabetes complications. J Nanobiotechnol. 2021a;19:1–18.

Zhao B, Zhang Q, Liang X, Xie J, Sun Q. Quercetin reduces inflammation in a rat model of diabetic peripheral neuropathy by regulating the TLR4/MyD88/NF-κB signalling pathway. Eur J Pharmacol. 2021b;912:174607.

Zhu CW, Lü H, Du LL, Li J, Chen H, Zhao HF, et al. Five blueberry anthocyanins and their antioxidant, hypoglycemic, and hypolipidemic effects in vitro. Front Nutr. 2023;10:1172982.

Harnessing Functional Foods for Diabetic Neuropathy Prevention and Management

11

Sandip Mondal, Swarnajeet Tripathy, Gopinath Halder, Debashri Mondal, Prasurjya Saikia, and Kaushik Modak

11.1 Introduction

An estimated 422 million people worldwide have diabetes, mostly living in low-and middle-income countries, and 1.5 million deaths are directly attributed to diabetes each year (Cando et al. 2024). Diabetic neuropathy is one of the most common complications of diabetes mellitus. The prevalence of neuropathy is a function of disease duration and a rough estimate, derived from several large clinical trials, is that about 50% of all patients with diabetes will ultimately develop neuropathy. Diabetic neuropathy is a type of nerve damage. This nerve damage results from chronic hyperglycemia. It often leads to significant morbidity and reduced quality of life (Chetty 2022; Feldman et al. 2019).

S. Mondal (✉)
Amity Institute of Pharmacy, Amity University Kolkata, Kolkata, West Bengal, India

S. Tripathy
School of Pharmacy and Life Sciences, Centurion University of Technology and Management, Balangir, Odisha, India

G. Halder
Department of Chemical Engineering, National Institute of Technology Durgapur, Durgapur, West Bengal, India

D. Mondal
Department of Zoology, Raiganj University, Uttar Dinajpur, Raiganj, West Bengal, India

P. Saikia
Himalayan University, Itanagar, Arunachal Pradesh, India

Faculty of Pharmaceutical Science, Assam down town University, Guwahati, Assam, India

K. Modak
Faculty of Pharmaceutical Science, Assam down town University, Guwahati, Assam, India

M. K. Gautam et al. (eds.), *Antioxidants and Functional Foods for Diabetic Neuropathy*, https://doi.org/10.1007/978-981-95-9326-2_11

Specifically Diabetic neuropathy can present in various forms. Symptoms of diabetic neuropathy range from mild tingling or numbness to severe pain and loss of motor function. Different types of nerves are affected by diabetic neuropathy. Affected nerves include sensory, motor and autonomic nerves. This causes a wide spectrum of clinical manifestations (Pop-Busui et al. 2022).

Prevalence of neuropathy is higher among individuals with type 2 diabetes. This is largely due to the longer duration of hyperglycemia. Hyperglycemia is often present before diagnosis. However, patients with type 1 diabetes are also at significant risk. This is especially relevant if glycemic control is not achieved early in the disease process (Mayeda et al. 2020; Elafros et al. 2022).

The pathophysiology of diabetic neuropathy presents a complex interplay of metabolic and vascular factors. Chronic hyperglycemia is the main trigger. This initiates a cascade of metabolic changes. These changes lead to nerve damage. One of the key mechanisms is the excessive formation of advanced glycation end-products (AGEs). AGEs result from non-enzymatic glycation of proteins and lipids (Méndez-Morales et al. 2022).

AGEs promote oxidative stress. They also lead to inflammation, both of which play critical roles in nerve damage. Hyperglycemia impairs mitochondrial function. This leads to energy deficits in neurons and exacerbates oxidative stress (Dewanjee et al. 2022). Alongside metabolic factors, vascular complications also contribute to diabetic neuropathy. Damage on a microvascular scale, termed nerve ischemia, results in reduced blood flow to peripheral nerves. It deprives them of essential nutrients and oxygen (Horton and Barrett 2021).

Diabetic neuropathy can be classified into several types. Classification is based on the affected nerves. Peripheral neuropathy is the most common form. It usually affects the feet and legs. It can cause sensory deficits, pain, and loss of motor function (Sloan et al. 2021). Next is autonomic neuropathy. This affects the autonomic nervous system. It leads to issues with heart rate, digestion, and bladder control (Ten Donkelaar et al. 2020). Then there's proximal neuropathy. It affects the hips, thighs or buttocks. It may lead to muscle weakness (Mullins and Kelsall 2023). Lastly, focal neuropathy affects specific nerves often in the head or torso. It causes sudden weakness or pain (Strand et al. 2022).

Each type presents unique challenges in management. This underscores the complexity of the condition. Diet plays a crucial role in preventing and managing diseases. The role of diet in disease prevention is significant, especially in diabetes and its complications. This role has received attention in recent years (Guo et al. 2020).

Treatments have always been the main focus of diabetic neuropathy management. These are intended to reduce symptoms and delay the course of the illness. Recognition is increasing. It's that nutrition plays an equally important role in the treatment and avoidance of this crippling illness. This has been the main emphasis. The relationship between diabetes and nutrition has been well established. Dietary interventions are at the heart of diabetes management (Garcia-Molina et al. 2020). Recent research has suggested that there are specific dietary patterns and foods. Particularly those rich in bioactive compounds. These may have direct benefits in preventing or delaying the onset of diabetic neuropathy (Barrea et al. 2023). These bioactive components include antioxidants and also, anti-inflammatory agents, vitamins and minerals. They have been shown to reduce oxidative stress and inflammation. These are key drivers of nerve damage in diabetes (Bhatti et al. 2022).

There's an emphasis on the synergy between nutrition and pharmacological treatments. While medications help to control blood glucose levels, dietary interventions provide a non-invasive and cost-effective approach (Ahmadian et al. 2023). They modulate disease processes at a metabolic level. This integrative approach involves diet and medication. It may offer more comprehensive management of diabetic neuropathy. It involves addressing the underlying causes of nerve damage (Smith et al. 2022).

Functional foods are foods that offer advantages beyond basic nourishment. They contain bioactive compounds. These can either maintain health or preclude illness. When referring to diabetic neuropathy, functional foods serve as a useful tool. They can stop damage to nerves. They contain bioactive components. These aim at the root cause of nerve injury. This does not hold true for conventional foods. Unlike such foods, functional foods are specially prepared. They are picked precisely because of their ability to promote good health (Dixit et al. 2023).

Functional foods fall into two categories: natural and fortified. Natural are those that contain bioactive compounds naturally. These compounds prevent disease. Examples are fruits, veggies, nuts, seeds, and fish. On the other hand, fortified functional foods are those that are infused with more nutrients. These nutrients can be naturally occurring or can be added. They provide health benefits. For example, cereal can be fortified. These cereals can have vitamins and omega-3 fatty acids, which nourish nerves and reduce inflammation (Essa et al. 2023).

Antioxidants are a key ingredient of functional foods. They are molecules that counteract free radicals. Free radicals are byproducts of bodily activities. They provoke oxidative pressure. Consequently, they are tied to nerve damage in diabetic individuals (Fatima et al. 2023). Antioxidants are plentiful in certain foods. Often these foods are fruits and veggies. Some other examples of plant-based foods also have them. These antioxidants include vitamins C and E. They also consist of polyphenols and flavonoids. They can be used to manage the metabolic pathways that elicit nerve degeneration (Akbari et al. 2022).

Then there are other vital bioactive elements that feature in functional foods. These are vitamins, minerals, and probiotics. To cite an example, magnesium is key. It plays a central role in nerve operation. Zinc Too cannot be left out. It plays an important role in the metabolism of glucose. Probiotics are another element. They may enhance gut health and boost gut-brain communication. This connection has been related to the progress of diabetic issues (Peng et al. 2020).

Probiotics are found in certain foods. These include yoghurt and kefir. These microorganisms are alive. They can provide health benefits when consumed in appropriate amounts. They might improve immune function. Additionally, they can dampen inflammation. This can help manage the deleterious effects of diabetes. To conclude, functional foods are a solid addition to standard pharmacological solutions. They can target metabolic and inflammatory mechanisms behind nerve damage. In this way, they can improve the general state of nerves. Consequently, functional foods might lower the impact of diabetic neuropathy (Hernandez-Ledesma and Martinez-Villaluenga 2021).

This book chapter has the goal of exploring the involvement of functional foods in the prevention and management of diabetic neuropathy. Diabetic neuropathy is a common diabetes complication. The chapter also seeks to foster an understanding of diabetic neuropathy's pathophysiology. It emphasizes the role of oxidative stress, inflammation, and vascular complications. These factors contribute to nerve damage. The chapter then delves into a growing recognition. It identifies this as dietary interventions functioning as supplements to pharmacological treatments. These interventions are used for managing diabetes-related complications. Next, the chapter defines functional foods. It examines their components. These include antioxidants, vitamins, and probiotics. The neuroprotective benefits of these components are also considered. Lastly, the chapter looks at the potential of functional foods. These foods include specific items like probiotics and omega-3 fatty acids. They can be used to mitigate diabetic neuropathy.

11.2 Pathophysiology of Diabetic Neuropathy

Diabetic neuropathy arises from several factors. One of the primary elements is chronic hyperglycemia. It leads to nerve damage mechanisms. One of these key elements involves oxidative stress. This type of oxidative stress originates from the excess of reactive oxygen species. These free radicals are a root cause of significant cellular damage. They result in inflammation and cause nerve degeneration (Lin et al. 2023).

Mitochondrial dysfunction is another contributor to nerve damage. Hyperglycemia disrupts mitochondrial function. This in turn disrupts the production of energy needed for survival of nerve cells. Ultimately, it leads to neurodegeneration too (Cheng et al. 2020).

AGEs, which are Advanced Glycation End-products, form an essential element. They form when proteins and lipids undergo nonenzymatic glycation. This occurs in the presence of high glucose levels. AGEs accumulate in nerve tissues. They worsen oxidative stress and inflammation. This in turn speeds up the process of nerve damage. Vascular complications also contribute. Take microvascular ischemia For instance. It reduces blood flow to peripheral nerves. This causes oxygen and nutrient deficiency. They further impair nerve function and promote degeneration (Khan et al. 2020; Kuzan 2021).

Clinical symptoms of diabetic neuropathy vary but are often sensory. They commonly include numbness and tingling or burning sensations in extremities. Motor symptoms occur too. These include muscle weakness and impaired coordination. Autonomic dysfunction can lead to gastrointestinal, cardiovascular, or bladder issues. If left untreated, these can bring about severe complications. Examples include foot ulcers, infections, and, in some cases, even amputations. This underscores the critical need for early intervention and management (Sloan et al. 2021;

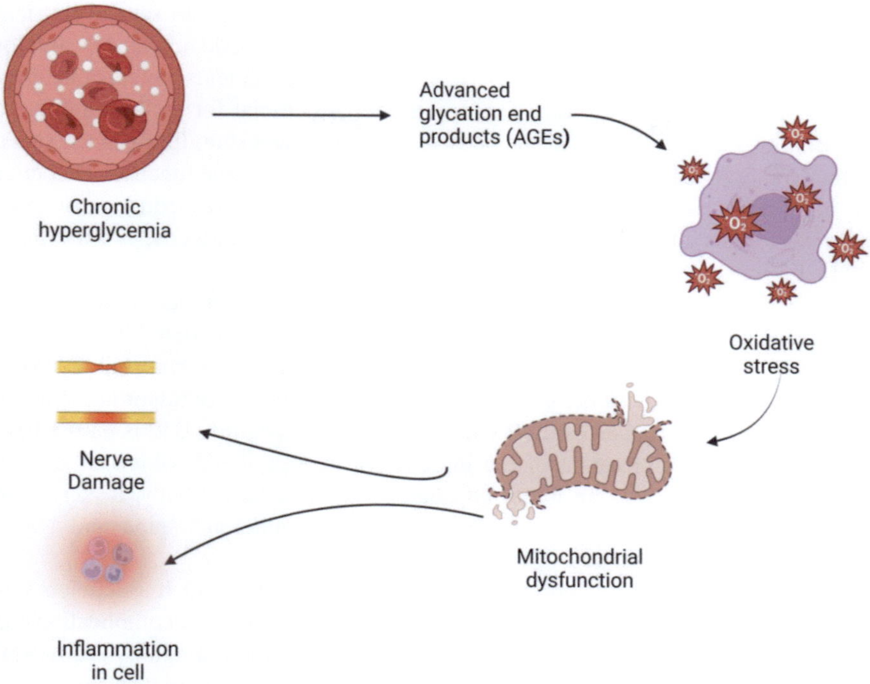

Fig. 11.1 Pathophysiology of diabetic neuropathy. (Created in BioRender. Mondal, S. (2025), https://BioRender.com/v56t710)

Anandhanarayanan et al. 2022). The general pathophysiology of diabetic neuropathy has been depicted in Fig. 11.1.

11.3 The Role of Nutrition in Diabetic Neuropathy

The role of nutrition in diabetic neuropathy cannot be overlooked. Nutrition plays a pivotal role in the onset and progression of diabetic neuropathy. Chronic hyperglycemia and nutrient imbalances critically influence nerve health. Hyperglycemia is a hallmark of poorly controlled diabetes. It directly and detrimentally affects nerve function. Chronic high blood sugar levels lead to excessive reactive oxygen species (ROS) and other toxic byproducts. Such high levels damage nerves structurally. This impairment affects the ability of nerves to transmit signals efficiently (Vincent et al. 2004).

Hyperglycemia is associated with inflammation and this inflammation continues to aggravate the process of nerve degeneration. This further impairs the health of nerve cells. Sensory and motor deficits characteristic of diabetic neuropathy become the net results of these processes (Vinik et al. 2020).

Diabetic neuropathy does not develop in isolation but due to several nutrient deficiencies. This holds true for the development of the diabetic neuropathy and for further progression of the condition. Most often diabetic patients have a deficiency in several micronutrients group. These nutrients are crucial for the optimal nerve functioning and the wellness of the body in general. Magnesium for instance, it is a critical mineral that is required in nerve transmission and muscle function. Diabetics are at a higher risk of developing a magnesium deficiency. The reduction or complete lack of this mineral creates a greater problem in terms of glucose control (Feng et al. 2020).

A similar scenario exists for zinc. It's critical for a well-functioning immune system and nerve repair. Zinc is normally deficient in diabetic patients. This, in turn, inhibits or delays nerve regeneration processes (Nguyên 2021; Chen et al. 2024).

Vitamins such as those belonging to the B-complex group are important in nerve maintenance repair processes. B12 vitamin is one such vitamin. B12 is particularly important in the preservation of the integrity of myelin sheath which covers the nerve fibers. If this vitamin is not adequately supplied to the body, it can cause demyelination and other forms of nerve damage (Calderón-Ospina and Nava-Mesa 2020).

Vitamin D deficiency is prevalent among many diabetics. Such a deficiency is closely linked to high levels of inflammation and other diabetic complications as well. It also causes a greater risk of diabetic complications. Neuropathy is a typical diabetic complication (Putz et al. 2022). Satisfactory intake of these nutrients is key. It is crucial in the prevention or retardation of nerve damage in diabetic patients.

Antioxidants and anti-inflammatory agents are important. They are significant in safeguarding nerves from the damaging effects of hyperglycemia. Antioxidants neutralise ROS. They reduce oxidative stress. This leads to minimal damage to nerves. Antioxidant-rich foods include berries, green leafy vegetables, and nuts. They help protect nerve function by shielding against oxidative damage. Such damage is often bad for nerve health (Fatima et al. 2023).

In addition to antioxidants, also present are anti-inflammatory agents. These can be found in foods like turmeric, ginger, and omega-3 fatty acids. These agents modulate the chronic inflammation. Inflammation contributes to nerve degeneration in diabetic neuropathy. It's good to take note of these types of agents (Shen et al. 2022).

Concluding appropriate nutrition is imperative. Prevention of nutrient deficiencies is of utmost relevance. Intake of antioxidants and anti-inflammatory substances is vital. This is crucial in managing diabetic neuropathy. Hyperglycemia and its nerve-damaging effects must be addressed (Yildiran et al. 2020).

It is equally important to ensure proper levels of essential nutrients. For those with diabetes, this can lead to an improvement in nerve health. It can also lower the risk of neuropathic complications (Smith et al. 2022).

11.4 Functional Foods: Definition, Classification, and Bioactive Components

The definition of functional food is food that has health benefits beyond basic nutrition. These foods are considered to have a positive impact on health. They play a role in disease prevention. Functional foods can be either a single ingredient or a composite food. We can either use naturally occurring ingredients or enhance them with health benefits (Gupta and Mishra 2021).

There are two ways to classify functional foods. The first method is the natural category. This includes fruits, vegetables, whole grains, and nuts. The other is the fortified category. This includes fiber-, vitamin-, and probiotic-enriched foods. An example of a natural functional food is an apple. It is high in fiber. An example of a fortified functional food is vitamin D-enriched milk. Its specific design aims to prevent conditions like osteoporosis in both children and adults (Muhammad Sajid et al. 2021).

Functional foods have bioactive components. These are naturally occurring compounds. These compounds can have an effect on health. An example of a bioactive ingredient is lycopene. Lycopene is found in tomatoes. It is known to have potential anticancer properties (Gengatharan et al. 2024).

Functional foods are becoming more popular. This is due to the recognition of their health benefits. Many people consume foods devoid of nutritional value in their diet. Functional foods address the need for health-promoting food options. They aim to fill nutritional gaps. Pretend. This helps to prevent diet-related health issues (Alongi and Anese 2021).

Functional foods are often misunderstood. Some people think they are different from normal foods. Recall that we still classify functional foods as foods. They are not medicine.

In summary, functional foods are a food type. They go beyond basic nutrition. They offer health-promoting benefits. This is due to bioactive components. These are naturally occurring compounds. They play a key role in preventing chronic diseases. An example is diabetes. Unlike conventional foods that just provide energy and nutrients, functional foods actively contribute to health. They promote specific physiological functions. They reduce the risk of certain diseases.

Functional foods can come from natural sources. They can also be fortified. Natural functional foods include fruits, vegetables, and whole grains. They also include plant-based oils. These are full of bioactive compounds. Examples include antioxidants, vitamins, and minerals. They contain healthy fats too.

Blueberries are a case in point. They are abundant in polyphenols and flavonoids. These substances help reduce oxidative stress and inflammation. Both are common contributors to diabetic neuropathy (Felgus-Lavefve et al. 2022). Fortified functional foods are products enhanced with nutrients. Examples include calcium-fortified orange juice and vitamin D-enriched milk. These foods aim to fill nutritional gaps. They provide specific health benefits. Such benefits include enhancing bone health and supporting immunity (Clark 2019; Mirmiran et al. 2014).

11.4.1 Classification Based on Bioactive Compounds

Functional foods can be classified. This is based on the bioactive compounds they contain. These bioactive components consist of polyphenols and flavonoids. They also have vitamins and minerals, prebiotics, and probiotics. There are dietary fibres too. Each offers unique mechanisms of action. These mechanisms support health. They also prevent disease. In diabetic neuropathy, bioactive compounds are relevant in particular categories.

11.4.2 Polyphenols and Flavonoids

Polyphenols are found in fruits, vegetables, tea, coffee, and dark chocolate. They have strong antioxidant properties. These neutralize reactive oxygen species (ROS) and reduce oxidative stress. This is a major factor in the development of diabetic neuropathy. Polyphenols also have anti-inflammatory properties. They help to reduce chronic inflammation that worsens nerve damage in diabetes. Flavonoids are a subgroup of polyphenols. They are abundant in foods like berries, citrus fruits, onions, and green tea. They play a crucial role in modulating glucose metabolism and improving insulin sensitivity (Rudrapal et al. 2022).

Vitamin D is also indispensable. This fat-soluble vitamin regulates inflammation. Vitamin D is vital to supporting immune function. Vitamin D deficiency is widespread among diabetics. It has been linked to heightened neuropathy risks. Foods such as fatty fish fortified dairy and mushrooms are great sources (Md Isa et al. 2023).

Zinc and Magnesium: These minerals impact nerve function. They also affect glucose regulation and immune response. Magnesium contributes immensely. It helps to regulate nerve transmission and muscle function. Zinc has a critical role in nerve repair and inflammation reduction. Good sources of both magnesium and zinc are leafy greens nuts, seeds, and legumes (Hoque et al. 2023).

Vitamins and minerals are essential. They are needed for various physiological processes. Their importance extends to nerve function, glucose metabolism, and immune response. When talking of diabetic neuropathy, certain vitamins and minerals are notably important.

- Vitamin B Complex: Thiamine, pyridoxine, and cobalamin are fundamental vitamins. This is in the context of nerve health. A deficit of any of these vitamins can result in nerve damage. It can also exacerbate neuropathy symptoms. Thiamine is crucial for energy metabolism. It also plays a role in nerve function. Vitamin B12 is vital for the health of the myelin sheath. This sheath protects nerve fibers.

Leafy greens, whole grains, and animal products are rich in these vitamins (Sarwar et al. 2021).

- Vitamin D: This is a fat-soluble vitamin. It has an active part in controlling inflammation. Vitamin D also supports immune function. Common in individuals with diabetes, vitamin D deficiency can potentially increase the chances of neuropathy. Fatty fish, fortified dairy products, and mushrooms can be excellent sources of vitamin D (Rak and Bronkowska 2018).
- Magnesium and Zinc: These minerals participate in nerve function. They also have a role in glucose regulation. Additionally, they affect immune responses. Magnesium has a regulatory function in nerve transmission. It also has a role in muscle function. Zinc is crucial in nerve repair and in minimizing inflammation. Rich sources of these minerals are foods like leafy greens, nuts, seeds, and legumes.

11.4.3 Prebiotics, Probiotics and Dietary Fibers

Gut health is given importance. This is in the management of diabetes; it is also for its complications. Gut-brain axis has bidirectional communication. This is an axis between the gut and nervous system. It plays a role in maintaining nerve health (Lu et al. 2024). Prebiotics and probiotic dietary fibers contribute. They help in the maintenance of gut health. It in turn supports nerve function and metabolic health.

- Prebiotics: Prebiotics are non-digestible fibers. They act as food for beneficial gut bacteria. They help maintain a healthy gut microbiota. This microbiota has been connected to improved insulin sensitivity. It can reduce inflammation. Foods rich in prebiotics exist. These include garlic, onions, leeks, asparagus, and bananas (Megur et al. 2022).
- Probiotics: Probiotics are beneficial live bacteria. They are found in fermented foods. Foods such as yoghurt, efir, sauerkraut, Sauerkraut and kimchi have these bacteria. These bacteria help balance gut microbiota (Mondal et al. 2022). They also reduce inflammation and improve glucose metabolism. This reduction in the risk of diabetic neuropathy is potentially significant.
- Dietary Fibers: Dietary fibers, especially soluble fibers, can help regulate blood glucose levels. They do this by slowing sugar absorption. An example would be preventing spikes in blood glucose levels. These spikes can contribute to nerve damage over time (Giuntini et al. 2022).

Fiber-rich foods are whole grains, legumes, fruits, and vegetables. All of these can contribute to a healthy diet and help manage blood glucose levels. Managing blood glucose is critical for maintaining overall health. It is especially important for people with diabetes.

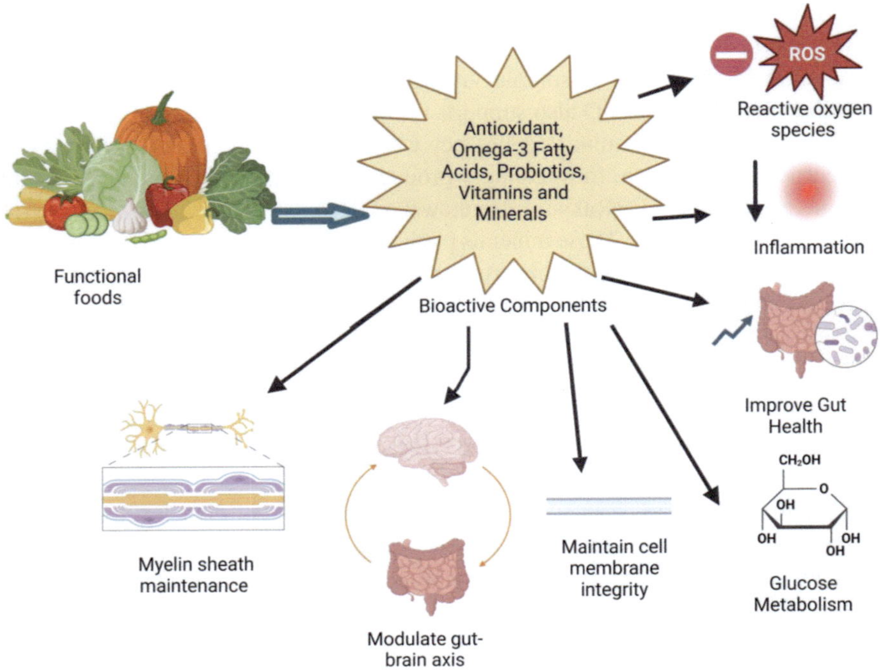

Fig. 11.2 Functional foods and different functions of bioactive components. (Created in BioRender. Mondal, S. (2025), https://BioRender.com/l61c706)

11.4.4 Mechanisms of Action of Bioactive Components

Bioactive components in functional foods are known for their health benefits. These benefits are a result of several mechanisms (Fig. 11.2). Particularly, these include glucose regulation, reduction of oxidative stress, and promotion of nerve health. All are crucial in preventing and managing diabetic neuropathy.

11.5 Modulation of Glucose Levels

Primary goal of managing diabetes is to maintain stable blood glucose levels. Functional foods with Dietary fibers, polyphenols, and flavonoids regulate glucose metabolism. They do so by improving insulin sensitivity and slowing the absorption of carbohydrates. For example, soluble fibers in oats, legumes, and fruits form a gel-like substance in the digestive tract. This slows the release of glucose into the bloodstream. It prevents postprandial glucose spikes. Those spikes can lead to complications like neuropathy. Polyphenols and flavonoids help modulate glucose transport. They improve insulin signaling too. This further supports glucose control (Fernandes et al. 2022).

11.6 Reduction of Oxidative Stress

Oxidative stress can be a big contributor to diabetic neuropathy development. Overproduction of reactive oxygen species (ROS) can happen. It Is In Response to Chronic Hyperglycemia. This can cause damage to nerve cells. It can impair their function.

Antioxidants are important in this situation. They can be in foods that are high in polyphenols. Berries, green tea, and dark chocolate are some examples. These antioxidants neutralize ROS. They reduce oxidative damage. They donate electrons to unstable ROS. This stabilizes them. This also prevents them from causing any further damage to cells.

The reduction in oxidative stress is important. It protects nerve cells. It supports overall nerve health. It can prevent damage to nerve cells. It can prevent a further decrease in nerve health (Caturano et al. 2023).

11.7 Anti-inflammatory Effects

Chronic inflammation is a key factor for the growth of diabetic neuropathy. Foods with anti-inflammatory bioactive compounds are functional. Such anti-inflammatory foods contain omega-3 fatty acids (curcumin from turmeric) and ginger. They play a crucial role in modulating inflammation (Shen et al. 2022).

Molecules known as pro-inflammatory cytokines drive inflammation. This contributes to nerve harm. Omega-3 fatty acids cut down the production of these molecules. They are found in fatty fish such as salmon and mackerel. These molecules drive inflammation and also contribute to nerve damage (Samanta et al. 2022).

Curcumin and ginger also help. They reduce the overall inflammatory burden on the body. They do so by inhibiting the activity of key inflammatory pathways. They lower chronic inflammation. This bioactive compound helps slow neuropathy progression and promotes nerve repair (Parsamanesh et al. 2018).

11.8 Promotion of Nerve Health

Regulating glucose is a property of these foods. They also serve as antioxidants. They have anti-inflammatory traits. Certain bioactive compounds go a step further. They actively support nerve health and regeneration.

Functioning of nerves requires B1, B6, and B12 vitamins. These are particularly significant for nerve function. Nerve cells function properly when provided with Vitamin B1. It helps convert glucose into energy (Sarwar et al. 2021).

Synthesis of neurotransmitters is supported by vitamin B6. Functioning of the myelin sheath is crucial for vitamin B12. Myelin sheath insulates nerve fibers. It facilitates efficient signal transmission (Baltrusch 2021).

A deficiency in these vitamins can lead to nerve damage. It can also exacerbate neuropathy symptoms. Adequate intake of vitamins is crucial. Such intake through functional foods can help. It helps maintain nerve health. It prevents further deterioration.

Functional foods are key in the prevention and management of diabetic neuropathy. They provide bioactive components. These components Moderate glucose levels reduce oxidative stress, lower inflammation, and promote nerve health. It is important to incorporate functional foods into the diet. Individuals with diabetes can then potentially reduce the neuropathy risk.

The individual can also improve their overall health. The inclusion of polyphenols, flavonoids, vitamins, minerals, prebiotics, probiotics, and dietary fibres in the diet is also vital. It is essential for supporting nerve function. It helps to prevent diabetes-related complications.

11.9 Functional Foods in the Prevention of Diabetic Neuropathy

Diabetic neuropathy, a common complication of diabetes, is primarily from chronic hyperglycemia. It leads to nerve damage through a couple of mechanisms. These include oxidative stress, inflammation, and impaired glucose metabolism. New studies have suggested that dietary interventions are beneficial. These interventions include functional foods. They can play an important role in preventing and managing diabetic neuropathy.

Functional foods are rich in bioactive components. These components include antioxidants, anti-inflammatory agents, polyphenols, vitamins, and minerals. They have shown potential to reduce nerve damage. They improve glucose metabolism and support nerve function.

Rich in antioxidants, foods are significant. They play a critical role in lowering oxidative stress. They protect against diabetic neuropathy progression.

Hyperglycemia damages nerves through one of the main mechanisms. It is known as oxidative stress. During times of high blood sugar, more reactive oxygen species (ROS) are created. The result is the oxidation of proteins, lipids, and DNA. As a result, nerve function can be impaired (Jomova et al. 2023). Nerve degeneration can also be promoted. Antioxidants have the ability to neutralize ROS. This protects nerves from oxidative damage. It also improves overall nerve health (Chun and Park 2020).

Those rich in antioxidants have another vital function. They play a key role in reducing oxidative stress. They also protect against diabetic neuropathy progression.

11.9.1 Antioxidant-Rich Foods

Oxidative stress is a key culprit in the nerve damage caused by hyperglycemia. When blood sugar is high, the body generates excess reactive oxygen species (ROS). These oxidize proteins, lipids, and DNA. This leads to impairment in nerve function and promotes nerve degeneration. ROS can be neutralized with antioxidants. This protects nerves from oxidative damage. It also improves nerve health overall (Preedy 2020).

Foods that have high antioxidant content are crucial in this battle. They help reduce oxidative stress. This also helps in protecting from the onset and slowing down the progression of diabetic nerve damage.

The Purpose of Antioxidants: Minimizing Oxidative Strain and Nervous Injuries
Oxidative stress results in cell damage and can lead to neuron death. Antioxidants are compounds that help mitigate such nerve injuries. These molecules protect against damage. By oxidizing agents. Antioxidants work by donating an electron to an oxidizing agent.

The oxidizing agent, known as a free radical, needs an electron. When the free radical receives an electron, it won't steal one from the cells. Otherwise, it would cause cell damage and lead to nerve injuries. Antioxidants complete this process by making the free radical stable.

Ginkgo biloba, vitamin E, coenzyme Q10, and many other antioxidants play this role. They neutralize the free radicals, preventing them from causing damage. This way, they reduce oxidative stress. Antioxidants exert their protective role both in the cells and in the nerves (Ekici-Günay 2020).

Antioxidants work by donating electrons to unstable ROS. This neutralizes the ROS. Before they can cause any damage to cellular structures. When we have diabetic patients, hyperglycemia often boosts the production of ROS. However, by including antioxidant-rich foods in the diet, we can decrease the oxidative stress quite significantly (Arabshomali et al. 2023).

The reduction in oxidative damage is beneficial. It protects nerves and also helps to restore normal cellular function. Furthermore, it can cut down on inflammation linked to oxidative stress. In the case of individuals with diabetic neuropathy, this is all the more important. Antioxidant-rich foods are crucial as they address a key mechanism of nerve damage. That mechanism is known as oxidative stress. It's important to keep this particular fact in mind (Arabshomali et al. 2023; Mukherjee et al. 2024).

11.9.2 Key Examples of Antioxidant-Rich Foods

Many foods boast high antioxidant content. Regularly consuming them can give diabetics crucial protective benefits. Table 11.1 outlines key functional food nutrients and their benefits for nerve health. Here are a few key examples of antioxidant-rich foods:

Table 11.1 Functional food nutrients and their role in nerve health

Nutrient	Food sources	Mechanism of action	Benefits for neuropathy	References
Polyphenols	Berries, green tea, dark chocolate	Strong antioxidant properties, reduce oxidative stress and inflammation	Protects nerve cells, improves blood flow to peripheral nerves	Mandal et al. (2024)
Omega-3 fatty acids	Fatty fish (salmon, mackerel), flaxseeds, walnuts	Anti-inflammatory properties, improve cell membrane fluidity	Promotes nerve repair, reduces pain and inflammation	Samanta et al. (2022)
Vitamin C	Citrus fruits (oranges, lemons), berries	Antioxidant properties, supports collagen synthesis	Reduces oxidative stress, enhances vascular health	Mandal et al. (2024)
Vitamin E	Nuts (almonds, sunflower seeds), avocados	Antioxidant that protects cell membranes from oxidative damage	Reduces nerve degeneration, protects nerve cells	Gorji et al. (2018)
Magnesium	Leafy greens (spinach, kale), nuts, legumes	Regulates nerve transmission and muscle function	Improves glucose metabolism, reduces nerve excitability	Feng et al. (2020)
Vitamin D	Fatty fish (salmon, tuna), fortified dairy, mushrooms	Regulates inflammation, supports immune function	Reduces chronic inflammation, lowers risk of nerve damage	Putz et al. (2022)
Probiotics	Yogurt, kefir, sauerkraut, kimchi	Modulate gut microbiota, enhance gut-brain communication	Reduces systemic inflammation, supports nerve health	Hernandez-Ledesma and Martinez-Villaluenga (2021)
Folate (vitamin B9)	Leafy greens (spinach, broccoli), beans, citrus fruits	Supports DNA repair and synthesis, reduces homocysteine levels	Protects nerve health, prevents vascular complications	Mironenko and Eliseeva (2019)
Thiamine (vitamin B1)	Whole grains, pork, fortified cereals	Provides energy to nerve cells, supports glucose metabolism	Prevents energy deficits in neurons, reduces neuropathy symptoms	Sarwar et al. (2021)
Curcumin	Turmeric	Potent anti-inflammatory and antioxidant effects	Alleviates chronic inflammation, reduces neuropathic pain	Parsamanesh et al. (2018)
Catechins	Green tea	Antioxidant and anti-inflammatory properties	Protects nerve cells, improves insulin sensitivity	Chen et al. (2022)

(continued)

Table 11.1 (continued)

Nutrient	Food sources	Mechanism of action	Benefits for neuropathy	References
Lycopene	Tomatoes, watermelon	Antioxidant properties, reduces oxidative stress	Protects against oxidative damage, supports nerve health	Gengatharan et al. (2024)

- Berries: blueberries, strawberries, raspberries, and blackberries. They are rich in polyphenols and vitamin C. Both are strong antioxidants (Mandal et al. 2024). Berry can reduce oxidative stress. They can also improve insulin sensitivity. Another benefit is the protection of nerve cells from damage.
- Dark Chocolate: Chocolate with high cocoa content. Rich in flavonoids. Flavonoids offer strong antioxidant properties. They have been proven to enhance blood flow to peripheral nerves. Flavonoids also reduce inflammation. They enhance nerve function (Shahanas et al. 2019).
- Green Tea: This beverage contains catechins. Catechins are a group of antioxidants. Research shows that catechins can minimize oxidative stress. They can also reduce inflammation. Studies indicate regular green tea consumption is associated with better blood sugar control. It also reduces nerve damage in diabetes.
- Nuts: almonds and walnuts, among others. Are high in Vitamin E. Vitamin E is an important antioxidant. It aids in the protection of nerve cells from oxidative damage. Nuts also contain good fats. This improves overall metabolic health. They also support nerve function.

11.9.3 Anti-inflammatory Foods

Chronic inflammation is a critical factor that contributes to the progression of diabetic neuropathy. Inflammation can make nerve damage worse. It boosts the production of pro-inflammatory cytokines. These are molecules that worsen oxidative stress. They also impair nerve repair mechanisms. Inclusion of anti-inflammatory foods in the diet helps. It can modulate the inflammatory response. This will slow the progression of neuropathy. It will also improve nerve health (Prabhakar 2021).

11.9.4 The Strategy of Reducing Inflammation to Slow Progression of Neuropathy

Inflammation is a natural immune reaction. Yet when it becomes chronic—as it often is in diabetic individuals—it can lead to further damage to nerves. It damages other tissues as well. Foods that are anti-inflammatory contain bioactive compounds. These can inhibit the activity of pro-inflammatory molecules. Cytokines are one such molecule. They reduce the overall inflammatory burden in the body. For people with diabetic neuropathy, consuming these foods helps alleviate nerve inflammation. This helps them have better nerve repair and function (Elbandy 2022).

11.9.5 Examples of Foods with Anti-inflammatory Properties

Certain foods are renowned for anti-inflammatory effects. These foods can be especially helpful for people with diabetes. This list includes turmeric. It has the active ingredient curcumin. Curcumin is noted for its potent anti-inflammatory properties. It inhibits inflammatory pathways. It reduces the production of pro-inflammatory cytokines. Pain is reduced by turmeric, and it enhances nerve function. These benefits of turmeric apply to diabetic neuropathy patients.

Ginger is another food with anti-inflammatory properties Ginger has gingerols and shogaols. These compounds have an anti-inflammatory effect. Regular use of ginger can lower inflammation. Oxidative stress is also reduced by ginger. Both of these factors promote diabetic neuropathy (Al Hroob et al. 2018).

Fatty fish have anti-inflammatory qualities. Salmon and mackerel are examples. Plant-based omega-3-rich sources are chia seeds and flaxseeds. Omega-3 fatty acids are rich in these foods. Omega-3s reduce pro-inflammatory cytokines. They also improve nerve function. These fatty acids are also good for the repair of damaged nerve tissues. Polyphenols and flavonoids are two types of bioactive compounds. They are found in numerous plant-based foods. These compounds provide neuroprotective effects. They also help in regulating glucose metabolism. This makes them vital for preventing and managing diabetic neuropathy.

There are bioactive compounds in polyphenols and flavonoids. These compounds protect nerves from damage. They achieve this by reducing oxidative stress. These compounds also modulate inflammation. Additionally, they enhance glucose regulation. These compounds enhance insulin sensitivity. They help in preventing spikes in blood sugar. Commonly, these spikes contribute to nerve damage. Polyphenols and flavonoids improve endothelial cell function (Li and Zhang 2023). This improves blood flow to the nerves. It also prevents ischemic damage, a common complication in diabetic neuropathy.

11.9.6 Principal Instances of Polyphenol and Flavonoid-Rich Foods

Flavonoids and polyphenols are primarily found in plant-based foods. Here are a few examples of food items rich in these important compounds:

- Blueberries: Blueberries are packed with plant chemicals. They offer copious amounts of these valuable compounds. Blueberries are a superfood because of their antioxidant properties. They help lower inflammation. Blueberries also support heart health.

- Dark Chocolate Dark chocolate is chockfull of beneficial antioxidants. It contains polyphenols and flavonoids. These nutrients strengthen blood vessels and lower blood pressure. These components also reduce the risk of heart disease. Dark chocolate can also help control blood sugar.
- Oranges: Oranges provide high amounts of vitamin C. They are rich in citrus flavonoids. These flavonoids are a form of antioxidant. They guard against cancer and heart disease. Oranges help reduce inflammation. They improve circulation too.
- Kale: Kale is a cruciferous vegetable. It contains glucosinolate. Glucosinolates have potent anti-cancer properties. They are rich in vitamins A, C, C and K. They also offer calcium. Kale is also a good source of fiber. It is great for heart health and digestion.

11.9.7 Key Examples of Dietary Fiber and Low Glycemic Index Foods

Dietary fiber is important for preventing many lifestyle-related diseases and ensuring a healthy lifestyle. The best sources of fiber include: fruits vegetables: whole grains and legumes. These are all high in fiber and low in saturated fat and cholesterol. They also provide essential nutrients, vitamins, and minerals. Moreover, high-fiber diets may prevent conditions such as constipation, diverticulosis, and hemorrhoids.

Low-glycemic-index (GI) foods are beneficial for those at risk for heart disease or diabetes. They help manage blood sugar levels. Foods with low GI include oatmeal, whole grains, vegetables, and some fruits. They digest slowly, reducing spikes in blood sugar. The ideal diet has a balanced supply of fiber and foods with low GI (Eleazu 2016).

It is important to include these food types in our daily diet. They are highly recommended due to their health benefits.

- Eating foods high in fiber can help lower cholesterol levels. It can also help reduce the risk of heart disease. Foods with high fiber content can help reduce blood sugar levels. This is also important for those with diabetes (McRae 2018).
- Whole grains possess a low glycemic index (GI). They help regulate blood sugar levels and prevent spikes in levels. This is important not only for diabetics but also for those at risk of diabetes. It is crucial to have balanced levels of blood sugar to maintain good health. They also help prevent heart disease and diabetes (Kaur et al. 2022).
- Apples are high in fiber. This fruit helps reduce cholesterol levels. The skin of an apple is the most fibrous part. Apples are also known to help with digestion. This reduces the risk of colon cancer (Josimuddin et al. 2022).
- Oatmeal is high in fiber. It is also low in saturated fat. Oatmeal can help with heart health. It also helps with weight control.

- Legumes are a rich source of fiber. They include lentils and beans. We can add them to almost any dish. They are low in fat. They are also filling.
- Kale is high in fiber and low in calories. It is rich in flavor. We can steam it or sauté it with some olive oil.
- Carrots are a great snack. They are high in fiber. They also contain vitamins such as vitamin A and K. This is good for eye health.
- Oranges are a low-GI food. They have fiber and vitamin C. This is good for immune support. They make a great healthy snack.
- Blueberries are high in fiber. They have antioxidants. They promote heart health. Blueberries are also good for brain health. They are a versatile snack.
- Avocado is a fruit high in fiber and good fat. It is low in GI. It is also a nutrient-rich superfood. It's a great addition to any dish.
- Chia seeds are a great source of fiber. They are nutrient-dense seeds. We can add them to smoothies or salads. They make a great snack.
- Quinoa is high in fiber and protein. It is a low-GI food. It is a good substitute for rice. It is also gluten-free. It is a versatile grain.
- Broccoli is a high-fiber food. It is packed with nutrients. They support digestive health. They reduce inflammation. Broccoli helps lower cholesterol levels.
- Lentils are a good source of fiber and protein. They are low in fat. They are versatile and can be added to soups, salads, and curries. They can be sprouted for added nutritional value. The fiber in lentils helps with digestion. It may reduce the risk of chronic diseases.
- Berries are high in fiber and antioxidants. They provide a great energy boost. They prevent inflammation. Blueberries and strawberries are top choices. They make a great snack.
- Almonds are high in fiber. They are rich in monounsaturated fats. Almonds can lower cholesterol levels. They are a great snack option.
- Chia seeds are fiber-rich. They are also a good source of protein. They promote healthy digestion. Chia seeds can help lower blood pressure. We can add them to smoothies or oatmeal. They make a great snack.
- Sweet potatoes are a good source of fiber. They are an excellent carbohydrate source. They are rich in vitamins and minerals. They promote digestion and heart health. They are a great snack option.
- Black beans are a cheap fiber source. They are a good source of protein. They are also rich in vitamins and minerals. They promote healthy digestion. They are a great snack option.

The above are just a few dietary fiber and low glycemic index samples. They provide health benefits and are easy to integrate into our diet. Each has its own unique taste and nutritional characteristics. They are affordable and can help enhance our meals. So incorporate these into our daily meals. Remember to always eat a balanced diet. This includes a wide variety of foods. This way, we can be on the path to better health. Optimize our diet by having a mix of high-fiber and low-glycemic-index foods. Remember that food is a central part of our lives. It doesn't have to be boring! Embrace variety and enjoy our meals.

11.9.8 Probiotics, Gut Health, and Diabetic Neuropathy

Diabetic neuropathy is a condition influenced by various metabolic, cardiovascular, vascular, inflammatory pathways. Recent research highlights the significant role of gut health in nerve health modulation. This is particularly through the gut–brain axis. This axis refers to bidirectional communication between the gut and central nervous system; the gut microbiota has influence. This influence is on brain and nerve health, immune function, as well as inflammatory responses. A growing body of evidence suggests the potential to use probiotics. This is for therapeutic benefits in managing diabetic neuropathy. Gut-Brain Axis and Nerve Health the gut-brain axis is an important pathway. It's how the gut microbiota talks to our nervous system. This affects both nerve health and inflammation. The gut microbiota is a varied microorganism community. It resides in the gastrointestinal tract. It also plays a critical role in maintaining metabolic balance. It also maintains immune function. This process is important for keeping the integrity of the intestinal barrier. Disruptions to the gut microbiota can happen such as dysbiosis. They can lead to an increase in systemic inflammation. They can also lead to oxidative stress. They can even lead to impaired glucose metabolism. All of these are critical factors in the development of diabetic neuropathy.

The gut-brain axis can influence neuroinflammatory pathways. It does this through modulation of gut bacteria. This can impact nerve health and overall neurological function. By restoring the balance of gut bacteria, the gut-brain axis can help reduce inflammation. It can also reduce nerve damage associated with diabetic neuropathy. Probiotics are living microorganisms that give health benefits when taken in the correct amounts. They function by bringing back the balance of gut microbiota. This enhances the gut-brain axis and improves gut health overall. Probiotic-rich functional foods are a fine source of helpful bacteria. Regular consumption of these foods can help moderate inflammation. They can also improve insulin sensitivity. They can also decrease the risk of diabetic complications. This includes neuropathy as well. Key Examples of Probiotic-Rich Functional Foods Functional foods are abundant in probiotics. They are linked to superior gut health and nerve function.

A couple of them include the following.

- Yoghurt: Yoghurt contains live cultures of Lactobacillus and Bifidobacterium species. They are known to support gut health. They can reduce inflammation. They also improve glucose metabolism. Regular consumption of probiotic-rich yoghurt can help to protect against nerve damage. This is beneficial for diabetic patients (Barengolts et al. 2019).
- Kefir: Kefir is a dairy product that is fermented. It contains many probiotics. These include *Lactobacillus, Bifidobacterium,* and yeast species. Kefir is known to boost gut microbiota diversity. It also enhances inflammatory markers. This is proven to be beneficial for individuals with diabetes (Bellikci-Koyu et al. 2019).
- Sauerkraut: Sauerkraut is a fermented cabbage product. It is rich in probiotics. Particularly *Lactobacillus plantarum*. This has been shown to improve gut

health. It can reduce oxidative stress. It can also modulate the immune system. This provides benefits for diabetic neuropathy patients (Ahmad et al. 2023).

11.9.9 Impact of Modulating Gut Microbiota

Modifying gut microbiota through probiotic consumption may have vast implications for diabetic complications. These could include neuropathy. Probiotics have the ability to mitigate systemic inflammation. They do this by triggering the production of anti-inflammatory cytokines. They also suppress pro-inflammatory pathways. Probiotics also aid in enhancing glucose metabolism. This is achieved by ratcheting up insulin sensitivity. Insulin sensitivity is crucial for preventing damage to nerves. This damage is usually caused by hyperglycemia.

Probiotics might also enhance the creation of short-chain fatty acids (SCFAs). SCFAs are known to promote nerve repair. They also protect against neuroinflammation. Thus, improving gut health can lead to these positive outcomes (Ikeda et al. 2022).

In conclusion, probiotics and a healthy gut Microbiota are becoming more important. They are now being acknowledged as crucial in the prevention and control of diabetic neuropathy. By altering the gut-brain axis, probiotics have the ability to decrease inflammation. They can improve insulin sensitivity and advance nerve health. This offers a hopeful dietary strategy for individuals with diabetes.

11.10 Omega-3 Fatty Acids and Their Neuropathy Prevention Role

Omega-3 fatty acids are crucial polyunsaturated fats. They are essential for maintaining overall health. Particularly, they are significant in reducing inflammation and preserving nerve function. For persons with diabetes, omega-3 fatty acids offer significant promise. They are effective in both preventing and managing neuropathy. These fatty acids counteract chronic inflammation and oxidative stress, both of which significantly contribute to nerve impairment (Bayram and Kızıltan 2024).

Significant within these fatty acids are eicosapentaenoic acid (EPA) and docosahexaenoic acid (DHA). These two fatty acids have neuroprotective effects. They operate through numerous mechanisms, offering a natural and effective approach. This method supports nerve health (Sun et al. 2018a).

11.10.1 Anti-inflammatory Neuroprotective Consequences

Neuroprotective impacts of omega-3 fatty acids are mainly ascribed to their robust anti-inflammatory traits. Omega-3s cut inflammation by downregulating the production of pro-inflammatory cytokines and promoting the synthesis of anti-inflammatory molecules. These molecules include resolvins and protectins (Calder

2010). Chronic inflammation is a key source of nerve damage in diabetic neuropathy. It exacerbates oxidative stress, mitochondrial dysfunction, and nerve cell degeneration. Omega-3s reduce inflammation. This offers protection to nerve cells from additional damage and improves nerve conduction. This can alleviate neuropathy's symptoms, such as pain, tingling, and numbness.

Apart from reducing inflammation, omega-3s provide tender nerve protection. They do this by improving cellular membrane integrity. They also promote neuronal repair. DHA is a crucial element of cell membranes in the nervous system. It aids in maintaining the fluidity and function of nerve cell membranes. This allows for enhanced communication between nerve cells and the nervous system. Omega-3s are therefore crucial for supporting nerve health. They also help in preventing the progression of diabetic neuropathy (Zhang et al. 2018).

11.10.2 Crucial Omega-3 Fatty Acid Sources

Omega-3 fatty acids are plentiful in marine and plant-based sources. This makes it possible to incorporate these fats into the diet through a variety of foods. Key dietary sources include:

- Fatty Fish: These include salmon, mackerel, sardines, and anchovies. They are rich in EPA and DHA. These are the most bioavailable forms of omega-3s. Regular consumption of these fatty fish is linked to reduced inflammation. It is also linked to improved nerve health in individuals with diabetes (Das et al. 2024).
- Flaxseeds and Chia Seeds: These plant-based sources provide alpha-linolenic acid (ALA). It is a precursor to EPA and DHA. ALA requires conversion to EPA and DHA. These seeds are still excellent sources of omega-3s. They are good for those who follow a plant-based diet (Saini et al. 2021).

11.10.3 Clinical Proof

Numerous clinical studies confirm the role of omega-3 fatty acids. They prevent and manage diabetic neuropathy. Research has shown that omega-3 supplementation can lower pain. It can also improve nerve conduction significantly. Patients with diabetic neuropathy have benefitted from this (Durán et al. 2022).

One study published in Diabetes Care highlights these benefits. Patients who supplemented with omega-3 fatty acids saw a drop in neuropathic pain. Their nerve function improved too. Other studies show omega-3s can improve insulin sensitivity. They can also lower inflammation markers. This extra protection against nerve damage is particularly useful for people with diabetes (Yorek 2018).

Another option is using supplementation. It's a strategy that individuals with diabetes can use. This helps maintain nerve health and stops the progression of neuropathy.

11.11 Functional Food Supplements for Diabetic Neuropathy

A promising approach for managing diabetic neuropathy is supplements derived from functional foods. This is due to their powerful bioactive parts. Supplements like curcumin have significantly proven their potential in reducing inflammation and also in lowering oxidative stress and nerve damage. These are major influences that can contribute to neuropathy progression, especially in diabetes (Sun et al. 2018b).

Other supplements like resveratrol and quercetin show great therapeutic potential. They also reduce inflammation and oxidative stress (Inchingolo et al. 2022). Alpha-lipoic acid (ALA) is another aid. It has demonstrated its potential in alleviating the symptoms of diabetes. All these supplements work by focusing on the underlying mechanisms of nerve damage. This means they can be an add-on to the usual pharmacological treatments (Rohilla et al. 2023).

The primary component is curcumin. It has shown to have potential in managing symptoms of diabetes. It lowers inflammation and oxidative stress. Another is resveratrol. It also reduces inflammation and oxidative stress. Quercetin has positive potential in diabetes. Alpha-lipoic acid, or Ala, is a significant aid. It has shown considerable potential in treating nerve damage. These compounds are naturally found in various plants, fruits, and vegetables. They deal with the root causes of nerve damage, offering an additional solution to traditional pharmacological treatments.

11.12 Derived Supplements from Functional Foods

The food industry has shown considerable interest in the production of functional foods. These are capable of reducing the risk of diseases and promoting good health. Manufacturers exploit advances in food technology to produce these foods. They incorporate additives or supplements in their formulations. These food products are highly popular for being able to enhance human health.

Functional food-based supplements refer to these products. They involve extracting ingredients from functional foods. This process takes place in a concentrated form. These concentrated forms are then used to prepare the supplements. Doctors often prescribe these supplements to help manage various health conditions.

The extraction process is often complex. For instance, if we look at supplements derived from turmeric, the active compound in turmeric is curcumin. Curcumin is known for its anti-inflammatory properties. It also has great antioxidant capabilities. In addition, it has the potential to prevent cancer. These properties make turmeric an important ingredient in functional foods (Sharifi-Rad et al. 2020).

Curcumin supplements are often used in the management of diseases like diabetes. These supplements have proven effective in managing diabetic neuropathy. Studies show that curcumin reduces inflammation. It also reduces oxidative stress. It is known that inflammation and oxidative stress are both involved in nerve damage (Park et al. 2021).

Functional foods are important in the management of diseases. These supplements have shown bioactive effects. They can play a significant role in enhancing health status. There is no doubt that functional food-based supplements are a fundamental part of medicinal food. These supplements offer many benefits. They can reduce the risk of chronic diseases. This is especially important in ageing populations (Adefegha 2018).

The rise in the use of functional food-based supplements is a noticeable trend. It's a promising development in the food and pharmaceutical industries. Not only do these supplements have potential to treat diseases, but they also have the potential to reduce the need for synthetic drugs. This will lessen our dependency on these drugs (Puri et al. 2022).

As a result, the demand for these supplements increases. With this rise comes the risk of unregulated and unapproved products. It's a substantial concern. There is an urgent need to regulate these supplements. This will ensure their quality and safety.

Both the food and pharmaceutical industries have a role in ensuring the safety of functional food-based supplements. They must join forces to create quality assurance systems. These systems will guarantee the quality and safety of the products.

To maximize the potential of these supplements, collaboration between both parties is crucial. This will ensure a standardized production process. It will also mean that consumers can trust and benefit from these products. What's more is that this will offer a new field of research for the scientific community. This field can investigate the efficacy of functional food-based supplements.

In conclusion, functional food-based supplements are an important aspect of managing diseases. Studies continue to highlight the efficacy of these supplements. Proper regulations and an increased focus on research are crucial. This will ensure the quality and safety of these supplements. It will also optimize their potential in improving human health.

11.13 Recipes Showcasing Functional Foods

Many dishes highlight the nutritional value of functional foods in today's food scene, from antioxidant-rich berries to whole grains and dark, leafy veggies. By providing more than just nourishment, these components improve our cookery and actively support our general well-being. Functional foods can address various health concerns. Conditions like heart disease, diabetes, and obesity. They also have the potential to boost mental health. By offering essential nutrients. Essential building blocks. This can give us an edge in the hectic world.

Functional foods are starting to gain popularity. They are part of a lifestyle choice or an approach towards a healthier life. But integrating them into a diet is not always easy. It is a challenge. We need to learn how to cook with these foods to change our habits (Vlaicu et al. 2023; Birch and Bonwick 2019). However, these foods can be both versatile and easy to incorporate. There are myriad recipes available such as online cookbooks, Food blogs, cooking shows etc. They all offer an abundance of ideas. For smoothies, salads, and soups to baked goods, stews, and sauces. The only limit is our own imagination.

A shift towards healthier eating habits is a journey. This journey can be both rewarding and challenging. It might be hard at times to choose a whole grain over a refined one. To pick a lentil or a chickpea over a meat. Yet with every small step we take, we can improve our health, our dietary habits and our overall well-being. It is therefore worth exploring this culinary adventure. Incorporating functional foods into our meals are a way to nurture our bodies and minds. It might be difficult at first. But as with any other skill, practice makes perfect. Preparing meals from scratch can be a therapeutic ritual—a time to unwind and connect with nature. By embracing functional food recipes, it also becomes an act of self-care, nourishing both body and soul with delicious, health-enhancing ingredients.

11.14 Challenges and Future Perspectives in Functional Foods for Diabetic Neuropathy

Development and use of functional foods the use is important for preventing diabetic neuropathy. It faces numerous scientific and regulatory hurdles. One major challenge is the complexity of verifying health claims linked with functional foods. Regulatory bodies like FDA and EFSA are involved. The bodies have stringent regulations for proving the effectiveness and food safety. It is Especially true for these foods when They are promoted for disease prevention (Díaz et al. 2020).

Moreover, consistency in Bioactive compound concentrations it's a Challenge in functional foods. It is tough because natural food sources and processing Methods differ Clinical trials have limitations too (Chakrabarti et al. 2018). Numerous studies examining effects of functional Foods or bioactive compounds on Diabetic neuropathy have constraints. These studies experience short durations and lack of control over Diet and lifestyle. They often have small sample sizes.

These restrictions create difficulties in creating strong evidence for long-term benefits of functional foods. They can help manage or prevent neuropathy. A lot of existing research focuses on certain populations. These populations may fail to reflect the broad diversity of diabetic patients. In the future, research should consider specific bioactive compounds such as polyphenols, flavonoids and omega-3 fatty acids (Kaur et al. 2024).

It should investigate their mechanisms for preventing diabetic neuropathy. We need to conduct long-term large-scale studies with diverse populations. They are crucial in better understanding the Potential of functional foods. Advancements in biotechnology and nutrigenomics may also help. They could aid in creating personalized Dietary interventions. This would tailor functional foods to individual needs. This approach could be more effective in preventing and managing diabetic neuropathy.

11.15 Conclusion

Functional foods have an important role in the prevention and management of diabetic neuropathy. This makes them a noteworthy supplement to traditional pharmacological treatments. They include nutrient-rich antioxidant fruits. Omega-3 fatty acids and fiber also contribute. Using these foods can benefit nerve health. It can reduce risks associated with diabetes. They offer bioactive compounds. These help regulate glucose levels, reduce oxidative stress, and alleviate inflammation. These factors are important in the pathophysiology of diabetic neuropathy.

We can't overemphasize the importance of a balanced diet. Such a diet is critical for those managing diabetes. Pharmacological interventions are beneficial. They help in controlling blood sugar levels. However, dietary strategies hold potential to enhance overall health. They can complement medical treatments as well. A well-rounded nutrition approach is beneficial. It doesn't only improve nerve function. It also promotes better metabolic outcomes. This contributes to a holistic strategy for management of diabetes.

Even so, additional exploration is necessary. We need to establish more robust clinical evidence. This is on the effectiveness of specific bioactive compounds. They are found in functional foods. We need large-scale and long-term studies. These are needed to explore the full potential of these. These studies are focused on diabetic neuropathy prevention. The research will be useful. It will help develop tailored dietary recommendations. It may also help in designing functional food products. These can significantly benefit individuals at risk of or suffering from diabetic neuropathy.

When both dietary interventions and rigorous scientific investigation are prioritized, we pave the way. It is for innovative solutions. These solutions can improve quality of life. Especially, they can benefit those affected by diabetes.

References

Adefegha SA. Functional foods and nutraceuticals as dietary intervention in chronic diseases; novel perspectives for health promotion and disease prevention. J Diet Suppl. 2018;15(6):977–1009.

Ahmad R, et al. Lactic acid bacteria fermented foods: impact on immune system and consequences over type 2 diabetes mellitus. J Appl Pharm Sci. 2023;13(6):18–56.

Ahmadian N, Manickavasagan A, Ali A. Comparative assessment of blood glucose monitoring techniques: a review. J Med Eng Technol. 2023;47(2):121–30.

Akbari B, et al. The role of plant-derived natural antioxidants in reduction of oxidative stress. Biofactors. 2022;48(3):611–33.

Al Hroob AM, et al. Ginger alleviates hyperglycemia-induced oxidative stress, inflammation and apoptosis and protects rats against diabetic nephropathy. Biomed Pharmacother. 2018;106:381–9.

Alongi M, Anese M. Re-thinking functional food development through a holistic approach. J Funct Foods. 2021;81:104466.

Anandhanarayanan A, et al. Diabetic neuropathies. In: Endotext [Internet]. MDText.com, Inc; 2022.

Arabshomali A, et al. Potential benefits of antioxidant phytochemicals in type 2 diabetes. Molecules. 2023;28(20):7209.
Baltrusch S. The role of neurotropic B vitamins in nerve regeneration. Biomed Res Int. 2021;2021:9968228.
Barengolts E, et al. The effect of probiotic yogurt on glycemic control in type 2 diabetes or obesity: a meta-analysis of nine randomized controlled trials. Nutrients. 2019;11(3):671.
Barrea L, et al. Comprehensive approach to medical nutrition therapy in patients with type 2 diabetes mellitus: from diet to bioactive compounds. Antioxidants. 2023;12(4):904.
Bayram SŞ, Kızıltan G. The role of omega-3 polyunsaturated fatty acids in diabetes mellitus management: a narrative review. Curr Nutr Rep. 2024;13(3):527–51.
Bellikci-Koyu E, et al. Effects of regular kefir consumption on gut microbiota in patients with metabolic syndrome: a parallel-group, randomized, controlled study. Nutrients. 2019;11(9):2089.
Bhatti JS, et al. Oxidative stress in the pathophysiology of type 2 diabetes and related complications: current therapeutics strategies and future perspectives. Free Radic Biol Med. 2022;184:114–34.
Birch CS, Bonwick GA. Ensuring the future of functional foods. Int J Food Sci Technol. 2019;54(5):1467–85.
Calder PC. Omega-3 fatty acids and inflammatory processes. Nutrients. 2010;2(3):355–74.
Calderón-Ospina CA, Nava-Mesa MO. B vitamins in the nervous system: current knowledge of the biochemical modes of action and synergies of thiamine, pyridoxine, and cobalamin. CNS Neurosci Ther. 2020;26(1):5–13.
Cando LFT, et al. Current status of diabetes mellitus care and management in the Philippines. Diabetes Metab Syndr Clin Res Rev. 2024;18:102951.
Caturano A, et al. Oxidative stress in type 2 diabetes: impacts from pathogenesis to lifestyle modifications. Curr Issues Mol Biol. 2023;45(8):6651–66.
Chakrabarti S, Guha S, Majumder K. Food-derived bioactive peptides in human health: challenges and opportunities. Nutrients. 2018;10(11):1738.
Chen B, et al. A comprehensive review on beneficial effects of catechins on secondary mitochondrial diseases. Int J Mol Sci. 2022;23(19):11569.
Chen B, et al. Cellular zinc metabolism and zinc signaling: from biological functions to diseases and therapeutic targets. Signal Transduct Target Ther. 2024;9(1):6.
Cheng H, et al. Mitochondrial dysfunction plays a key role in the development of neurodegenerative diseases in diabetes. Am J Physiol Endocrinol Metab. 2020;318(5):E750–64.
Chetty L. Prevalence, traditional medicine use and co-morbidities among type 2 diabetes mellitus in outpatients-a cross sectional hospital-based survey in KwaZulu-Natal. Durban University of Technology; 2022.
Chun H, Park Y. Oxidative stress and diabetic neuropathy. In: Diabetes. Elsevier; 2020. p. 13–23.
Clark N. Nancy Clark's sports nutrition guidebook. Human Kinetics; 2019.
Das P, et al. Advances in therapeutic applications of fish oil: a review. Meas Food. 2024;13:100142.
Dewanjee S, et al. Altered glucose metabolism in Alzheimer's disease: role of mitochondrial dysfunction and oxidative stress. Free Radic Biol Med. 2022;193:134–57.
Díaz LD, Fernández-Ruiz V, Cámara M. An international regulatory review of food health-related claims in functional food products labeling. J Funct Foods. 2020;68:103896.
Dixit V, et al. Functional foods: exploring the health benefits of bioactive compounds from plant and animal sources. J Food Qual. 2023;2023(1):5546753.
Durán AM, et al. Dietary omega-3 polyunsaturated fatty-acid supplementation upregulates protective cellular pathways in patients with type 2 diabetes exhibiting improvement in painful diabetic neuropathy. Nutrients. 2022;14(4):761.
Ekici-Günay N. Ginkgo biloba extract as an antioxidant in nerve regeneration. In: Pathology. Elsevier; 2020. p. 235–45.
Elafros MA, et al. Towards prevention of diabetic peripheral neuropathy: clinical presentation, pathogenesis, and new treatments. Lancet Neurol. 2022;21(10):922–36.
Elbandy M. Anti-inflammatory effects of marine bioactive compounds and their potential as functional food ingredients in the prevention and treatment of neuroinflammatory disorders. Molecules. 2022;28(1):2.

Eleazu CO. The concept of low glycemic index and glycemic load foods as panacea for type 2 diabetes mellitus; prospects, challenges and solutions. Afr Health Sci. 2016;16(2):468–79.
Essa MM, et al. Functional foods and their impact on health. J Food Sci Technol. 2023;60:820–34.
Fatima MT, et al. The role of dietary antioxidants in type 2 diabetes and neurodegenerative disorders: an assessment of the benefit profile. Heliyon. 2023;9(1):e12698.
Feldman EL, et al. Diabetic neuropathy. Nat Rev Dis Prim. 2019;5(1):1–18.
Felgus-Lavefve L, et al. The effects of blueberry phytochemicals on cell models of inflammation and oxidative stress. Adv Nutr. 2022;13(4):1279–309.
Feng J, et al. Role of magnesium in type 2 diabetes mellitus. Biol Trace Elem Res. 2020;196(1):74–85.
Fernandes I, et al. The role of nutraceutical containing polyphenols in diabetes prevention. Metabolites. 2022;12(2):184.
Garcia-Molina L, et al. Improving type 2 diabetes mellitus glycaemic control through lifestyle modification implementing diet intervention: a systematic review and meta-analysis. Eur J Nutr. 2020;59(4):1313–28.
Gengatharan A, Che Zahari C-N-M, Mohamad N-V. Exploring lycopene: a comprehensive review on its food sources, health benefits and functional food applications. Curr Nutr Food Sci. 2024;20(8):914–31.
Giuntini EB, Sardá FAH, de Menezes EW. The effects of soluble dietary fibers on glycemic response: an overview and futures perspectives. Foods. 2022;11(23):3934.
Gorji N, Moeini R, Memariani Z. Almond, hazelnut and walnut, three nuts for neuroprotection in Alzheimer's disease: a neuropharmacological review of their bioactive constituents. Pharmacol Res. 2018;129:115–27.
Guo Y, et al. The role of nutrition in the prevention and intervention of type 2 diabetes. Front Bioeng Biotechnol. 2020;8:575442.
Gupta E, Mishra P. Functional food with some health benefits, so called superfood: a review. Curr Nutr Food Sci. 2021;17(2):144–66.
Hernandez-Ledesma B, Martinez-Villaluenga C. Current advances for development of functional foods modulating inflammation and oxidative stress. Academic Press; 2021.
Hoque M, et al. Comprehensive guide to vitamin and mineral sources with their requirements. Indiana J Agric Life Sci. 2023;3(6):23–31.
Horton WB, Barrett EJ. Microvascular dysfunction in diabetes mellitus and cardiometabolic disease. Endocr Rev. 2021;42(1):29–55.
Ikeda T, et al. Short-chain fatty acid receptors and gut microbiota as therapeutic targets in metabolic, immune, and neurological diseases. Pharmacol Ther. 2022;239:108273.
Inchingolo AD, et al. Effects of resveratrol, curcumin and quercetin supplementation on bone metabolism – a systematic review. Nutrients. 2022;14(17):3519.
Jomova K, et al. Reactive oxygen species, toxicity, oxidative stress, and antioxidants: chronic diseases and aging. Arch Toxicol. 2023;97(10):2499–574.
Josimuddin S, Kumar M, Rastogi H. A review on nutritional and medicinal value of malus domestica with various activity. Int J Health Sci. 2022;6(S4):7251–65.
Kaur J, et al. Insights into the latest advances in low glycemic foods, their mechanism of action and health benefits. J Food Meas Charact. 2022;16:1–14.
Kaur M, et al. Macroalgae-derived bioactive compounds for functional food and pharmaceutical applications—a critical review. Crit Rev Food Sci Nutr. 2024;65:1–23.
Khan R, et al. Advanced glycation end products: formation, role in diabetic complications, and potential in clinical applications. In: The eye and foot in diabetes. IntechOpen; 2020.
Kuzan A. Toxicity of advanced glycation end products. Biomed Rep. 2021;14(5):1–8.
Li H, Zhang Q. Research progress of flavonoids regulating endothelial function. Pharmaceuticals (Basel). 2023;16(9):1201.
Lin Q, et al. Oxidative stress in diabetic peripheral neuropathy: pathway and mechanism-based treatment. Mol Neurobiol. 2023;60(8):4574–94.
Lu S, et al. The communication mechanism of the gut-brain axis and its effect on central nervous system diseases: a systematic review. Biomed Pharmacother. 2024;178:117207.

Mandal B, Das R, Mondal S. Anthocyanin: a potential phytochemical candidate for the amelioration of non-alcoholic fatty liver disease. Ann Pharm Fr. 2024;82:373–91. Elsevier.
Mayeda L, et al. Glucose time in range and peripheral neuropathy in type 2 diabetes mellitus and chronic kidney disease. BMJ Open Diabetes Res Care. 2020;8(1):e000991.
McRae MP. Dietary fiber intake and type 2 diabetes mellitus: an umbrella review of meta-analyses. J Chiropr Med. 2018;17(1):44–53.
Md Isa Z, Amsah N, Ahmad N. The impact of vitamin D deficiency and insufficiency on the outcome of type 2 diabetes mellitus patients: a systematic review. Nutrients. 2023;15(10):2310.
Megur A, et al. Prebiotics as a tool for the prevention and treatment of obesity and diabetes: classification and ability to modulate the gut microbiota. Int J Mol Sci. 2022;23(11):6097.
Méndez-Morales S, et al. Diabetic neuropathy: molecular approach a treatment opportunity. Vasc Pharmacol. 2022;143:106954.
Mirmiran P, Bahadoran Z, Azizi F. Functional foods-based diet as a novel dietary approach for management of type 2 diabetes and its complications: a review. World J Diabetes. 2014;5(3):267–81.
Mironenko A, Eliseeva T. Vitamin B9–description, benefits, effects on the body and best sources. J Healthy Nutr Diet. 2019;4(10):88–100.
Mondal S, et al. Application of probiotic bacteria for the management of fish health in aquaculture. In: Bacterial fish diseases. Elsevier; 2022. p. 351–78.
Muhammad Sajid A, et al. Functional foods and human health: an overview. In: Muhammad Sajid A, Muhammad Haseeb A, editors. Functional foods. IntechOpen: Rijeka; 2021.
Mukherjee S, et al. Therapeutic effect of targeted antioxidant natural products. Discov Nano. 2024;19(1):1–28.
Mullins BM, Kelsall N. Anatomy of the lumbar and sacral plexuses and lower limb peripheral neuropathies. Surgery (Oxford). 2023;41(4):193–9.
Nguyên B. Nutrition and its effects on immunity. 2021.
Park H, et al. Effects of curcumin treatment in a diabetic neuropathic pain model of rats: involvement of c-Jun N-terminal kinase located in the astrocytes and neurons of the dorsal root ganglion. Pain Res Manag. 2021;2021:8787231.
Parsamanesh N, et al. Therapeutic potential of curcumin in diabetic complications. Pharmacol Res. 2018;136:181–93.
Peng M, et al. Effectiveness of probiotics, prebiotics, and prebiotic-like components in common functional foods. Compr Rev Food Sci Food Saf. 2020;19(4):1908–33.
Pop-Busui R, et al. Diagnosis and treatment of painful diabetic peripheral neuropathy. American Diabetes Association; 2022.
Prabhakar O. Inflammatory biomarkers as a part of diagnosis in diabetic peripheral neuropathy. J Diabetes Metab Disord. 2021;20(1):869.
Preedy VR. Diabetes: oxidative stress and dietary antioxidants. Academic Press; 2020.
Puri V, et al. A comprehensive review on nutraceuticals: therapy support and formulation challenges. Nutrients. 2022;14(21):4637.
Putz Z, et al. Vitamin D in the prevention and treatment of diabetic neuropathy. Clin Ther. 2022;44(5):813–23.
Rak K, Bronkowska M. Immunomodulatory effect of vitamin D and its potential role in the prevention and treatment of type 1 diabetes mellitus—a narrative review. Molecules. 2018;24(1):53.
Rohilla M, et al. Discussing pathologic mechanisms of diabetic retinopathy & therapeutic potentials of curcumin and β-glucogallin in the management of diabetic retinopathy. Biomed Pharmacother. 2023;169:115881.
Rudrapal M, et al. Dietary polyphenols and their role in oxidative stress-induced human diseases: insights into protective effects, antioxidant potentials and mechanism(s) of action. Front Pharmacol. 2022;13:806470.
Saini RK, et al. Omega-3 polyunsaturated fatty acids (PUFAs): emerging plant and microbial sources, oxidative stability, bioavailability, and health benefits—a review. Antioxidants. 2021;10(10):1627.
Samanta C, et al. Omega-3 fatty acid and its protective effect against cancer and cancer-related complication. J Pharm Res Int. 2022;34:51–62.

Sarwar MF, Sarwar MH, Sarwar M. Deficiency of vitamin B-complex and its relation with body disorders. In: B-complex vitamins-sources, intakes and novel applications. IntechOpen; 2021. p. 79–100.
Shahanas E, et al. Health benefits of bioactive compounds from cocoa (Theobroma cacao). Agric Rev. 2019;40(2):143–9.
Sharifi-Rad J, et al. Turmeric and its major compound curcumin on health: bioactive effects and safety profiles for food, pharmaceutical, biotechnological and medicinal applications. Front Pharmacol. 2020;11:01021.
Shen C-L, et al. Bioactive compounds for neuropathic pain: an update on preclinical studies and future perspectives. J Nutr Biochem. 2022;104:108979.
Sloan G, Selvarajah D, Tesfaye S. Pathogenesis, diagnosis and clinical management of diabetic sensorimotor peripheral neuropathy. Nat Rev Endocrinol. 2021;17(7):400–20.
Smith S, et al. Prevention and management strategies for diabetic neuropathy. Life. 2022;12(8):1185.
Strand N, et al. Small fiber neuropathy. Curr Pain Headache Rep. 2022;26(6):429–38.
Sun GY, et al. Docosahexaenoic acid (DHA): an essential nutrient and a nutraceutical for brain health and diseases. Prostaglandins Leukot Essent Fat Acids. 2018a;136:3–13.
Sun J, et al. Role of curcumin in the management of pathological pain. Phytomedicine. 2018b;48:129–40.
Ten Donkelaar HJ, et al. The autonomic nervous system. In: Clinical neuroanatomy: brain circuitry and its disorders. Springer; 2020. p. 669–710.
Vincent AM, et al. Oxidative stress in the pathogenesis of diabetic neuropathy. Endocr Rev. 2004;25(4):612–28.
Vinik AI, Casellini C, Neumann S. Diabetes and the nervous system. In: Diabetes complications, comorbidities and related disorders. Springer; 2020. p. 275–353.
Vlaicu PA, et al. Designing nutrition for health-incorporating dietary by-products into poultry feeds to create functional foods with insights into health benefits, risks, bioactive compounds, food component functionality and safety regulations. Foods. 2023;12(21):4001.
Yildiran H, Macit MS, Özata Uyar G. New approach to peripheral nerve injury: nutritional therapy. Nutr Neurosci. 2020;23(10):744–55.
Yorek MA. The potential role of fatty acids in treating diabetic neuropathy. Curr Diab Rep. 2018;18:1–10.
Zhang AC, et al. Omega-3 polyunsaturated fatty acid supplementation for improving peripheral nerve health: protocol for a systematic review. BMJ Open. 2018;8(3):e020804.

Clinical Evidence: Functional Foods in Diabetic Neuropathy Management

12

Rakesh Das, Tamal Mondal, Puttaswamy Madhusudan, Tiyas Saha, Koustav Dutta, and Ananta Choudhury

12.1 Introduction

Diabetes Mellitus is a metabolic disorder characterized by chronic hyperglycemia and insulin-related defects, affecting lipid, protein, and carbohydrate metabolism. It has varying symptoms based on the type and duration of the disease, and if not managed properly, can lead to organ dysfunction and even death. Complications include cardiovascular, retinopathy, neuropathy, and nephropathy. Diabetes has been referred to as the "epidemic of the century" due to its high incidence and prevalence rates. The World Health Organization (1999) estimated diabetes as the seventh

R. Das
School of Pharmaceutical Technology, School of Health and Medical Sciences, Adamas University, Kolkata, West Bengal, India

T. Mondal
Department of Botany, Hiralal Mazumdar Memorial College for Women, Dakshineswar, Kolkata, India

P. Madhusudan (✉)
Department of Civil and Environmental Engineering, Environmental Materials Laboratory, Hanyang University, Seoul, South Korea
e-mail: madhu2010@hanyang.ac.kr

T. Saha
Radiant Institute of Pharmaceutical Sciences, Department of Pharmaceutics, Kishanganj, Bihar, India

K. Dutta
Department of Pharmaceutical Technology, School of Health and Medical Sciences, Adamas University, Kolkata, West Bengal, India

A. Choudhury
Faculty of Pharmaceutical Science, Assam down town University, Guwahati, Assam, India

M. K. Gautam et al. (eds.), *Antioxidants and Functional Foods for Diabetic Neuropathy*, https://doi.org/10.1007/978-981-95-9326-2_12

major cause of mortality in 2016, with 2.2 million deaths linked to increased blood sugar levels in 2012 (Kharroubi and Darwish 2015; Wong et al. 2017; Deshmukh and Jain 2015; Santaguida et al. 2005).

Diabetes mellitus, an autoimmune disease caused by insufficient insulin secretion by pancreatic beta cells, results in insulin insufficiency and affects 25% of diabetics. Despite being more common in children, this disease can also strike adults because of genetic reasons. Cumulative damage includes conditions including neuropathy, coronary heart disease, and nephropathy. Nutrient-dense foods may reduce the risk of cardiovascular disease, dyslipidemia, type 2 diabetes, and cancer. By lowering oxidative stress, cholesterol, inflammation, and elevated insulin resistance, they improve insulin sensitivity and decrease insulin tolerance (John et al. 2020; Cassidy et al. 2018; Dey et al. 2020).

By inhibiting enzymes involved in the digestion of carbohydrates, decreasing glucose absorption, and boosting insulin secretion, biologic peptides found in dietary supplements aid in the treatment and management of type 2 diabetes. Phenolic chemicals found in blackberries increase the amount of glycogen in cells and encourage HepG2 cells to consume glucose, two processes that are critical for maintaining mitochondrial membrane potential. By lowering blood glucose levels, insulin production, and glucose intake, flavonoids—which are found in many fruits, nuts, cereals, chocolate, and plants—have the potential to prevent diabetes. Quercetin, a flavonoid with antihyperglycemic effects, regulates the insulin-dependent mitogen-activated protein kinase (MAPK) pathway, enhances glucose absorption, increases glutathione peroxidase, regulates NF-κB signaling, and prevents pancreatic β-cell death. It is present in berries, onions, apples, seeds, flowers, and leaves (Gadang et al. 2011; Oboh et al. 2014).

Prebiotics are substances are inedible fiber-containing materials found in food that enhance the growth and function of gut bacteria. Common probiotics, such as Bifidobacterium and Lactobacillus, are found in dietary supplements and nutritious foods such fruit juices, ice cream, cheese, and baby formula. Synbiotics, a mix of prebiotics and probiotics, increase the number of lactobacilli and bifidobacteria, which prolongs the life of probiotics and maintains the equilibrium of gut microorganisms (Mofleh et al. 2013; Koh et al. 2014; Chawla et al. 2016; Jin and Liu 2021; Liu et al.).

It is being established that probiotics and synbiotics enhance NO, GSH, and TAC levels despite diminishing TNF-α, MDA, and CRP levels in diabetics. It is being discovered that Bifidobacteria, an important probiotics in the gut of humans, enhances insulin resistance. Administration via probiotics boosted the adhesion effectiveness of insulin in type 2 diabetes mellitus (T2DM) and lowered β-cell mortality in diabetic mice. Fruits, vegetables, and herb include edible polysaccharides that have anti-inflammatory, anti-cancer, anti-diabetic, and anti-oxidative properties (Lan-Pidhainy and Wolever 2011; Thule and Umpierrez 2014; Li et al. 2022).

12.2 Diabetes: A Metabolic Disorder

Type 2 diabetes is a condition of metabolism characterized by elevated sugar levels, resistance to insulin, β-cell dysfunction, and impaired insulin production (Kharroubi and Darwish 2015; Wong et al. 2017) Type 2 diabetes commonly leads to metabolic difficulties, such as oxidative stress, inflammation, vascular endothelial dysfunction, decreased lipid and lipoprotein metabolism, and hypertension (Deshmukh and Jain 2015; Melendez-Ramirez et al. 2010). The long-term pathological illnesses such as micro- and macro-vascular penalties involving neuropathy, retinopathy, and nephropathy, as well as a reduced quality living and higher mortality rate, can be carried on by these abnormal metabolic processes (Kang et al. 2020; Ozougwu et al. 2013). The cure of type 2 diabetes demands clinical nutrition intervention (MNT), involving nutritional lipid and cholesterol standards, food exchanging lists, glycemic index and load, energy and nutrient estimation, carbohydrate tally, and proper dietary planning (Granato et al. 2020; John et al. 2020); since there is little research to back up those recommendations, it is challenging to measure the beneficial effects of several micronutrients for controlling sugar levels in patients with diabetes (Cassidy et al. 2018; Dey et al. 2020; Gadang et al. 2011; Oboh et al. 2014; Mofleh et al. 2013). Evidently, the curative and medicinal properties of foodstuffs that might boost the efficiency of type 2 diabetes diet management may be excluded from the MNT pathway.

12.2.1 The Composition of Whole Grains and Their Effects on Diabetics

Foods based on grains and cereals are essential for the nutrition of individuals, especially for those with type 2 diabetes. Whole-grind bread and porridge items derived from barley, oats, rye, and whole wheat are essential components of a diet that emphasizes healthful foods. Results show that in individuals with the condition, dietary carbohydrate modification increases insulin responsiveness, reduces adipocyte cell size, and alters adipose tissue genes implicated in insulin pathway signaling (Koh et al. 2014) In contrary to blanched grains, entire wheat grains include complex polysaccharides that are impermeable in water, as well as tocopherols, phenolic acids, carotenoids, phytates, and phytoestrogens. Such compounds are thought to provide protection against a variety of malignancies, obesity, diabetes, high blood pressure metabolic syndrome, and cardiovascular problems (Chawla et al. 2016; Jin and Liu 2021; Lan-Pidhainy and Wolever 2011; Salehi et al. 2012) Research indicates that both WGs and cereal-based foods help diabetics by increasing their insulin sensitivity, increasing the activity of their pancreatic β-cells, hiding more insulin, and effectively controlling their blood sugar levels (Thule and Umpierrez 2014; Mishra and Mishra 2021). The main digestible and easily absorbed fibers in WGs, inulin and β-glucan, have hypoglycemic and hypolipidemic effects. However, they additionally serve as prebiotics in the gastrointestinal tract and alter the microbiota of the intestinal tract by promoting the establishment and activity of bifidobacteria and lactic acid bacteria, which increases the use of energy (Li et al. 2018; Zhang et al. 2009; Loo et al. 2013). Higher intakes of whole grain cereals, including fiber, bran, and germ, are associated with fewer deaths from circulatory and all-cause illnesses in studies that follow up with patients who have diabetes over an extended period of time. Blood pressure, serum homocysteine, triglycerides, LDL-C, oxidative and inflammatory states, endothelial functioning, and several other atherosclerosis risk factors for illnesses may all be altered by regular usage of WGs supplements (Yang et al. 2021). Since of its high fiber content, rye, a grain that is widely consumed in Northern and Eastern Europe, has been classified as a functional grain since it slows down the absorption and digestion of dietary carbs. When intestinal fermentation in rye products leads to an increase in molecules, β-cells release more insulin. Rye for breakfast reduces ghrelin and satisfaction, promotes gastrointestinal breakdown, and immediately increases insulin and glucose sensitivity.

Items made from oat meal are also thought to be excellent sources of carbohydrates for diabetics since they are rich in soluble fiber, antioxidants, and bioactive compounds. In diabetic individuals, oat products improve glycemic, insulinemic, and lipidemic responses and actively reduce postprandial glycemia. Barley and its derivatives have been found to aid with glucose tolerance and insulin resistance. It has been demonstrated that barley and its byproducts improve glucose tolerance and insulin resistance in pre-diabetic people. Furthermore, the grain and its constituents are thought to improve insulin and blood sugar regulation. Fiber, potassium, magnesium, phenolic acids, α-tocopherols, carotenoids, and antioxidants are all abundant in wheat bran and other whole wheat products. Because brown rice and its near cousins have a lower GI and a much lower amount of glucose than white rice, they have recently been studied as nutritious foods (Zhang et al. 2009).

In already diabetic individuals, barley and its derivatives are being shown to improve insulin sensitivity and sugar tolerance. White rice and other refined carbs can be used with barley to lower blood glucose levels. Additionally, barley products contain anti-inflammatory, antioxidant, and hypolipidemic qualities. Barley improves liver characteristics, lowers cholesterol, and raises adiponectin and fatty acid oxidation in experimental types of diabetes (Reusch and Manson 2017; Patel and Dhanabal 2013; Jung and Choi 2014; Li et al. 2022; El-Sayed et al. 2023).

Some research has shown that whole grain wheat and associated substances improve insulin and glucose metabolism control. Dietary fiber, potassium, magnesium, phenolic acids, α-tocopherols, carotenoids, and antioxidants are all abundant in whole grain bran and whole grain products. The bran and germ components are responsible for many of the aforementioned benefits. A substantial source of fiber, lignans, phenolic acid, and alkylresorcinol, wheat bran might raise lipid abnormalities, cardiovascular risk factors, postprandial glycemic response, and glycosylated hemoglobin in diabetics. The grain germ's high concentration of phytosterols, benzoquinone, flavonoids, and non-digestible oligosaccharides makes it an efficient inflammatory and antioxidant agent (Banerjee and Maulik 2002; Bakar et al. 2009; Khoo et al. 2017; Hanhineva et al. 2010; Tian et al. 2016).

Given that brown rice and its cousins require less insulin and have a lower glycemic index than white rice, they are beneficial foods. In individuals with type 2 diabetes, they include dietary fiber, vitamins, minerals, and bioactive substances that can improve glycemic control, liver issues, endothelial cell dysfunction, dyslipidemia, and abdominal obesity. γ-orizanol, found in brown rice, can prevent type 2 diabetes, change high-fat diets, and increase β-cell function. Both hatched and pre-germinated brown rice have special ingredients that have potent anti-diabetic actions. Pectin, β-glucan, tocopherols, and 31% fiber are among the nutrients found in bran rice, which is the leftover grain from brown rice. It may improve HDL-C while lowering total cholesterol, LDL-C, and hemoglobin glycosylated (Fang 2014; Wallace et al. 2015; Wu et al. 2006; Su and Chien 2007; Borges et al. 2010; Manach et al. 2005).

Peas, beans, lentils, peanuts, and other legumes are significant suppliers of dietary protein, oligosaccharides, resistant starches, non-digestible carbohydrates, and bioactive components. Frequent consumption of beans reduces the risk of cardiovascular disease, type 2 diabetes, and obesity. α-amylase regulators peptides, which are found in legumes, alter the postcolonial glycemic response, control lipid metabolism, and reduce the metabolism and absorption of caloric foods. They additionally include a variety of phenolic acids that can stop the blood flow and side effects of angiotensin II.

Beans are another important legume grain because of their high quantities of fiber, phytate, ω3 fatty acids, antioxidants, and phenolic compounds. Because of their similar hypoglycemic effect to anti-diabetic drugs, they can help regulate weight, lower the postprandial glycemic response, and improve dyslipidemia. Soybeans are rich in unique phytoestrogens that help with insulin sensitivity, glycemic control, renal function, and dyslipidemia. Regular use of soybean products can help diabetic patients better manage their dyslipidemia.

Other anti-obesity properties of legumes include their thermogenic properties, ability to induce satiety, and their ability to mediate gene expression and release of key adipocytokines. Regular consumption of these functional meals may help control a number of aspects of type 2 diabetes (Milenkovic et al. 2019; Bakuradze et al. 2012; Forester and Waterhouse 2009; Williamson and Clifford 2010; Kofuji et al. 2012; Whent et al. 2012; Stevenson et al. 2012; Ross et al. 2004).

Nuts like almonds, pistachios, walnuts, and hazelnuts that are rich in antioxidant vitamins as well as proteins might help lower the risk of developing heart disease. Consuming nuts can help people with type 2 diabetes improve their insulin obstruction, lower their postprandial glycemic response, reduce oxidative stress, normalize their lipoprotein and cholesterol levels, and improve the quality of their diet overall. Nuts also change hunger neurotransmitters, reduce inflammation, and help control weight. Feeding nuts also increases the level in plasma of adiponectin, decreases high-sensitivity C-reactive protein levels, and lowers soluble inflammatory and cardiovascular risk markers. Nuts also positively affect the functioning of endothelial cells due to their high L-arginine and antioxidant levels (Yao et al. 2012; Geil and Anderson 1994; Preuss 2009; Thompson et al. 2012; Anderson and Bush 2011).

12.3 Wealthy Fruits and Vegetables, Which Are Full in Phytochemicals, Provide Essential Nutrients and Health Benefits

Fruits and vegetables are essential for maintaining health and avoiding chronic illnesses due to their high dietary fiber, vitamin, and phytochemical content (Azadbakht et al. 2003). Regular consumption of fruits and vegetables improves glycemic control, lowers HbA1c and triglyceride levels, increases antioxidant defense, minimizes oxidative stress, reduces the risk of diabetic retinopathy, and lessens the burden of carotid atherosclerosis in diabetes patients (Bhathena and Velasquez 2002; Gilbert and Liu 2013; Ronis et al. 2009; Yang et al. 2011). Eating a range of vegetables and fruits is recommended because these foods are rich in micronutrients and bioactive compounds. The vibrant hue of these fruits and vegetables reflects the pigmentation of the phytochemicals that comprise most of their nutraceuticals. Watermelon, grapefruit, tomatoes, and other fruits and vegetables that are rich in phytonutrients have beneficial effects on lipid metabolism, vascular function, and diabetic diseases. Anthocyanin-rich fruits also have antioxidant and hypoglycemic properties.

12.4 Exceptionally Valuable Functional Foods and Bioactive Substances for Diabetic Patients

It is outside the purview of this article to examine all of the numerous natural foods, nutraceuticals, or bioactive chemicals that may be regarded as functional components and have a positive impact on the treatment of diabetes. Some of these possible functional foods, along with dairy products, probiotics, fish meat, green tea, and spices, are included in Table 12.1.

12.5 Diabetic Neuropathy

Diabetic peripheral neuropathy, which affects up to 60% of individuals with type 2 diabetes, is associated with major morbidity, including amputations, anxiety, depression, poor quality of life, and irregularities in gait (Josef and Raj 2011; Kles and Vinik 2006; Wu et al. 2007). The condition damages the final sections of peripheral nerves by affecting the small fibers that transmit temperature, light touch, and pain. As neuropathy increases, it affects large fibers that regulate muscle tone and reflexes, which can lead to problems with walking and balance. In 15–30 percent of cases, discomfort is felt. Plant-based dietary interventions frequently improve glycemic control and other factors associated with type 2 diabetes and its effects. In a study comparing a low in fat plant-based diet with a more conventional diet, the plant-based group's HbA1c (percent hemoglobin A1c) dropped by 1.2 points, whereas the control group's declined by 0.4 points. Glucose control appears to have a major impact on the likelihood of issues (Dobretsov et al. 2007; Yokoyama et al.

Table 12.1 The bioactive substances and useful qualities of several healthful fruits and vegetables are investigated

References	Potential functional characteristics of diabetes	Principal phytochemicals and bioactive ingredients	Vegetables and fruits
Mattes and Dreher (2010), Heber (2004), Tanaka et al. (2013)	Postprandial glycemia, lipid peroxidation, free radical generation, serum total capacity for antioxidants, vascular damage prevention, and dyslipidemia are all improved by dietary carbohydrate uptake.	Quercetin, epicatechin, soluble fiber, phlordizin, procyanidins, P-cumaric acid, chlorogenic acid, and gallic acid	Red apple, apple peel, apple and its by products
Takahashi et al. (2012), Chan et al. (2013), Ali and Agha (2009)	Increased insulin resistance, postprandial oxidative stress, lipid peroxidation, better dyslipidemia, blood sugar regulation, and inhibition of α-glucosidase and α-amylase activity. Activity of antioxidant enzymes Systolic blood pressure as a whole The total antioxidant capacity of plasma Adipocyte lipolysis, inflammation, and modulation of receptors that activate peroxisome proliferator-activated	α-carotene, β-carotene, lutein, delphinidins, pelargonidins, ciyanidins, tannins, ellagitanins, anthocyanins, hydroxy-cinnamic acid, and catechins	Berries; cranberry, blackberry, black raspberry, blueberry, red raspberry, strawberries
Martins et al. (2007), Setorki et al. (2009), Boyer and Liu (2004), Chai et al. (2012), Hyson (2011)	↓ oxidative stress ↓ cytokine production ↓ PPARγ activation ↓ hypoglycemia ↓ HbA1c, promote the development of metabolic syndrome and fatty liver features, alleviate lipid disorders, and have anti-inflammatory properties (inhibit cyclooxygenase).	Quercetin, hydroxy-cinnamic acid, lutein, xanthine, gallic acid, phenolic acids, carotenoids, melatonin, anthocyanins, and β-carotene	Cherries
Kaume et al. (2012), Seymour et al. (2011), Suh et al. (2011), Apostolidis et al. (2006)	↓ Hyperglycemia, lessen the metabolic issues brought about by it, enhance dyslipidemia, delay the development of nephropathy, lower inflammation, raise superoxide dismutase, glutathione reductase, glutathione peroxidase, and lipid peroxidation	Isothiocyanates, anthocyanins (red cabbage), carotenoids, lutein, β-carotene	Cabbage, Cauliflower

(continued)

Table 12.1 (continued)

References	Potential functional characteristics of diabetes	Principal phytochemicals and bioactive ingredients	Vegetables and fruits
Banihani et al. (2013), Taj Eldin et al. (2010), Gabler et al. (2006), Ashraf et al. (2005), Chanet et al. (2012), Assini et al. (2013)	Inflammatory cytokines ↓ lipid peroxidation ↓ oxidative stress and inhibiting stress-sensitive signaling pathways ↓ endothelial macrophage activation ↓ platelet hyperactivity and aggregation, improving vascular function ↓ dietary lipid digestion, improving dyslipidemia	naringenin, hesperidin, β-carotene, lutein, xanthine, α-cryptoxanthin, β-cryptoxanthin, and phytosterols	Citrus fruits
Bergman et al. (2001), Moser et al. (2011)	↓ Free radical generation, lipid peroxidation, bile acid binding, cholesterol excretion, lipid profile enhancement, and plasma total antioxidant capacity	Lutein, betaine, violaxanthine, opioid peptides (rubisculins), *P*-cumaric acid, ferulic acid	Spinach
Yadav et al. (2010), Yoshinari et al. (2009)	Elevated insulinemic and glycemic response ↓ systemic inflammation ↓ risk factors for cardiovascular disease.	Pectin, oleic and linolenic acids, carotenoids	Pumpkin
Utsunomiya et al. (2005)	Reduce hyperglycemia and dyslipidemia adiponectin, antioxidant, and anti-inflammatory properties	Anthocyanins, flavonoids, fiber, polyphenols, and chlorogenic acid	Plums
Lucas et al. (2011), Masibo and He (2009), Li et al. (2018)	In order to avoid postprandial glycemia, stop α-amylase. Preventing diabetic nephropathy, lowering lipid peroxidation, improving dyslipidemia, and producing glycogen	Gallic acid, caffeic acid, carotenoids, quercetin, kampferol, tannins, catechins, and mangiferin	Mango
Al-Farsi and Lee (2008), Saafi et al. (2011), Zangiabadi et al. (2011)	Reducing lipid peroxidation, increasing antioxidant enzymes, and protecting the liver and kidneys from oxidative injury are all advantages of preventing diabetic neuropathy.	Dietary fiber, polyphenols, acid cinnamic, melatonin	Date fruit

2014a, b; Ferdowsian et al. 2010). A randomized, regulated pilot study was conducted to explore the hypothesis that a plant-based dietary intervention could reduce the pain related to diabetic neuropathy. Clarifying the mechanisms by which a dietary intervention can lead to clinical changes or pinpointing the precise elements of the intervention responsible for any observed impact were not the study's objectives (Barnard et al. 2004; Smith et al. 2006).

12.6 Diabetic Neuropathy-Based Clinical Evidential Study

The few of investigation, volunteers to have been diagnosed with type 2 diabetes, have experienced severe diabetic neuropathy for at least 6 months, and be between the ages of 18 and 65. Exclusions included vegetarianism, pregnancy, unstable mental or physical health problems, alcohol and recreational drug use, vitamin B12 deficiency, and unwillingness to participate. Participants were evaluated for B12 deficiency using a blood level test for B12 and a serum methylmalonic acid test. A screening interview was used to determine eligibility, and it was presumed that a patient had painful diabetic neuropathy if they reported having pain in their hands or feet. The Ethical and Independent Review Services approved the study, and each participant filled out a permission form (Crane and Sample 1994). 35 people were divided into the intervention as well as control groups using a distribution sequence created from a table with random numbers. In the first copying, the allocation was concealed using sealed envelopes, and assessments of the Michigan Neuropathy Screening Instrument and Neuropathy Impairment Score-Lower Limb were carried out by a blinded, independent physician who was impartial toward the study premise (Savage et al. 1994; Bastyr III et al. 2005; Bax et al. 1996; Bril 1999).

12.7 An Interventional Study Design with Protocol

Some investigation was to look at how a diet consisting of plants affected those with type 2 diabetes. A diet that excluded animal products, restricted fat consumption, and prioritized items with a low glycemic index was prescribed to the intervention group. Fruits, grains, legumes, and vegetables were all part of the diet. A daily vitamin B12 supplement was also administered to the participants. In contrast, the control group did not undergo any dietary changes but received the same supplement every day. The supplement was meant to act as a placebo for the control group and prevent B12 insufficiency in the intervention group. At baseline, 10 weeks, and 20 weeks, assessments were carried out. Participants were instructed to continue taking their diabetic medications as prescribed, but to heed the recommendations of their individual doctors regarding medication use (Sun et al. 2005).

12.8 Notable Dependent Variables with Evidence

A certified dietician assessed dietary modifications using two-day records. The arterial pressure was obtained on three separate occasions, body weight was measured using a digital scale, and blood glucose, HbA1c, and plasma lipid contents were tested using Quest Diagnostics' standard protocols. We combined the two remaining measures.

Pin-prick and contact pressure were measured using the Neuropathy Impairment Score-Lower Limb, and sensory perception was measured using the Michigan Neuropathy Screening Instrument physical examination. The electrochemical conductivity of the skin, which is linked to pain, was tested using a Sudoscan device, which evaluates each foot and hand and gauges the activity of the sudomotor neurons. The patient's responses were recorded for further analysis (Bax et al. 1996; Bril 1999; Sun et al. 2005).

The research employed a variety of scales for assessing pain and sensory symptoms, including the visual analog "worst pain" scale, the global impression scale, the Short Form McGill Pain Questionnaire, the Michigan Neuropathy Screening Instrument questionnaire, the Neuropathy Total Symptom Score, the weekly pain diary, and the Norfolk Quality of Life Questionnaire (Kamper et al. 2009; Graham et al. 1980; Kremer and Atkinson Jr. 1981; Vinik et al. 2005, 2008). In addition to filling out other questionnaires independently, volunteers were asked to use a scale that they completed with an interviewer to rate the level of pain they had felt during the preceding 2 weeks.

Perceived pain improvement is measured using the patient's world impression of progress investigation, which uses a scale from "no change" to "a great deal better". Pain diaries quantify pain over time using monthly ratings for average, worst, and nocturnal pain (Tanenberg et al. 2011). Depression and mood are assessed using the Beck Inventory of Depression (BDI) and the Center for Epidemiological Studies Depression Updated Scale, which measure the quantity and intensity of symptoms (Beck et al. 1996; Macrodimitris and Endler 2001).

12.9 Statistical Interpretation

The study aimed to ascertain the effects of a plant-based diet on pain in diabetic neuropathy because there was not enough of prior controlled trials and research on the subject. The exploratory study was open to anyone who met the eligibility circumstances (Smith et al. 2006; Crane and Sample 1994). Baseline descriptive data were applied to body weight, levels of cholesterol, and pain scales. Two-tailed Student's t-tests were used for regularly distributed data, while Wilcoxon signed-rank tests were used for irregularly distributed variables. Chi-squared testing was used to categorical variables. An alpha of 0.05 was adopted for all statistical tests. Baseline values for meaningful outcome variables are incorporated as covariates in the multivariate analysis of covariance. Values from the prior reporting period were carried forward for missing data. Dropouts were presumed to have reached their initial body weight.

Of the 71 individuals, 35 had been selected for the study's execution between November 2012 and January 2014. The majority of participants were female, with a mean age of 57 years and higher levels of education. The intervention group's higher Neuropathy Impairment Score Lower Limb (NIS-LL) score showed more severe neuropathy. At baseline, both the intervention and control groups had been given medication to manage severe diabetic neuropathy symptoms. The sole measure that exhibited a statistically significant difference between the two groups was the Neuropathy Impairment Score Lower Limb. The research investigation ended on time since there were no significant negative consequences (Fig. 12.1).

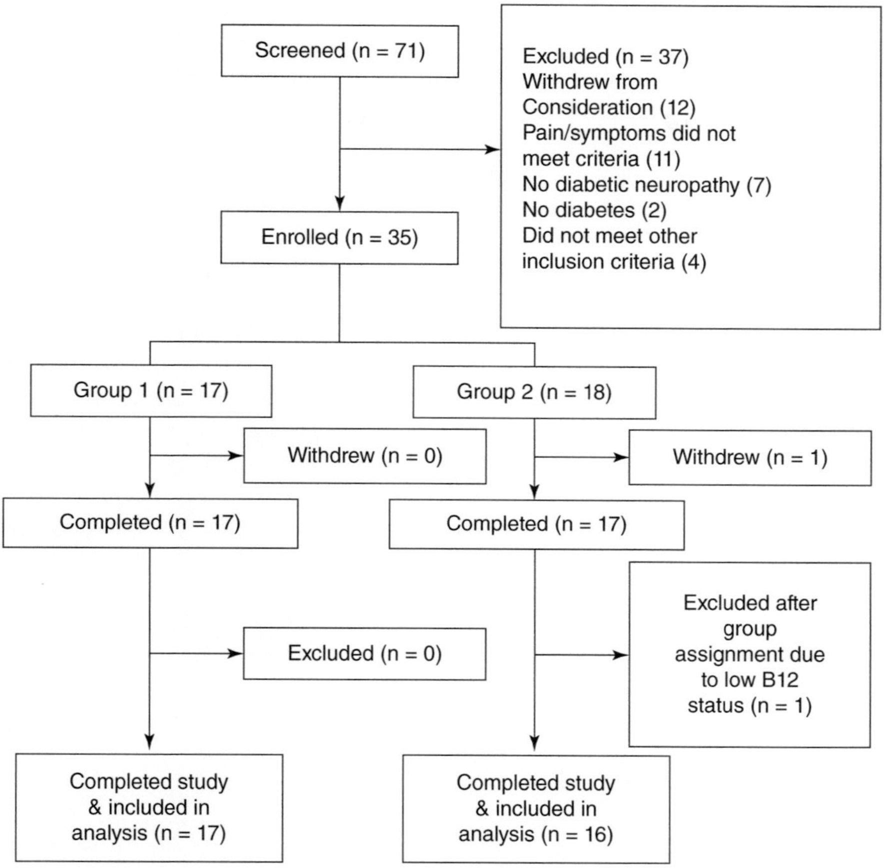

Fig. 12.1 A randomized controlled experiment examining that diet affects chronic pain from diabetic neuropathy

12.10 Adherence to Diet

At the halfway point and endpoint evaluations, 13 out of the trial's 17 intervention-group humans avoided from all animal products. Eight of them admitted to eating a low-fat diet at each point in time, implying as no more than 25% of their calories were from fat. Three of them followed a low-fat diet throughout one of the two periods. Two people in one record were fully compliant with the low-fat recommendations, though they ingested little amounts of animal-based foods. There was no data acquired about the application of vitamin B12 supplements.

12.11 Adjustments to Clinical Measurements

A per one case study, the objective of the research was to explore the impacts of an intervention on cholesterol levels and mood. Total cholesterol climbed in the control group and declined in the intervention group as more individuals discontinued the consumption of lipid-lowering medications and fewer expanded their use. Systolic and diastolic blood pressure reduced in both groups with no apparent distinctions. Six people from the intervention group and one participant from the control group had their blood pressure medicines altered over the course of 20 weeks.

The intervention group also had substantial improvements in pain and neuropathy symptoms over the course of 20 weeks. The average foot conductance of the control group demonstrated a drop, whereas the intervention group revealed a non-significant increase in sudomotor nerve function. Despite the intervention group's quality of life substantially rose, there was no statistically significant difference amongst the groups. The intervention group displayed significant improvements on 7/10 items on the Neuropathic Pain Scale; however, these differences were not statistically significant across groups.

Despite improvements in handling symptoms in both groups, these advantages were not considered of statistical significance. Although neither finding was statistically significant, the Neuropathy Impairment Score-Lower Limb and the MNSI-PA both revealed major improvements in the intervention group.

During the course of the investigation, one of the people in the intervention group started to use neuropathy medications, while two others stopped. The results did not change when the analysis was restricted to only participants who did not change their symptom management medication, with a single exception that changes in the overall Quality of Life score were no longer significant within the intervention group.

12.12 Modifications in Neuropathy Symptoms and Perceived Pain

The investigation examined the alterations in an assortment of people's perceptions of pain and neuropathy symptoms. The intervention group observed a non-significant boost in sudomotor nerve function, while the control group observed a decrease. The sensory subscore of the Short Form McGill Pain Questionnaire, that gauges pain, showed a substantial reduction. The intervention group's quality of life substantially improved considering being aware that there was no statistically significant difference comparing the groups. The intervention group exhibited significant improvements on seven out of the ten criteria on the Neuropathic Pain Scale; however, these differences were not statistically significant across groups. Increases were seen in both groups, although they were not statistically significant. Although it dropped for the control group, the MNSI-PA substantially rose for the intervention group. The research study also included a control group that commenced and ceased taking pregabalin. Conclusions showed that changes in the overall Quality of Life score were no longer noteworthy within the intervention group. The study suggests that modifying symptom-management prescriptions may improve the therapy of neuropathy symptoms.

12.13 Description of Clinical Case Evidence

This 20-week nutrition strategy was developed to enhance the lives of diabetics who struggle with severe neuropathy. A substantial decrease in corpulence, body mass index, and HbA1c have been observed in the intervention group, although electrophysiological conductance of skin in the foot diminished in the placebo group but maintained steady in the group receiving the treatment. Additionally, the intervention group possessed increases in the Neuropathy Total Symptom Score and self-reported worst pain.

The outcomes coincided with before that, less large research that demonstrated a dietary intervention could mitigate the discomfort and manifestations of neuropathy. In the Crane and Sample research, 21 people with type 2 diabetes and painful neuropathy received part in a residential lifestyle intervention program which featured exercise and a low-fat, low-calorie, plant-based diet lasting 25 days. The intervention group reported full resolution of searing pain and higher touch perception, while the remainder of the respondents reported just minor symptom relief. Additional benefits included weight loss, lowered blood lipid levels, and a decreased need for cardiovascular and blood glucose medications.

Smith along with colleagues looked at 32 people without diabetes who had peripheral neuropathy and a decreased tolerance to glucose. Participants received dietary and fitness coaching during the one-year the study. Improvements in body weight, cholesterol, and glucose tolerance were seen together with substantial improvements in intraepidermal nerve fiber density and foot sweat volume. The result was no statistically significant improvement in pain. The intervention in the present study consisted of weekly nutritional counseling meetings and was more intensive but have a shorter duration.

One or more of the methods by which that the low-fat plant-based diet reduces neuropathy pain may be via improved insulin sensitivity, resulting in to better glucose management. Diabetes neuropathy has been linked to obesity, dyslipidemia, and hypertension, all of which can be treated with a diet high in vegetables and fruits. The low-fat demand was poorly committed to, whilst the plant-based element of the diet was strongly committed to the low-fat criterion.

Medication modifications to the therapy group may have disguised a few improvements in blood pressure, blood lipids, and glucose control. The majority of individuals in the intervention group altered their diabetes medicines largely for reasons of hypoglycemia. Pain and other symptoms may have shifted as a result of changes in symptom management drugs throughout the course of the 20 weeks.

The neuropathy total symptom score revealed an important decrease in the control group's symptoms and self-reported more difficult pain (visual analog pain). Regarding the scope of the improvement, it's plausible that the B12 supplement—which was designed to function as a placebo—had a positive impact on both groups.

The positive aspects of this pilot investigation involve a control group, a randomized design, and an adequate period for recording clinical adjustments. However, it has a number of flaws notably the inability to make distinctions between the weekly lessons and the effect of the food. Additional investigation is necessary to identify the particular channels via which the intervention might result in benefits in physical health and symptoms.

12.14 Extra Biological Elements and Practical Diabetic Functional Foods

Analyzing every one of the numerous natural foods, nutraceuticals, or biologically active substances that may be termed functional ingredients and have an advantageous impact on diabetes treatment lies outside the boundaries of this article. Along with dairy products, probiotics, fish meat, green tea, and spices, Table 12.2 lists some of these possibly functional foods (Parvin et al. 2014).

Table 12.2 Beneficial functional meals' bioactive ingredients and functional characteristics

Potential functional characteristics of diabetes	Key bioactive ingredients and nutritional supplements	Functional foods
Improve the traits of the metabolic disorder, change the gut flora, regulate appetite and fullness, change adiponectin and adipocytokines, and initiate β-oxidation, lipolysis, and thermogenesis. Body mass index and obesity ↑ dietary fat excretion ↓ antioxidant capacity ↓ lipid peroxidation ↓ HbA1c ↑ inflammatory markers, oxidative stress, hypolipidemia, and antithrombotic effects Patients with diabetes can alter their defenses and tolerance to insulin.	α- and β-lactorphines, lactoferrin, lactoferricin, α-lactalbumin, β-lactoglobulin, growth factors, conjugated linoleic acids, lactic acid bacteria, bifidobacteria, calcium, vitamin B, immunoglobulines, bioactive proteins including casein and whey, and bioactive peptides	Dairy products and probiotics
Better control of obesity and weight ↓ inflammation and insulin resistance ↓ cardiovascular disease ↓ glycemic control ↓ proteinuria ↓ oxidative stress, inhibit and induce lipolysis, induce PPARα and PPARβ ↓ thermogenesis and energy expenditure, inhibit angiotensin converting enzyme, and control elevated blood pressure.	Bioactive peptides, antioxidants, taurine, selenium, and ω3 fatty acids (docosahexaenoic and eicosapentaenoic acids)	Fish and seafood
Regulate the breakdown of cholesterol, stop LDL oxidation, protect the vascular endothelium from atherogenesis, stop accumulation of platelets, stop atherosclerosis from developing, activate PPARγ, and lessen subclinical inflammation.	Oleic acid, ω_3 fatty acids, Flavonoids, cinnamic acid, benzoic acid, lignans, cumaric acid, ferulic acid, tocopherols, carotenoids, oleuropein, oleocanthal	Olive oil

(continued)

Table 12.2 (continued)

Potential functional characteristics of diabetes	Key bioactive ingredients and nutritional supplements	Functional foods
Boost glycemic control, trigger catalase and superoxide dismutase to induce lipid peroxidation, and support the body's natural antioxidant defenses. ↑ Gluconeogenesis Thermogenesis, adipocyte proliferation and differentiation, pro-inflammatory cytokines, the breakdown of collagen and fibrosis, cardiovascular muscles protection, restriction of lipogenic enzymes, lipid and adipose tissue metabolism control, satiety, insulin sensitivity, glycogen content, and monocyte chemotactic protein-1	Chlorophyll, carotenoids, pectin, epigallocatechin-3-gallat, polyphenols, phenolic acids, catechins, and plant sterols	Green tea
Heightened gluconeogenesis and glycolysis, improved periphery glucose absorption, insulin sensitivity, hypoglycemic and hypolipidemic effects, and antioxidant and anti-inflammatory properties	Procyanidins B-2, epicatechin, coumarin, cinnamaldehyde, and cinnamic acid	Cinnamon
Protect inflammatory enzymes with the value as nuclear hormone κB, lipoxygenase, and cyclooxygenase-2; prevent postprandial glycemic response and proteinuria by blocking α-glucosidase and α-amylase activity; prevent diabetes cataracts by activating PPARγ to regulate the metabolism of carbohydrates and lipids.	Bioactive peptide turmerin, 2-hydroxy methyl anthraquinone, β-sitosterol, stigmasterol, and curcuminoids	Turmeric
Minimize the effects of oxidative stress and offer defense against oxidative harm. Enhanced urea and blood creatinine levels, atherogenic lipoprotein levels, tissue in the kidney lipid peroxidation, α-glucosidase activity suppression, better carbohydrate absorption and digestion, and liver protection against oxidative damage brought on by diabetes	Gallic acid, phenolic acid, anthocyanins, flavonoids, and tannins	Sumac

12.15 Conclusion

In overall, dietary interventions high in whole grains, legumes, and nuts are vital for the best possible management of Type 2 Diabetes Mellitus since these foods improve insulin sensitivity and decrease the risk of serious consequences. The importance of diet to controlling diabetes is emphasized by the observation humans can drastically improve their metabolic health and overall happiness of life by emphasizing nutrient-dense meals and taking probiotics and prebiotics. In a nutshell, many studies indicate that a plant-based diet may assist with regulate type 2 diabetes and reduce the symptoms of diabetic neuropathy. Participants' body weight, HbA1c levels, and general quality of life seemed significantly improved. Even though the results were below statistical significance on all regulations they nonetheless highlight how crucial nutritional strategies are for improving the health of people with diabetes. The trials also demonstrate how dietary therapies, especially plant-based, low-fat diets, may help people with diabetes and neuropathy in their legs manage their metabolic condition and reduce the symptoms of their neuropathy. Although more investigation is required to elucidate the precise processes and impacts of these treatments, the encouraging results highlight the significance of dietary and lifestyle modifications in the management of problems associated with diabetes.

References

Al-Farsi MA, Lee CY. Nutritional and functional properties of dates: a review. Crit Rev Food Sci Nutr. 2008;48(10):877–87. https://doi.org/10.1080/10408390701724264.

Ali MM, Agha FG. Amelioration of streptozotocin-induced diabetes mellitus, oxidative stress and dyslipidemia in rats by tomato extract lycopene. Scand J Clin Lab Invest. 2009;69(3):371–9. https://doi.org/10.1080/00365510802658473.

Anderson JW, Bush HM. Soy protein effects on serum lipoproteins: a quality assessment and meta-analysis of randomized, controlled studies. J Am Coll Nutr. 2011;30:79–91. https://doi.org/10.1080/07315724.2011.10719947.

Apostolidis E, Kwon YI, Shetty K. Potential of cranberry-based herbal synergies for diabetes and hypertension management. Asia Pac J Clin Nutr. 2006;15(3):433–41.

Ashraf R, Aamir K, Shaikh AR, Ahmed T. Effects of garlic on dyslipidemia in patients with type 2 diabetes mellitus. J Ayub Med College Abbottabad. 2005;17(3):60–4.

Assini JM, Mulvihill EE, Huff MW. Citrus flavonoids and lipid metabolism. Curr Opin Lipidol. 2013;24(1):34–40. https://doi.org/10.1097/MOL.0b013e32835c07fd.

Azadbakht L, Shakerhosseini R, Atabak S, Jamshidian M, Mehrabi Y, Esmaill-Zadeh A. Beneficiary effect of dietary soy protein on lowering plasma levels of lipid and improving kidney function in type II diabetes with nephropathy. Eur J Clin Nutr. 2003;57:1292–4. https://doi.org/10.1038/sj.ejcn.1601688.

Bakar MFA, Mohamed M, Rahmat A, Burr SA. Phytochemicals and antioxidant activity of different parts of bambangan (*Mangifera pajang*) and tarap (*Artocarpus odoratissimus*). Food Chem. 2009;113(2):479–83. https://doi.org/10.1016/j.foodchem.2008.07.082.

Bakuradze T, Tausend C, Galan J, Neubauer L, van der Woude H, Appel KE, et al. Antioxidant-rich beverages reduce the formation of 8-oxo-2′-deoxyguanosine in humans. Free Radic Biol Med. 2012;52(3):536–43. https://doi.org/10.1016/j.freeradbiomed.2011.11.004.

Banerjee SK, Maulik SK. Effect of garlic on cardiovascular disorders: a review. Nutr J. 2002;1:4. https://doi.org/10.1186/1475-2891-1-4.

Banihani S, Swedan S, Alguraan Z. Pomegranate and type 2 diabetes. Nutr Res. 2013;33(5):341–8. https://doi.org/10.1016/j.nutres.2013.03.003.

Barnard ND, Scialli AR, Turner-McGrievy G, Lanou AJ. Acceptability of a low-fat vegan diet compares favorably to a step II diet in a randomized, controlled trial. J Cardpulm Rehabil. 2004;24(4):229–35.

Bastyr EJ III, Price KL, Bril V. Development and validity testing of the neuropathy total symptom score-6: questionnaire for the study of sensory symptoms of diabetic peripheral neuropathy. Clin Ther. 2005;27(8):1278–94.

Bax G, Fagherazzi C, Piarulli F, Nicolucci A, Fedele D. Reproducibility of Michigan Neuropathy Screening Instrument (MNSI): a comparison with tests using the vibratory and thermal perception thresholds. Diabetes Care. 1996;19(9):904–5.

Beck A, Steer R, Brown G. Manual for the Beck depression inventory-II. Psychological Corporation; 1996.

Bergman M, Varshavsky L, Gottlieb HE, Grossman S. The antioxidant activity of aqueous spinach extract: chemical identification of active fractions. Phytochemistry. 2001;58(1):143–52. https://doi.org/10.1016/s0031-9422(01)00137-6.

Bhathena SJ, Velasquez MT. Beneficial role of dietary phytoestrogens in obesity and diabetes. Am J Clin Nutr. 2002;76:1191–201. https://doi.org/10.1093/ajcn/76.6.1191.

Borges G, Degeneve A, Mullen W, Crozier A. Identification of flavonoid and phenolic antioxidants in black currants, blueberries, raspberries, red currants, and cranberries. J Agric Food Chem. 2010;58(7):3901–9. https://doi.org/10.1021/jf902061e.

Boyer J, Liu RH. Apple phytochemicals and their health benefits. Nutr J. 2004;3:5. https://doi.org/10.1186/1475-2891-3-5.

Bril V. NIS-LL: the primary measurement scale for clinical trial endpoints in diabetic peripheral neuropathy. Eur Neurol. 1999;41(1):8–13.

Cassidy YM, Mc Sorley EM, Allsopp PJ. Effect of soluble dietary fibre on postprandial blood glucose response and its potential as a functional food ingredient. J Funct Foods. 2018;46:423–39. https://doi.org/10.1016/j.jff.2018.05.019.

Chai SC, Hooshmand S, Saadat RL, Payton ME, Brummel-Smith K, Arjmandi BH. Daily apple versus dried plum: impact on cardiovascular disease risk factors in postmenopausal women. J Acad Nutr Diet. 2012;112(8):1158–68. https://doi.org/10.1016/j.jand.2012.05.005.

Chan HT, Yiu KH, Wong CY, Li SW, Tam S, Tse HF. Increased dietary fruit intake was associated with lower burden of carotid atherosclerosis in Chinese patients with type 2 diabetes mellitus. Diabet Med. 2013;30(1):100–8. https://doi.org/10.1111/j.1464-5491.2012.03764.x.

Chanet A, Milenkovic D, Manach C, Mazur A, Morand C. Citrus flavanones: what is their role in cardiovascular protection? J Agric Food Chem. 2012;60(36):8809–22. https://doi.org/10.1021/jf300669s.

Chawla A, Chawla R, Jaggi S. Microvascular and macrovascular complications in diabetes mellitus: distinct or continuum? Indian J Endocrinol Metabol. 2016;20(4):546–51. https://doi.org/10.4103/2230-8210.183480.

Crane MG, Sample C. Regression of diabetic neuropathy with total vegetarian (vegan) diet. J Nutr Med. 1994;4(4):431–9.

Deshmukh CD, Jain A. Diabetes mellitus: a review. Int J Pure Appl Biosci. 2015;3(3):224–30.

Dey D, Saha S, Narendhirakannan RT. Role of dietary phytochemicals in the prevention of diabetes mellitus: a review. Food Front. 2020;1(1):52–66. https://doi.org/10.1002/fft2.11.

Dobretsov M, Romanovsky D, Stimers JR. Early diabetic neuropathy: triggers and mechanisms. World J Gastroenterol. 2007;13(2):175–91.

El-Sayed M, Younes HA, Mohamed A. Plant bioactives and diabetes mellitus: role of dietary polyphenols and anthocyanins in the management of type 2 diabetes. Biomed Pharmacother. 2023;157:113992. https://doi.org/10.1016/j.biopha.2022.113992.

Fang J. Classification of fruits based on anthocyanin types and relevance to their health effects. Nutrition. 2014;30(11–12):1246–52. https://doi.org/10.1016/j.nut.2014.02.018.

Ferdowsian HR, Barnard ND, Hoover VJ, Katcher HI, Levin SM, Green AA, et al. A multicomponent intervention reduces body weight and cardiovascular risk at a GEICO corporate site. Am J Health Promot. 2010;24(6):384–7.

Forester SC, Waterhouse AL. Metabolites are key to understanding health effects of wine polyphenolics. J Nutr. 2009;139(9):1824S–31S. https://doi.org/10.3945/jn.109.108340.

Gabler NK, Osrowska E, Imsic M, Eagling DR, Jois M, Tatham BG, Dunshea FR. Dietary onion intake as part of a typical high fat diet improves indices of cardiovascular health using the mixed sex pig model. Plant Foods Hum Nutr. 2006;61(4):179–85. https://doi.org/10.1007/s11130-006-0030-8.

Gadang V, Hettiarachchy N, Horax R, Chen P, Johnson M. Proximate composition and functional properties of protein isolate from cowpea (Vigna unguiculata L. Walp.). J Food Sci. 2011;76(2):C175–81. https://doi.org/10.1111/j.1750-3841.2010.02034.x.

Geil PB, Anderson JW. Nutrition and health implications of dry beans: a review. J Am Coll Nutr. 1994;13:549–58. https://doi.org/10.1080/07315724.1994.10718446.

Gilbert ER, Liu D. Anti-diabetic functions of soy isoflavone genistein: mechanisms underlying its effects on pancreatic β-cell function. Food Funct. 2013;4:200–12. https://doi.org/10.1039/c2fo30199g.

Graham C, Bond SS, Gerkovich MM, Cook MR. Use of the McGill pain questionnaire in the assessment of cancer pain: replicability and consistency. Pain. 1980;8(3):377–87.

Granato D, Barba FJ, Bursać Kovačević D, Lorenzo JM, Cruz AG, Putnik P. Functional foods: product development, technological trends, efficacy testing, and safety. Annu Rev Food Sci Technol. 2020;11:93–118. https://doi.org/10.1146/annurev-food-032519-051708.

Hanhineva K, Törrönen R, Bondia-Pons I, Pekkinen J, Kolehmainen M, Mykkänen H, Poutanen K. Impact of dietary polyphenols on carbohydrate metabolism. Int J Mol Sci. 2010;11(4):1365–402. https://doi.org/10.3390/ijms11041365.

Heber D. Vegetables, fruits and phytoestrogens in the prevention of diseases. J Postgrad Med. 2004;50:145–9.

Hyson DA. A comprehensive review of apples and apple components and their relationship to human health. Adv Nutr. 2011;2(5):408–20. https://doi.org/10.3945/an.111.000513.

Jin T, Liu L. The Wnt signaling pathway effector TCF7L2 and type 2 diabetes mellitus. Mol Endocrinol. 2021;34(1):45–58. https://doi.org/10.1210/me.2020-00102.

John T, Samuel B, Abolaji O, Folashade O, Oyetooke A, Oluwatosin F. Functional foods and bioactive compounds: roles in the prevention, treatment and management of neurodegenerative diseases. GSC Biol Pharm Sci. 2020;11(2):297–313. https://doi.org/10.30574/gscbps.2020.11.2.0143.

Josef B, Raj S. Pharmacognostic and phytochemical properties of *Ficus carica* Linn: an overview. Int J Pharma Tech Res. 2011;3:8–12.

Jung UJ, Choi MS. Obesity and its metabolic complications: the role of adipokines and the relationship between obesity, inflammation, insulin resistance, dyslipidemia and nonalcoholic fatty liver disease. Int J Mol Sci. 2014;15(4):6184–223. https://doi.org/10.3390/ijms15046184.

Kamper SJ, Maher CG, Mackay G. Global rating of change scales: a review of strengths and weaknesses and considerations for design. J Manual Manipulat Ther. 2009;17(3):163–70.

Kang GG, Francis N, Hill R, Waters D, Blanchard C, Santhakumar AB. Dietary polyphenols and gene expression in molecular pathways associated with type 2 diabetes mellitus: a review. Int J Mol Sci. 2020;21(1):140. https://doi.org/10.3390/ijms21010140.

Kaume L, Howard LR, Devareddy L. The blackberry fruit: a review on its composition and chemistry, metabolism and bioavailability, and health benefits. J Agric Food Chem. 2012;60(23):5716–27. https://doi.org/10.1021/jf203318p.

Kharroubi AT, Darwish HM. Diabetes mellitus: the epidemic of the century. World J Diabetes. 2015;6(6):850–67. https://doi.org/10.4239/wjd.v6.i6.850.

Khoo HE, Azlan A, Tang ST, Lim SM. Anthocyanidins and anthocyanins: Colored pigments as food, pharmaceutical ingredients, and the potential health benefits. Food Nutr Res. 2017;61(1):1361779. https://doi.org/10.1080/16546628.2017.1361779.

Kles KA, Vinik AI. Pathophysiology and treatment of diabetic peripheral neuropathy: the case for diabetic neurovascular function as an essential component. Curr Diabetes Rev. 2006;2(2):131–45.

Kofuji K, Aoki A, Tsubaki K, Konishi M, Isobe T, Murata Y. Antioxidant activity of β-glucan. ISRN Pharm. 2012;2012:125864. https://doi.org/10.5402/2012/125864.

Koh GCH, Peacock SJ, van der Poll T, Wiersinga WJ. The impact of diabetes on the pathogenesis of sepsis. Eur J Clin Microbiol Infect Dis. 2014;33(4):481–9. https://doi.org/10.1007/s10096-013-2007-4.

Kremer E, Atkinson JH Jr. Pain measurement: construct validity of the affective dimension of the McGill Pain Questionnaire with chronic benign pain patients. Pain. 1981;11(1):93–100.

Lan-Pidhainy X, Wolever TM. The hypoglycemic effect of fat and protein is not attenuated by insulin resistance. Am J Clin Nutr. 2011;94(2):389–95. https://doi.org/10.3945/ajcn.110.005363.

Li X, Zhang B, Huang J, He Y, Zhang Y, Yang L. Phytochemicals and biological studies of plants from the genus Morus. J Food Drug Anal. 2018;26(1):1–15. https://doi.org/10.1016/j.jfda.2017.05.008.

Li Y, Ding Y, Minireview, Tian Y. Dietary polyphenols and type 2 diabetes: from mechanisms to human studies. Foods. 2022;11(3):296. https://doi.org/10.3390/foods11030296.

Loo AY, Jain K, Darah I. Antioxidant and radical scavenging activities of the root of *Ficus deltoidea* var. *kunstleri*. BMC Complement Altern Med. 2013;13:243. https://doi.org/10.1186/1472-6882-13-243.

Lucas EA, Li W, Peterson SK, Brown A, Kuvibidila S, Perkins-Veazie P, Clarke SL, Smith BJ. Mango modulates body fat and plasma glucose and lipids in mice fed a high-fat diet. Br J Nutr. 2011;106(10):1495–505. https://doi.org/10.1017/S0007114511002066.

Macrodimitris S, Endler N. Coping, control and adjustment in Type II diabetes. Health Psychol. 2001;20(3):208–16.

Manach C, Williamson G, Morand C, Scalbert A, Rémésy C. Bioavailability and bioefficacy of polyphenols in humans. I. Review of 97 bioavailability studies. Am J Clin Nutr. 2005;81(1 Suppl):230S–42S. https://doi.org/10.1093/ajcn/81.1.230S.

Martins MJ, Negrão MR, Azevedo I. Watermelon: the value of higher plasma arginine concentrations. Nutrition. 2007;23(6):517. https://doi.org/10.1016/j.nut.2007.03.007.

Masibo M, He Q. Mango bioactive compounds and related nutraceutical properties: a review. Food Rev Intl. 2009;25(4):346–70.

Mattes RD, Dreher ML. Nuts and healthy body weight maintenance mechanisms. Asia Pac J Clin Nutr. 2010;19:137–41.

Melendez-Ramirez LY, Richards RJ, Cefalu WT. Complications of type 1 diabetes. Endocrinol Metab Clin. 2010;39(3):625–40. https://doi.org/10.1016/j.ecl.2010.05.009.

Milenkovic D, Deval C, Dubray C, Mazur A, Morand C, Michalski M-C. Modulatory impact of food polyphenols on gut microbiota: a critical review of in vivo human evidence. Trends Food Sci Technol. 2019;92:178–92. https://doi.org/10.1016/j.tifs.2019.08.007.

Mishra S, Mishra M. Management of diabetes mellitus by various natural products: a short review. Plant Arch. 2021;21(1):270–5.

Mofleh AY, Ismail Z, Abdul Rahman NS, Mahmood AA, Albayaty M. Antioxidant and cytoprotective activities of the hydroalcoholic extract of Labisia pumila var. alata in vitro. BMC Complement Altern Med. 2013;13:175. https://doi.org/10.1186/1472-6882-13-175.

Moser B, Szekeres T, Bieglmayer C, Wagner KH, Mišík M, Kundi M, Zakerska O, Nersesyan A, Kager N, Zahrl J, et al. Impact of spinach consumption on DNA stability in peripheral lymphocytes and on biochemical blood parameters: results of a human intervention trial. Eur J Nutr. 2011;50(8):587–94. https://doi.org/10.1007/s00394-011-0167-6.

Oboh G, Ademosun AO, Bello F. Comparative effects of local coagulants on the nutritive and antinutritive composition of tofu. Food Sci Nutr. 2014;2(4):395–401. https://doi.org/10.1002/fsn3.112.

Ozougwu JC, Obimba KC, Belonwu CD, Unakalamba CB. The pathogenesis and pathophysiology of type 1 and type 2 diabetes mellitus. J Physiol Pathophysiol. 2013;4(4):46–57. https://doi.org/10.5897/JPAP2013.0001.

Parvin M, Zahra B, Fereidoun A. Functional foods-based diet as a novel dietary approach for management of type 2 diabetes and its complications: a review. World J Diabetes. 2014;5(3):267–81. https://doi.org/10.4239/wjd.v5.i3.267.

Patel DK, Dhanabal SP. Role of phytoconstituents in the management of diabetes mellitus. Int J Herbal Med. 2013;1(2):1–10.

Preuss HG. Bean amylase inhibitor and other carbohydrate absorption blockers: effects on diabesity and general health. J Am Coll Nutr. 2009;28:266–76. https://doi.org/10.1080/07315724.2009.10719781.

Reusch JEB, Manson JE. Management of type 2 diabetes in 2017: getting to goal. JAMA. 2017;317(10):1015–6. https://doi.org/10.1001/jama.2017.0242.

Ronis MJ, Chen Y, Badeaux J, Badger TM. Dietary soy protein isolate attenuates metabolic syndrome in rats via effects on PPAR, LXR, and SREBP signaling. J Nutr. 2009;139:1431–8. https://doi.org/10.3945/jn.109.107029.

Ross AB, Kamal-Eldin A, Åman P. Dietary alkylresorcinols: absorption, bioactivities, and possible use as biomarkers of whole-grain wheat- and rye-rich foods. Nutr Rev. 2004;62:81–95. https://doi.org/10.1111/j.1753-4887.2004.tb00029.x.

Saafi EB, Louedi M, Elfeki A, Zakhama A, Najjar MF, Hammami M, Achour L. Protective effect of date palm fruit extract (*Phoenix dactylifera* L.) on dimethoate induced-oxidative stress in rat liver. Exp Toxicol Pathol. 2011;63(5):433–41. https://doi.org/10.1016/j.etp.2010.03.002.

Salehi A, Gunnerud U, Muhammed SJ, Östman E, Holst JJ, Björck I, Ahrén B. The insulinogenic effect of whey protein is partially mediated by a direct effect of amino acids and GIP on β-cells. Nutr Metabol. 2012;9:48. https://doi.org/10.1186/1743-7075-9-48.

Santaguida PL, Balion C, Hunt D, Morrison K, Gerstein H, Raina P, Booker L, Yazdi H. Diagnosis, prognosis, and treatment of impaired glucose tolerance and impaired fasting glucose. Evid Rep Technol Assess (Summ). 2005;128:1–11.

Savage DG, Lindenbaum J, Stabler SP, Allen RH. Sensitivity of serum methylmalonic acid and total homocysteine determinations for diagnosing cobalamin and folate deficiencies. Am J Med. 1994;96(3):239–46.

Setorki M, Asgary S, Eidi A, Rohani AH, Esmaeil N. Effects of apple juice on risk factors of lipid profile, inflammation and coagulation, endothelial markers and atherosclerotic lesions in high cholesterolemic rabbits. Lipids Health Dis. 2009;8:39. https://doi.org/10.1186/1476-511X-8-39.

Seymour EM, Tanone II, Urcuyo-Llanes DE, Lewis SK, Kirakosyan A, Kondoleon MG, Kaufman PB, Bolling SF. Blueberry intake alters skeletal muscle and adipose tissue peroxisome proliferator-activated receptor activity and reduces insulin resistance in obese rats. J Med Food. 2011;14(12):1511–8. https://doi.org/10.1089/jmf.2010.0292.

Smith AG, Russell J, Feldman EL, Goldstein J, Peltier A, Smith S, et al. Lifestyle intervention for pre-diabetic neuropathy. Diabetes Care. 2006;29(6):1294–9.

Stevenson L, Phillips F, O'Sullivan K, Walton J. Wheat bran: its composition and benefits to health, a European perspective. Int J Food Sci Nutr. 2012;63:1001–13. https://doi.org/10.3109/09637486.2012.687366.

Su MS, Chien PJ. Antioxidant activity, anthocyanins, and phenolics of rabbiteye blueberry (*Vaccinium ashei*) fluid products as affected by fermentation. Food Chem. 2007;104(1):182–7. https://doi.org/10.1016/j.foodchem.2006.11.024.

Suh JH, Romain C, González-Barrio R, Cristol JP, Teissèdre PL, Crozier A, Rouanet JM. Raspberry juice consumption, oxidative stress and reduction of atherosclerosis risk factors in hypercholesterolemic golden Syrian hamsters. Food Funct. 2011;2(7):400–5. https://doi.org/10.1039/c1fo10047e.

Sun Y, Lai MS, Lu CJ. Effectiveness of vitamin B12 on diabetic neuropathy: systematic review of clinical controlled trials. Acta Neurol Taiwanica. 2005;14(2):48–54.

Taj Eldin IM, Ahmed EM, Elwahab HMA. Preliminary study of the clinical hypoglycemic effects of *Allium cepa* (red onion) in type 1 and type 2 diabetic patients. Environ Health Insights. 2010;4:71–7. https://doi.org/10.4137/EHI.S5540.

Takahashi K, Kamada C, Yoshimura H, Okumura R, Iimuro S, Ohashi Y, Araki A, Umegaki H, Sakurai T, Yoshimura Y, et al. Effects of total and green vegetable intakes on glycated hemoglobin A1c and triglycerides in elderly patients with type 2 diabetes mellitus: the Japanese Elderly Intervention Trial. Geriatr Gerontol Int. 2012;12(Suppl 1):50–8. https://doi.org/10.1111/j.1447-0594.2011.00812.x.

Tanaka S, Yoshimura Y, Kawasaki R, Kamada C, Tanaka S, Horikawa C, Ohashi Y, Araki A, Ito H, Akanuma Y, et al. Fruit intake and incident diabetic retinopathy with type 2 diabetes. Epidemiology. 2013;24:204–11. https://doi.org/10.1097/EDE.0b013e318281725e.

Tanenberg RJ, Irving GA, Risser RC, Ahl J, Robinson MJ, Skljarevski V, et al. Duloxetine, pregabalin, and duloxetine plus gabapentin for diabetic peripheral neuropathic pain management in patients with inadequate pain response to gabapentin: an open-label, randomized, noninferiority comparison. Mayo Clin Proc. 2011;86(7):615–26.

Thompson SV, Winham DM, Hutchins AM. Bean and rice meals reduce postprandial glycemic response in adults with type 2 diabetes: a cross-over study. Nutr J. 2012;11:23. https://doi.org/10.1186/1475-2891-11-23.

Thule PM, Umpierrez GE. Sulfonylureas: a new look at old therapy. Curr Diab Rep. 2014;14:473. https://doi.org/10.1007/s11892-014-0473-0.

Tian C, Ye X, Zhang R, Long J, Ren W, Ding S, Yang J. Mulberry leaf polyphenols and fiber induce synergistic antiobesity and display a modulation effect on gut microbiota and metabolites. Nutrients. 2016;8(5):315. https://doi.org/10.3390/nu8050315.

Utsunomiya H, Yamakawa T, Kamei J, Kadonosono K, Tanaka S. Anti-hyperglycemic effects of plum in a rat model of obesity and type 2 diabetes, Wistar fatty rat. Biomed Res. 2005;26(4):193–200. https://doi.org/10.2220/biomedres.26.193.

Vinik EJ, Hayes RP, Oglesby A, Bastyr E, Barlow P, Ford-Molvik SL, et al. The development and validation of the Norfolk QOL-DN, a new measure of patients' perception of the effects of diabetes and diabetic neuropathy. Diabetes Technol Ther. 2005;7(3):497–508.

Vinik EJ, Paulson JF, Ford-Molvik SL, Vinik AI. German-translated Norfolk quality of life (QOL-DN) identifies the same factors as the English version of the tool and discriminates different levels of neuropathy severity. J Diabetes Sci Technol. 2008;2(6):1075–86.

Wallace TC, Giusti MM, The Anthocyanins Committee. Anthocyanins. In: Galanakis CG, editor. Nutraceutical and functional food components: effects of innovative processing techniques. Academic Press; 2015. p. 221–31. https://doi.org/10.1016/B978-0-12-404709-9.00014-4.

Whent M, Huang H, Xie Z, Lutterodt H, Yu L, Fuerst EP, Morris CF, Yu LL, Luthria D. Phytochemical composition, anti-inflammatory, and antiproliferative activity of whole wheat flour. J Agric Food Chem. 2012;60:2129–35. https://doi.org/10.1021/jf203807w.

Williamson G, Clifford MN. Colonic metabolites of berry polyphenols: the missing link to biological activity? Br J Nutr. 2010;104(Suppl 3):S48–66. https://doi.org/10.1017/S0007114510003946.

Wong CY, Al-Salami H, Dass CR. Potential of insulin nanoparticle formulations for oral delivery and diabetes treatment. J Control Release. 2017;264:247–75. https://doi.org/10.1016/j.jconrel.2017.09.003.

World Health Organization (1999) Definition, diagnosis and classification of diabetes mellitus and its complications.

Wu X, Beecher GR, Holden JM, Haytowitz DB, Gebhardt SE, Prior RL. Concentrations of anthocyanins in common foods in the United States and estimation of normal consumption. J Agric Food Chem. 2006;54(11):4069–75. https://doi.org/10.1021/jf060300l.

Wu SC, Wrobel JS, Armstrong DG. Assessing the impact of pharmacologic intervention on the quality of life in diabetic peripheral neuropathic pain and fibromyalgia. Pain Med. 2007;8(Suppl. 2):S33–42.

Yadav M, Jain S, Tomar R, Prasad GB, Yadav H. Medicinal and biological potential of pumpkin: an updated review. Nutr Res Rev. 2010;23(2):184–90. https://doi.org/10.1017/S0954422410000107.

Yang B, Chen Y, Xu T, Yu Y, Huang T, Hu X, Li D. Systematic review and meta-analysis of soy products consumption in patients with type 2 diabetes mellitus. Asia Pac J Clin Nutr. 2011;20:593–602.

Yang Y, Zhao L, Li W, Duan G, Liu L. Traditional Chinese medicine in the treatment of patients infected with 2019-new coronavirus (SARS-CoV-2): a review and perspective. Int J Biol Sci. 2021;17(6):1341–51. https://doi.org/10.7150/ijbs.45538.

Yao Y, Cheng XZ, Wang LX, Wang SH, Ren G. Major phenolic compounds, antioxidant capacity and antidiabetic potential of rice bean (*Vigna umbellata* L.) in China. Int J Mol Sci. 2012;13:2707–16. https://doi.org/10.3390/ijms13032707.

Yokoyama Y, Barnard ND, Levin SM, Watanabe M. Vegetarian diets and glycemic control in diabetes: a systematic review and meta-analysis. Cardiovasc Diagn Ther. 2014a;4(5):373–82.

Yokoyama Y, Nishimura K, Barnard ND, Takegami M, Watanabe M, Sekikawa A, et al. Vegetarian diets and blood pressure: a meta-analysis. JAMA Intern Med. 2014b;174(4):577–87.

Yoshinari O, Sato H, Igarashi K. Anti-diabetic effects of pumpkin and its components, trigonelline and nicotinic acid, on Goto-Kakizaki rats. Biosci Biotechnol Biochem. 2009;73(5):1033–41. https://doi.org/10.1271/bbb.80805.

Zangiabadi N, Asadi-Shekaari M, Sheibani V, Jafari M, Shabani M, Asadi AR, Tajadini H, Jarahi M. Date fruit extract is a neuroprotective agent in diabetic peripheral neuropathy in streptozotocin-induced diabetic rats: a multimodal analysis. Oxidative Med Cell Longev. 2011;2011:976948. https://doi.org/10.1155/2011/976948.

Zhang M, Lv XY, Li J, Xu ZG, Chen L. The characterization of high-fat diet and multiple low-dose streptozotocin induced type 2 diabetes rat model. Exp Diabetes Res. 2009;2009:1–9. https://doi.org/10.1155/2009/969231.

Part IV

Specific Antioxidants and Their Efficacy

13 Vitamin E: Mechanisms of Action and Benefits in Diabetic Neuropathy

Swetlana Gautam, Khalid Orayj, Md Sayeed Akhtar, Fauzia Tabassum, Sadaf Farooqui, and Nihal AlMuraikhi

13.1 Introduction

High blood sugar levels are a hallmark of diabetes. In type 2 diabetes, target tissues such as muscle and adipose tissue lose their capacity to react to insulin, while in type 1 diabetes, the pancreatic β-cells that produce insulin are destroyed. Unlike type 2 diabetes, which usually grows more common as individuals age, type 1 diabetes frequently first appears in infant and adult (Schwarz et al. 2009). An increase in oxygen-produced free radicals leads to lipid peroxidation, a severe form of oxidative stress associated with diabetes. Free radicals of harmful oxygen have been linked to the development of diabetes mellitus and its consequences for both the micro and macrovascular systems. The metabolism experiences oxidative stress due to an imbalance brought on by increased production and/or reduced clearance of these free radicals, which ultimately results in tissue damage (Jain and Jain 2012).

The most prevalent diabetic neuropathic condition, diabetic peripheral neuropathy (DPN), also known as distal symmetrical polyneuropathy (DSPN), can strike about 50% of people with type 2 diabetes mellitus (T2DM) within 10 years. It has been demonstrated that antioxidants can avoid the problems associated with diabetes.

S. Gautam (✉)
Institution of Pharmacy and Paramedical Sciences, Dr Bhimrao Ambedkar University, Agra, Uttar Pradesh, India

K. Orayj · M. S. Akhtar
Department of Clinical Pharmacy, College of Pharmacy, King Khalid University, Alfara, Abha, Saudi Arabia

F. Tabassum · N. AlMuraikhi
Department of Pharmacology, Vision College, Riyadh, Saudi Arabia

S. Farooqui
Department of Clinical Pharmacy, College of Pharmacy, Prince Sattam Bin Abdulaziz University, Al-Kharj, Saudi Arabia

M. K. Gautam et al. (eds.), *Antioxidants and Functional Foods for Diabetic Neuropathy*, https://doi.org/10.1007/978-981-95-9326-2_13

Diabetes is a particularly attractive candidate for antioxidant supplements because it provides a good model of persistent oxidative damage. The results showed that the reduction of antioxidants and elevated blood sugar levels were significantly correlated. The development of the problems was significantly influenced by this depletion, and antioxidant treatment (vitamins E and C) might reduce this risk (Ng et al. 2020). Reduced levels of vitamins C and E in diabetics can lead to neuropathy. In diabetic rats, vitamin E treatment maintained the cellular bodies of the neurones, which had a neurotrophic effect. Pharmaceutical therapy with vitamin E supplementation also affects diabetic people' defective nerve conduction. Hypomagnesaemia and low vitamin B6 levels have been shown in diabetic neuropathy patients, but not in diabetic individuals without neuropathy. As a result, antioxidant vitamins and minerals that help diabetic patients reduce oxidative stress by enhancing glycemic management and/or increasing antioxidant activity may be suggested as an adjuvant treatment to help diabetic patients with neuropathy symptoms (Farvid et al. 2011).

The human body's intricate network of antioxidant defense and repair mechanisms regulates the harm that reactive oxygen species (oxidative damage) inflict to bimolecular structures. Additionally, diet is a source of certain antioxidants. Vitamin E, a fat-soluble vitamin that aids in reducing lipid damage from oxygen free radicals, is one of the most well-characterized of them. Targeting the lipids in membranes or lipoproteins with highly reactive species causes lipid peroxidation, which sets off a chain reaction. Vitamin E inhibits the chain-breaking process of lipid peroxidation and stops this chain reaction (Jain and Jain 2012). In nature, vitamin E may be found in vegetables, plants, and plant oils. Tocopherols make up the majority of common food oils, such as peanut, soybean, maize, and olive oil, whereas tocotrienols are few or non-existent. Compared to tocopherol, which is widely available, tocotrienol is present in the plant world and in food in very tiny concentrations (Peh et al. 2016). The largest concentration of tocotrienols may be found in some vegetable oils, such as palm oil, rice bran oil, and annatto seeds. Research has indicated that the absorption of vitamin E depends on β-hydroxylase and α-tocopherol transport protein (αTTP). Vitamin E is packaged and transported across the liver to other tissues with the help of αTTP (Sontag and Parker 2002).

In humans, tocotrienol's plasma half-life was calculated to be between 2 and 4 h (Yap et al. 2001). In a double-blind, placebo clinical research, taking 250 mg of tocotrienol supplements daily for 8 weeks increased plasma α-tocotrienol levels by 0.8 μM in human volunteers (O'byrne et al. 2000). As a fat-soluble substance, tocotrienol has a restricted bioavailability. Recent research has shown that tocotrienols are effective in preventing cancer, heart disease, stroke, and neurodegeneration due to their antioxidant, anticholesterolemic, anti-inflammatory, and anti-cancer qualities (Chin et al. 2016).

13.2 Associations Between Vitamin E and Diabetes

The primary cause of long-term complications in type II diabetes, such as kidney damage, retinopathy, and neuropathy, is inadequate glycemic control. Type 2 diabetes and oxidative stress have a complicated interaction, with hyperglycemia and hyperinsulinemia promoting increased generation of reactive oxygen species. This impacts the person's general health by impairing insulin signaling, β-cell function, and raising hemoglobin glycation (Xu et al. 2014).

Research has indicated that the concentrations of plasma alpha-tocopherol are lower in individuals with diabetes than in those without the condition, and they seem to be considerably lower in those who have complications such as microangiopathy (Martín-Gallán et al. 2003). There exists a correlation between the risk of diabetes and physiological vit. E. The risk of type II diabetes was specifically 46% lower for individuals with the greatest tertile of serum a-tocopherol than for those with the lowest tertile. Another study by Mayer-Davis et al. discovered that a higher risk was linked to low vitamin E levels of both forms of diabetes, although high plasma a-tocopherol may be preventive against type II diabetes (Mayer-Davis et al. 2002).

13.3 Mechanisms of Action of Vitamin E in Diabetic Neuropathy

Vitamin E is a good candidate for research on how it protects aging-related illnesses, especially those where reactive oxygen species (ROS) are a prominent factor, because it is the body's primary fat-soluble antioxidant. Vitamin E is soluble in lipids and belongs to the same physiologically active group of compounds as RRR-α-tocopherol. This includes α-, β-, γ-, and δ-tocopherol, as well as the four tocotrienols: α, β, γ, and δ-tocotrienol. It's crucial to remember that although most vit. Although α-tocopherol is used in E research, additional tocopherol and tocotrienol stereoisomers and their metabolites also seem to have strong biological effects. It seems that many of these characteristics are unrelated to vitamin E's antioxidant capacities. For instance, it has been demonstrated that α-tocopherol affects cell signaling via blocking protein kinase C (PKC). Since tocopherols have been demonstrated to have the ability to bind to the transcription factor peroxisome proliferator-activated receptor-γ (PPAR-γ), vitamin E has also been linked to gene regulation activities (Abd El-Aal et al. 2018).

13.3.1 Antioxidant Properties of Vitamin E

It exists as tocopherol and tocotrienol, two forms of the lipophilic antioxidant. As an antioxidant that occurs naturally, vitamin E can aid in protecting cells from oxidative stress. Vitamin E's ability to reduce the risk of hyperglycemia and its effects when combined with other diabetic treatments were the focus of several studies. Vit. E lowers hepatic lipid peroxidation in streptozotocin-induced diabetes, according to a number of studies conducted on in vivo models. Lipid peroxide levels may have increased as a result of an alteration in antioxidant status. Antioxidant enzymes including SOD, CAT, and GPX decline with diabetes (Fardoun 2007). Because of this, taking 440 mg per kg of body weight of vit. E orally once every week for a duration of 1 month has shown positive results in decreasing hyperoxide levels and increasing antioxidant enzyme activity (Tuzcu and Baydas 2006).

Protein glycation changes that are expected to happen in diabetes may be avoided by vitamin E. High glucose concentrations cause hemoglobin to become glycosylated, a significant diabetes marker that may be avoided with vitamin E. By inhibiting oxidative stress, vitamin E can help lower the HbA1C level. Vitamin E has been demonstrated to reduce protein glycation by inhibiting the development of MDA (malondialdehyde), which is a factor in the glycation of proteins in diabetics (Mondal et al. 2025).

In biological systems, structure dictates function. Since oxidative stress usually causes specific residues to oxidize, resulting in structural changes and functional deficits, this statement is especially pertinent to proteins. Protein oxidation is elevated among people with diabetes. (Telci et al. 2000). as well as in diabetic model animals. It has been demonstrated that antioxidant treatments lower the amounts of oxidized protein in rats (Gautam et al. 2015). Protein oxidation is commonly quantified by derivatizing proteins with 2,4 dinitrophenylhydrazine (DNPH) and then detecting the resultant carbonyl content using immunochemical methods. Vitamin E mitigates this process.

13.3.2 Transcriptional Regulation of Antioxidant Defense Genes

In addition to acting as a main antioxidant, vitamin E also has an impact on the body's natural antioxidant defense system. The components of this system include enzymes that aid in transport, detoxification, and antioxidant defense. The transcription factor responsible for these enzymes' regulation is nuclear factor 2 (Nrf2). Typically, Nrf2 expression is rather low in the steady state. However, during stress, Nrf2 half-life increases along with the synthesis of other proteins and enzymes whose genes contain upstream Nrf2-sensitive regions, or so-called antioxidant response component (Kobayashi and Yamamoto 2006). Vitamin E's stimulation of this pathway may provide an indirect means of protection against stress brought on by high blood sugar levels, because studies of diabetes and its effects have shown the preventive advantages of Nrf2.

A-tocopherol may have a minor effect on the generation of antioxidant enzymes by lowering the degradation of proteins and mRNA brought on by oxidative stress. Vitamin E has little effect on catalase or SOD1 activity, although it does influence mRNA stability. More mRNA stabilization, not transcription stimulation, is probably to blame for this. The antioxidant HO-1 is expressed in greater amounts in the glomeruli of rats with diabetes compared to the control group. Treatment with vit. E or the antioxidant normalizes this result by lowering the amounts of HO-1 mRNA and protein (Gautam et al. 2012).

13.3.3 Preventing Tissue Injury in Diabetics

Vitamin E has a critical role in the onset and course of type I diabetes, but it may also have protective benefits on the pancreas when combined with other antioxidants. Type 2 diabetes is caused by pathological changes in the pancreas, namely in the β-cells that secrete insulin. Together with the oxidative stress, the number of β-cells in the pancreas that release insulin is significantly reduced. Through increased β-cell death and disruption in the transcriptional control of insulin via PDX-1 (pancreatic and duodenal homeobox 1) and MafA, oxidative stress seems to gradually reduce insulin output over time (Gangwar et al. 2014). Because it is an antioxidant, vitamin E enhances pancreatic physiology in diabetics, which may enhance the functional results of diabetes in animal models (Sena et al. 2008). It is possible that vitamin E protects the pancreas, but it may do so by working in tandem with other antioxidants to produce these benefits.

13.3.4 Neuroprotective Effects of Vitamin E

In diabetic rats, vit. E supplements enhance nerve conduction, whereas diets deficient in vitamin E cause oxidative stress and the depletion of sympathetic and sensory neurones. A lack of vitamin E exacerbates diabetic nerve conduction, which mostly affects sensory nerves rather than motor ones. Compared to controls, diabetics have a lower plasma vitamin E to lipid ratio, particularly in those who have neuropathy. In diabetics with neuropathy, this impact is more noticeable (Ziegler et al. 2004). Since individuals with diabetic neuropathy are likely to experience a quicker depletion of vitamin E in their tissues than diabetic patients without neuropathy, maintaining optimal vitamin E status may be crucial for them. Diabetes has a substantial impact on the CNS, as evidenced by its association with an increased risk of dementia. The diabetic rat brain loses membrane fluidity as a result of oxidative stress and the associated peroxidation of long-chain fatty acids; antioxidant therapy may help restore this fluidity. This fluidity is necessary for normal neuronal transmission. Rats with diabetes that received vitamin E showed enhanced brain membrane fluidity, higher synthesis of antioxidant enzymes, and decreased oxidative stress. Moreover, it has been suggested that tocotrienols have unique neuroprotective properties (Skalska et al. 2008).

13.4 Clinical Evidence of Vitamin E in the Management of Diabetes

A literature review investigating the connection with antioxidant nutrients and type 2 diabetes suggests that vitamin E supplementation "may be a valuable strategy for controlling diabetes complications" (Balbi et al. 2018). The many forms of vitamin E were combined in this meta-analysis, making it challenging to identify whether a form of vitamin E's beneficial or harmful effects was specific to that form.

Furthermore, the main articles occasionally do not discriminate between RRR-alpha-TOH and all-rac-alpha-TOH. However, every vitamin E vitamin inhibits lipid peroxidation and quenches free radicals. Hence, it is plausible that the antioxidant characteristics that all members of the vitamin E family possess play a role in their beneficial impacts on managing T2D problems. Remarkably, in T2D patients, elevated lipid peroxidation is linked to inadequate glycemic management (Gunawardena et al. 2019).

Antioxidants such as taurine or NAC have been shown to help prevent the generation of additional ROS in people with diabetes. To illustrate how diabetes enhanced the formation of ROS, the glomeruli of obese rats treated with 2',7' dichlorofluorescin diacetate (DCFH-DA) fluoresced more than those of control rats (Gautam et al. 2012). When vitamin E or the antioxidant probucol, which is mostly used to prevent LDL oxidation in atherosclerosis, were given to diabetic rats, this rise in fluorescence was reduced. In particular, Rhee et al. (2005) discovered that the livers of streptozotocin-induced diabetic rats had considerably greater amounts of superoxide radicals than the livers of control rats when superoxide radical formation was examined. In a dose-dependent way, vitamin E treatment decreased the quantities of liver superoxide radicals.

A clinical study examining the impact of ascorbate and vitamin E on male adults T2D patients using metformin was carried out by Abd El-Aal et al. Those who took vitamin E, ascorbate, or both supplements for 90 days showed improvements in blood sugar management, according to the research. Future research should examine the advantages of vitamin E supplementation for young individuals with type 2 diabetes or prediabetes, according to the study's conclusions. It also implies that those with antioxidant deficiencies who may react significantly to antioxidant supplements or activity if they have elevated OxS levels (Abd El-Aal et al. 2018).

There are few treatment options available for prediabetes, particularly for children. Recent random controlled pilot research that examined the impacts of T3 derived from annatto seeds seems encouraging. Neither dietary supplements nor pharmacological treatments for glucose control were administered to the prediabetic adult research group. Glycemic control significantly improved after 12 weeks of consuming 300 mg per day of T3 in comparison to the controls who were taking a placebo. To enhance glycemic control, T3s are thought to upregulate the expression of GLUT4, insulin receptor substrate 1, and AKT signaling, as well as activate peroxisome proliferator-activated receptor-gamma (PPAR-gamma) (Suleman et al. 2022).

13.5 Vitamin E and Diabetic Neuropathy: Clinical Evidence

An interventional experiment was conducted in order to assess the effectiveness and safety of using vit. E and Eve primrose oil together to treat diabetic neuropathy. Eighty individuals with painful neuropathy who had been identified as having type 2 diabetes participated in the experiment for a year. The study participants had biochemical and clinical assessments at baseline. Every day, they were also given

500–1000 mg of eve primrose oil and 400 mg of vitamin E. They were then evaluated every 2 weeks for 3 months to see whether their symptoms had subsided and whether there had been any adverse effects. The primary end measure was the reduction of neuropathy symptoms. Of the study population, 70 patients (88%) reported having their neuropathic pains relieved. For the treatment of mild to moderate diabetic neuropathy, vitamin E and Eve primrose combined are helpful (Gautam et al. 2014).

The possible advantages of vitamin E intake for diabetic neuropathy were examined in a 1998 study by Tutuncu NB et al. The study recruited twenty-one people with type 2 diabetes and randomly assigned them to either 900 mg of the intervention or a placebo for the duration of 6 months. In terms of nerve conduction, the vitamin E group's performance improved (Tütüncü et al. 1998). In 2014, Rajanandh MG et al. conducted another investigation with comparable findings. This randomized controlled trial had ninety-two DPN patients who were allocated to receive either vit. E (test sample) or pregabalin (control). It was shown that both the overall pain score and the random blood glucose level decreased in the intervention group. (Rajanandh et al. 2014). A research by Kuhad A et al. examined the effects of tocotrienol in conjunction with insulin and DPN on diabetic rats, in addition to the previously mentioned investigations. Thus, in addition to reducing oxidative stress and inflammatory cytokines, tocotrienol combined with insulin relieves neuropathic pain in diabetic rats (Kuhad and Chopra 2009).

13.6 Future Directions in Vitamin E Research

There has been some research on the use of vitamin E to treat DPN symptoms, however the results are conflicting and the data is small. Although the precise process is not entirely known, a number of plausible theories have been put forth. Antioxidant vitamin E aids in scavenging free radicals, which can damage cells and tissues. One possible source of this stress is oxidative stress, which has been connected to diabetes. Because vitamin E reduces oxidative stress, it can help prevent or postpone the formation of diabetic peripheral neuropathy (DPN) and shield neurones from harm. Its anti-inflammatory qualities may lessen the inflammation of nerve tissue and ease the symptoms of neuropathic pain. Vitamin E can enhance blood flow and maintain nerve function by increasing blood flow, but diabetes can harm peripheral nerves by decreasing blood supply (Kim et al. 2006).

Improvements in nerve function and decreased pain have been seen in certain trials including vitamin E administration in DPN patients (Murthy et al. 2013). Some research, however, has shown contradictory or no outcomes at all. In JAMA Neurology, for instance, a more extensive clinical study conducted in 2018 revealed no appreciable effect of oral mixed tocotrienols for DPN patients (Hor et al. 2018). Finding out the specific molecular pathways by which vitamin E protects against diabetic neuropathy is one of the most important topics for future study. The effects of vitamin E on pathways associated with nerve damage, such as chronic inflammation, oxidative stress, and mitochondrial breakdown, require more investigation.

Innovative targets and pathways impacted by vitamin E might be found by using advanced omics technologies, such as proteomics, metabolomics, and genomes. The creation of more specialized treatment approaches may result from this.

The limited generalizability of current studies on Vitamin E and diabetic neuropathy is attributed to the small and homogenous patient groups included. Large-scale, multi-center clinical studies including a variety of people with differing ages, ethnic origins, and stages of diabetic neuropathy should be the main emphasis of future research. With a more thorough understanding of the safety and effectiveness of vitamin E across a wider range of demographics, this approach may result in more individualized therapy choices.

While initial studies have shown promising outcomes, further study is required to ascertain vitamin E's long-term safety and efficacy in the treatment of diabetic neuropathy. High-dose vitamin E usage for an extended period of time may be associated with hazards, such as bleeding difficulties and drug interactions. To track these dangers and evaluate the long-term advantages of vitamin E therapy, long-term research is necessary.

Finally, since diet and exercise are major lifestyle treatments that help manage diabetes and its complications, vitamin E therapy should be explored in conjunction with these strategies in future studies. Examining how lifestyle choices affect vitamin E's efficacy in diabetic neuropathy may result in more comprehensive therapeutic strategies that enhance patient outcomes. In conclusion, although vitamin E shows promise in treating symptoms of neuropathy associated with diabetes, more investigation is required in a number of areas. The full therapeutic benefits of vitamin E might be realized by addressing these future areas, which would result in better care and an enhanced quality of life for those who suffer from diabetic neuropathy.

13.7 Conclusion

The therapy of diabetes and its consequences, especially diabetic neuropathy, has shown vitamin E to be a viable therapeutic agent. Because of its strong antioxidant characteristics, it helps lessen oxidative stress, which is crucial for the onset and advancement of diabetic neuropathy. Because it reduces inflammation, protects nerve cells, and may even enhance nerve function, a comprehensive method of reducing the adverse impacts of diabetes affecting the neurological system is provided by vitamin E. Despite encouraging results from preclinical and clinical studies, vitamin E's maximum effectiveness in the treatment of diabetes and diabetic neuropathy has not yet been realized. To elucidate its mechanisms of action, optimize dose and distribution, and assess its long-term safety and efficacy, more study is required. Furthermore, knowing the genetic and epigenetic variables that affect each person's reaction to vitamin E may open the door to more individualized therapies.

In conclusion, vitamin E has a great deal of promise for use as a therapeutic tool against diabetes and its aftereffects. Its function will need to be more thoroughly established via additional study and well-planned clinical trials, which might lead to better outcomes and a higher standard of living for diabetic neuropathy sufferers.

References

Abd El-Aal A, Abd El-Ghffar EA, Ghali AA, Zughbur MR, Sirdah MM. The effect of vitamin C and/or E supplementations on type 2 diabetic adult males under metformin treatment: a single-blinded randomized controlled clinical trial. Diabetes Metab Syndr Clin Res Rev. 2018;12(4):483–9.

Balbi ME, Tonin FS, Mendes AM, Borba HH, Wiens A, Fernandez-Llimos F, Pontarolo R. Antioxidant effects of vitamins in type 2 diabetes: a meta-analysis of randomized controlled trials. Diabetol Metab Syndr. 2018;10:1–12.

Chin KY, Pang KL, Soelaiman IN. Tocotrienol and its role in chronic diseases. In: Anti-inflammatory nutraceuticals and chronic diseases; 2016. p. 97–130.

Fardoun RZ. The use of vitamin E in type 2 diabetes mellitus. Clin Exp Hypertens. 2007;29(3):135–48.

Farvid MS, Homayouni F, Amiri Z, Adelmanesh F. Improving neuropathy scores in type 2 diabetic patients using micronutrients supplementation. Diabetes Res Clin Pract. 2011;93(1):86–94.

Gangwar M, Gautam MK, Sharma A, Tripathi YB, Goel RK, Nath G. Antioxidant capacity and radical scavenging effect of Mallotus philippinensis fruit extract on human erythrocytes: an in vitro study. Sci World J. 2014;2014:279451.

Gautam MK, Gupta A, Vijaykumar M, Rao CV, Goel RK. Studies on the hypoglycemic effects of Murraya paniculata Linn. Extract on alloxan-induced oxidative stress in diabetic and non-diabetic models. Asian Pac J Trop Dis. 2012;2(S1):S186–91.

Gautam MK, Purohit V, Agarwal M, Singh A, Goel RK. In-vivo healing potential of Aegle marmelos in excision, incision and dead space wound models. Sci World J. 2014;2014:740107.

Gautam MK, Gangwar M, Singh SK, Goel RK. Effects of Azadirachta indica on vascular endothelial growth factor and cytokines in diabetic deep wound. Planta Med. 2015;81(8):713–21.

Gunawardena HP, Silva R, Sivakanesan R, Ranasinghe P, Katulanda P. Research article poor glycaemic control is associated with increased lipid peroxidation and glutathione peroxidase activity in type 2 diabetes patients. Oxid Med Cell Longev. 2019;2019:9471697.

Hor CP, Fung WY, Ang HA, Lim SC, Kam LY, Sim SW, et al. Efficacy of oral mixed tocotrienols in diabetic peripheral neuropathy: a randomized clinical trial. JAMA Neurol. 2018;75(4):444–52.

Jain AB, Jain VA. Vitamin E, its beneficial role in diabetes mellitus (DM) and its complications. J Clin Diagn Res. 2012;6(10):1624.

Kim HK, Kim JH, Gao X, Zhou JL, Lee I, Chung K, Chung JM. Analgesic effect of vitamin E is mediated by reducing central sensitization in neuropathic pain. Pain. 2006;122(1–2):53–62.

Kobayashi M, Yamamoto M. Nrf2–Keap1 regulation of cellular defense mechanisms against electrophiles and reactive oxygen species. Adv Enzym Regul. 2006;46(1):113–40.

Kuhad A, Chopra K. Tocotrienol attenuates oxidative–nitrosative stress and inflammatory cascade in experimental model of diabetic neuropathy. Neuropharmacology. 2009;57(4):456–62.

Martín-Gallán P, Carrascosa A, Gussinyé M, Domínguez C. Biomarkers of diabetes-associated oxidative stress and antioxidant status in young diabetic patients with or without subclinical complications. Free Radic Biol Med. 2003;34(12):1563–74.

Mayer-Davis EJ, Costacou T, King I, Zaccaro DJ, Bell RA. Plasma and dietary vitamin E in relation to incidence of type 2 diabetes: the insulin resistance and atherosclerosis study (IRAS). Diabetes Care. 2002;25(12):2172–7.

Mondal NS, Islam A, Gautam MK, Mondal S, Ahmed M, Jamir SI, Khongwir LD, Mahajon B. Unraveling the mechanisms of diabetic wounds: insights into pathogenesis and advanced treatment strategies. Curr Diabetes Rev. 2025;22(1):113–36.

Murthy S, Gautam MK, Goel S, Purohit V, Sharma H, Goel RK. Evaluation of in-vivo wound healing activity of Bacopa monniera on different wound model in rats. Biomed Res Int. 2013;2013:972028.

Ng YT, Phang SCW, Tan GCJ, Ng EY, Botross Henien NP, M. Palanisamy UD, et al. The effects of tocotrienol-rich vitamin E (Tocovid) on diabetic neuropathy: a phase II randomized controlled trial. Nutrients. 2020;12(5):1522.

O'byrne D, Grundy S, Packer L, Devaraj S, Baldenius K, Hoppe PP, et al. Studies of LDL oxidation following α-, γ-, or δ-tocotrienyl acetate supplementation of hypercholesterolemic humans. Free Radic Biol Med. 2000;29(9):834–45.

Peh HY, Tan WD, Liao W, Wong WF. Vitamin E therapy beyond cancer: tocopherol versus tocotrienol. Pharmacol Ther. 2016;162:152–69.

Rajanandh MG, Kosey S, Prathiksha G. Assessment of antioxidant supplementation on the neuropathic pain score and quality of life in diabetic neuropathy patients–a randomized controlled study. Pharmacol Rep. 2014;66(1):44–8.

Rhee SJ, Jeong YC, Choi JH. Effects of vitamin E on phospholipase A2 activity and oxidative damage to the liver in streptozotocin-induced diabetic rats. Ann Nutr Metab. 2005;49(6):392–6.

Schwarz PE, Li J, Lindstrom J, Tuomilehto J. Tools for predicting the risk of type 2 diabetes in daily practice. Horm Metab Res. 2009;41(02):86–97.

Sena CM, Nunes E, Gomes A, Santos MS, Proença T, Martins MI, Seiça RM. Supplementation of coenzyme Q10 and α-tocopherol lowers glycated hemoglobin level and lipid peroxidation in pancreas of diabetic rats. Nutr Res. 2008;28(2):113–21.

Skalska S, Kyselova Z, Gajdosikova A, Karasu C, Stefek M, Stolc S. Protective effect of stobadine on NCV in streptozotocin-diabetic rats: augmentation by vitamin E. Gen Physiol Biophys. 2008;27(2):106.

Sontag TJ, Parker RS. Cytochrome P450 ω-hydroxylase pathway of tocopherol catabolism: novel mechanism of regulation of vitamin E status. J Biol Chem. 2002;277(28):25290–6.

Suleman F, Khan DA, Pervez MA, Aamir M. Effects of delta-tocotrienol supplementation on glycaemic control in individuals with prediabetes: a randomized controlled study. JPMA. 2022;72(1):4–7.

Telci A, Cakatay U, Kayali R, Erdoğan C, Orhan Y, Sivas A, Akcay T. Oxidative protein damage in plasma of type 2 diabetic patients. Horm Metab Res. 2000;32(01):40–3.

Tütüncü NB, Bayraktar M, Varli K. Reversal of defective nerve conduction with vitamin E supplementation in type 2 diabetes: a preliminary study. Diabetes Care. 1998;21(11):1915–8.

Tuzcu M, Baydas G. Effect of melatonin and vitamin E on diabetes-induced learning and memory impairment in rats. Eur J Pharmacol. 2006;537(1–3):106–10.

Xu R, Zhang S, Tao A, Chen G, Zhang M. Influence of vitamin E supplementation on glycaemic control: a meta-analysis of randomised controlled trials. PLoS One. 2014;9(4):e95008.

Yap SP, Yuen KH, Wong JW. Pharmacokinetics and bioavailability of α-, γ-and δ-tocotrienols under different food status. J Pharm Pharmacol. 2001;53(1):67–71.

Ziegler D, Sohr CG, Nourooz-Zadeh J. Oxidative stress and antioxidant defense in relation to the severity of diabetic polyneuropathy and cardiovascular autonomic neuropathy. Diabetes Care. 2004;27(9):2178–83.

Polyphenols: Sources, Types and Benefits in Diabetic Neuropathy

14

Anirbita Ghosh, Anagha Ranade, Rabinarayan Acharya, and Sharad D. Pawar

Abbreviations

SOD	Superoxide dismutase
CAT	Catalase
GPx	Glutathione peroxidase
GSH	Reduced glutathione
NADPH	Nicotinamide Adenine Dinucleotide Phosphate Hydrogen
ROS	Reactive oxygen species
AGE	Advanced glycation end products
RAGE	Receptor for Advanced glycation end products
TNF-α	Tumour necrosis factor—alpha
STZ	Streptozotocin
NO	Nitric oxide

14.1 Diabetic Neuropathy: Pathophysiological Insight

The initiation of the polyol pathway, oxidative stress, accumulation of advanced glycation end products (AGEs) in various body tissues including the neurons, and inflammation are among the numerous mechanisms implicated in diabetic neuropathy. Prolonged hyperglycaemia, like in other tissues, decreases the expression and

A. Ghosh · S. D. Pawar (✉)
Department of Pharmacology, Central Ayurveda Research Institute (CARI), Kolkata, West Bengal, India

A. Ranade · R. Acharya
Central Council for Research in Ayurvedic Sciences (Ministry of Ayush, Government of India), New Delhi, India

M. K. Gautam et al. (eds.), *Antioxidants and Functional Foods for Diabetic Neuropathy*, https://doi.org/10.1007/978-981-95-9326-2_14

activity of the antioxidant enzymes SOD (superoxide dismutase), CAT (catalase), GPx (glutathione peroxidase) and GSH (reduced glutathione) in neurons. One of the significant mechanisms leading to diabetic neuropathy is the polyol pathway. In this process, NADPH (nicotinamide adenine dinucleotide phosphate hydrogen) and aldose reductase are involved in the metabolism of glucose, which produces fructose and sorbitol. An excess of NADPH is used, sorbitol accumulates, fructose is produced in large amounts, and the redox balance goes negative. The reduction in NADPH caused by the polyol pathway results in a decrease in glutathione synthesis causing an accumulation of oxidative stress in the neurons (Ratan et al. 2024). Transcript factors like B-cell lymphoma 2 (Bcl-2), which are essential for cell survival and functionality, also regulate protein expression, may become less abundant as a result of abnormal production of reactive oxygen species like hydrogen peroxide and superoxide ions. Nevertheless, studies have demonstrated that pro-apoptotic protein activities, including as PARP (poly (ADP-ribose) polymerase), BAX (Bcl-2-associated X protein) and c-Jun N-terminal kinase, are elevated in diabetic neuropathy patients (Redza-Dutordoir and Averill-Bates 2021).

Under diabetic conditions, abnormally high glucose levels surge the flux of glucose through the hexosamine pathway. The enzyme glutamine amidotransferase (GFAT) transforms fructose-6-phosphate (from glycolysis) into glucosamine-6-phosphate in the hexosamine pathway. Uridine diphosphate N-acetylglucosamine (UDP-GlcNAc), a necessary substrate for the O-linked glycosylation (O-GlcNAcylation) of proteins, is created through further metabolism of glucosesamine-6-phosphate. Increased UDP-GlcNAc from elevated glucose levels causes aberrant O-GlcNAcylation of certain proteins, changing those proteins' functions. The neurodegenerative alterations in diabetic neuropathy are facilitated by this excessive modification, which affects transcription factors, communication pathways and enzyme activity in neurons and glial cells. Activation of the hexosamine pathway may compromise mitochondrial damage (Coussement et al. 2020).

The AGE pathway, which converts fructose-6-phosphate to glyceraldehyde-3-phosphate, is activated by persistent hyperglycaemia. The Maillard reaction is responsible for the non-enzymatic glycation of proteins, lipids and peptides with reducing sugars. This results in the cross-linking of important proteins, which damages cells. Additionally, the attachment of these proteins to cell surface receptors, such as RAGE, triggers inflammatory processes and neurotrophic function (Fig. 14.1) (Ratan et al. 2024).

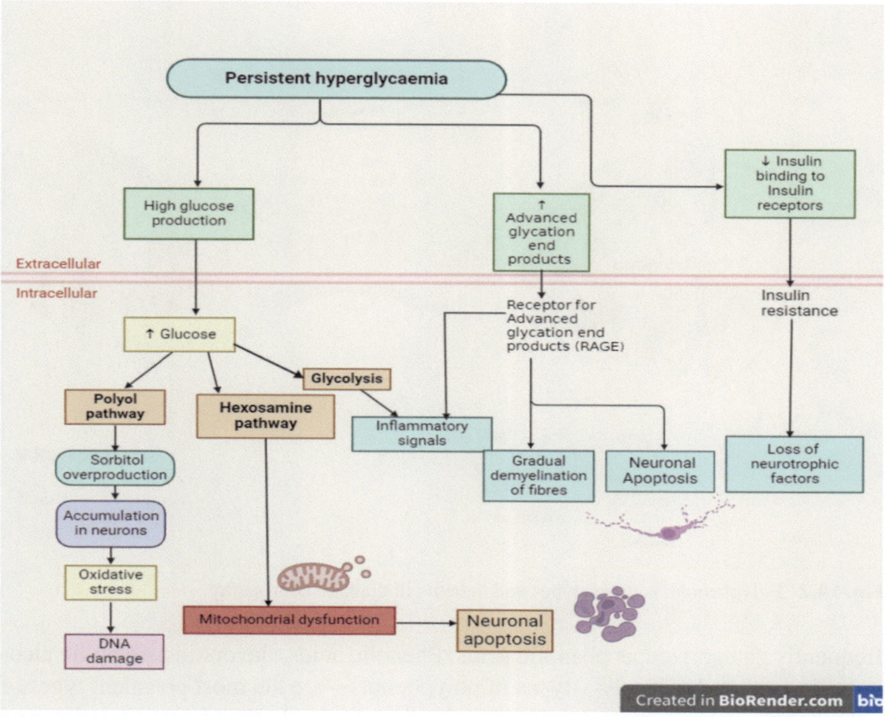

Fig. 14.1 Diagrammatic representation of pathophysiology of diabetic neuropathy. (Created with BioRender.com)

14.2 Polyphenols: A Brief Introduction

Foods based on plants naturally include chemicals called polyphenols, which exist in various intricate forms. The primary factors piqueing the interest of researchers and consumers in polyphenols are their antioxidant qualities, high dietary abundance and potential contribution to the prevention of a number of ROS mediated illnesses, including cancer, cardiovascular disease and neurological disorders (Rudrapal et al. 2022). They are found in many dietary sources, including green tea, grapes, red fruits, vegetables, coffee and wine. The scientific description of polyphenols and their history show that these naturally occurring plant-derived products were previously known as "vegetable tannins" due to their widespread use in the extraction of plant extracts used to turn animal skins into leather. This early application of polyphenolic plant extracts is the subject of the first definition of "plant polyphenols" in the scientific literature (Quideau et al. 2011). Since the phenolic ring is the most fundamental monomer in polyphenols, phenolic alcohols are

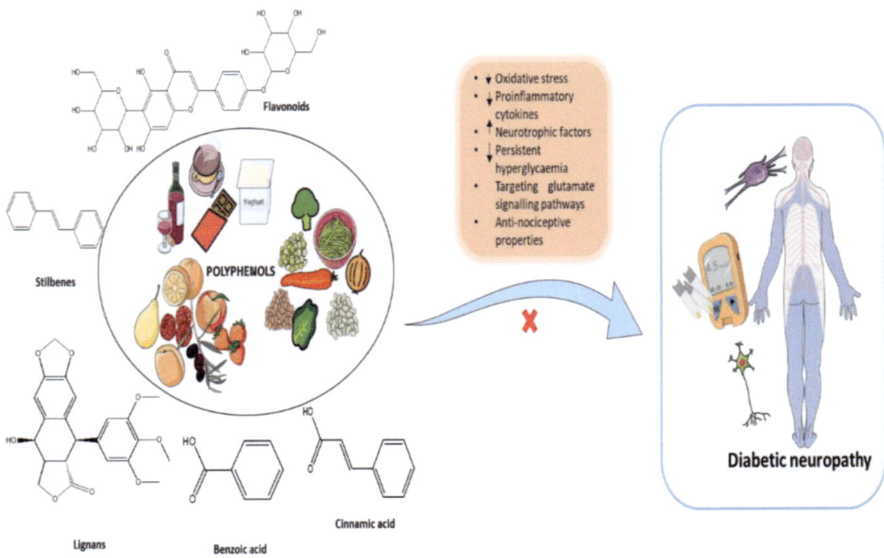

Fig. 14.2 Polyphenols: sources, types and benefits in diabetic neuropathy

frequently categorized as phenolic acids. Phenolic acids, flavonoids, phenolic alcohols, lignans, and stilbenes—types of polyphenols—are the most prevalent types of polyphenols, which are categorized according to the basis of the phenolic ring (Abbas et al. 2017). Some commonly known bioactive flavonoids include quercetin, kaempferol, catechins, epicatechins, procyanidins. Phenolic acids including caffeic, chlorogenic and gallic acids are some of the most common polyphenolic compounds with proven therapeutic roles in treating various diseases. Plant polyphenols have been defined historically based on protein precipitation and structural traits, but in recent years, this classification has undergone a significant revision to include structural traits and biosynthesis pathways (Fig. 14.2) (Belščak-Cvitanović et al. 2018).

14.3 Types and Structures of Polyphenols

14.3.1 Flavonoids

Flavonoids are a diverse group of phytochemicals, which are found in our common diet like fruits, vegetables, herbs and some beverages like tea and wine. They possess antioxidant, antidiabetic, anti-inflammatory and neuroprotective attributes, making them therapeutically essential, particularly in preventing or managing chronic diseases like diabetic neuropathy (Manzoor et al. 2020).

Structure and Classification of Flavonoids Flavonoids consist of 15 carbon atoms, which are arranged in a C6-C3-C6 configuration. This comprises two benzene rings linked by a three-carbon chain that forms a closed pyran ring (Dias et al.

2021). They are further divided into classes based on variations in the structure of the pyran ring (Nkembo 2023):

Name	Structure
Flavonols (e.g. kaempferol)	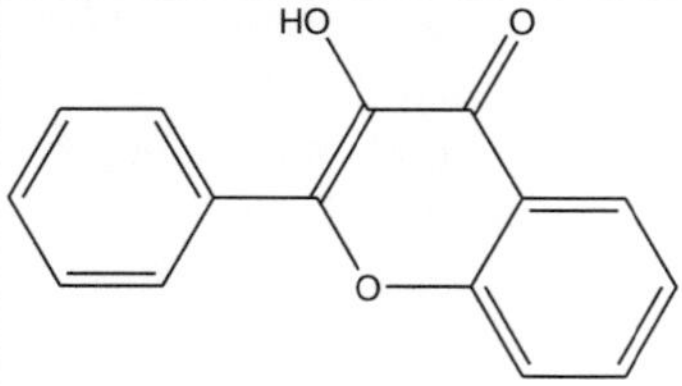
Flavones (e.g. luteolin)	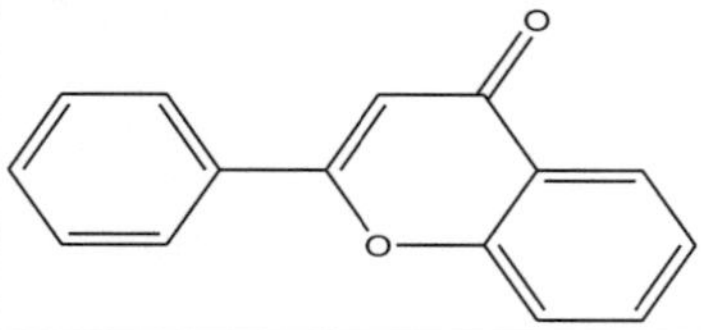
Flavanones (e.g. hesperidin)	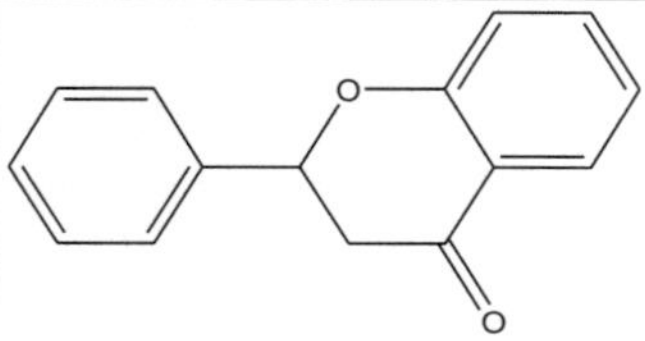
Flavanols (Catechins) (e.g. epigallocatechin gallate [EGCG])	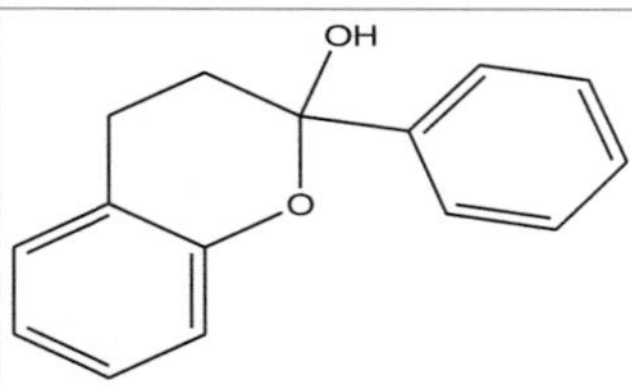
Anthocyanins (e.g. delphinidin)	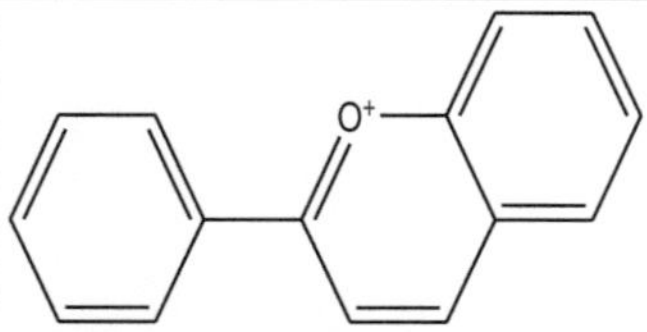
Isoflavones (e.g. daidzein)	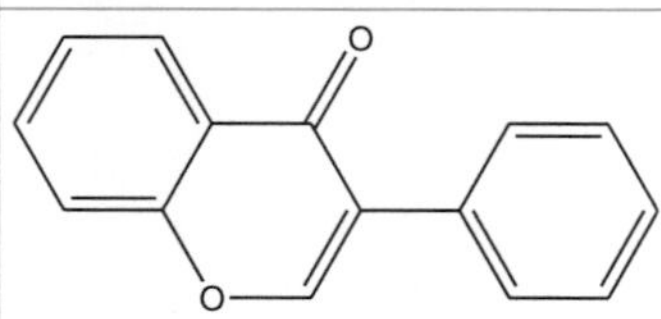

14.3.2 Phenolic Acids

Phenolic acids comprise polyphenolic acids existing in both free and conjugated forms. Among their many biological attributes are neuroprotective, free radical scavenging, anti-cancer and anti-inflammatory effects, which bear significance in human health. Phenolic acids abound in fruits, grains, vegetables and seeds (Rahman et al. 2021; Rashmi and Negi 2020). Besides health promoting effects, they help in defence mechanisms against UV radiation and pathogens.

Structure and Classification of Phenolic Acids In phenolic acids, a phenolic ring is attached to a carboxylic acid group. Their biosynthesis metabolic pathways are the shikimate pathway and the phenylpropanoid pathway. These pathways lead to the formation of essential phenolic compounds that help in plant defence, growth and pigmentation (Kumar and Goel 2019). Structurally, they are categorized into two main classes based on the arrangement of the carbon atoms in their structure:

1. *Benzoic Acid Derivatives:* These are simple phenolic acids with a C6-C1 skeleton. They include:

Name	Structure
Gallic acid	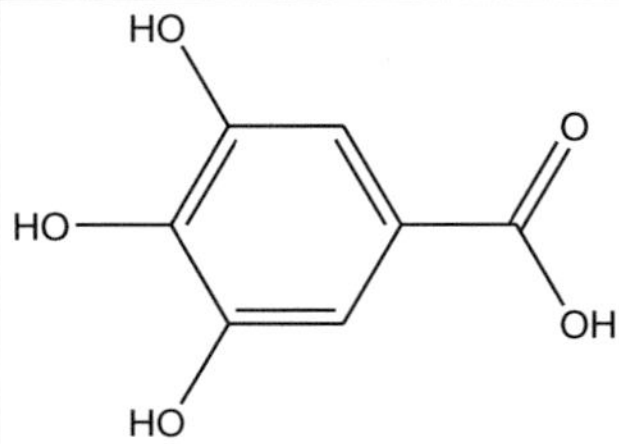
Vanillic acid	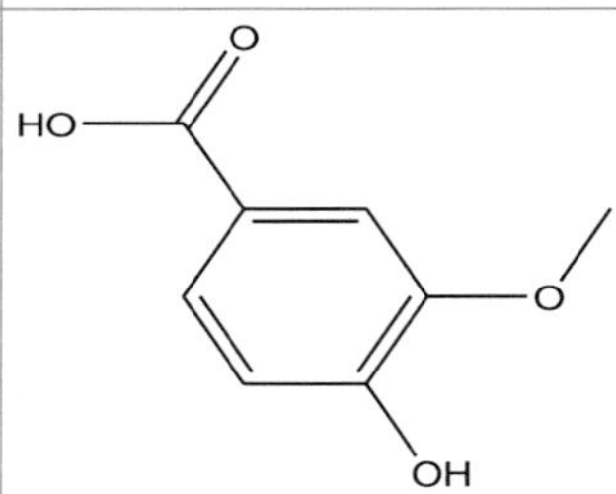
Syringic acid	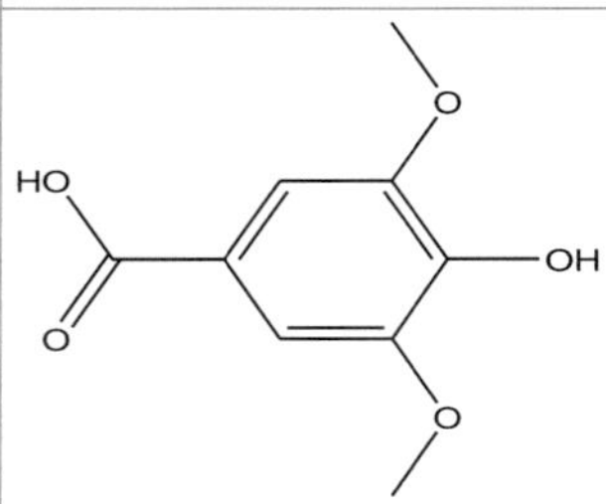
Protocatechuic acid	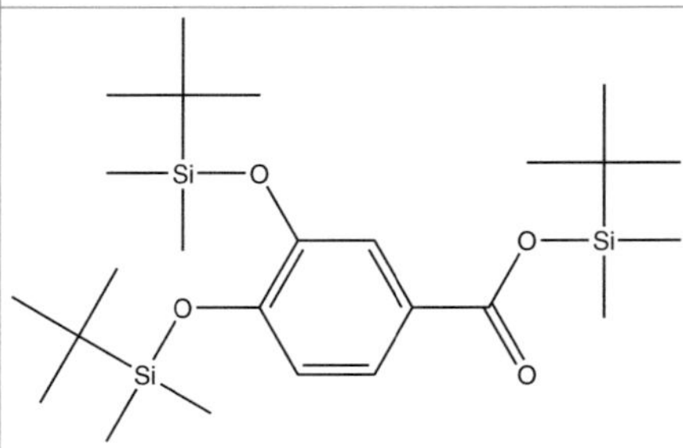

2. *Cinnamic Acid Derivatives:* These have a C6-C3 skeleton and are typically more complex than benzoic acid derivatives. The most common cinnamic acid derivatives include (Marchiosi et al. 2020):

Name	Structure
Caffeic acid	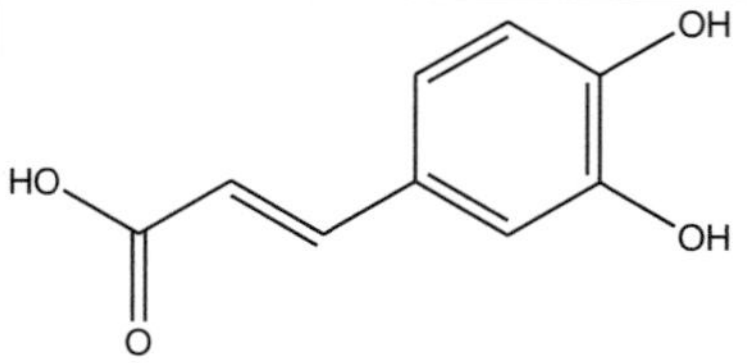
Ferulic acid	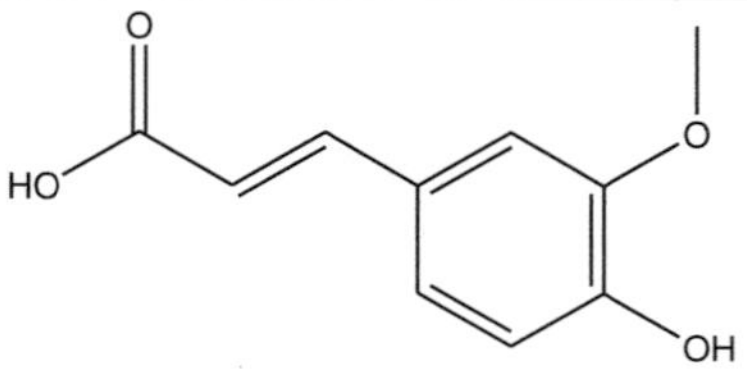
p-Coumaric acid	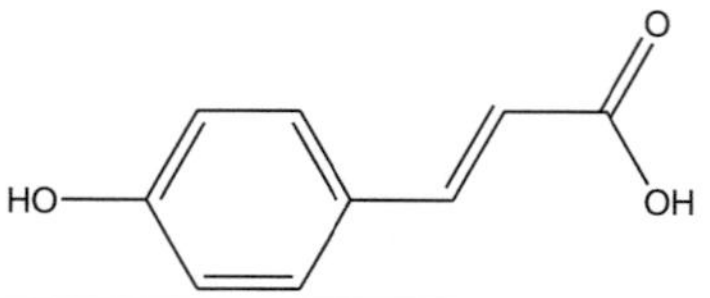
Sinapic acid	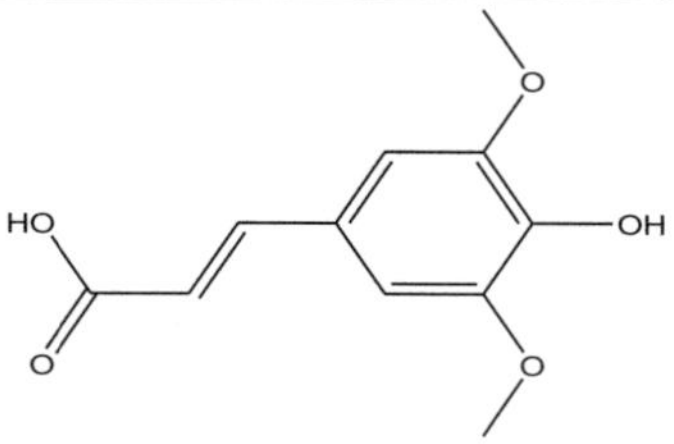

14.3.3 Stilbenes

Stilbenes comprise a group of polyphenols with a C6-C2-C6 structure supported by 1,2-diphenylethylene. They provide protection against plant infections, UV radiation and injury. They possess biological activities, including antioxidant, anti-inflammatory, cardioprotective, neuroprotective properties. Among the biologically important stilbenes, resveratrol is the most studied and well-known for its health-promoting effects. Stilbenes function as phytoalexins, having antimicrobial properties to protect plants from pathogen attacks (Rudrapal et al. 2022; Pecyna et al. 2020).

Chemical Structure of Stilbenes Stilbenes comprise of cyclic conjugated rings (two) joined by an ethylene bridge. Their structure is simple but flexible, allowing for different isomers depending on the position of the hydroxyl or other functional groups. Stilbenes mainly exist in two isomeric forms:

1. *Trans-isomers*: The more stable and bioactive form, commonly found in nature
2. *Cis-isomers*: Less stable and typically formed during UV exposure or under certain processing conditions

Resveratrol exists both in cis- and trans-isomeric forms, but the trans-resveratrol isomer is biologically more active and is the most studied (Al-Khayri et al. 2023). Examples of some health promoting stilbenes with structures (El Khawand et al. 2018):

Name	Structure
Resveratrol	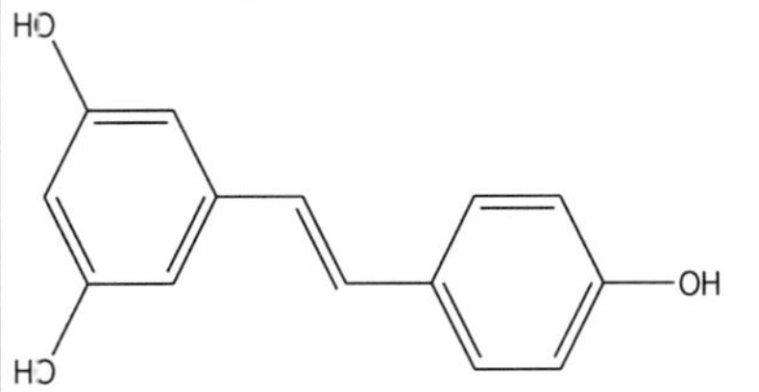
Pterostilbene	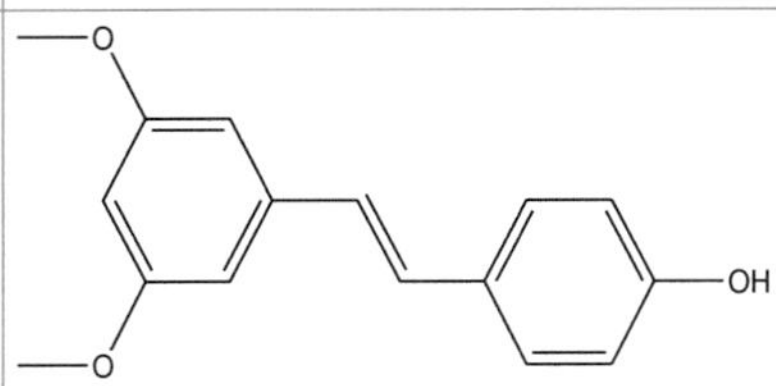
Piceatannol	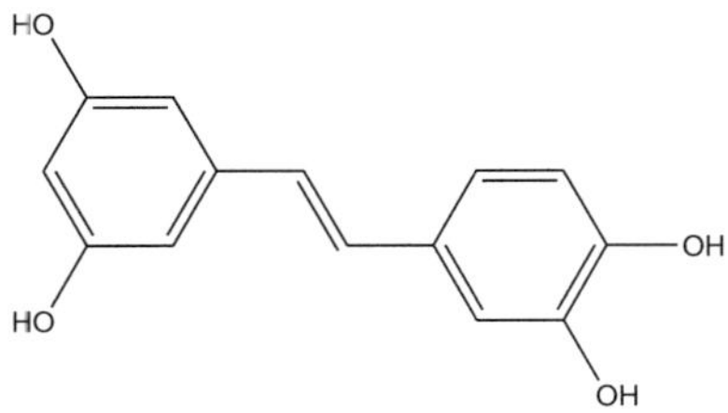

14.3.4 Lignans

Lignans are a class of phenolic compounds widely found in plants, especially in seeds, whole grains, vegetables and fruits. In these compounds, a two-carbon bridge (C6–C2–C6) connects two phenyl groups linked through a β-β' bond (Rocchetti et al. 2022). Lignans are part of a broader group of phytochemicals known as phytoestrogens, due to their ability to modulate estrogenic activity in mammals. These compounds have gained scientific importance owing to their appreciably high biological efficacy, resulting in antioxidant, anti-diabetic, anticancer, anti-inflammatory and cardioprotective effects (Plaha et al. 2022).

Chemical Structure of Lignans Lignans comprise of two phenylpropane units (C6-C3), biosynthetically obtained from the phenylpropanoid pathway in plants. Their structure is based on a dimeric phenylpropanoid framework, and the complexity arises from various hydroxylations, methylations and glycosylations. Lignans can be classified into two major groups based on their chemical structure (Suzuki and Umezawa 2007):

(a) *Classical Lignans*: Formed by the conjugation of two cinnamyl alcohol derivatives.
    Secoisolariciresinol, matairesinol, lariciresinol and pinoresinol
(b) *Non-classical Lignans*: These include norlignans characterized by C6-C2 structure and neolignans (where phenylpropanoid units are joined by bonds other than β-β').

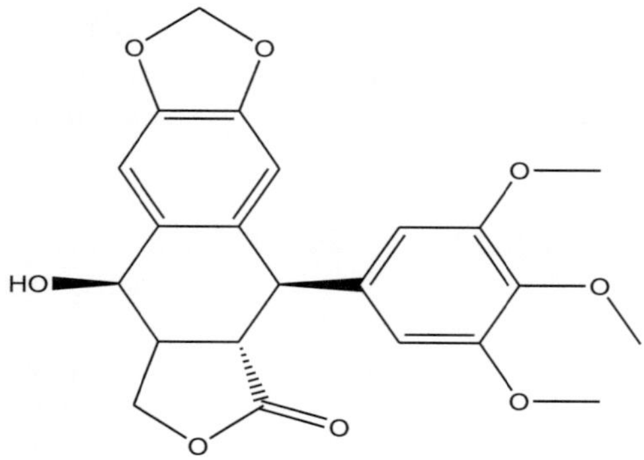

General structure of lignan

Common Lignans

- *Secoisolariciresinol*: Found in flaxseeds and known for its role in forming mammalian lignans
- *Matairesinol*: Found in sesame seeds, nuts and whole grains, it can be metabolized into enterolactone and enterodiol in the gut.
- *Pinoresinol and Lariciresinol*: Found in whole grains and vegetables (Nadeem et al. 2020).

14.4 Sources of Polyphenols

14.4.1 Sources of Flavonoids

Flavonoid chemicals are naturally occurring substances that are derived from plants and can be found in various plant components. Vegetables utilize flavonoids to support their growth and defence against plaque. Dietary flavonoids are so named because they are also widely present in meals and drinks that come from plants, including fruits, vegetables, tea, chocolate and wine (Panche et al. 2016).

Dietary Sources Hesperidin, hesperetin, diosmin, naringin, naringenin rutin, nobiletin, quercetin, tangeretin and other dietary flavonoids are abundant in citrus fruits and constitute a significant supply of these chemicals. A variety of citrus fruits, including pomelos, bergamots, grapefruits, lemons, limes, mandarins, oranges, and pomelos, contain these flavonoids. Citrus flavonoids possess antioxidant, anti-inflammatory and radical scavenging properties (Mahmoud et al. 2019). In addition to pome fruits (apples and pears), berries, wine, tea, cocoa and nuts, flavonoids like flavan-3-ols exist in stone fruits and legumes. It is seen that there are species-specific variations in composition of the fruits. Their sensory qualities specifically astringency and bitterness profiles are influenced by the presence of flavan-3-ols. Other flavonoids like procyanidins, (+)-catechin and (~)-epicatechin abound in *Theobroma cacao*, commonly known as cocoa, and its by-products. Conversely, high concentrations of (–)-epigallocatechin-3-O-gallate, (–)-epicatechin-3-O-gallate and (–)-epigallocatechin are found in *Camellia sinensis* leaves, used in green tea (Kanon et al. 2023).

Medicinal Plants Containing Flavonoids Flavonoids present in *Bauhinia variegata* L. demonstrated antioxidant qualities by the neutralization of radicals and iron binding (Mishra et al. 2013). Medicinal plants like *Gingko biloba* abound in flavonoids like Quercetin, Isorhamnetin and Kaempferol (Das et al. 2022). *Echinacea purpurea* is rich in Quercetin and rutin, which possess immune-boosting and anti-inflammatory properties (Li et al. 2016b; Hou et al. 2010). *Hypericum perforatum* (St. John's Wort) contains hyperoside and rutin used in treating anxiety and neuronal damage, thus providing neuroprotection (Oliveira et al. 2016). Few medicinal plants with their flavonoid compounds are listed below in the table that have shown a beneficial effect against diabetic neuropathy (Table 14.1).

Table 14.1 Flavonoid compounds from natural sources with reported activity against diabetic neuropathy

Sl. No.	Flavonoid compounds	Sources	Parts derived from	Activity	References (Year)
1.	Quercetin	*Allium cepa* Lam.	Leaves	Diabetic neuropathy	Dureshahwar et al. (2017)
2.	Kaempferol	*Eruca sativa*	Seeds	Diabetic neuropathy	Kishore et al. (2018)
3.	Epigallocatechin-3-gallate	*Camelia sinensis* (Green tea)	Leaves	Diabetic neuropathy	Chen et al. (2022)
4.	Genistein	Soya bean *Glycine max*	Beans	Diabetic neuropathy	Valsecchi et al. (2011)
5.	Curcumin	*Curcuma longa*	Rhizome	Diabetic neuropathy	Banafshe et al. (2014)
6.	Diosmin	Citrus fruits	Fruits skin	Diabetic neuropathy	Das et al. (2022)

14.4.2 Sources of Phenolic Acids

Dietary Sources Cereals such as wheat, oats, rice and corn contain the most abundant phenolic acid called Ferulic acid. It contributes to the antioxidant properties of these grains (Shahidi et al. 2022). Chlorogenic acid is found in apples, pears, plums, cherries and coffee beans (Wianowska and Gil 2019). P-coumaric acid and caffeic acid are found in high concentrations in vegetables like tomatoes, carrots, spinach and potatoes (Liu 2013). Rosmarinic acid and caffeic acid are common in herbs such as rosemary, thyme, oregano and basil (Yashin et al. 2017). A tabular summary of phenolic acids, their sources from plants, their biological activities against diabetic neuropathy and neuropathic pain and references is given in Table 14.2.

14.4.3 Sources of Stilbenes

Information regarding the overall intake of stilbenes is limited because of the difficulty of analysing the various stilbenes' occurrence in their natural environments, both qualitatively and quantitatively (Kluska et al. 2023). Despite being formed in plants via the phenylalanine ammonia lyase and shikimate pathways, they are often produced in less amounts than polyphenols, which are also mostly produced by the same routes. In contrast to other polyphenol families that are more prevalent, such as flavonoids, this makes the study of stilbenes challenging. Several plant sources, in

Table 14.2 Phenolic acids from medicinal plants and their biological activities in diabetic neuropathy

Phenolic acid	Sources	Parts derived from	Biological activity	Reference (year)
Ellagic acid, gallic acid and chebulic acid	*Emblica officinalis* (Amla)	Leaves	Diabetic neuropathy and anti-diabetic	Prananda et al. (2023) Tiwari et al. (2011)
Gallic acid, ellagic acid and chebulic acid	*Terminalia chebula* (Haritaki)	Fruits	Neuropathic pain	Bulbul et al. (2022)
Rosmarinic acid, ursolic acid, gallic acid and vanillic acid	*Ocimum sanctum* (Tulsi)	Leaves	Anti-diabetic and neuropathic pain	Muthuraman et al. (2008)
Gallic acid, caffeic acid, protocatechuic acid	*Curcuma longa* (Turmeric)	Rhizomes	Diabetic neuropathy and neuroprotective	Nagilla and Reddy (2014), Chumroenphat et al. (2021)
Gallic acid, ellagic acid	*Azadirachta indica* (Neem)	Leaves	Neuroprotective and diabetic neuropathy	Kandhare et al. (2017), Gupta et al. (2017)
Sinapic acid	*Emblica officinalis* (Amla)	Fruits	Neuroprotective and antihyperglycaemic	Chaikul et al. (2021), Vineela and Thakur (2020), Alaofi (2020)

contrast to grapes and peanuts, are high in stilbenes but are either locally confined or infrequently used as food. For instance, *Polygonum cuspidate* is utilized in Asia and Europe for culinary or medicinal purposes, but in Europe, it is used to extract resveratrol or piceid for use as a nutraceutical or dietary supplement (El Khawand et al. 2018).

Dietary Sources Grapes and red wine contain one of the greatest quantities of stilbenes, which have shown to be protective against neurodegenerative diseases. Stilbenes production can genetically differ based on the variety of grapes and can be affected by agricultural and environmental conditions (Farhan and Rizvi 2023). Grains and wine derivatives have one of the greatest amounts of stilbenes, and this amount can be highly changeable. Stilbenes have also been found in quantifiable amounts in beers (Chiva-Blanch et al. 2011). It has been reported that the blanch water of almonds contained piceatanol and oxyresveratrol in the quantities of 0.91–2.55 g/100 g. Stilbenes are also reported to be useful in diabetes as it is reported that alpha-glucosidase activity can be inhibited by stilbene glycosides found in rhubarb. In early-stage diabetes, piceatannol can lower plasma glucose levels and improve glucose absorption (Oluwole et al. 2022). Here is Table 14.3 summarizing some potent stilbenes along with their sources and potential against diabetic neuropathy and neurological diseases.

Table 14.3 Plant-derived stilbenes, their sources, plant parts and neuroprotective effects in diabetic neuropathy

Sl No.	Stilbene compound	Plant source	Plant part	Biological activity against diabetic neuropathy/ neuropathic pain	References
1.	Resveratrol	*Vitis vinifera* (Grapevine)	Seeds, Skins	Diabetic neuropathy	Amin et al. (2024)
2.	Piceatannol	*Rheum emodi* (Indian Rhubarb)	Rhizomes	Anti-diabetic, neuroprotective and antioxidant	Matsuda et al. (2001), Al-Jaber et al. (2024)
3.	Kuwanon C	*Curdrania tricuspidata*	Roots	Neuroprotective and anti-inflammatory	Ko et al. (2020)
4.	Mulberrofuran G and K, isomulberrofuran G	*Polygonum cuspidatum*		Neuroinflammatory and neuroprotective	Teka et al. (2022)
5.	Pterostilbene	*Pterocarpus marsupium, Vitis vinifera*	Leaves	Antidiabetic and neuroprotective	Wang and Chen (2014), Nieoczym et al. (2021)
6.	Pinosylvin	*Triticum turgidum ssp.*	Whole grains	Neuroprotective and antioxidant	Al-Khayri et al. (2023)
7.	Cis-3,30 -dihydroxy-5-methoxystilbene, Thunalbene, Batatasin III	*Pholidota cantonensis*	Grass	Neuroprotective	Al-Khayri et al. (2023)

14.4.4 Sources of Lignans

Dietary Sources With lignan production 75–800 times higher than that of cereals, legumes, other oil seeds, vegetables and fruit, flaxseeds are primarily the richest known source of lignans (9–30 mg/g). A major dietary lignan found in flaxseeds is called secoisolariciresinol (2,3-bis (3-methoxy-4-hydroxybenzyl) butane-1,4-diol). Herbacetin diglucoside (HD), ferulic acid glucoside (FeAG), p-coumaric acid glucoside (CouAG) and HMG make up flaxseed lignan (Soleymani et al. 2020). Sesame seeds carry sesamin in concentrations ranging from 0.1% to 0.5%. Pinoresinol and lariciresinol make up the majority of the about 29 mg/100 g of lignan found in sesame seeds, which is relatively high (Andargie et al. 2021). The regular consumption range of the most common lignans in Western countries, matairesinol and secoisolariciresinol, is from 2 to 74 μg and 70–1000 μg, respectively (Soleymani et al. 2020).

Sources from Medicinal Plants A special kind of lignan called sauchinone is taken from *Saururus chinensis* (Lour.) Baill roots. It possesses a range of biological actions, including immune-suppressive, cytoprotective, hepatoprotective and anti-inflammatory properties (Li et al. 2014). These properties can be helpful in treating metabolic disorders. Lignans derived from various medicinal plants have vast therapeutic applications. Here is Table 14.4 summarizing some potent lignans along with their sources and potential against diabetic neuropathy and neuropathic pain.

Table 14.4 Lignan phytochemicals investigated for diabetic neuropathy and peripheral neuropathic pain

Sl No.	Lignan compounds	Plant sources	Plant parts	Activity against diabetic neuropathy/ neurotoxicity	References
1.	Magnolol	*Magnolia officinalis*	Root and stem bark	Diabetic peripheral neuropathy	Wu et al. (2006), Yang et al. (2022)
2.	Honokiol	*Magnolia officinalis*	Stem bark	Diabetic neuropathy	Bibi et al. (2023)
4.	Arctigenin	*Forsythia viridissima*	Fruits Seeds	Peripheral Neuropathic pain, Diabetic neuropathy	Yi et al. (2019), Medras et al. (2023)
5.	*Eucommia ulmoides* lignans	*Eucommia ulmoides* Oliv		Glaucoma induced neurodegeneration	Li et al. (2016a, b)
6.	Secoisolariciresinol diglycoside	Flaxseed	Seeds	Diabetic neuropathy	Hu et al. (2015)
7.	Phyllanthin, Niranthin	*Phyllanthus amarus*	Aerial parts	Neuropathic pain	Pereira et al. (2016), Kassuya et al. (2003)

14.5 Polyphenol-Rich Healthy Meals

14.5.1 Main Course

It has been reported that consumption of 48 g of whole grains each day can reduce the risk of developing type 2 diabetes. This activity may be attributed to the presence of phenolic acids like ferulic acid, p-coumaric acid and sinapic acid (Shahidi et al. 2022). Low fat, vegetarian meal consumption for about 2 weeks has been found to be helpful in alleviating the pain in 17 of 21 patients of diabetic neuropathy. A 1 day meal for low-fat vegetarian diet can include polyphenol-rich items like cooked oatmeal with half teaspoon of cinnamon and no added fat, a banana for breakfast. One can take cucumber, mango and spinach salad for lunch. Dinner may include lentil-spinach soup or one cup broccoli with oil-free dressing (Trapp and Barnard 2010). Oats consist of polyphenols like phenolic acids, flavonoids and anthracamide, with ferulic acid being the most abundant. P-coumaric acid and coffee phenols are also present in it (Tang et al. 2023). Phenolic acid derivatives are present in fruits and vegetables, such as p-coumaric acid abounds in spinach (Septembre-Malaterre et al. 2018). Broccoli abounds in flavonoids like kaempferol and quercetin (Gasmi et al. 2024).

14.5.2 Snacks or Side Dish

Snacks may include hummus comprising of one-fourth cup chickpeas and tahini with tomatoes (Trapp and Barnard 2010). Tomatoes contain naringenin chalcone, the primary polyphenol, at up to 18.2 mg 100 g^{-1} fresh weight. The primary flavonol and one of the most significant flavonoids found in tomatoes is quercetin. With a content ranging from 1.4 to 3.3 mg 100 g^{-1} fresh weight, chlorogenic acid is the primary phenolic acid present. Cultivated tomatoes include trace amounts of other flavonols, such as myricetin and kaempferol (Martí et al. 2016). Dry legumes like chickpeas are a staple meal that are rich in flavonoids and polyphenols. While the total flavonoid content of chickpeas drastically decreased, the total polyphenol content was dramatically raised by pressure cooking (53%), open-pan boiling (64%) and microwave heating (increased doubly) (Hithamani and Srinivasan 2014). These polyphenols have been discussed earlier in this chapter to be helpful in diabetic neuropathy.

14.5.3 Beverages

Blueberry juice intake can be helpful as blueberry anthocyanins have been reportedly shown to prevent capillary albumin filtration, a condition associated diabetic neuropathy (Hameed et al. 2020). Numerous water-soluble polyphenols, such as phenolic acids, stilbenes, tannins, flavanols and anthocyanins, are found in wines. Red wines have more polyphenols than white wines because of wine processing.

Because of its phenolic composition, red wine has a higher antioxidant capacity than white wine. Flavonoids make up the majority of phenolic red wines, with 1450 mg/L for young wines and 1285 mg/L for mature ones. Wines are known to have neuroprotective properties due to their polyphenols (Basli et al. 2012). Resveratrol in red wine protects against diabetic neuropathy by various mechanisms discussed earlier. Green tea intake may also be helpful as green tea extract has been used in treating diabetic neuropathy in a randomized controlled trial (Essmat and Hussein 2021).

14.6 Ayurvedic Implications of Polyphenols in Diabetic Neuropathy

"Triphala" is a widely known formulation in Ayurveda comprising of polyphenol-rich dried fruits of *Emblica officinalis (Amalaki)*, *Terminalia belerica (Behada)* and *Terminalia chebula (Haritaki)* in equal concentrations (1:1:1) (Mukherjee et al. 2022). *Emblica officinalis* is rich in phenolic acids like ellagic acid, gallic acid and chebulic acid. *Terminalia belerica* abounds in phenolic acids like gallic acid, ellagic acid and chebulic acid. Triphala churna has been reported to show neuroprotective effects in case of streptozotocin-induced diabetic neuropathy in rats. In this study, Triphala churna therapy dramatically reduced blood glucose levels, increased body weight and enhanced motor nerve conduction velocity in diabetic rats. Additionally, it controlled the levels of circulating cytokines and reduced allodynia and thermal and mechanical hyperalgesia. It is likely to have treated diabetic neuropathy by reducing oxidative stress, inhibition of inflammatory cytokines and higher neural growth factor (NGF) expression in rats (Suryavanshi et al. 2021).

In a clinical study, 3 g of "Bhumyamalaki" powder and 10 g of "Atibala" root decoction were administered twice daily for 30 days to 33 patients with diabetic neuropathy. Both medications substantially lessened symptoms and restored diminished sensory perception based on recordings made by the Neuropathy Analyzer machine of vibration, cold and hot sensations (Patel et al. 2011). Atibala (*Abutilon indicum* Linn.) is rich in various polyphenolic compounds like caffeic acid, vanillin, p-coumaric acid, etc. (Suryawanshi and Umate 2020).

"Bhunimbadi churna," another Ayurvedic powder formulation, has shown to treat diabetic neuropathy in a clinical study. Its polyphenol-rich ingredients include Pippali (*Piper longum*), Murva (*Marsdenia tenacissima*), Patola (*Trichosanthes dioica*), Nisha (*Curcuma longa*), Bhunimba (*Andrographis paniculata*), Nimba (*Azadirachta indica*), Triphala (*Terminalia chebula*, *Terminalia bellirica*, *Emblica officinalis*), Padmak (*Prunus cerasoides*) and Ativisha (*Aconitum heterophyllum*) (Kumar et al. 2022; Parveen and Baruah 2019). It was shown that Bhunimbadi churna was useful in lowering diabetic neuropathy signs and symptoms, particularly subjective sensory complaints. Electrical investigations showed that peripheral nerves were in better general health. The experimental medication appears to promote remyelination since the velocity of transmission of electric impulses through the nerves increased to normal or almost normal velocity. There was a noticeable drop in serum cholesterol, postprandial blood sugar and fasting blood sugar (Kumar et al. 2022).

A clinical investigation was carried out on a 61-year-old patient of diabetic peripheral neuropathy at Bharati Ayurved Hospital who was on conventional anti-diabetic medications for 10 years for type II diabetes. He had symptoms of pain, numbness, and burning feeling and was prescribed Gabapentin to treat neuropathy. "Vasanta Kusumakar Rasa" tablet (125 mg) was taken by the patient orally twice a day at 6 a.m. and after lunch. It was taken with one teaspoon of ghee for 3 months along with Padaabhyanga (foot massage) with hair oiling at night. This formulation provides nourishment to the cells and improves blood circulation, thus enhancing the sensory functions in diabetic neuropathy patients. It reduces aldose reductase levels, relieves oxidative stress brought on by hyperglycaemia, and helps to lessen oxidative stress. It is helpful in lowering diabetic neuropathy and rectifying the metabolic imbalance brought on by diabetes mellitus, even though it does not need to have a strong hypoglycaemic effect (Katkar and Mahajan 2023). Vasanta Kusumakar Rasa has also been shown to reduce blood glucose level in dose-dependent manner and intestinal glucose uptake in Streptozotocin and high-fat-diet induced diabetes in Sprague Dawley rats. It also improved insulin, glycated haemoglobin, HOMA-IR, and lipid profiles and showed organo-protective actions (Gaidhani et al. 2023). Vasanta Kusumakar Rasa comprises metals and polyphenol containing herbs. It consists of processed Gold, processed Tin, processed Silver, processed Lead, processed Mica, processed Coral, processed Pearl, processed Iron, red sulfide of Mercury, taken with cow milk, and decoction of herbs like *Saccharum officinarum*, *Adhatoda vasica*, *Santalum album*, *Vetiveria zizanioides*, *Pavonia odorata*, *Curcuma longa* and *Musa paradisiaca*. Fresh juices of *Nelumbium speciosum* and *Jasminum officinale* are also present (Singh et al. 2024). It can be an effective option for treatment of diabetic neuropathy.

Nishamlaki is an Ayurvedic formulation consisting of the fruits of Amla (*Emblica officinalis*) and dried rhizomes of Haldi (*Curcuma longa*) in a 6:1 ratio which has been shown to treat diabetic neuropathy in a preclinical study. Type 2 diabetic Wistar rats were administered 0.9 g/kg and 1.8 mg/kg of Nishamlaki along with honey. The study found that Nishamlaki improved thermal analgesia and cold allodynia, reduced lipid levels (serum cholesterol and serum triglyceride) and improved antioxidant enzymes function (increased SOD and Catalase and reduced malonaldehyde levels) of the sciatic nerves in the diabetic rats. The treatment group produced a dose-dependent increase in the pain threshold as compared to the untreated group in tail-immersion and hot plate assays (Dawane et al. 2016).

A 67-year-old male diabetic patient with diabetic peripheral neuropathy was successfully managed with a combination of Ayurvedic oral medication and external therapies. The oral medication included "Balaguduchi ksheera Kashaya" (consisting of *Sida cordifolia* and *Tinospora cordifolia* both containing polyphenols) in the dose of 100 ml taken two times after food, "Ksheerbala" 101 Capsule (1 capsule two times after food) and "Nishamalaki churna" (5 g twice after food), which is rich in polyphenols. The external therapies involved Udvartana, Abhyanga, Patrapotala sveda and Ksheeravasthi. The patient showed significant improvement in symptoms and quality of life after treatment, with complete recovery of ataxia, pinprick sensation, numbness, tingling sensation and weakness (Rao et al. 2020; Ahmed et al. 2018; Khan et al. 2020).

Basti chikitsa is an Ayurvedic therapy, which involves administering medicated liquids like herbal oils or decoctions into the rectum (medicated enema therapy). It comes under Panchakarma therapy, used to balance the Vata dosha (responsible for movement and communication in the body) contributing to Prameha (Diabetes). Basti Dravya stimulates the Central Nervous System (CNS) and acts on the receptors in the enteric nervous system, which results in the release of the necessary hormones. The CNS regulates the insulin secretion and pancreatic output, thus maintaining glucose homeostasis, an important factor in diabetic and neural complications. Till date there has been no study of polyphenol-rich Basti chikitsa formulations in treating diabetic neuropathy. The extent of absorption of polyphenols through this drug delivery system also remains in question. Research should be conducted as they can be an effective option due to the properties present in them (Suma et al. 2021).

14.7 Therapeutic Potential of Polyphenols in Diabetic Neuropathy

14.7.1 General Mechanism of Action

Polyphenols, especially flavonoids, stilbenes and phenolic acids, are effectively proven against diabetic neuropathy as they possess antioxidant, anti-inflammatory and neuroprotective activities. Polyphenols have also shown effective neuro-protection. For example, a study was conducted to investigate the role of apigenin on inhibiting trans-dedifferentiation of Schwann cells during peripheral nerve degeneration. Sciatic nerve explants were cultured in the presence or absence of apigenin, and the effects of apigenin on myelin fragmentation, axonal degradation and Schwann cell proliferation were evaluated using immunohistochemistry and Western blotting. Apigenin significantly inhibited myelin fragmentation, axonal degradation and Schwann cell proliferation in sciatic nerve explants. It decreased the expression of LAMP1 and p75 NTR, inhibited Schwann cell proliferation and thereby effectively delayed peripheral nerve degeneration (Kim et al. 2019). According to ethnobotanical research, some 800 plant species have been traditionally utilized to cure the diabetic disease. Among the many antidiabetic phytoconstituents present in plants are polyphenols and their in vivo constituents which function as active signalling agents. For example, HPLC analysis in the microwave-assisted as well as Soxhlet-assisted extract of a polyherbal formulation called Mehani revealed the presence of eight polyphenols with gallic acid being the predominant compound. The polyphenol-rich formulated showed $80.3 \pm 1.12\%$ and $85.3 \pm 1.86\%$ of alpha amylase inhibitory activity and $78.7 \pm 0.15\%$ and $80.0 \pm 0.63\%$ of alpha glucosidase inhibitory activity depicting its antidiabetic efficacy. The hydroalocoholic extracts obtained through Soxhlet and microwave-assisted

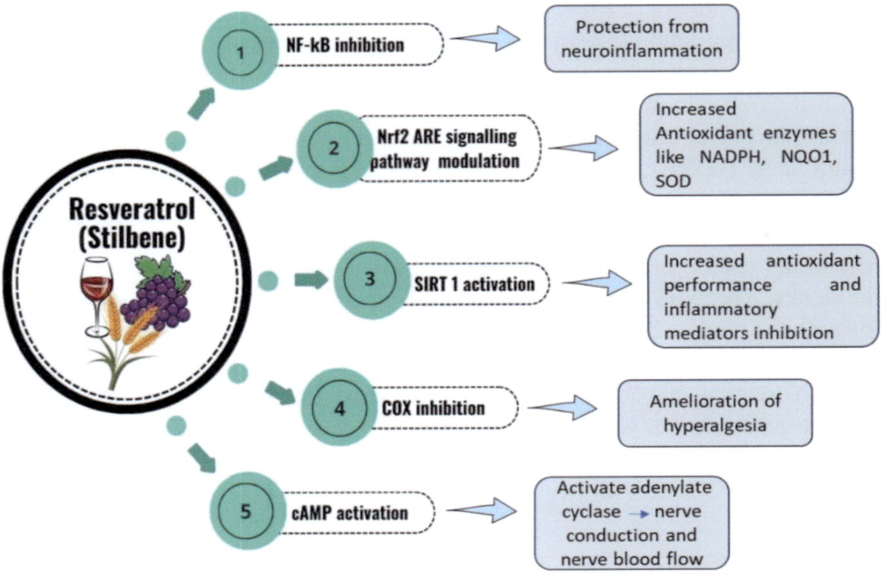

Fig. 14.3 Schematic representation of mode of action of resveratrol, a significant stilbene against diabetic neuropathy

extraction procedures exhibited the highest activities in terms of DPPH-free radical scavenging assay (93.8 ± 0.04%, 94.85 ± 0.04%) and hydrogen peroxide assay (88.7 ± 0.9%, 87.2 ± 1.4%) (Venkateswaran et al. 2021) (Fig. 14.3).

The mechanisms through which polyphenols act in treating diabetic neuropathy are multifaceted, for instance:

(a) *Reduction of oxidative stress*

Oxidative stress is a major contributor to nerve damage in diabetic neuropathy. Hyperglycaemia leads to the production of oxidative stress-induced reactive oxygen species (ROS). This results in reduced GSH levels, sodium dismutase, glutathione S-transferase, decreased glutathione peroxidase (GPx), oxidations of leukocytes and catecholamines, and increased mitochondrial breakdown and ischemia. Polyphenols act as potent antioxidants, neutralizing ROS and protecting neurons from oxidative damage. Polyphenols enhance the activity of endogenous antioxidant enzymes (such as superoxide dismutase, catalase, and glutathione peroxidase), which neutralize free radicals. This decreases lipid peroxidation and protein oxidation in the peripheral (Ye et al. 2022). For example, in Streptozotocin-induced diabetic rat models, apocynin and curcumin increased the SOD level and decreased high levels of malonaldehyde (MDA) and spinal Hydrogen peroxide. Curcumin reportedly reverses the overexpression of phagocyte NADPH oxidase subunits, suppressing the activation of spinal NADPH oxidases, the primary enzymes responsible for producing (Fig. 14.4) (Zhao et al. 2014).

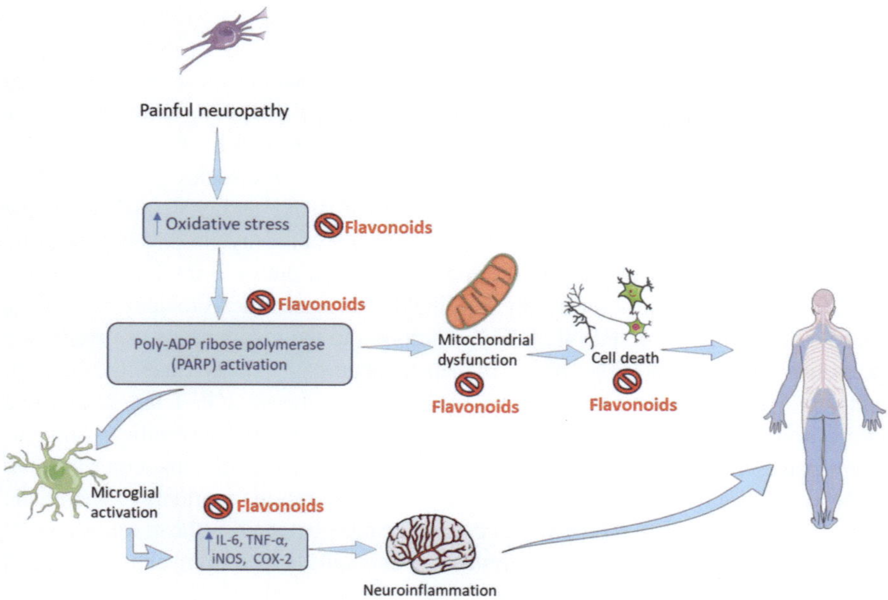

Fig. 14.4 Schematic representation of mode of action of flavonoids against painful neuropathy

(b) *Anti-inflammatory effects*

Persistent hyperglycaemia, oxidative stress and build-up of advanced glycation end products (AGEs) stimulate pro-inflammatory factors production like nuclear factor kappa B (NF-κB) and cytokines like TNF-α and IL-6. NF-κB is a redox-sensitive transcriptional factor triggered by hyperglycaemia, oxidative stress and pro-inflammatory cytokines. In a streptozotocin (STZ)-induced diabetes model, ischemia reperfusion damage triggered the overexpression of NF-κB in the diabetic Schwann and sciatic endothelial cells. Also, overexpression of intercellular adhesion molecule-1 (ICAM-1) and extensive permeation of monocyte macrophages in comparison to controls reported that NF-κB activation was responsible for the enhanced inflammatory response in the diabetic nerves (Wang et al. 2006). Also, the pro-inflammatory response and neuropathic alterations caused by diabetes are linked to the cytokine TNF-α, which is produced from NF-κB, inducing COX-2 overexpression and mitogen-activated protein kinase (MAPK) activation (Pop-Busui et al. 2016). Systemic inflammation and recurrent episodes of oxidative/nitrosative stress, endoplasmic and mitochondrial stress, and cumulative cellular damage are caused by hyperglycaemia, loss of insulin signalling, insulin resistance, dyslipidaemia and lipid metabolism abnormalities (Albers and Pop-Busui 2014). Polyphenols like resveratrol have been shown to mitigate diabetes-associated neuropathy through inhibition of the NF-κB cascade and also minimize the levels of TNF-α, IL-6 and COX-2 in the experimental rats. This reduces inflammation in the nerves, preventing further damage (Fig. 14.3) (Kumar and Sharma 2010)

(c) *Neuroprotective effects via nerve growth factors*

Diabetic neuropathy leads to depletion of neuronal growth factors like brain-derived neurotrophic factor (BDNF) and nerve growth factor (NGF), resulting in impaired nerve regeneration. It leads to decrease in the levels of ciliary neurotrophic factor, insulin-like growth factors-1 (IGF-1), neurotrophin-3/4/5, and BDNF in muscles (Rozanska et al. 2020). Polyphenols help upregulate the expression of BDNF and NGF, assisting nerve regeneration and repair. IGF-1 helps in the maintenance of the myelin sheath. In diabetic neuropathy, demyelination can occur, and IGF-1 helps preserve the integrity of the myelin sheath. Resveratrol reportedly can shield cells against neurotoxicity by causing the release of GDNF and BDNF. IGF-1 contributes to the maintenance of myelin, the protective sheath around nerve fibres. In diabetic neuropathy, demyelination can occur, and IGF-1 helps preserve the integrity of the myelin sheath. IGF-1 has insulin-like effects and can improve glucose metabolism. By improving insulin sensitivity and glucose uptake in cells, IGF-1 may help alleviate the hyperglycaemia that contributes to diabetic neuropathy (Kabir et al. 2021). Polyphenol like morin reportedly increased the IGF-1 and NGF levels in sciatic nerves in diabetic rats compared to the negative control group (AlSharari et al. 2014). Also, mulberry flavonoids helped in myelin degradation inhibition and upgradation of axonal function (Song-Tao et al. 2014).

(d) *Inhibition of Advanced Glycation End Products (AGEs)*

AGEs are formed when glucose binds to proteins, leading to nerve damage in diabetes. The release of proteases by macrophages in response to glycated myelin makes it susceptible to phagocytosis, which may lead to demyelination of the nerve cells. Significant axonal cytoskeletal proteins including tubulin, neurofilament and actin have also been found to be modified by AGE, which has been linked to axonal atrophy/degeneration and impaired axonal transport. Observational studies have suggested a decrease in relative muscular strength in patients with type 2 diabetes with increased AGE levels in their blood (Mengstie et al. 2022). Glycation affects neuron activity by two distinct processes: (a) glycation of neuronal protein, which results in functional loss; and (b) binding of AGE to nerve cell surfaces, which stimulates RAGE (receptor for AGE signalling) to activate inflammatory response that exacerbates neuronal injury. Thus, AGEs accumulate in peripheral nerves and activate RAGE, contributing to oxidative stress and inflammation (Bhat et al. 2017). Polyphenols like resveratrol, catechins, ferulic acid, sinapic acid and syringic acid block AGEs production and also block RAGE receptor activation, thereby reducing the AGEs-mediated inflammatory and oxidative responses (Yeh et al. 2017).

(e) *Activation of Sirtuins (SIRT1)*

SIRT1 is a protein that regulates cellular stress resistance, mitochondrial biogenesis and inflammation. Polyphenols, particularly resveratrol, are known SIRT1 activators. Resveratrol alters the structure of SIRT1 and enhances binding with its substrates, importantly p65 (RelA). p65 is associated with the NF-kB complex that controls the activation of leukocytes and inflammatory cytokines. It is mentioned earlier in this chapter that inflammation is connected with the development of diabetic neuropathy. By lowering the expression of inflammatory factors such TNFa, IL-1b, IL-6, metalloprotease (MMP)-1, MMP-3 and Cox-2, SIRT1 stimulated by resveratrol suppresses the acetylation of P65. Apart from resveratrol, Quercetin, Fisetin (found in strawberries), Curcumin (in *Curcuma longa*) and Berberine stimulate the SIRT1 activation. By activating SIRT1, polyphenols help regulate glucose metabolism, reduce oxidative stress, and protect neurons from hyperglycaemia-induced damage. SIRT1 activation also suppresses inflammation and enhances the repair mechanisms of nerve cells (Iside et al. 2020; Ayissi et al. 2014). Resveratrol has been found in the seeds and skin of grapes, which has helped in treating diabetic neuropathy. *Allium cepa* leaves containing quercetin have been useful in treating diabetic neuropathy. Ayurvedic medicinal plants like *Phyllanthus niruri* have been reported to activate Sirt1 thereby alleviating hyperglycaemia and also diabetic polyneuropathy in another study (Swargiary et al. 2024; Patel et al. 2022). In Ayurveda, variations in vata and vatarakta can lead to a variety of neurological system-related illnesses, including memory loss and synaptic transmission impairment. Ayurvedic antioxidant products including *Syzygium aromaticum* may aid in scavenging free radicals and preventing nerve illnesses., *Syzygium aromaticum*, due to its ability to calm Vata activates nerves by activation of SIRT1, thereby providing neuroprotection (Fig. 14.5) (Shekhar et al. 2018).

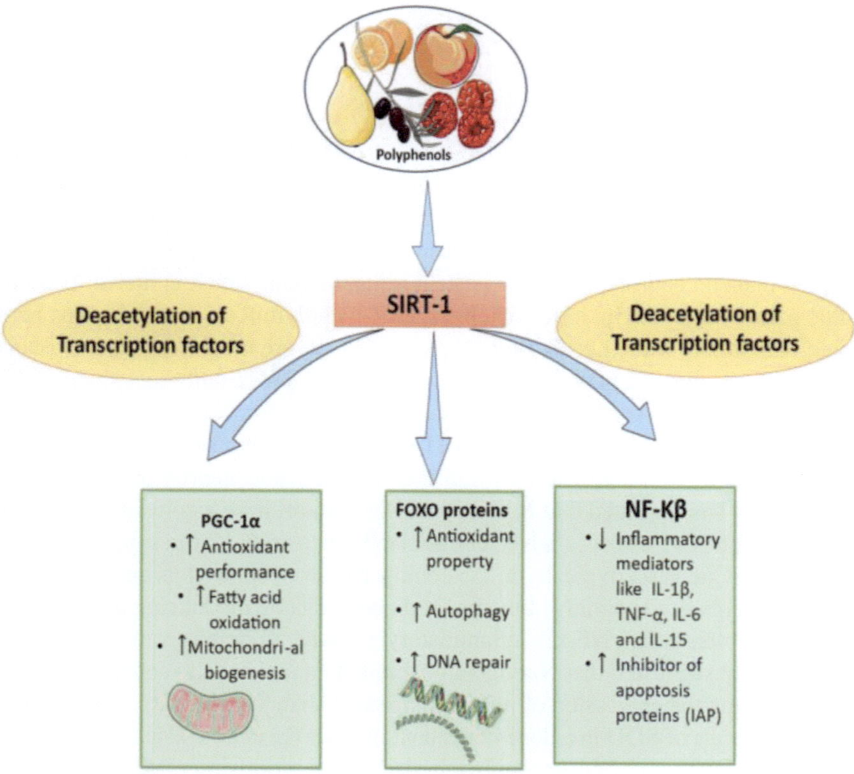

Fig. 14.5 Mode of action of Sirtuins against diabetic neuropathy

(f) *Modulation of the glutamate signalling pathway*

In diabetic neuropathy, abnormal levels of extracellular glutamate can overstimulate glutamate receptors, especially NMDA (N-methyl-D-aspartate) receptors, leading to excessive calcium influx into neurons. This results in glutamate excitotoxicity leading to neuronal injury and death, causing neuropathic pain in diabetes (Palazzo et al. 2024). Polyphenols such as resveratrol and quercetin can inhibit NMDA receptor overactivation by reducing oxidative stress and restoring cellular redox balance, thus protecting neurons from glutamate-induced excitotoxicity. By reducing the overexpression of glutamate receptors, particularly NMDA receptors, polyphenols protect neurons from excessive calcium influx, which can lead to mitochondrial dysfunction and neuronal death. For example, resveratrol inhibits the phosphorylation of NR2B, a subunit of NMDA receptors, which can attenuate glutamate excitotoxicity.

Glutamate is cleared from the synaptic cleft by glutamate transporters like GLT-1 (excitatory amino acid transporter 2) on astrocytes. Diabetic neuropathy-associated loss of function of these transporters leads to impaired glutamate clearance and increased excitotoxicity (Magdaleno Roman and Chapa González 2024). Polyphenols upregulate the expression and activity of glutamate transporters. For instance, resveratrol has been shown to enhance GLT-1 expression in astrocytes, thereby improving glutamate uptake and reducing extracellular glutamate levels (Girbovan and Plamondon 2015).

(g) *miRNAs as targets*

Sensitization of miR-25 in db/db mice is associated with improved neurological function in case of diabetic neuropathy. This is linked to the decline in the amount of reactive oxygen species (ROS) produced in diabetic peripheral nerves. This is mediated by PKC-α phosphorylation. It is also stated that there is a strong correlation between diabetic neuropathy and persistent inflammation, which is linked to neuropathic pain. MiR-190a-5p, miR-155 chr11, miR-9 chr2, miR-25 and miR-146a regulate inflammation, while Let-7i, miR-29b and miR-29c are involved in neurovascular dysfunctions in diabetic neuropathic patients, causing increased fibronectin. MiR-155 is claimed to exhibit its analgesic activity in neuropathy-associated pain by targeting factors like NF-κB (Malakoti et al. 2024). Consequently, it has been found that resveratrol inhibits miR-155 to exert its anti-inflammatory and neuroprotective effects. Also, a study has indicated that MiR-126 and resveratrol have a correlation, and resveratrol's protective effect is adversely affected when miR-126 is suppressed (Xin et al. 2018.). Quercetin, typically in the conjugated form, showed its neuroprotective activity and enhanced cognitive function in diabetic rats by modulating the activation of the miR-146a/miR-9/NF-κB inflammatory pathway (Ebrahimpour et al. 2020). Another study suggested that Naringenin exhibited its antidiabetic activity in rats by downregulating the expression of miR-224-3p and enhancing the mRNA and protein expression levels of SOD in a dose-dependent manner (Li et al. 2021).

There is a plethora of polyphenol-rich plants with promising effects against diabetic complications including diabetic neuropathy as summarized in Table 14.5.

Table 14.5 Summary of studies to evaluate the potency of polyphenols against diabetic neuropathy

Sl No.	Source plant name	Polyphenolic compounds	Aim of the study	Subjects involved	Model used	Results	Mechanism of action	References
1.	*Camelia sinensis* (Green tea)	Epigallocatechin-3-O-gallate	Effect on green tea polyphenol in maintaining spinal IGF-1/IGF1R signalling-mediated diabetic neuropathy	Swiss albino mice	STZ-induced diabetic rats	Reduced neuroinflammation and pain-like behaviours	Upregulation of spinal IGF-1/IGF1R signalling: Decrease in interleukin 1β (IL-1β), tumour necrosis factor-α (TNF-α) levels	Chen et al. (2022)
2.		Epigallocatechin-3-O-gallate	Effect on diabetes-induced neuropathic pain	Male Swiss albino rats	STZ-induced diabetic rats	Reduction in the latency time in the hot plate and tail immersion tests	Improved serum levels of insulin, lipids and NO: Enhanced SOD, CAT, GSH-Px and GSH levels in the brain tissues	Abo-Salem et al. (2020)
3.	*Boesenbergia rotunda*	Panduratin A, Pinostrobin, Pinocembrin, Cardamonin, Boesenbergin A and Alpinetin	Effects of Boesenbergia rotunda on inflammation and nociceptive responses diabetes peripheral neuropathic rats using polyphenol extract	Male Sprague Dawley rats	STZ-induced diabetic rats	Increased the mechanical pain threshold, duration; decrease in the inflammatory phase paw licking time	Decreasing blood glucose concentration and inflammatory mediators	Wang et al. (2022)

(continued)

Table 14.5 (continued)

Sl No.	Source plant name	Polyphenolic compounds	Aim of the study	Subjects involved	Model used	Results	Mechanism of action	References
4.		Hydroxytyrosol	Effect of hydroxytyrosol on oxidative stress and mitochondrial function on mouse brain	Male db/db mice	Normal rats	Neurons survival and mitochondrial respiratory chain complex is enhanced	Activation of AMPK, Sirtuin 1 and PPARγ coactivator 1α (PGC-1α)	Zheng et al. (2015)
5.		Naringin	Neuroprotective effect of naringin by modulation		STZ-induced diabetic rats	Enhanced Nociceptive threshold, tail-flick delay, reduced mechanotactile allodynia and oxidative-nitrosative stress	Reduced TNF-α level; Increased SOD levels	Kandhare et al. (2012)
6.		6-Methoxyflavanone	Amelioration of mechanical allodynia and vulvodynia in the streptozotocin-induced diabetic neuropathic pain	Female Sprague Dawley rats	STZ-induced diabetic rats	Anti-nociceptive activity: mechanical allodynia and vulvodynia on rats	Opioid (delta and kappa) and GABA-A receptor agonist	Akbar et al. (2016)
7.	*Selinum vaginatum*	Chlorogenic acid	Neuroprotective action of Chlorogenic acid isolated from *Selinum vaginatum in* diabetic rats	Adult Wistar rats of either sex	STZ-induced diabetic rats	Amelioration of neuropathic pain	Reduced TNF-α, neural nitrite in the sciatic nerve and LPO level; increased Na+K+ATPase level (membrane-bound inorganic phosphate) and SOD, levels	Saraswat et al. 2020

8.	*Cola nitida*	Caffeine	Examine the effect of acetone extract of Cola nitida in the spatial memory and brain Na+/ K+-ATPase activity of streptozotocin (STZ)-induced diabetic female Wistar rats.	Female Wistar rats	STZ-induced diabetic rats	Elevated brain Na^+/K^+ ATPase activity		Imam-Fulani et al. (2018)
9.	Apple peel extract		Activity of apple peel extract against peripheral diabetic neuropathy	Adult male Sprague-Dawley rats	High fat diet and STZ inducing diabetes	Analgesic effect in neuropathic rats Also improved paw withdrawal latency in diabetic animals.	Increased SOD, GSH and decreased MDA levels	Kamdi et al. (2021)
10.	Eugenol	*Syzygium aromaticum*	Investigate the effects of diabetes on nerve blood flow, oxygen delivery, and oxidative stress, and to explore the potential therapeutic effects of various antioxidants and compounds on diabetic neuropathy	Adult male Sprague-Dawley rats	STZ inducing diabetes	Restoration of impairments in the conduction velocities of motor and sensory nerves, increased blood flow to the nerves	Inhibition of TNF-α and (NF)kB cascade and advanced glycation end products (AGEs) formation	Nangle et al. (2006)

(continued)

Table 14.5 (continued)

Sl No.	Source plant name	Polyphenolic compounds	Aim of the study	Subjects involved	Model used	Results	Mechanism of action	References
11.		Ferulic acid	Evaluation of the potential of Ferulic acid against diabetic neuropathy in rats	Adult male Sprague-Dawley rats	STZ-induced diabetic rats	Improved peripheral nerve function	Enhanced insulin secretion and sensitivity; α-glucosidase inhibition activity; increased Bcl-2 levels, lowered Bax levels and caspase-3 activity in the diabetic rats; increased activity of Na+-K+ ATPase, and decreased activity of COX-2	Dhaliwal et al. (2020)
12.		Resveratrol	Evaluate the effects of resveratrol on TRPV4-mediated diabetic peripheral neuropathy (DPN), oxidative stress, and apoptosis in STZ-induced diabetic mice.	C57BL/6j mice	STZ-induced diabetic mice	Relief from neuropathic pain	Inhibition of TRPV4 channels, elevation of vitamin A, β-carotene, and vitamin E in the brain, erythrocytes, plasma, liver and kidney, reduction of oxidative stress, mitochondrial dysfunction and apoptosis in dorsal root ganglia	Amin et al. (2024)

13.		Isorhamnetin	Examine the neuroprotective effects of isorhamnetin in STZ-induced diabetic rats	Male Wistar rats	STZ-induced diabetes	Reduced pain, blood glucose levels, and increased body weight significantly compared to control.	Inhibition of astroglial activation, acetyl-cholinesterase activity, oxidative stress, apoptosis, and inflammation	Jamali-Raeufy et al. (2019)
14.	*Heterotheca inuloides*	7-Hydroxy-3,4-dihydrocadalin	Examine the neuroprotective effects of 7-hydroxy-3,4-dihydrocadalin in streptozotocin-induced diabetic rodents	Male Wistar rats	STZ-induced diabetes	Reduced formalin evoked hyperalgesia and allodynia	Decrease in MDA concentration, activation of 5-HT receptors, inhibition of guanylyl cyclase enzyme	Rocha-González et al. (2014)
15.	*Symphyocladia latiuscula*	*Symphyocladia latiuscula* polyphenols	Investigation of *Symphyocladia latiuscula* polyphenols in attenuating diabetic peripheral neuropathy	Wistar rats	STZ-induced diabetes	Improvement of nerve conduction velocity and muscle grip strength	Inhibition of aldose reductase and pro-inflammatory cytokines, restoration of NaþKþ-ATPase activity	Samaddar and Koneri (2019)
16.	*Ecklonia cava*	*Ecklonia cava* polyphenols	Exploration of the neuroprotective activity of *Ecklonia cava in* diabetic peripheral neuropathy	Wistar rats	STZ-induced diabetes	Improved nerve conduction velocity, compound muscle action potential, muscle grip strength and relief from neuropathic pain	Inhibition of aldose reductase and pro-inflammatory cytokines, restoration of NaþKþ-ATPase activity	Samaddar and Koneri (2019)

(continued)

Table 14.5 (continued)

Sl No.	Source plant name	Polyphenolic compounds	Aim of the study	Subjects involved	Model used	Results	Mechanism of action	References
17.	Rice bran oil	Oryzanol	Investigating the role of oryzanol from crude rice bran oil in treating diabetic neuropathy	Wistar rats of either sex	STZ-induced diabetes (45 mg/kg)	Increase in pain threshold levels in hot plate and tail immersion tests	Inhibition of $Na^{+}K^{+}$ATPase, MDA and Nitrite levels	Ghatak and Panchal (2012)
18.	*Bacopa monnieri*	Bacosine	Investigating the role of Bacosine from Bacopa monnieri in treating diabetic neuropathy	Male Wistar rats	STZ-Nicotinamide induced diabetes	Increase in pain threshold levels in hot plate and tail immersion tests	Decrease in interleukin 1β (IL-1β), tumour necrosis factor-α (TNF-α), Hb1Ac, Nitrite and MDA levels Increase in SOD, GSH and serum insulin levels	Kishore et al. (2017)
19.		Curcumin	Randomized double-blind placebo- controlled clinical trial of nano curcumin in lessening diabetic sensorimotor polyneuropathy	Diabetic patients(80)		Reduction in severity of pain of neuropathy	Reduction in glycated haemaglobin, total neuropathic score, fasting blood glucose, total reflex score	Asadi et al. (2019)

STZ Streptozotocin, *SOD* Sodium dismutase, *GSH* Reduced glutathione, *NO* Nitric oxide, *MDA* Malonaldehyde

14.8 Challenges in This Context

14.8.1 Safety Profile

Coming to the safety profile, resveratrol has no toxic effects so it can be used as a complementary and alternative medicine in treating neuropathic and inflammatory pain. Thus, in an animal trial, Longevinex, a resveratrol supplement possessed zero side effects, suggested positive outcomes (Sangeetha et al. 2013). Three flavonoid polyphenols—quercetin, EGCG (Epigallocatechin-3-gallate) and C3G—were assessed in a study for their safety profile. After being taken orally, these polyphenols were found in plasma and had a moderate-to-high antioxidant capacity. Their profiles varied, nevertheless, in terms of direct neurotoxicity, brain accessibility and capacity to shield brain neurons from oxidative damage. Although EGCG demonstrated a higher potential for neuroprotection and less disruption of the blood-brain barrier (BBB), it did exhibit some neurotoxicity. Although C3G merely prevented necrosis and disrupted the BBB, it showed less direct neurotoxicity. It was also reported that EGCG exhibited mitochondrial collapse on mice hepatocytes and tea polyphenols have been found to cause hepatic damage in mice after 24 h. The worst result was that of quercetin. Also, diabetic mice may exhibit cardiotoxicity if high dosages of EGCG are administered intraperitoneally (Silva and Pogačnik 2020). A study assessed the pharmacokinetics and safety of long-term use of green tea polyphenol or polyphenol E in Fitzpatrick skin types II or III in healthy men and women. Excess gas, stomachache, heartburn, headache, dizziness and muscle soreness were mild side effects of 800 mg of EGCG once daily or 400 mg twice daily in the treatment group. Although an 800 mg daily dose of EGCG was safe, well-tolerated and increased systemic EGCG by more than 60%, it did not offer any defence against UV-induced erythema (Chow et al. 2003).

14.8.2 Production Procedures

It comes with great difficulties in establishing a uniform procedure for obtaining all the required metabolites from plant sources owing to the variations in structure and composition of polyphenols in different plant materials and the fluctuations of plant matrix. The majority of laboratories choose traditional extraction techniques since they are inexpensive and simple to employ. Additionally, research has demonstrated that traditional techniques like maceration, percolation and soxhlet extraction have inadequate effectiveness and environmental issues because they require huge volumes of organic solvent. A recovery step is frequently included in the traditional extraction process, which is followed by the lengthy evaporation process for extract concentration (Alara et al. 2021). Factors influencing the efficiency of the extraction processes include the kind of solvent, plant to solvent ratio, extraction time, temperature and agitation rate. To overcome these problems, newer techniques such as microwave assisted extraction have been developed. The total phenolic content of certain plant extracts was found to increase with semi-continuous microwave-assisted extraction and increased stirring using larger magnetic stirrers with lower energy production and enhanced anti-oxidant potential of the plant extracts. Other methods like ultrasound-assisted, simultaneous ultrasound-microwave, supercritical fluid, microwave-assisted extractions are novel techniques with better production rates (Ofosu et al. 2020).

Metabolic engineering can boost the synthesis of desired metabolites by modifying and improving genetic and regulatory mechanisms in living things. Microbes with metabolic engineering are becoming more and more popular in food, medicinal and nutraceutical applications. Their commercial manufacturing is severely hampered by their declined yield, detrimental effects to cell growth or reduced host susceptibility to particular chemicals, and costly raw material procurement (Lyu et al. 2019).

14.9 Future Perspectives

Notable Indian medicinal plants like *Mentha piperita, Datura metel, Andrographis paniculata, Eclipta prostrata, Luffa acutangula, Withania somnifera* are rich in polyphenols. They are of Ayurvedic importance too in treatment of chronic diseases. These plants have reportedly shown activities against diabetes and diabetic complications if not specifically against diabetic neuropathy. Their mechanism of action in treating diabetes and related complications and the presence of polyphenols can be considered for future research against diabetic neuropathy. Many of the conventional drugs used in diabetic neuropathy are used to mitigate the symptoms but come with side effects like dizziness, drowsiness, peripheral oedema, etc. Therefore, polyphenolic therapy singly or combined with the conventional therapies can prove to be quite beneficial in treating diabetic neuropathy.

Moreover, diabetic neuropathy in Ayurveda parlance refers to complications arising due to further vitiated vataj avastha in diabetic patients resulting in respective systemic presentations. The typical management includes the use of lipid-based dosage forms of medicinal plants (medicated ghee) per oral and also medicated oils for topical and per rectal applications. These unique dosage forms can be taken up for their presence of polyphenols and can be considered for future research against diabetic neuropathy along with their mode of administration as per the dosha status of the patient as drug delivery also plays a major role. Thus, research on reported medicinal plants from various traditional medicine systems along with their unique drug delivery system can definitely provide new insights in the management of diabetic neuropathy.

Plants that can be considered in lipid-based formulations/preparations in diabetic neuropathy management from Ayurveda texts are described in Table 14.6.

Table 14.6 Medicinal plant materials used in the study: botanical identification and plant parts

Sr. No.	Sanskrit name	Botanical name	Part used
1.	Priyangu	*Calicarpa macrophylla*	Pushpa
2.	Lajjalu	*Mimosa pudica*	Whole plant
3.	Dhataki	*Woodfordia fruticosa*	Pushpa, bark
4.	Punnaga	*Calophyllum inophyllum*	Bark, seed, fruit
5.	Nagakesara	*Mesua ferrea*	Flowers
6.	Chandana	*Santalum album*	Heartwood, oil
7.	Kuchnadana	*Caesalpinia sappan*	Heartwood
8.	Mocharasa	*Bombax malabaricum*	Exudate
9.	Rasanjana	*Berberis aristata*	Stem bark preparation
10.	Kumbha	*Careya arborea*	Stem
11.	Padmakesara	*Nelumbo nucifera*	Stamen
12.	Manjishtha	*Rubia cordifolia*	Root
13.	Duralabha	*Fagonia cretica*	Whole plant
14.	Aralu	*Ailanthus excelsa*	Bark, root
15.	Madhuk	*Glycyrrhiza glabra*	Stolons
16.	Lodhra	*Symplocos racemose*	Bark
17.	Bilva	*Aegele marmelos*	Raw Fruit pulp
18.	Punarnava	*Boerhavia diffusa*	Root
19.	Danti	*Balipospermum montanum*	Root
20.	Shati	*Hedychium spicatum*	Root/rhizome
21.	Amalaki	*Phyllanthus emblica*	Fruit
22.	Haritaki	*Terminalia chebula*	Fruit
23.	Shirish	*Albizzia lebbeck*	Bark, seed
24.	Pippali	*Piper longum*	Fruit /spike
25.	Vacha	*Acorus calamus*	Rhizome
26.	Bharangi	*Clerodendrum serratum*	Root
27.	Shunthi	*Zingiber officinale*	Rhizome
28.	Vidanga	*Embelia ribes*	Fruits
29.	Kashmari	*Gmelina arborea*	Root, fruit
30.	Palash	*Butea monosperma*	Bark, flower

14.10 Conclusion

Polyphenols can certainly prove to be an effective and promising option pertaining to managing and treating diabetic neuropathy. Polyphenols have been reported to improve the oxidative state, reduce the release of proinflammatory mediators, enhance the survival of neurons by supporting the mitochondrial function and inhibiting apoptotic pathways and improve nerve regeneration by promoting the growth of new nerve fibres according to various in vitro studies using various cell cultures, in vivo studies employing animal models and limited clinical studies. Thus, it is essential to gather more evidence-based data for its potent implications in translational research.

References

Abbas M, Saeed F, Anjum FM, Afzaal M, Tufail T, Bashir MS, et al. Natural polyphenols: an overview. Int J Food Prop. 2017;20(8):1689–99.

Abo-Salem OM, Ali TM, Harisa GI, Mehanna OM, Younos IH, Almalki WH. Beneficial effects of (−)-epigallocatechin-3-O-gallate on diabetic peripheral neuropathy in the rat model. J Biochem Mol Toxicol. 2020;34(8):e22508.

Ahmed H, Juraimi AS, Swamy MK, Ahmad-Hamdani MS, Omar D, Rafii MY, et al. Botany, chemistry, and pharmaceutical significance of Sida cordifolia: a traditional medicinal plant. In: Anticancer plants: properties and application: volume 1; 2018. p. 517–37.

Akbar S, Subhan F, Karim N, Shahid M, Ahmad N, Ali G, et al. 6-Methoxyflavanone attenuates mechanical allodynia and vulvodynia in the streptozotocin-induced diabetic neuropathic pain. Biomed Pharmacother. 2016;84:962–71.

Alaofi AL. Sinapic acid ameliorates the progression of streptozotocin (STZ)-induced diabetic nephropathy in rats via NRF2/HO-1 mediated pathways. Front Pharmacol. 2020;11:1119.

Alara OR, Abdurahman NH, Ukaegbu CI. Extraction of phenolic compounds: a review. Curr Res Food Sci. 2021;4:200–14.

Albers JW, Pop-Busui R. Diabetic neuropathy: mechanisms, emerging treatments, and subtypes. Curr Neurol Neurosci Rep. 2014;14:1–11.

Al-Jaber HI, Shakya AK, Al-Qudah MA, Barhoumi LM, Abu-Sal HE, Hasan HS, et al. Piceatannol, a comprehensive review of health perspectives and pharmacological aspects. Arab J Chem. 2024;17:105939.

Al-Khayri JM, Mascarenhas R, Harish HM, Gowda Y, Lakshmaiah VV, Nagella P, et al. Stilbenes, a versatile class of natural metabolites for inflammation—an overview. Molecules. 2023;28(9):3786.

AlSharari SD, Al-Rejaie SS, Abuohashish HM, Aleisa AM, Parmar MY, Ahmed MM. Ameliorative potential of morin in streptozotocin-induced neuropathic pain in rats. Trop J Pharm Res. 2014;13(9):1429–36.

Amin GSM, Marouf BH, Namiq HS, Salih JM. Impact of resveratrol and pharmaceutical care on Type 2 diabetes mellitus and its neuropathic complication: a randomized placebo controlled clinical trial. J Clin Pharm Ther. 2024;2024(1):7739710.

Andargie M, Vinas M, Rathgeb A, Möller E, Karlovsky P. Lignans of sesame (Sesamum indicum L.): a comprehensive review. Molecules. 2021;26(4):883.

Asadi S, Gholami MS, Siassi F, Qorbani M, Khamoshian K, Sotoudeh G. Nano curcumin supplementation reduced the severity of diabetic sensorimotor polyneuropathy in patients with type 2 diabetes mellitus: a randomized double-blind placebo-controlled clinical trial. Complement Ther Med. 2019;43:253–60.

Ayissi VBO, Ebrahimi A, Schluesenner H. Epigenetic effects of natural polyphenols: a focus on SIRT1-mediated mechanisms. Mol Nutr Food Res. 2014;58(1):22–32.

Banafshe HR, Hamidi GA, Noureddini M, Mirhashemi SM, Mokhtari R, Shoferpour M. Effect of curcumin on diabetic peripheral neuropathic pain: possible involvement of opioid system. Eur J Pharmacol. 2014;723:202–6.

Basli A, Soulet S, Chaher N, Mérillon JM, Chibane M, Monti JP, Richard T. Wine polyphenols: potential agents in neuroprotection. Oxidative Med Cell Longev. 2012;2012(1):805762.

Belščak-Cvitanović A, Durgo K, Huđek A, Bačun-Družina V, Komes D. Overview of polyphenols and their properties. In: Polyphenols: properties, recovery, and applications. Woodhead Publishing; 2018. p. 3–44.

Bhat S, Mary S, Giri AP, Kulkarni MJ. Advanced glycation end products (AGEs) in diabetic complications. In: Mechanisms of vascular defects in diabetes mellitus; 2017. p. 423–49.

Bibi T, Bano S, ud Din F, Ali H, Khan S. Preparation, characterization, and pharmacological application of oral Honokiol-loaded solid lipid nanoparticles for diabetic neuropathy. Int J Pharm. 2023;645:123399.

Bulbul MRH, Chowdhury MNU, Naima TA, Sami SA, Imtiaj MS, Huda N, Uddin MG. A comprehensive review on the diverse pharmacological perspectives of *Terminalia chebula* Retz. Heliyon. 2022;8(8):e10220.

Chaikul P, Kanlayavattanakul M, Somkumnerd J, Lourith N. *Phyllanthus emblica* L.(amla) branch: a safe and effective ingredient against skin aging. J Tradit Complement Med. 2021;11(5):390–9.

Chen X, Le Y, Tang SQ, He WY, He J, Wang YH, Wang HB. Painful diabetic neuropathy is associated with compromised microglial IGF-1 signaling which can be rescued by green tea polyphenol EGCG in mice. Oxidative Med Cell Longev. 2022;2022(1):6773662.

Chiva-Blanch G, Urpi-Sarda M, Rotchés-Ribalta M, Zamora-Ros R, Llorach R, Lamuela-Raventós RM, et al. Determination of resveratrol and piceid in beer matrices by solid-phase extraction and liquid chromatography–tandem mass spectrometry. J Chromatogr A. 2011;1218(5):698–705.

Chow HS, Cai Y, Hakim IA, Crowell JA, Shahi F, Brooks CA, et al. Pharmacokinetics and safety of green tea polyphenols after multiple-dose administration of epigallocatechin gallate and polyphenol E in healthy individuals. Clin Cancer Res. 2003;9(9):3312–9.

Chumroenphat T, Somboonwatthanakul I, Saensouk S, Siriamornpun S. Changes in curcuminoids and chemical components of turmeric (*Curcuma longa* L.) under freeze-drying and low-temperature drying methods. Food Chem. 2021;339:128121.

Coussement P, Bauwens D, Peters G, Maertens J, De Mey M. Mapping and refactoring pathway control through metabolic and protein engineering: the hexosamine biosynthesis pathway. Biotechnol Adv. 2020;40:107512.

Das R, Lami MS, Chakraborty AJ, Mitra S, Tallei TE, Idroes R, et al. *Ginkgo biloba*: a treasure of functional phytochemicals with multimedicinal applications. Evid Based Complement Alternat Med. 2022;2022(1):8288818.

Dawane JS, Pandit VA, Bhosale MSK, Khatavkar PS. Evaluation of effect of nishamalaki on STZ and HFHF diet induced diabetic neuropathy in wistar rats. J Clin Diagn Res. 2016;10(10):FF01.

Dhaliwal J, Dhaliwal N, Akhtar A, Kuhad A, Chopra K. Beneficial effects of ferulic acid alone and in combination with insulin in streptozotocin induced diabetic neuropathy in Sprague Dawley rats. Life Sci. 2020;255:117856.

Dias MC, Pinto DC, Silva AM. Plant flavonoids: chemical characteristics and biological activity. Molecules. 2021;26(17):5377.

Dureshahwar K, Mubashir M, Une HD. Quantification of quercetin obtained from *Allium cepa* Lam. leaves and its effects on streptozotocin-induced diabetic neuropathy. Pharm Res. 2017;9(3):287.

Ebrahimpour S, Esmaeili A, Dehghanian F, Beheshti S. Effects of quercetin-conjugated with superparamagnetic iron oxide nanoparticles on learning and memory improvement through targeting microRNAs/NF-κB pathway. Sci Rep. 2020;10(1):15070.

El Khawand T, Courtois A, Valls J, Richard T, Krisa S. A review of dietary stilbenes: sources and bioavailability. Phytochem Rev. 2018;17(5):1007–29.

Essmat A, Hussein MS. Green tea extract for mild-to-moderate diabetic peripheral neuropathy a randomized controlled trial. Complement Ther Clin Pract. 2021;43:101317.

Farhan M, Rizvi A. The pharmacological properties of red grape polyphenol resveratrol: clinical trials and obstacles in drug development. Nutrients. 2023;15(20):4486.

Gaidhani SN, Ala S, Reddy V, Kumar YS, Jamadagni S, Selvam T, Kumar KA. Antidiabetic activity of Vasant Kusumakar ras in streptozotocin and high fat diet induced type 2 diabetes mellitus in Sprague Dawley rats. J Nat Remed. 2023;23:521.

Gasmi A, Gasmi Benahmed A, Shanaida M, Chirumbolo S, Menzel A, Anzar W, et al. Anticancer activity of broccoli, its organosulfur and polyphenolic compounds. Crit Rev Food Sci Nutr. 2024;64(22):8054–72.

Ghatak SB, Panchal SS. Protective effect of oryzanol isolated from crude rice bran oil in experimental model of diabetic neuropathy. Rev Bras. 2012;22:1092–103.

Girbovan C, Plamondon H. Resveratrol downregulates type-1 glutamate transporter expression and microglia activation in the hippocampus following cerebral ischemia reperfusion in rats. Brain Res. 2015;1608:203–14.

Gupta NK, Srivastva N, Puri S, Bubber P, Puri V. Neuroprotective potential of *Azadirachta indica* leaves in diabetic rats. Asian J Pharm Clin Res. 2017;10(4):243–8.

Hameed A, Galli M, Adamska-Patruno E, Krętowski A, Ciborowski M. Select polyphenol-rich berry consumption to defer or deter diabetes and diabetes-related complications. Nutrients. 2020;12(9):2538.

Hithamani G, Srinivasan K. Bioaccessibility of polyphenols from wheat (*Triticum aestivum*), sorghum (*Sorghum bicolor*), green gram (*Vigna radiata*), and chickpea (*Cicer arietinum*) as influenced by domestic food processing. J Agric Food Chem. 2014;62(46):11170–9.

Hou CC, Chen CH, Yang NS, Chen YP, Lo CP, Wang SY, et al. Comparative metabolomics approach coupled with cell-and gene-based assays for species classification and anti-inflammatory bioactivity validation of Echinacea plants. J Nutr Biochem. 2010;21(11):1045–59.

Hu P, Mei QY, Ma L, Cui WG, Zhou WH, Zhou DS, et al. Secoisolariciresinol diglycoside, a flaxseed lignan, exerts analgesic effects in a mouse model of type 1 diabetes: engagement of antioxidant mechanism. Eur J Pharmacol. 2015;767:183–92.

Imam-Fulani AO, Sanusi KO, Owoyele BV. Effects of acetone extract of *Cola nitida* on brain sodium-potassium adenosine triphosphatase activity and spatial memory in healthy and streptozotocin-induced diabetic female Wistar rats. J Basic Clin Physiol Pharmacol. 2018;29(4):411–6.

Iside C, Scafuro M, Nebbioso A, Altucci L. SIRT1 activation by natural phytochemicals: an overview. Front Pharmacol. 2020;11:1225.

Jamali-Raeufy N, Baluchnejadmojarad T, Roghani M. Isorhamnetin exerts neuroprotective effects in STZ-induced diabetic rats via attenuation of oxidative stress, inflammation and apoptosis. J Chem Neuroanat. 2019;102:101709.

Kabir MT, Tabassum N, Uddin MS, Aziz F, Behl T, Mathew B, et al. Therapeutic potential of polyphenols in the management of diabetic neuropathy. Evid Based Complement Alternat Med. 2021;2021(1):9940169.

Kamdi SP, Raval A, Nakhate KT. Effect of apple peel extract on diabetes-induced peripheral neuropathy and wound injury. J Diabetes Metab Disord. 2021;20:119–30.

Kandhare AD, Raygude KS, Ghosh P, Ghule AE, Bodhankar SL. Neuroprotective effect of naringin by modulation of endogenous biomarkers in streptozotocin induced painful diabetic neuropathy. Fitoterapia. 2012;83(4):650–9.

Kandhare AD, Mukherjee AA, Bodhankar SL. Neuroprotective effect of *Azadirachta indica* standardized extract in partial sciatic nerve injury in rats: evidence from anti-inflammatory, antioxidant and anti-apoptotic studies. EXCLI J. 2017;16:546.

Kanon A, Carroll A, Lomiwes D. Flavanols: Epigallocatechin Gallate, green tea, cocoa, clinical trial, functional foods. In: Handbook of food bioactive ingredients: properties and applications. Cham: Springer International Publishing; 2023 p. 1–35.

Kassuya CA, Silvestre AA, Rehder VLG, Calixto JB. Anti-allodynic and anti-oedematogenic properties of the extract and lignans from *Phyllanthus amarus* in models of persistent inflammatory and neuropathic pain. Eur J Pharmacol. 2003;478(2–3):145–53.

Katkar S, Mahajan M. Management of diabetic peripheral neuropathy with yoga, ayurved medicine and diet: a case study. Purana. 2023;LXV:153.

Khan TA, Ipshita AH, Mazumdar RM, Abdullah ATM, Islam GMR, Rahman MM. Bioactive polyphenol profiling and in-vitro antioxidant activity of *Tinospora cordifolia* Miers ex Hook F and Thoms: a potential ingredient for functional food development. Bangladesh J Sci Indus Res. 2020;55(1):23–34.

Kim M, Jung J, Jeong NY, Chung HJ. The natural plant flavonoid apigenin is a strong antioxidant that effectively delays peripheral neurodegenerative processes. Anat Sci Int. 2019;94(4):285–94.

Kishore L, Kaur N, Singh R. Bacosine isolated from aerial parts of *Bacopa monnieri* improves the neuronal dysfunction in Streptozotocin-induced diabetic neuropathy. J Funct Foods. 2017;34:237–47.

Kishore L, Kaur N, Singh R. Effect of Kaempferol isolated from seeds of *Eruca sativa* on changes of pain sensitivity in Streptozotocin-induced diabetic neuropathy. Inflammopharmacology. 2018;26:993–1003.

Kluska M, Jabłońska J, Prukała W. Analytics, properties and applications of biologically active stilbene derivatives. Molecules. 2023;28(11):4482.
Ko W, Yoon CS, Kim KW, Lee H, Kim N, Woo ER, et al. Neuroprotective and anti-inflammatory effects of Kuwanon C from Cudrania tricuspidata are mediated by heme oxygenase-1 in HT22 hippocampal cells, RAW264. 7 macrophage, and BV2 microglia. Int J Mol Sci. 2020;21(14):4839.
Kumar N, Goel N. Phenolic acids: natural versatile molecules with promising therapeutic applications. Biotechnol Rep. 2019;24:e00370.
Kumar A, Sharma SS. NF-κB inhibitory action of resveratrol: a probable mechanism of neuroprotection in experimental diabetic neuropathy. Biochem Biophys Res Commun. 2010;394(2):360–5.
Kumar A, Wasnik K, Mahto RR. Effect of Bhunimbadi Churna in management of uncontrolled Type 2 diabetes mellitus-a case study. J Ayurveda Integ Med Sci. 2022;7(11):232–7.
Li B, Lee YJ, Kim YC, Yoon JJ, Lee SM, Lee YP, et al. Sauchinone from *Saururus chinensis* protects vascular inflammation by heme oxygenase-1 induction in human umbilical vein endothelial cells. Phytomedicine. 2014;21(2):101–8.
Li CP, Qiu GZ, Liu B, Chen JL, Fu HT. Neuroprotective effect of lignans extracted from *Eucommia ulmoides* Oliv. on glaucoma-related neurodegeneration. Neurol Sci. 2016a;37:755–62.
Li Y, Yao J, Han C, Yang J, Chaudhry MT, Wang S, et al. Quercetin, inflammation and immunity. Nutrients. 2016b;8(3):167.
Li SH, Wang MS, Ke WL, Wang MR. Naringenin alleviates myocardial ischemia reperfusion injury by enhancing the myocardial miR-126-PI3K/AKT axis in streptozotocin-induced diabetic rats. Exp Ther Med. 2021;22(2):1–9.
Liu RH. Health-promoting components of fruits and vegetables in the diet. Adv Nutr. 2013;4(3):384S–92S.
Lyu X, Lee J, Chen WN. Potential natural food preservatives and their sustainable production in yeast: terpenoids and polyphenols. J Agric Food Chem. 2019;67(16):4397–417.
Magdaleno Roman JY, Chapa González C. Glutamate and excitotoxicity in central nervous system disorders: ionotropic glutamate receptors as a target for neuroprotection. Neuroprotection. 2024;2:137.
Mahmoud AM, Hernandez Bautista RJ, Sandhu MA, Hussein OE. Beneficial effects of citrus flavonoids on cardiovascular and metabolic health. Oxidative Med Cell Longev. 2019;2019(1):5484138.
Malakoti F, Mohammadi E, Akbari Oryani M, Shanebandi D, Yousefi B, Salehi A, Asemi Z. Polyphenols target miRNAs as a therapeutic strategy for diabetic complications. Crit Rev Food Sci Nutr. 2024;64(7):1865–81.
Manzoor A, Dar IH, Bhat SA, Ahmad S. Flavonoids: health benefits and their potential use in food systems. In: Functional food products and sustainable health; 2020. p. 235–56.
Marchiosi R, dos Santos WD, Constantin RP, de Lima RB, Soares AR, Finger-Teixeira A, et al. Biosynthesis and metabolic actions of simple phenolic acids in plants. Phytochem Rev. 2020;19:865–906.
Martí R, Roselló S, Cebolla-Cornejo J. Tomato as a source of carotenoids and polyphenols targeted to cancer prevention. Cancer. 2016;8(6):58.
Matsuda H, Morikawa T, Toguchida I, Park JY, Harima S, Yoshikawa M. Antioxidant constituents from rhubarb: structural requirements of stilbenes for the activity and structures of two new anthraquinone glucosides. Bioorg Med Chem. 2001;9(1):41–50.
Medras ZJ, Mostafa YM, Ahmed AA, El-Sayed NM. Arctigenin improves neuropathy via ameliorating apoptosis and modulating autophagy in streptozotocin-induced diabetic mice. CNS Neurosci Ther. 2023;29(10):3068–80.
Mengstie MA, Chekol Abebe E, Behaile Teklemariam A, Tilahun Mulu A, Agidew MM, Teshome Azezew M, et al. Endogenous advanced glycation end products in the pathogenesis of chronic diabetic complications. Front Mol Biosci. 2022;9:1002710.

Mishra A, Sharma AK, Kumar S, Saxena AK, Pandey AK. *Bauhinia variegata* leaf extracts exhibit considerable antibacterial, antioxidant, and anticancer activities. Biomed Res Int. 2013;2013(1):915436.

Mukherjee PK, Banerjee S, Gupta BD, Kar A. Evidence-based validation of herbal medicine: translational approach. In: Evidence-based validation of herbal medicine. Elsevier; 2022 p. 1–41.

Muthuraman A, Diwan V, Jaggi AS, Singh N, Singh D. Ameliorative effects of *Ocimum sanctum* in sciatic nerve transection-induced neuropathy in rats. J Ethnopharmacol. 2008;120(1):56–62.

Nadeem M, Taj Khan I, Khan F, Ajmal Shah M, Silva AS, Nabavi SF, et al. Lignans and flavonolignans. In: Silva AS, Nabavi SF, Saeedi M, Nabavi SM, editors. Recent advances in natural products analysis; 2020. p. 98–116.

Nagilla B, Reddy KP. Neuroprotective and antinociceptive effect of curcumin in diabetic neuropathy in rats. Int J Pharm PharmSci. 2014;6(5):131–8.

Nangle MR, Gibson TM, Cotter MA, Cameron NE. Effects of eugenol on nerve and vascular dysfunction in streptozotocin-diabetic rats. Planta Med. 2006;72(06):494–500.

Nieoczym D, Socała K, Zelek-Molik A, Pieróg M, Przejczowska-Pomierny K, Szafarz M, et al. Anticonvulsant effect of pterostilbene and its influence on the anxiety-and depression-like behavior in the pentetrazol-kindled mice: behavioral, biochemical, and molecular studies. Psychopharmacology. 2021;238:3167–81.

Nkembo AT. Flavonoids' classification and their applications as nutraceuticals. In: Flavonoids and anti-aging. CRC Press; 2023. p. 23–44.

Ofosu FK, Daliri EBM, Elahi F, Chelliah R, Lee BH, Oh DH. New insights on the use of polyphenols as natural preservatives and their emerging safety concerns. Front Sustain Food Syst. 2020;4:525810.

Oliveira AI, Pinho C, Sarmento B, Dias AC. Neuroprotective activity of *Hypericum perforatum* and its major components. Front Plant Sci. 2016;7:1004.

Oluwole O, Fernando WB, Lumanlan J, Ademuyiwa O, Jayasena V. Role of phenolic acid, tannins, stilbenes, lignans and flavonoids in human health–a review. Int J Food Sci Technol. 2022;57(10):6326–35.

Palazzo E, Marabese I, Ricciardi F, Guida F, Luongo L, Maione S. The influence of glutamate receptors on insulin release and diabetic neuropathy. Pharmacol Ther. 2024;263:108724.

Panche AN, Diwan AD, Chandra SR. Flavonoids: an overview. J Nutr Sci. 2016;5:e47.

Parveen R, Baruah H. An ayurvedic approach to the diagnosis and management of diabetic neuropathy. AYUHOM. 2019;6(2):53–9.

Patel K, Patel M, Gupta SN. Effect of Atibalamula and Bhumyamalaki on thirty-three patients of diabetic neuropathy. Ayu. 2011;32(3):353–6.

Patel MV, Patel MM, Patel KB, Chhayani PV, Mittwede M, Scheidbach D, Gupta SN. A randomized placebo-compared study on the efficacy of classical ayurvedic pharmaceutical form versus aqueous alcoholic extracts of *Phyllanthus niruri* Linn. Plus *Sida cordifolia* Linn. in patients of diabetic sensory polyneuropathy. J Ayurveda Integ Med. 2022;13(3):100619.

Pecyna P, Wargula J, Murias M, Kucinska M. More than resveratrol: new insights into stilbene-based compounds. Biomolecules. 2020;10(8):1111.

Pereira RG, Garcia VL, Rodrigues MVN, Martínez J. Extraction of lignans from *Phyllanthus amarus* Schum. & Thonn using pressurized liquids and low pressure methods. Sep Purif Technol. 2016;158:204–11.

Plaha NS, Awasthi S, Sharma A, Kaushik N. Distribution, biosynthesis and therapeutic potential of lignans. 3 Biotech. 2022;12(10):255.

Pop-Busui R, Ang L, Holmes C, Gallagher K, Feldman EL. Inflammation as a therapeutic target for diabetic neuropathies. Curr Diab Rep. 2016;16:1–10.

Prananda AT, Dalimunthe A, Harahap U, Simanjuntak Y, Peronika E, Karosekali NE, et al. *Phyllanthus emblica*: a comprehensive review of its phytochemical composition and pharmacological properties. Front Pharmacol. 2023;14:1288618.

Quideau S, Deffieux D, Douat-Casassus C, Pouységu L. Plant polyphenols: chemical properties, biological activities, and synthesis. Angew Chem Int Ed. 2011;50(3):586–621.

Rahman MM, Rahaman MS, Islam MR, Rahman F, Mithi FM, Alqahtani T, et al. Role of phenolic compounds in human disease: current knowledge and future prospects. Molecules. 2021;27(1):233.

Rao SK, Indu S, Kumar PP, Nair PG, Radhakrishnan P. Management of diabetic peripheral neuropathy through Ayurveda. J Ayurveda Case Rep. 2020;3(1):30–4.

Rashmi HB, Negi PS. Phenolic acids from vegetables: a review on processing stability and health benefits. Food Res Int. 2020;136:109298.

Ratan Y, Rajput A, Pareek A, Pareek A, Kaur R, Sonia S, et al. Recent advances in biomolecular patho-mechanistic pathways behind the development and progression of diabetic neuropathy. Biomedicine. 2024;12(7):1390.

Redza-Dutordoir M, Averill-Bates DA. Interactions between reactive oxygen species and autophagy: special issue: death mechanisms in cellular homeostasis. Biochim Biophys Acta Mol Cell Res. 2021;1868(8):119041.

Rocchetti G, Gregorio RP, Lorenzo JM, Barba FJ, Oliveira PG, Prieto MA, et al. Functional implications of bound phenolic compounds and phenolics–food interaction: a review. Compr Rev Food Sci Food Saf. 2022;21(2):811–42.

Rocha-González HI, Ramírez-Aguilar M, Granados-Soto V, Reyes-García JG, Torres-López JE, Huerta-Cruz JC, Navarrete A. Antineuropathic effect of 7-hydroxy-3, 4-dihydrocadalin in streptozotocin-induced diabetic rodents. BMC Complement Altern Med. 2014;14:1–12.

Rozanska O, Uruska A, Zozulinska-Ziolkiewicz D. Brain-derived neurotrophic factor and diabetes. Int J Mol Sci. 2020;21(3):841.

Rudrapal M, Khairnar SJ, Khan J, Dukhyil AB, Ansari MA, Alomary MN, et al. Dietary polyphenols and their role in oxidative stress-induced human diseases: insights into protective effects, antioxidant potentials and mechanism (s) of action. Front Pharmacol. 2022;13:806470.

Samaddar S, Koneri R. Polyphenols of marine red macroalga *Symphyocladia latiuscula* ameliorate diabetic peripheral neuropathy in experimental animals. Heliyon. 2019;5(5):e01781.

Sangeetha MK, Vallabi DE, Sali VK, Thanka J, Vasanthi HR. Sub-acute toxicity profile of a modified resveratrol supplement. Food Chem Toxicol. 2013;59:492–500.

Saraswat N, Sachan N, Chandra P. Anti-diabetic, diabetic neuropathy protective action and mechanism of action involving oxidative pathway of chlorogenic acid isolated from *Selinum vaginatum* roots in rats. Heliyon. 2020;6(10):e05137.

Septembre-Malaterre A, Remize F, Poucheret P. Fruits and vegetables, as a source of nutritional compounds and phytochemicals: changes in bioactive compounds during lactic fermentation. Food Res Int. 2018;104:86–99.

Shahidi F, Danielski R, Rhein SO, Meisel LA, Fuentes J, Speisky H, et al. Wheat and rice beyond phenolic acids: genetics, identification database, antioxidant properties, and potential health effects. Plants. 2022;11(23):3283.

Shekhar S, Yadav Y, Singh AP, Pradhan R, Desai GR, Dey AB, Dey S. Neuroprotection by ethanolic extract of *Syzygium aromaticum* in Alzheimer's disease like pathology via maintaining oxidative balance through SIRT1 pathway. Exp Gerontol. 2018;110:277–83.

Silva RF, Pogačnik L. Polyphenols from food and natural products: neuroprotection and safety. Antioxidants. 2020;9(1):61.

Singh AD, Chawda MB, Kulkarni YA. Cardioprotective effects of 'Vasant Kusumakar Rasa,'a herbo-metallic formulation, in type 2 diabetic cardiomyopathy in rats. Cardiovasc Toxicol. 2024;24(9):942–54.

Soleymani S, Habtemariam S, Rahimi R, Nabavi SM. The what and who of dietary lignans in human health: special focus on prooxidant and antioxidant effects. Trends Food Sci Technol. 2020;106:382–90.

Song-Tao M, Dong-lian L, Jing-jing D, Yan-juan P. Protective effect of mulberry flavonoids on sciatic nerve in alloxan-induced diabetic rats. Braz J Pharm Sci. 2014;50(4):765–71.

Suma J, Meghana N, Shalini C. A review on basti chikitsa in prameha: a critical analysis. Int J Res Ayurveda Pharm. 2021;12:110–4. https://doi.org/10.7897/2277-4343.1205157.

Suryavanshi SV, Barve K, Addepalli V, Utpat SV, Kulkarni YA. Triphala Churna—a traditional formulation in ayurveda mitigates diabetic neuropathy in rats. Front Pharmacol. 2021;12:662000.

Suryawanshi VS, Umate SR. A review on phytochemical constituents of Abutilon indicum (Link) sweet-an important medicinal plant in Ayurveda. Plant Sci. 2020;3(3):15–9.

Suzuki S, Umezawa T. Biosynthesis of lignans and norlignans. J Wood Sci. 2007;53:273–84.

Swargiary D, Kashyap B, Sarma P, Ahmed SA, Gurumayum S, Barge SR, et al. Free radical scavenging polyphenols isolated from *Phyllanthus niruri* L. ameliorates hyperglycemia via SIRT1 induction and GLUT4 translocation in in vitro and in vivo models. Fitoterapia. 2024;173:105803.

Tang Y, Li S, Yan J, Peng Y, Weng W, Yao X, et al. Bioactive components and health functions of oat. Food Rev Intl. 2023;39(7):4545–64.

Teka T, Zhang L, Ge X, Li Y, Han L, Yan X. Stilbenes: source plants, chemistry, biosynthesis, pharmacology, application and problems related to their clinical application-a comprehensive review. Phytochemistry. 2022;197:113128.

Tiwari V, Kuhad A, Chopra K. *Emblica officinalis* corrects functional, biochemical and molecular deficits in experimental diabetic neuropathy by targeting the oxido-nitrosative stress mediated inflammatory cascade. Phytother Res. 2011;25(10):1527–36.

Trapp CB, Barnard ND. Usefulness of vegetarian and vegan diets for treating type 2 diabetes. Curr Diab Rep. 2010;10:152–8.

Valsecchi AE, Franchi S, Panerai AE, Rossi A, Sacerdote P, Colleoni M. The soy isoflavone genistein reverses oxidative and inflammatory state, neuropathic pain, neurotrophic and vasculature deficits in diabetes mouse model. Eur J Pharmacol. 2011;650(2–3):694–702.

Venkateswaran MR, Jayabal S, Murugesan S, Periyasamy S. Identification of polyphenolic contents, in vitro evaluation of antioxidant and antidiabetic potentials of a polyherbal formulation-Mehani. Nat Prod Res. 2021;35(16):2753–7.

Vineela S, Thakur SR. Sinapic acid ameliorates Acrylamide induced neurotoxicity through inhibition of oxidative stress in rats. Res J Pharm Technol. 2020;13(12):6009–16.

Wang YJ, Chen RJ. Pterostilbene protection and bladder cancer cells. In: Cancer. Academic Press; 2014. p. 271–81.

Wang Y, Schmeichel AM, Iida H, Schmelzer JD, Low PA. Enhanced inflammatory response via activation of NF-κB in acute experimental diabetic neuropathy subjected to ischemia–reperfusion injury. J Neurol Sci. 2006;247(1):47–52.

Wang P, Wen C, Olatunji OJ. Anti-inflammatory and antinociceptive effects of *Boesenbergia rotunda* polyphenol extract in diabetic peripheral neuropathic rats. J Pain Res. 2022;15:779–88.

Wianowska D, Gil M. Recent advances in extraction and analysis procedures of natural chlorogenic acids. Phytochem Rev. 2019;18:273–302.

Wu YT, Lin LC, Tsai TH. Simultaneous determination of honokiol and magnolol in *Magnolia officinalis* by liquid chromatography with tandem mass spectrometric detection. Biomed Chromatogr. 2006;20(10):1076–81.

Xin Y, Zhang H, Jia Z, Ding X, Sun Y, Wang Q, Xu T. Resveratrol improves uric acid-induced pancreatic β-cells injury and dysfunction through regulation of miR-126. Biomed Pharmacother. 2018;102:1120–6.

Yang J, Wei Y, Zhao T, Li X, Zhao X, Ouyang X, et al. Magnolol effectively ameliorates diabetic peripheral neuropathy in mice. Phytomedicine. 2022;107:154434.

Yashin A, Yashin Y, Xia X, Nemzer B. Antioxidant activity of spices and their impact on human health: a review. Antioxidants. 2017;6(3):70.

Ye D, Fairchild TJ, Vo L, Drummond PD. Painful diabetic peripheral neuropathy: role of oxidative stress and central sensitisation. Diabet Med. 2022;39(1):e14729.

Yeh WJ, Hsia SM, Lee WH, Wu CH. Polyphenols with antiglycation activity and mechanisms of action: a review of recent findings. J Food Drug Anal. 2017;25(1):84–92.

Yi JM, Shin S, Kim NS, Bang OS. Neuroprotective effects of an aqueous extract of *Forsythia viridissima* and its major constituents on oxaliplatin-induced peripheral neuropathy. Molecules. 2019;24(6):1177.

Zhao WC, Zhang B, Liao MJ, Zhang WX, He WY, Wang HB, Yang CX. Curcumin ameliorated diabetic neuropathy partially by inhibition of NADPH oxidase mediating oxidative stress in the spinal cord. Neurosci Lett. 2014;560:81–5.

Zheng A, Li H, Xu J, Cao K, Li H, Pu W, et al. Hydroxytyrosol improves mitochondrial function and reduces oxidative stress in the brain of db/db mice: role of AMP-activated protein kinase activation. Br J Nutr. 2015;113(11):1667–76.

Alpha-Lipoic Acid: Mechanisms and Clinical Evidence in Diabetic Neuropathy

15

Biresh Kumar Sarkar, K N Anitha, Lalchand D. Devhare, S Umadevi, Geeta N. Lodhi, and Durga Prasad Thammisetty

15.1 Introduction

Peripheral nerve damage in diabetics who have had other reasons ruled out is a feature of diabetic neuropathy. Depending on the kind of nerve fibres affected, symptoms might change. One of three primary neurological conditions is more likely to occur in people with diabetes: extensive clinical neuropathy with distal sensory and autonomic symptoms, subclinical neuropathy with abnormalities by electrodiagnostic and quantitative testing, or localised syndromes (Vallianou et al. 2009).

Many risk factors, such as the length of diabetes, inadequate glycaemic control, and obesity, might contribute to diabetic neuropathy. Oxidative stress, inflammation, cell damage, hyperglycaemia, and mitochondrial malfunction might hasten it. Nerve cells undergo programmed cell death due to oxidative stress brought on by hyperglycaemia (Hsieh et al. 2023). Reactive oxygen and nitrogen species (RONS)

B. K. Sarkar
Central Ayurveda Research Institute, CCRAS, Ministry of AYUSH, Kolkata, India

K. N. Anitha (✉)
Department of Pharmacology, Government College of Pharmacy, Bangalore, Karnataka, India

L. D. Devhare
Nagpur College of Pharmacy, Wanadongri, Maharashtra, India

S. Umadevi
Department of Pharmaceutics, School of Pharmaceutical Sciences Vels Institute of Science Technology and Advanced Studies, Pallavaram, Chennai, India

G. N. Lodhi
Taywade College of Pharmacy, Koradi Nagpur, Maharashtra, India

D. P. Thammisetty
Department of Pharmacy Practice, Sri Padmavathi School of Pharmacy (Autonomous), Tirupati, Andhra Pradesh, India

M. K. Gautam et al. (eds.), *Antioxidants and Functional Foods for Diabetic Neuropathy*, https://doi.org/10.1007/978-981-95-9326-2_15

are linked to vascular tone signalling, insulin synthesis, hypoxia-inducible factor activation, cell proliferation, differentiation, and migration. Because free radicals can create secondary radicals, which can cause toxicity and oxidative stress, redox balance is crucial (Gautam et al. 2012).

The endogenous immune system maintains redox equilibrium through the action of glutathione peroxidase, catalase, and SOD. A rise in oxidative stress results from the accumulation and overpowering of endogenous molecules by reactive oxygen species (ROS). Neural pain following a stroke, spinal nerve damage, neuropathy caused by diabetes, and chemotherapy-induced neuropathy are among the painful neuropathies connected to oxidative stress (Viana et al. 2022). Because of its potent antioxidant qualities, ALA was recently investigated as a potential therapy for neuropathic pain. In a diabetic neuropathy rat model, ALA increased neuronal blood flow, reduced oxidative stress, and enhanced distal nerve transmission. The results of clinical studies on neuropathic complaints are encouraging (Agathos et al. 2018).

Studies on diabetic patients have shown that alpha-lipoic acid can protect peripheral neurones against ischaemia and increase the conduction acceleration of motor nerves in clinical diabetic neuropathy. It also raises the quantity of decreased glutathione (GSH), an antioxidant that occurs naturally and functions in redox-dependent activities with lipoic acid. The strong lipid-based free radical scavenger of peripheral nerves, alpha-lipoic acid, also promotes fibre regeneration and nerve growth factor stimulation. Based on how it affects nerve function, its therapeutic effectiveness is assessed (Vallianou et al. 2009).

15.2 Alpha-Lipoic Acid: A Brief Overview

The potent antioxidant properties of ALA were discovered in the 1980s. It is the only antioxidant soluble in both fat and water. It is present in potatoes, broccoli, skeletal muscle, liver, and other foods that both humans and animals produce. Nutritional supplements containing ALA typically contain either R-ALA alone or a combination of racemic of ALA's R and S isomers (Vallianou et al. 2009). When linoleic acid is broken down, ALA and a medium-chain triglyceride are created. This compound has the molecular weight of 206.32556 and the formula C8H14O2S2. Both R- (+)-ALA and S- (−)-ALA are the two ALA stereoisomers. The coenzyme of oxoglutarate dehydrogenase, R- (+)-ALA, may be found in the natural world in plants like spinach and broccoli as well as in the hearts, liver, and kidneys of tiny mammals. Tissue oxidation is either reduced or delayed in these cases. Preparations for medical use include both stereoisomers in a racemic combination. In comparison to the S- (−) stereoisomer, the R- (+) stereoisomer absorbs drugs more quickly and thoroughly after delivery. Since the former can be more effective than the latter, studies are being done to see if a compound's overall pharmacological properties can be affected by the latter stereoisomer, necessitating higher R- (+) quantities in pharmaceuticals that are marketed commercially. The reason for this is that the latter may hinder the compound's overall benefits (Carlson et al. 2007).

One kind of coenzyme is ALA necessary for the mitochondria to produce energy. It has significant antioxidant qualities and affects the physiology of the entire body. ALA is found in trace levels in almost every meal and is used as a medicinal component and dietary supplement. Well absorbed, A-lipoic acid diffuses well in intra- and extracellular environments and crosses the BBB. It has been applied to a number of oxidative stress models, including heavy metal and mushroom poisoning, diabetes, ischemia-reperfusion damage, cataracts, and neurodegenerative diseases. α-lipoic acid side effects might include moderate skin reactions, nausea, and vomiting (Evans and Goldfine 2000).

Chronic pain and migraine are two conditions for which α-lipoic acid treatment is supported by clinical evidence. According to a study, ingesting 600 mg of ALA orally every day for 3 months dramatically decreased the overall response rate of migraine patients and the number of migraine crises by 30%. People with peripheral neuropathic sciatic pain from a herniated disc had notable improvements in their symptoms and electromyography findings over 60 days. Individuals who took α-lipoic acid on a regular basis reported using fewer analgesics (71% versus 45.5%). When individuals with chronic cervical and lower back pain received 600 mg/day of ALA, and those with carpal tunnel syndrome received 300 mg/day, similar results were shown (Agathos et al. 2018).

15.3 Mechanisms of Action of Alpha-Lipoic Acid

After therapy, diabetics with neuropathic pain may benefit from ALA's antioxidant qualities. It neutralises ROS, repairs damage and stops oxidants from forming. Additionally, it increases blood flow to the nerves, which may lessen neuropathic pain. ALA can bind with other antioxidants or enhance their antioxidant activity. Its antioxidant function could be crucial in reducing neuropathic pain (Gautam et al. 2015).

ALA reduces lipid peroxidation, restores glutathione levels, and activates antioxidant enzymes while increasing peripheral nerve blood flow, glucose absorption, metabolism, and nerve conduction velocity (NCV). It lowers hyperalgesia, improves neuroprotection against reperfusion damage, decreases lipid oxidation, raises adenosine triphosphate activity, repairs neuropeptide shortages in the spinal cord, and inhibits NF-kB activation in peripheral neurones (Papanas and Ziegler 2014).

15.3.1 Antioxidant and Anti-inflammatory Properties

ALA provides the basis for the activity of pyruvate dehydrogenase and α-ketoglutarate dehydrogenase, whereas pyruvate's oxidative decarboxylation to acetyl-CoA connects glycolysis and the citric acid cycle. Potent antioxidants such as dihydrolipoic acid and ALA restore vit. C and Vit. E and increase GSH (Laher 2011).

By lowering oxidised forms of various antioxidant agents and influencing signalling pathways like insulin and NF-kB, ALA/DHLA functions as a biological antioxidant and metal chelator. Both DHLA and ALA inhibit the production of protein carbonyls, and only DHLA has the ability to restore the oxidised versions of vitamins E and C as well as neutralise free radicals. CoQ10, ubiquinone in its oxidised state, is reduced by DHLA and can also lower the alpha-tocopheroxyl radical (Bast and Haenen 2003). ALA is a pleiotropic drug that has antioxidant and anti-inflammatory qualities. It works by changing NF-kB to achieve these effects. Indeed, in several cell types and tissues, NF-kB regulates cytokines, such as IL-1β and IL-6. In brain models, IL-1β and IL-6 are modulated in a DNA methylation-dependent manner, opening the door to research on the epigenetic processes activated by ALA (Dinicola et al. 2017).

15.3.2 Effects on Mitochondrial Function

Supplementing with ALA is a practical approach for the activity of mitochondria in the context of inflammation-induced NO in chronic inflammatory diseases. ATP production was increased, mitochondrial capacity was increased, and mitochondrial enzyme activity was partially recovered. ALA may now have the ability to modify the S-nitrosylation of proteins and control the manifestation of certain proteins impacted by S-nitrosylation pathways. Since GSH is an antioxidant that shields mitochondrial complexes from damage caused by NO, it is hypothesised that this novel molecular pathway of ALA may be through the control of GSH levels, which were also examined. ALA supplementation via S-nitrosylation control can mitigate the reduced mitochondrial enzyme activity, which is essential for energy generation, caused by nitrosative stress. All of these mitochondrial enzyme complexes showed a large decrease in S-nitrosylation with the introduction of ALA, which also successfully restored the mitochondrial enzyme activity that had been suppressed by excess NO while producing a considerable increase in ATP (Hiller et al. 2016).

15.3.3 Modulation of Glucose Metabolism

Although there are significant changes between the two isomeric forms of ALA, it generally affects glucose absorption, glycogen synthesis, and insulin metabolism. According to research, ALA stimulates the muscle of obese Zucker rats to express more insulin receptor substrate 1 (IRS1) and facilitates its binding to the p85 regulatory subunit of PI3K. By raising local transporter levels, ALA is also thought to attract glucose transporter type 4 (GLUT4) to the skeletal muscle's sarcolemma, improving glucose absorption. Studies on cell cultures confirm that ALA stimulates the insulin-signalling pathway, which causes GLUT1 and GLUT4 to translocate. Additionally, ALA promotes insulin sensitivity by triggering AMPK and decreasing lipid build-up in skeletal muscle (Serhiyenko et al. 2018).

15.4 Clinical Evidence for Alpha-Lipoic Acid in Diabetic Neuropathy

Numerous advantages of ALA have been demonstrated in diabetic neuropathy. It improves endoneurial blood flow and NCV, reduces lipid peroxidation in the brain and sciatic nerve, and aids in the restoration of normal redox-sensitive nuclear transcription factor (NF)-κB dysfunction. It also improves myo-inositol content, the breakdown of calories, and glucose absorption in the endoneurium. Additionally, alpha-lipoic acid provides neurotrophic support by restoring neuropeptide deficits. It reduces diabetes-induced mechanical and thermal hyperalgesia, suppresses apoptosis in diabetic rat hippocampus tissue, protects peripheral nerves prevents ischemia-reperfusion injury and enhances antioxidant enzyme activity in the sciatic nerve (Ziegler 2004).

15.4.1 Clinical Trial Overview

ALA has been shown to enhance microcirculation in individuals with diabetic polyneuropathy and nitric oxide-mediated, endothelium-dependent vasodilation in diabetic patients. A 1999 study found that ALA works well for distal sensory nerves with modest ischemia-reperfusion injury (Mitsui et al. 1999). Researchers have investigated the neuroprotective benefits of ALA well, and these effects are attributed to decreased oxidative stress and increased microcirculation. Nevertheless, there has been conflicting evidence regarding the best way to administer the medication and how much is needed (Khan et al. 2022).

In order to assess the safety and efficacy of several therapies, including a 600 mg (i. v.) of ALA given over 3 weeks, Ziegler et al. performed a meta-analysis of four studies: ALADIN I, ALADIN III, SYDNEY, and NATHAN II. The therapy was safe and considerably lessened painful neuropathic symptoms, according to the results (Ziegler et al. 2004).

Researchers conducted a meta-analysis in 2012 to evaluate the safety as well as efficacy of treatments, including the intravenous injection of 300–600 mg of ALA per day for 2–4 weeks. The findings indicated significant improvements in both the NCS result and positive neuropathic symptoms (Han et al. 2012). The impact of Methylcobalamin (MC) individually vs MC + ALA therapy in patients with DPN was subsequently evaluated by the study team using a second meta-analysis. Their findings showed that daily therapy of MC (500–1000 mg, i.v. or i.m.) and ALA (300–600 mg, i.v. administration) for 2–4 weeks was connected with a better positive result in the nerve conduction investigation without suffering any negative consequences (Xu et al. 2013 Abubaker et al. conducted a thorough assessment in 2022 and found that ALA use alone did not considerably lessen diabetic patients' neuropathic discomfort. It did, however, have some impact on neuropathic pain (Abubaker et al. 2022).

After administering 600 mg intravenously of ALA every day for 3 weeks, diabetic individuals with neuropathy had less pain, paresthesias, and numbness. After

3 weeks, the same medication improved neuropathic symptoms and impairments in 1258 individuals, according to a recent meta-analysis (Ziegler et al. 2004). In diabetic patients, acute ALA infusion enhanced nitric oxide-mediated endothelium-dependent vasodilation, and in patients with diabetic polyneuropathy, it enhanced microcirculation (Gautam et al. 2014).

Alpha-lipoic acid therapy taken orally for 5 weeks reduced neuropathic indications and impairments in 187 patients with diabetic polyneuropathy. One of the main indications for neuropathic foot ulcers is deficits. This is a good discovery, therefore. According to the SYDNEY 2 trial, the optimal risk-benefit proportion is achieved with a dose of 600 mg once a day. (Ziegler et al. 2006). Approximately 21% of patients had side effects, primarily nausea, from the 1200 mg daily dosage. Compared to the ALADIN I and II studies, which employed the exact dosage of ALA, the seven-month ALADIN III experiment revealed a somewhat higher incidence of individuals taking 600 mg of ALA or a control. Although there was no discernible variation in the subjective assessment of symptoms across the groups, alpha-lipoic acid therapy was linked to enhanced nerve function (Vallianou et al. 2009).

According to the ISLAND Study, alpha-lipoic acid, either alone or in conjunction with imbesartan, considerably enhanced the brachial artery's endothelium-dependent flow-mediated vasodilation by 44% and 75%, respectively, after 4 weeks when compared to placebo treatment. Plasminogen activator-1 and interleukin-6 plasma concentrations decreased as a result of this impact, indicating that alpha-lipoic acid may lessen endothelial dysfunction (Vallianou et al. 2009). In streptozotocin-diabetic rats, alpha-lipoic acid treatment dramatically decreased von Willebrand factor, fibrinogen, and factor VII, suggesting that it has anti-inflammatory and antithrombotic qualities.

According to the study, at different dosages, alpha-lipoic acid efficiently suppresses the expression of molecules that bind to cells, particularly ICAM-1 and VCAM-1 (Roy et al. 1998). These findings may help prevent arteriosclerosis and other inflammatory illnesses, as well as treat them. The medication has an excellent safety profile, according to clinical and postmarketing monitoring studies. However, further research is required to evaluate the neurophysiological and therapeutic characteristics of alpha-lipoic acid (Magis et al. (2007)).

An additional study was carried out to evaluate the effects of oral ALA on diabetic peripheral neuropathy, LDL-C, HDL-C, and glycaemic control. Patients with a form of diabetes mellitus (DM) undergoing treatment at the outpatient internal medicine and diabetes clinics at Benha University Hospital were the subjects of an interventional, prospective study. The patient was treated with ALA for 3 months while they had diabetic peripheral neuropathy. Measurements of peripheral neuropathy symptoms, nerve conduction velocities, autonomic neuropathy cardiovascular (CV) tests, and posterior tibial nerve cross-section area were taken both before and after therapy. The findings suggested that administering ALA orally to diabetic individuals may improve their peripheral neuropathy, HbA1c, and LDL- and HDL-C levels (Elbadawy et al. 2021).

15.4.2 Dosage and Administration

ALA can be administered intravenously and taken as an oral supplement. Research suggests that it has potential therapeutic benefits for a number of diseases, such as neuropathy, ischemic-perfusion damage, and type 1 and type 2 diabetic mellitus (DM).

Oral Research indicates that giving individuals with type 1 and type 2 diabetes mellitus a daily oral dose of 300 mg of ALA for 3 months preserves and enhances their functional eyesight. Nevertheless, ALA can be taken orally for up to 6 months at a dosage of 600–1800 mg per day.

Intravenous injection The research supports the use of 600 mg IV therapy per day to treat neuropathic deficits and positive neuropathic sensations (Gębka et al. 2014).

After being consumed orally, α-LA is taken up by the gut and goes to the brain (where it may be able to pass across the blood–brain barrier without restriction). Dietary supplements or other sources aside, α-LA is converted to DHLA (hydrolipidic acid in reduced form) and metabolised in the liver to produce other metabolites including tetranorliponate and bisnorliponate, which are then eliminated by the kidneys. The cellular transport of α-LA has been associated with many mechanisms, including sodium-dependent transport, a transmembrane protein generated by the SLC5A6 gene that also facilitates the translocation of other vitamins and cofactors. Also, in charge of α-LA intestinal absorption are both transporters (Skibska et al. 2023).

15.4.3 Comparison with Other Therapies

A strong antioxidant with extensive research on its possible use in the management of diabetic neuropathy is ALA. The three major ways that ALA promotes nerve health are by lowering oxidative stress, strengthening mitochondrial activity, and increasing glucose absorption. In contrast to other treatments, ALA provides a distinct mode of action. For example, the drugs gabapentin and pregabalin, which are frequently prescribed for neuropathic pain, function by altering neurotransmitters in the central nervous system, but they frequently cause adverse effects such as sleepiness and vertigo. Another medication that effectively reduces pain is duloxetine, a serotonin-norepinephrine reuptake inhibitor that can also raise blood pressure and induce nausea. Another choice is tricyclic antidepressants, which are useful but can have serious adverse effects, including drowsiness and heart problems. Contrarily, at greater dosages, ALA mostly causes gastrointestinal pain and is typically well-tolerated with little adverse effects. By treating the oxidative stress and mitochondrial dysfunction that underlie diabetic neuropathy, ALA delivers systemic advantages in contrast to topical therapies like capsaicin, which alleviate localised pain by depleting substance but frequently result in a burning sensation. Opioids are occasionally used to treat extremely painful conditions, but ALA is a better long-term option due to ALA's lack of dependence risk and other significant adverse effects. Likewise, although acetyl-L-carnitine, a further mitochondrial enhancer, has certain advantages similar to those of ALA, its application is restricted to the regeneration of nerve tissue. Overall, because of its neuroprotective properties, ALA is a useful treatment for diabetic neuropathy. Depending on the needs of each patient, it may also be coupled with other therapies in an efficient manner (Deli et al. 2014).

15.5 Safety and Tolerability of Alpha-Lipoic Acid

15.5.1 Adverse Effects

The skin was the site of the most commonly reported adverse medication responses, mostly allergic reactions such as urticaria, pruritus, and dermatitis. Headaches, perspiration, and heat flushes were examples of whole-body symptoms. After parenteral injection, two slight anaphylactoid reactions along with a serious reaction, including laryngospasm, were noted. The most often reported gastrointestinal problems included nausea, vomiting, abdominal discomfort, and diarrhoea. Elevations of liver enzymes were seen in certain cases; however, they were primarily associated with concurrent illnesses, which may potentially account for these results. Vertigo and dizziness were among the central and peripheral nervous system symptoms. Only in the first year of observation were psychiatric problems listed as unexplained symptoms, such as worry, agitation, and sleep abnormalities. There were three documented occurrences of hypoglycaemia responses (Ziegler 2004).

There have been extremely rare reports of hypoglycaemia and increased liver enzymes. However, these events cannot be definitively linked to ALA. On rare occasions, ALA may also cause a little decrease in thyroid hormones or a severe case of laryngospasm. Lastly, it is not recommended during pregnancy or breastfeeding because its safety is uncertain (Papanas and Ziegler 2014).

15.5.2 Drug Interactions

Gleiter et al. examined the relationship between thioctic acid taken orally and meals in six healthy male and female participants (Gleiter et al. 1996). Following a 12-h fast, the participants took 600 mg of thioctic acid tablets after eating a typical meal of eggs, bacon, and butter. Compared to fasting, the study discovered that giving thioctic acid after a meal reduced AUC values, Cmax values, and tmax times. Since there was no discernible difference in the pharmacokinetic behaviour of the two isomers, it is advised to take the drug empty-handed. In people with type 2 diabetes, the trial also looked at how oral thioctic acid interacted with traditional treatment using glibenclamide or acarbose. There were no noticeable pharmacokinetic medication interactions (Gleiter et al. 1999).

15.5.3 Long-Term Use Considerations

In clinical practice, thioctic acid tablets and ampoules have usually satisfactory tolerability, according to post-marketing monitoring data concerning the drug's usage in Germany. According to monitoring data from an electronic record of 362 doctors, between April 1992 and March 1995, 112 spontaneous adverse drug reactions were recorded in 78 individuals, even though between 3 lacs to 4.3 lacs patients were treated each year. There was no correlation between cumulative thioctic acid doses and adverse drug events, even when the usage of larger quantities of medications increased (Rathmann et al. 1998).

15.6 Alpha-Lipoic Acid in Combination Therapies

In order to improve therapy results for diabetic neuropathy, alpha-lipoic acid (ALA) is frequently employed in combination treatments that target many disease processes. ALA's primary contribution to the combo therapy is its potent antioxidant effect, as oxidative stress and the activity of mitochondria are two crucial factors in the onset of diabetic neuropathy and its course. The effectiveness of a rehabilitation program in conjunction with resveratrol, cholecalciferol, acetyl-L-carnitine, and ALA in treating sciatica in young individuals resulting from herniated discs was assessed in research. The study focused on the improvement of pain resolution, postural modifications, pain medication use, and overall quality of life. Treatment for sciatica using a combination of rehabilitation and anti-inflammatory, analgesic, and antioxidant supplements reduces pain and promotes better postural stability (Scaturro et al. 2023).

Researchers also aimed to ascertain if Diabetic Neuropathy (DN) could be successfully treated with a single tablet that contained Superoxide Dismutase, Vitamin B12 (B12), Acetyl L-Carnitine, and Alpha Lipoic Acid. Cardiovascular Autonomic Reflex Tests (CARTs), the Michigan Neuropathy Screening Instrument Questionnaire and Examination (MNSIQ and MNSIE), and the vibration perception threshold (BIO) were used. All peripheral neuropathy indicators, including SNAP and SNCV, pain, and quality of life perception, were improved in patients with DMT2 when all four components were combined into a single pill and used for a full year. However, CARTs and MNSIE were not affected (Didangelos et al. 2020).

In treating diabetic peripheral neuropathy, research comparing epalrestat with α-lipoic acid (ALA) discovered that the dual treatment was more successful and efficient than the ALA and epalrestat monotherapies. Of the 1894 DPN patients, 473 were in the ALA group, 557 were in the epalrestat group, and 864 were in the (ALA + epalrestat) group (Zhao et al. 2018).

In treating diabetic peripheral neuropathy (DPN), research comparing methylcobalamin (MC) and lipoic acid (LA-MC) discovered that, when compared to MC therapy, LA-MC administration for 2–4 weeks improves NCV and neuropathic complaints (Xu et al. 2013).

Furthermore, ALA's advantages can be increased by combining it with dietary changes and other lifestyle alterations like better glycaemic management. ALA enhances the benefits of these additional therapies by lowering oxidative stress and enhancing nerve health, resulting in a more comprehensive strategy for addressing diabetic neuropathy.

15.7 Future Perspectives and Research Directions

An effective treatment for diabetic neuropathy is ALA. Crucially, it offers three more advantages over analgesic medications already approved for use in the treatment of symptomatic pain: this therapy, which is aetiology orientated, improves muscle strength, sensory deficits, and other symptoms, including paraesthesia, numbness, and pain. It is also more accepted and acts more quickly. ALA's efficacy and absence of significant side effects make it superior to popularly used analgesics such as duloxetine, which has subsequently been connected to a 16% withdrawal rate due to resistance. One significant disadvantage of ALA and any other analgesic medication used in painful DSPN is the short duration of every single controlled clinical study focused on neuropathic pain or symptoms (Lunn et al. 2014).

Currently, ALA is known for its antioxidant qualities and its ability to treat chronic pain and disorders involving central sensitisation. A powerful neutraliser of free radicals and metal chelate form, ALA also has opioid effects associated with decreased sensitisation of the CNS and peripheral elements supporting NP. These effects have been shown in studies with an approximate dosage of 10 mg ALA/kg for 30 days. There haven't been any trials, nevertheless, to show that people with NP continue to benefit from therapy or retain it when it ends (Mantovani et al. 2003).

ALA was investigated using several dosages and delivery methods. There is no standard dosage for ALA, and no amount has been found to be superior. In a similar vein, there is insufficient proof that one administration method is better than another. Importantly, research has not been done on the effectiveness of ALA when combined with other widely used medications (such as more recent medicines like tapentadol or more established ones like duloxetine). It is yet to be shown, but theoretically, given the many methods of action, one can hypothesise an additive effect. Thus, further knowledge in this area is needed. Further research is needed to determine whether gains in neurologic function are associated with a decrease in oxidative stress markers. A recent study on the new methylcobalamin, L-methylfolate, and pyridoxal-5′-phosphate combination (Metanx) demonstrated such a link (Fonseca et al. 2013). would support the pathophysiology-based therapeutic justification for ALA. Lastly, further information on the ideal and/or maximum length of therapy with this drug would be interesting to know.

15.7.1 Potential for New Therapeutic Approaches and Emerging Research

There has been a trend in the current situation towards a rise in the quantity of patents that have been issued and filed that centre on ALA formulations for diverse medicinal uses. Original research publications that outline several methods for creating ALA formulations, such as lipid-, polymer-, and metal-based nanoparticles, also looked at new treatment techniques. These novel approaches sought to improve the pharmacological and pharmacokinetic profile of ALA in order to augment its medicinal qualities. A total of 167 patents that were filed, granted, and are still in effect over the decade between 2011 and 2020 were noted. Due to its large number of pending patents, the United States is an uncommon market for products based on ALA. With 110 patents overall, the bulk of the nation's patents centre on mixing ALA with other medications in pharmaceutical formulations; others, however, to enhance ALA's medicinal properties, focus on cyclodextrins and liposome-based nanogels (Viana et al. 2022).

Apart from the conventional dosage forms already available for purchase, the application of nanocarriers is primarily impacted by ALA's substantial hepatic breakdown during first-pass metabolism and limited solubility in the gastrointestinal system (Salehi et al. 2019). Several strategies, such as metal, polymer, as well as lipid-based nanoparticles, have been studied during the last 5 years.

15.7.1.1 Metal-Based Nanoparticles

Crescenzo et al. investigated how gold nanoparticles (GNP) may improve ALA delivery by employing NaBH4. The optimised formulation, which is employed in Parkinson's disease research, showed great biocompatibility with human neuronal cell type SH-SY5Y and released the drug content in about an hour. This shows that nanoparticles made up of metals are a cutting-edge technology that needs more research (Di Crescenzo et al. 2016). Furthermore, utilising the SH-SY5Y cell line for Parkinson's disease, Piersimoni et al. synthesised GNP coated with ALA and examined its antioxidant qualities. By using antioxidant processes, they concluded that GNP-ALA protects the SH-SY5Y in living cells from ROS damage while maintaining the physiological properties of the cell membrane (Piersimoni et al. 2020).

15.7.1.2 Polymer-Based Nanoparticles

Additionally, polymeric matrix-based methods have been studied. Examples include nanospheres and nanofibers composed of polymers with a long half-life, such as poly (lactic-co-glycolic acid) and polycaprolactone. The nephrotoxicity of gentamicin was reduced by Aljaeid and El-Moselhy using ALA as a physiologically active additive. These two compounds were placed within the nanoparticles to do this, and the results were examined in vivo in rabbits. Compared to rabbits given a vehicle, rabbits treated with gentamicin exhibited higher levels of biochemical indicators of kidney injury. Yet, there was no discernible difference among the animals in the experiment who got nanoparticles containing both gentamicin and ALA and the biochemical markers of the control group. This suggests that ALA provided protection against nephrotoxicity caused by gentamicin (Aljaeid and El-Moselhy 2018). Haidar et al. investigated the use of poly nanofibers to administer atorvastatin and ALA to treat peripheral nerve injury. Chitosan nanoparticles were preloaded with atorvastatin. The first hour saw the release of almost 83% of the ALA. Compared to dosage forms that included only one active ingredient, in vivo studies on male Sprague-Dawley rats demonstrated that combining atorvastatin and ALA successfully recovered motor function (Haidar et al. 2020).

15.7.1.3 Lipid-Based Nanocarriers

In order to treat ALA, researchers have created a micellar system that is patent patent-protected and uses lipid-based techniques. The combination of ALA and carnosine had more neuroprotective benefits than carnosine by itself. When coupled with stearyl poly (20) oxyethylene ether and inorganic metal salts. These included enhanced pigmentation brought on by UV radiation, thicker epidermis, acceleration of keratinocyte proliferation and differentiation in response to dosage, and superior skin permeability (Viana et al. 2022).

15.8 Conclusion

In both experimental and clinical diabetes, oxidative stress is a major factor in nerve damage. It has been demonstrated that the natural antioxidant ALA inhibits glycation processes, stops beta-cell death, enhances glucose absorption, restores vitamin levels, and improves brain function. It is a potentially effective therapeutic strategy for the avoidance and management of long-term diabetes mellitus because of its critical function in mitochondrial bioenergetic activities. In models of diabetic neuropathy, peripheral nerve constriction damage, and autoimmune encephalomyelitis, ALA has also been demonstrated to alleviate neuropathic pain syndromes by modifying oxidative stress pathways and reducing nociception. Clinical trials including patients with severe diabetic neuropathy, diabetic sensorimotor polyneuropathy, multiple sclerosis-related neuropathic pain, and chronic lumbosacral radicular pain have demonstrated its analgesic benefits, which are backed by antioxidant pathways.

It has been demonstrated that ALA protects peripheral nerves from ischaemia in rats and increases motor-NCV in experimental diabetic neuropathy. ALA therapy raises GSH both in vivo and in vitro. GSH is a significant endogenous antioxidant that appears to be a major player in the redox-dependent processes of a variety of cellular targets, along with lipoic acid. Oxidative stress has been linked to endoneurial oxidative damage and impaired nerve blood supply because diabetes has been linked to increased generation and/or poor clearance of ROS. Nevertheless, more research utilising an improved technique and suitable demographic, dosage, and duration selection may offer proof of the latent therapeutic potential of ALA as well as its possible effects on other diabetes problems, including diabetic autonomic neuropathy.

Plant-based ALA has demonstrated promise in preclinical and clinical studies as a possible adjunctive therapy for neuropathy, particularly in diabetic polyneuropathies. Though further study is required to properly characterise its therapeutic usage and clinical feasibility, recent patent filings underscore the importance of for pharmacological optimisation of potential ALA-based medicines.

References

Abubaker SA, Alonazy AM, Abdulrahman A. Effect of alpha-lipoic acid in the treatment of diabetic neuropathy: a systematic review. Cureus. 2022;14(6):e25750.

Agathos E, Tentolouris A, Eleftheriadou I, Katsaouni P, Nemtzas I, Petrou A, et al. Effect of α-lipoic acid on symptoms and quality of life in patients with painful diabetic neuropathy. J Int Med Res. 2018;46(5):1779–90.

Aljaeid BM, El-Moselhy MA. Loading of gentamicin and alpha lipoic acid on a biodegradable polymer for more effective and less nephrotoxic formula. Int J Pharmacol. 2018;14(6):796–801.

Bast A, Haenen GR. Lipoic acid: a multifunctional antioxidant. Biofactors. 2003;17:207.

Carlson DA, Smith AR, Fischer SJ, Young KL, Packer L. The plasma pharmacokinetics of R-(+)-lipoic acid administered as sodium R-(+)-lipoate to healthy human subjects. Altern Med Rev. 2007;12(4):343.

Deli G, Bosnyak E, Pusch G, Komoly S, Feher G. Diabetic neuropathies: diagnosis and management. Neuroendocrinology. 2014;98(4):267–80.

Di Crescenzo A, Cacciatore I, Petrini M, D'Alessandro M, Petragnani N, Del Boccio P, et al. Gold nanoparticles as scaffolds for poor water soluble and difficult to vehiculate antiparkinson codrugs. Nanotechnology. 2016;28(2):025102.

Didangelos T, Karlafti E, Kotzakioulafi E, Kontoninas Z, Margaritidis C, Giannoulaki P, Kantartzis K. Efficacy and safety of the combination of superoxide dismutase, alpha lipoic acid, vitamin B12, and carnitine for 12 months in patients with diabetic neuropathy. Nutrients. 2020;12(11):3254.

Dinicola S, Proietti S, Cucina A, Bizzarri M, Fuso A. Alpha-lipoic acid downregulates IL-1β and IL-6 by DNA hypermethylation in SK-N-BE neuroblastoma cells. Antioxidants. 2017;6(4):74.

Elbadawy AM, Abd Elmoniem RO, Elsayed AM. Alpha lipoic acid and diabetes mellitus: potential effects on peripheral neuropathy and different metabolic parameters. Alexandria J Med. 2021;57(1):113–20.

Evans JL, Goldfine ID. α-Lipoic acid: a multifunctional antioxidant that improves insulin sensitivity in patients with type 2 diabetes. Diabetes Technol Ther. 2000;2(3):401–13.

Fonseca VA, Lavery LA, Thethi TK, Daoud Y, DeSouza C, Ovalle F, et al. Metanx in type 2 diabetes with peripheral neuropathy: a randomized trial. Am J Med. 2013;126(2):141–9.

Gautam MK, Gangwar M, Nath G, Rao CV, Goel RK. In-vitro antibacterial activity on human pathogens and total phenolic, flavonoid contents of Murraya paniculata (L.) leaves. Asian Pac J Trop Biomed. 2012;2(S2):S1660–4.

Gautam MK, Purohit V, Agarwal M, Singh A, Goel RK. In-vivo healing potential of Aegle marmelos in excision, incision and dead space wound models. Sci World J. 2014;2014:740107.

Gautam MK, Gangwar M, Singh SK, Goel RK. Effects of Azadirachta indica on vascular endothelial growth factor and cytokines in diabetic deep wound. Planta Med. 2015;81(8):713–21.

Gębka A, Serkies-Minuth E, Raczyńska D. Effect of the Administration of Alpha-Lipoic Acid on contrast sensitivity in patients with Type 1 and Type 2 diabetes. Mediat Inflamm. 2014;2014(1):131538.

Gleiter CH, Schug BS, Hermann R, Elze M, Blume HH, Gundert-Remy U. Influence of food intake on the bioavailability of thioctic acid enantiomers. Eur J Clin Pharmacol. 1996;50:513–4.

Gleiter S, Freudenthaler E, Potthast S, et al. Lack of interaction between thioctic acid, glibenclamide and acarbose. Br J Clin Pharmacol. 1999;48(6):819–25.

Haidar MK, Timur SS, Kazanci A, Turkoglu OF, Gürsoy RN, Nemutlu E, et al. Composite nanofibers incorporating alpha lipoic acid and atorvastatin provide neuroprotection after peripheral nerve injury in rats. Eur J Pharm Biopharm. 2020;153:1–13.

Han T, Bai J, Liu W, Hu Y. Therapy of endocrine disease: a systematic review and meta-analysis of α-lipoic acid in the treatment of diabetic peripheral neuropathy. Eur J Endocrinol. 2012;167(4):465–71.

Hiller S, DeKroon R, Hamlett ED, Xu L, Osorio C, Robinette J, et al. Alpha-lipoic acid supplementation protects enzymes from damage by nitrosative and oxidative stress. Biochim Biophys Acta Gen Subj. 2016;1860(1):36–45.

Hsieh RY, Huang IC, Chen C, Sung JY. Effects of oral alpha-lipoic acid treatment on diabetic polyneuropathy: a meta-analysis and systematic review. Nutrients. 2023;15(16):3634.

Khan H, Singh TG, Dahiya RS, Abdel-Daim MM. α-Lipoic acid, an organosulfur biomolecule a novel therapeutic agent for neurodegenerative disorders: an mechanistic perspective. Neurochem Res. 2022;47(7):1853–64.

Laher I. Diabetes and alpha lipoic acid. Front Pharmacol. 2011;2:15229.

Lunn MP, Hughes RA, Wiffen PJ. Duloxetine for treating painful neuropathy, chronic pain or fibromyalgia. Cochrane Database Syst Rev. 2014;2014(1):CD007115.

Magis D, Ambrosini A, Sándor P, Jacquy J, Laloux P, Schoenen J. A randomized double-blind placebo-controlled trial of thioctic acid in migraine prophylaxis. Headache. 2007;47(1):52–7.

Mantovani G, Macciò A, Madeddu C, Mura L, Gramignano G, Lusso MR, et al. The impact of different antioxidant agents alone or in combination on reactive oxygen species, antioxidant enzymes and cytokines in a series of advanced cancer patients at different sites: correlation with disease progression. Free Radic Res. 2003;37(2):213–23.

Mitsui Y, Schmelzer JD, Zollman PJ, Mitsui M, Tritschler HJ, Low PA. Alpha-lipoic acid provides neuroprotection from ischemia-reperfusion injury of peripheral nerve. J Neurol Sci. 1999;163(1):11–6.

Papanas N, Ziegler D. Efficacy of α-lipoic acid in diabetic neuropathy. Expert Opin Pharmacother. 2014;15(18):2721–31.

Piersimoni ME, Teng X, Cass AE, Ying L. Antioxidant lipoic acid ligand-shell gold nanoconjugates against oxidative stress caused by α-synuclein aggregates. Nanoscale Adv. 2020;2(12):5666–81.

Rathmann W, Haastert B, Delling B, Gries FA, Giani G. Postmarketing surveillance of adverse drug reactions: a correlational study approach using multiple data sources. Pharmacoepidemiol Drug Saf. 1998;7(1):51–7.

Roy S, Sen CK, Kobuchi H, Packer L. Antioxidant regulation of phorbol ester-induced adhesion of human Jurkat T-cells to endothelial cells. Free Radic Biol Med. 1998;25(2):229–41.

Salehi B, Berkay Yılmaz Y, Antika G, Boyunegmez Tumer T, Fawzi Mahomoodally M, Lobine D, et al. Insights on the use of α-lipoic acid for therapeutic purposes. Biomolecules. 2019;9(8):356.

Scaturro D, Vitagliani F, Tomasello S, Sconza C, Respizzi S, Letizia Mauro G. Combined rehabilitation with alpha lipoic acid, acetyl-L-carnitine, resveratrol, and Cholecalciferolin discogenic sciatica in young people: a randomized clinical trial. Medicina. 2023;59(12):2197.

Serhiyenko V, Serhiyenko L, Suslik G, Serhiyenko A. Alpha-lipoic acid: mechanisms of action and beneficial effects in the prevention and treatment of diabetic complications. MOJ Publ Health. 2018;7(4):174–8.

Skibska B, Kochan E, Stanczak A, Lipert A, Skibska A. Antioxidant and anti-inflammatory effects of α-lipoic acid on lipopolysaccharide-induced oxidative stress in rat kidney. Arch Immunol Ther Exp. 2023;71(1):16.

Vallianou N, Evangelopoulos A, Koutalas P. Alpha-lipoic acid and diabetic neuropathy. Rev Diab Stud. 2009;6(4):230.

Viana MDM, Lauria PSS, Lima AAD, Opretzka LCF, Marcelino HR, Villarreal CF. Alpha-lipoic acid as an antioxidant strategy for managing neuropathic pain. Antioxidants. 2022;11(12):2420.

Xu Q, Pan J, Yu J, Liu X, Liu L, Zuo X, et al. Meta-analysis of methylcobalamin alone and in combination with lipoic acid in patients with diabetic peripheral neuropathy. Diabetes Res Clin Pract. 2013;101(2):99–105.

Zhao M, Chen JY, Chu YD, Zhu YB, Luo L, Bu SZ. Efficacy of epalrestat plus α-lipoic acid combination therapy versus monotherapy in patients with diabetic peripheral neuropathy: a meta-analysis of 20 randomized controlled trials. Neural Regen Res. 2018;13(6):1087–95.

Ziegler D. Thioctic acid for patients with symptomatic diabetic polyneuropathy: a critical review. Treat Endocrinol. 2004;3:173–89.

Ziegler D, Nowak H, Kempler P, Vargha P, Low PA. Treatment of symptomatic diabetic polyneuropathy with the antioxidant α-lipoic acid: a meta-analysis. Diabet Med. 2004;21(2):114–21.

Ziegler D, Ametov A, Barinov A, Dyck PJ, Gurieva I, Low PA, et al. Oral treatment with α-lipoic acid improves symptomatic diabetic polyneuropathy: the SYDNEY 2 trial. Diabetes Care. 2006;29(11):2365–70.

Co-enzyme Q10 in Diabetic Neuropathy: Mechanisms of Action and Clinical Evidence

16

Lalchand D. Devhare, Yamjala Ganesh Kumar, Sukirti Priya, Amit Verma, Manish Dubey, and T. Naga Aparna

16.1 Introduction

A microvascular side effect of diabetes, diabetic neuropathy (DN) can seriously harm many parts of the peripheral nervous system, including dorsal root ganglia (DRG) neurones and Schwann cells. DN, the most common type of diabetes with substantial morbidity, early mortality, and a worse quality of life, affects around 50% of those with diabetes (Galeshkalami et al. 2019).

Painful diabetic neuropathy (PDN) is a common symptom of peripheral nerve damage that affects as much as 1/3 of diabetic patients. This crippling chronic ailment mostly affects the feet and legs and is typified by searing pain, tingling, and paresthetic and electric shock feelings. It also has a nocturnal aggravation. These excruciating symptoms typically worsen the patient's quality of life overall and are associated with psychiatric illnesses, including anxiety and depression, as well as significant sleep disturbances. Pharmaceuticals are commonly utilised for the symptomatic therapy of PDN in addition to strict glycaemic control and risk factor

L. D. Devhare (✉)
Nagpur College of Pharmacy, Wanadongri, Maharashtra, India

Y. G. Kumar
Sree Dattha Institute of Pharmacy, Rangareddy, Telangana, India

S. Priya
Central University of South Bihar, Fatehpur, Bihar, India

A. Verma
School of Pharmacy, Chandigarh University, Lucknow, Uttar Pradesh, India

M. Dubey
Vananchal College of Science (Pharmacy), Babu Dinesh Singh University, Garhwa, Jharkhand, India

T. N. Aparna
G. Pulla Reddy College of Pharmacy, Hyderabad, Telangana, India

M. K. Gautam et al. (eds.), *Antioxidants and Functional Foods for Diabetic Neuropathy*, https://doi.org/10.1007/978-981-95-9326-2_16

management. On the other hand, symptomatic therapies for PDN are often not very well tolerated and only partially successful. Thus, researchers have recently focused a great deal of emphasis on disease-modifying therapies that have the potential to change the underlying processes responsible for PDN in addition to relieving pain (Alam et al. 2020).

It has been proposed that the development and advancement of DN are significantly influenced by mitochondrial damage as well as inflammatory actions that ensue, despite the fact that the precise pathogenetic processes underlying the condition have not yet been determined. Given these pathogenic pathways, several anti-inflammatory and antioxidant medications have been tried to treat diabetic neuropathy patients (Mondal et al. 2025).

An essential part of the mitochondrial electron transport chain, which produces energy, is CoQ10. It has potent antioxidant and anti-inflammatory qualities. By recycling and rebuilding endogenous antioxidants like vit. E and C, it indirectly lessens tissue oxidative stress damage and directly scavenges reactive oxygen species (ROS). By preventing nuclear factor kappa B (NFkB), a major contributor to inflammation, from activating, CoQ10 may also have anti-inflammatory effects. This implies that CoQ10 may be a long-term, safe treatment for a number of illnesses marked by mitochondrial malfunction, oxidative stress, and inflammation (Villalba et al. 2010). In neuropathic pain caused by sciatic and spinal nerve ligation, CoQ10 has demonstrated antinociceptive properties in experimental investigations. Additionally, promising findings from a number of experimental investigations have been published about the possible effectiveness of CoQ10 in the treatment of neuropathic pain brought on by diabetes mellitus (Amini et al. 2022).

16.2 Biochemical Properties of Co-enzyme Q10

16.2.1 Structure and Function

The compound 2,3-dimethoxy,5-methyl,6-polyisoprene parabenzoquinone is known as coenzyme Q. Ten isoprene units, each with five carbons, or a total of fifty carbons, make up the polyisoprene chain of coenzyme Q10, which is present in humans. The all-trans polyisosoprene is responsible for its liking for the inner layer of cell walls. The methyl group and the two methoxy groups both affect how selectively an enzyme act. The fully substituted quinone ring inhibits further interactions with thiol compounds in cells, including glutathione, thioredoxin, and thioctic acid. The group is represented by the quinone ring. Quinone reduction results in the creation of proton and electron carriers (Rauchova et al. 1995).

Coenzyme Q is dispersed throughout the cell layers. The enzymes that regulate coenzyme Q oxidation reduction have well-characterized protein attachment area in mitochondria. Furthermore, coenzyme Q floats inside the membranes' phospholipid bilayer. The all-trans polyisoprene chain may float in the free form and extend in a lengthy straight shape due to the CH3 groups incorporated in the lipid's fatty acid chains. Additionally, evidence suggests that the polyisoprenoid chain might occasionally fold into a more dense, smaller shape. It is believed that the lipid bilayer can be stabilised by the isoprenoid chain. The quinone head group exists in two different forms: oxidised (quinone) and reduced (quinol). Most membranes include known enzymes that may oxidise quinol and reduce quinone.

16.2.2 Role in Mitochondrial Electron Transport Chain

Coenzyme Q is a crucial component of the biological machinery that generates ATP, the energy required for muscular contraction and other key biological processes. The orientated enzymatic protein complexes have a bond with the coenzyme Q. It undergoes oxidation, releasing protons into the outer membrane while accumulating protons and electrons within the mitochondrial membrane. An essential enzyme in the aerobic chain, CoQ10 transfers electrons from complex II to complex III along with protein I and complex II. When it receives electrons from such complexes, it stays in its decreased phase as ubiquinol; when it sends electrons to complex III, it reverts to its oxidized form as ubiquinone (Casagrande et al. 2018; Gautam et al. 2014).

Another aspect of proton transport in the outer most layer of the cell involves coenzyme Q. In this case, coenzyme Q helps the Na/H antiport to initiate the H_2 interchange through the barrier. Coenzyme Q is part of the cell membrane electron transport system, which is responsible for moving electrons from the cytoplasm to external electron acceptors like iron or oxygen using NADH. Activating this mechanism significantly raises the proton outflow via the H/Na antiport (Crane 2001).

16.2.3 Antioxidant Mechanisms

Coenzyme Q serves as a major free radical scavenger by being situated in membranes near to unsaturated lipid chains. In various membranes, the concentration of CoQ ranges from three to thirty times that of tocopherol. Since coenzyme Q is mostly located in cell walls in the quinol state, it can be a very strong antioxidant. The decreased levels of coenzyme Q in plasma and endomembrane are known to be sustained by at least three enzymes. The three enzymes in concern are NADPH coenzyme Q reductase, NADH cytochrome b5 reductase, and NADH/NADPH oxidoreductase (DT diaphorase) (Villalba and Navas 2000). When there is oxidative stress brought on by a diet deficient in selenium and tocopherol, membrane coenzyme Q levels are significantly elevated.

When decreased, CoQ10 functions as an antioxidant, preventing oxidation of cellular membranes, blocking lipid peroxidation, helping to recycle α-tocopherol and indirectly stabilising Ca^{2+} channels to avoid calcium overload (Sugiyama et al. 1980).

Because it is necessary for the manufacture of ATP, CoQ10 is also needed for mitochondrial bioenergy, which affects every cell in the body and is thus necessary for good health. It is helpful in neutralising reactive oxygen species, or free radicals, because of its redox characteristic. CoQ10, the body's sole naturally occurring liposoluble antioxidant, may participate in redox processes to shield DNA and proteins from oxidative degradation and lipid peroxidation. By protecting calcium excess, it also obliquely preserves Ca^{2+} channel stability. The enzyme works on lipid peroxidation by either securing free radicals or converting the α-tocopheryl radical to α-tocopherol (Kagan et al. 1990; Murthy et al. 2013). It functions similarly to vit. E, however unlike CoQ10, vit. E is not extracted internally and is only obtained from diet and hepatic storage.

16.2.4 Synthesis and Regulation in the Human Body

Normal CoQ10 levels are sustained in healthy persons by two mechanisms: endogenous synthesis via the mevalonate cycle and exogenous synthesis through food consumption. In the mevalonate cycle of endogenous synthesis, acetyl-CoA is the initial substrate. The end products include Triglyceride, CoQ10, and Long-chain mostly unsaturated organic compound the latter of which is necessary for the glycosylation of proteins. Through this process, the enzyme prenyltransferase synthesises the isoprenoid side chain of CoQ10, which then goes through another condensation with 4-hydroxybenzoate. CoQ10 enters the body through the exogenous pathway in its oxidised state and is then transformed at the RBC level into a decreased state. It is present in various foods in trace amounts in nature, but is most commonly found in dark veggies like spinach, legumes like broccoli, grains like soy and peanuts, oleaginous fruits like almonds and nuts, and red meats like liver and heart. It is also present in some fish like sardines and mackerel (Casagrande et al. 2018).

16.3 Mechanisms of Action of Co-enzyme Q10 in Diabetic Neuropathy

Hyperglycaemia, hypoxia, and inflammation all increase the production of reactive oxygen species, which leads to the development of diabetic neuropathy. In addition to decreased insulin and insulin growth factor (IGF) activity, oxidative stress is primarily responsible for the pathogenic process. In theory, Diabetic neuropathy may be treated with coenzyme Q10, an antioxidant having bioenergetic, anti-inflammatory, and neuronal apoptosis prevention qualities (Shi et al. 2013).

Inflammation, hypoxia, and hyperglycaemia all increase the Producing of ROS, which leads to the progress of DN. The pathogenic process is mostly attributed to oxidative stress, together with reduced insulin and IGF activity. It participates in the process of catalysing the electron transfer from glycerol-3-phosphate to CoQ10. The β cell's physiological function is affected by G3PD activity. G3PD was shown to be deficient in pancreatic islets from either cadaveric human islet from type 2 diabetes patients or rat models of the disease. It was hypothesised that elevated CoQ10 levels in the inner membrane of the mitochondria might improve the reduced G3PD activity in the pancreatic islet (Shen and Pierce 2015).

16.3.1 Reduction of Oxidative Stress and Free Radical Damage

All tissues contain CoQ10, which shields cell membranes from free radical-induced lipoperoxidation. Its main antioxidant is ubiquinol, which is CoQ10 in a decrease form. CytB5 and NQO1 are two examples of oxidoreductases that take part in the CoQ10 redox cycle in cell membranes. This makes it possible to recover the active form of CoQ10, and ubiquinol can interact with α-tocopherol and vit. C radicals to lessen oxidative stress (Sifuentes-Franco et al. 2022).

16.3.2 Modulation of Inflammatory Pathways

CoQ10 has been shown to have positive benefits via direct and indirect anti-inflammatory mechanisms. When administered supplements, CoQ10 has been demonstrated to change the function of TNF-α and IL-1 in individuals with diabetic nephropathy (Mazidi et al. 2018). Additionally, research indicates that CoQ10 functions through redox-dependent pathways to prevent apoptosis and reduce inflammation, as seen by the reduction in plasma levels of TNF-α, IL-6, and CRP after taking a CoQ10 supplement (Heidari et al. 2018). By inhibiting the expression of the NF-Kb gene, CoQ10 can lower the production of pro-inflammatory cytokines. During infection, oxidising agents and pathogen lipoperoxides trigger a signalling cascade in monocytes that releases TNF-alpha and activates the NF-kB protein, hence lowering inflammation. Because of its ability to scavenge radicals, CoQ10's antioxidant capability can inhibit the signalling pathway. Additional research indicates that the anti-inflammatory properties of CoQ10 could be linked to adiponectin; raising adiponectin levels by supplementation raises CoQ10 levels, which in turn lowers TNF-α-mediated inflammatory response (Farsi et al. 2016).

16.3.3 Effects on Cellular Apoptosis and Nerve Regeneration

The production of CoQ10 reaches its peak around the age of 25, after which it begins to drop with ageing and may change depending on the tissue. The production of CoQ10 in heart tissue at age 65, for example, is around half that of heart tissue at age 25. CoQ10 is important for both normal neuronal activity and the risk of developing brain diseases because many neurological disorders are caused by mitochondrial disorders and free radical-induced oxidative stress, and normal brain function necessitates a high energy intake (20% of total body energy consumption). Glia and neurones alike contain coenzyme Q10 (CoQ10), which is present throughout the brain. In cultured human brain cells, it has been shown that additional CoQ10 restores depleted levels (Mantle et al. 2021).

16.4 Clinical Evidence and Studies on Co-enzyme Q10 in Diabetic Neuropathy

CoQ10 may directly scavenge ROS and may also recycle and regenerate other natural antioxidants, including vit. E and C, which can help avoid or reduce damage from oxidative stress on tissues. Given these routes of action, CoQ10, either alone or as an adjuvant, has been shown as a safe and effective long-term treatment for a number of illnesses, particularly those marked by inflammation, oxidative stress, and mitochondrial dysfunction. According to research, taking a CoQ10 supplement may help control blood sugar levels. The potential application of CoQ10 in the treatment of specific chronic pain syndromes has been the subject of recent scientific studies. Additionally, promising findings from a number of experimental investigations have been published about the possible effectiveness of CoQ10 in the treatment of neuropathic pain brought on by diabetes mellitus (Amini et al. 2022).

16.4.1 Overview of Preclinical Studies

Positive preclinical results have been found in studies on the neuroprotective benefits of Co-enzyme Q10 (CoQ10) in the treatment of diabetic neuropathy. Both animal and in vitro cellular models of diabetes have been used in research. We looked at the well-known T2D model, leptin receptor-deficient db−/db− mice. The findings imply that sensory abnormalities in the late-stage diabetic db−/db− mice may be related to decreased PLCβ3 expression and reactive oxygen species. Antioxidant CoQ10 supplementation, both early and long-term, may be a potential therapy strategy for type 2 diabetic neuropathy (Shi et al. 2013).

In a different research, the antioxidant coenzyme Q10 (CoQ10) is assessed for its preventive and antinociceptive properties against neuropathic pain in a diabetic rat model caused by diabetes. By preventing oxidative stress and suppressing proinflammatory factors, CoQ10 may offer protection against inflammation. According to these findings, administering CoQ10 could be a relatively safe, high-reward strategy for diabetic neuropathy prevention or treatment (Zhang et al. 2013).

Researchers are exploring Coenzyme Q10 (CoQ10) as a possible therapy for DN pain. They evaluated the inhibition effects of CoQ10 and the overexpression of downstream cytokines, NF-Kb, MAPK, and TLR4. These elements are thought to play a significant role in the onset of neuropathic pain. Because of its antioxidant and free-radical scavenger properties, CoQ10 may reduce oxidative stress in the CNS and PNS. The findings suggest that CoQ10 may be a safe, long-term treatment for DN (Zhang et al. 2013).

To evaluate the possible beneficial impacts of Coenzyme Q10 in combination with Pyridoxine for the management of diabetic neuropathic pain caused by streptozotocin (STZ), more research was done. By blocking high levels of glycated haemoglobin, hyperlipidaemia, oxidative-nitrosative stress, and membrane-bound inorganic phosphate enzyme, the researchers discovered that pyridoxine and coenzyme Q10 not only decreased high blood sugar levels but also markedly decreased neuropathic pain. This, in turn, lessens allodynia and hyperalgesia (Visnagri et al. 2012).

16.4.2 Human Clinical Trials

Clinical research on the helpful benefits of CoQ10 supplementation in treating individuals with chronic pain disorders have yielded positive, but limited, results. Numerous clinical investigations have suggested that CoQ10 therapy may help fibromyalgia sufferers have better quality of life and less pain. In this context, a research was conducted with the premise that combining coenzyme Q10 (CoQ10) with pregabalin might aid patients with painful diabetic neuropathy (PDN) by reducing their symptoms. Reports indicate that in comparison to the control with pregabalin treatment, the CoQ10 + pregabalin treatment significantly decreased pain. The mean pain NRS score reductions for the two groups at the end of week two did not vary significantly. However, between weeks four and eight, patients receiving CoQ10 + pregabalin saw a considerably higher mean pain NRS score drop than those taking placebo + pregabalin (Amini et al. 2022).

The study found that administering 100 mg/day of CoQ10 to women with T2D reduced oxidative stress and FBS levels while increasing total antioxidant capacity (TAC), catalase enzyme activity, and quantitative insulin sensitivity check index. Another study has reported that T2DM patients who took 200 mg of CoQ10 daily saw a substantial reduction in their HbA1c values, indicating better glycaemic management. Blood LDL-C levels also significantly decreased, which may protect T2DM patients from atherosclerotic plaques and the resulting cardiovascular problems (Samimi et al. 2024).

An additional study was carried out to assess the effects of adding L-carnitine and coenzyme Q10 to current oral antidiabetic drugs on type 2 diabetic patients' lipid profiles, lipoprotein levels, and glycaemic control. According to the findings, each of the CoQ10 and L-carnitine groups saw a significant decrease in FBG, total lipids, LDL-c, and lipoprotein (a) after 8 weeks of treatment. Additionally, the HbA1c% of the coenzyme Q10 group's levels were noticeably less than the control groups (Mohammed-Jawad et al. 2014).

Another 12-week clinical study that was placebo-controlled examined how Coenzyme Q10 (CoQ10) affected the lipid profile, diabetic neuropathy, blood glucose, and oxidative stress among those with T2D. The average insulin tolerance and TAC concentration of the Q10 group rose considerably after the experiment in comparison to the placebo group, the results showed. Following the experiment, the treatment group's C-reactive protein (hsCRP) dramatically dropped in comparison to the control group (Fakhrabadi et al. 2014).

16.4.3 Meta-Analyses and Systematic Reviews

Based on a comprehensive review of 40 randomized controlled trials, taking CoQ10 supplements dramatically reduced fasting levels of insulin, glucose, HbA1c, and HOMA-IR. This suggests that taking CoQ10 supplements may help with glycaemic management. The findings revealed a "U"-shaped dose–response slope for CoQ10 intake and result markers of glycaemic management, suggesting that 100–200 mg/day of CoQ10 is more effective in lowering fasting blood sugar level, HbA1c, HOMA-IR, and insulin concentrations (Liang et al. 2022).

The researchers assess CoQ10's effectiveness as well as safety for individuals with T2DM in a different a literature review. To evaluate coenzyme Q10's benefits on patients with T2D in comparison to control or western treatment, data from randomised controlled studies were gathered. There were thirteen studies with a total of 765 subjects. The outcome shown that coenzyme Q10 may help individuals with T2DM with glycaemic management, TG reduction, and HDL-C enhancement (Zhang et al. 2018).

16.5 Co-enzyme Q10 Dosage and Administration in Neuropathy Management

16.5.1 Optimal Dosing Strategies

There are several types of CoQ10-containing commercial goods on the market that come in powder, pill, or oil form, with varying dosages that indicate various levels of bioavailability. The products can be found in reduced or oxidised forms. Compared to powders, pills, capsules, or oil powder solutions, soluble CoQ10 formulations have higher bioavailability and are absorbed more quickly (Schulz et al. 2006). It has been demonstrated that ubiquinol is more readily absorbed when comparing the solubilised forms of ubiquinone and ubiquinol. Numerous clinical studies, including those including the widest spectrum of disorders, have shown that negative effects were more common with concentrations over 1200 mg per day, with ranges between 22 and 400 mg per day being deemed safe (Kumar et al. 2013; Casagrande et al. 2018).

16.5.2 Bioavailability and Absorption Challenges

The recommended daily dosage of CoQ10 from food is 2.5–5 mg, and the gastrointestinal tract absorbs only about 10% of the consumed amount because of the enzyme's poor water solubility and a large molecular weight. This is insufficient to meet the body's needs when there is a redox imbalance (Kumar et al. 2009). The decreased ubiquinol form of CoQ10 makes up around 95% of the plasma in healthy persons. As with fats, CoQ10 is absorbed from the food slowly and preferentially because of its elevated molecular weight and hydrophobic properties. 1–2 h after oral ingestion, plasma CoQ10 levels begin to rise. Their peak concentration happens in 7–8 h, and their half-life may last up to 34 h. CoQ10 is primarily taken up in the small intestine and then transferred to the liver to form the lipoprotein complex. The liver absorbs the CoQ10 carried by the chylomicrons and incorporates it into Low-density lipoprotein, which has 58% of the molecule, and high-density lipoprotein, which has 26%. The heart, lungs, kidneys, spleen, and adrenal glands are among the tissues that subsequently get CoQ10 (Miles 2007).

16.5.3 Potential Side Effects and Contraindications

It's safe to use CoQ10 supplements. High dosages were used in several clinical trials, but no side effects were reported that were severe enough to jeopardise the treatment. Pregnant lady, nursing mothers, teenager should take the enzyme with caution as Its impact throughout these times has not been entirely established. Studies have shown that less than 1% of individuals experience gastrointestinal side effects, such as headaches, allergic skin rashes, diarrhoea, vomiting, and nausea (Fuke et al. 2000). This medicine has the potential to cause hypotension and antiplatelets, thus individuals using it should be closely watched. Because the mevalonate route is used to synthesise both CoQ10 and cholesterol, several studies have found decreased CoQ10 levels following CoQ10 treatment in conjunction with HMG-CoA reductase inhibitors (statins). Serum CoQ10 concentrations may drop by as much as 54% (Ghirlanda et al. 1993). It has been demonstrated that the amount of CoQ10 decrease while taking statins in combination is dose-related and reversible when therapy is stopped. It has been suggested that this decrease might be the root of the statin side effects; therefore, as long as CoQ10 supplementation is carefully controlled, it may serve as a potential mediator treatment while on statin therapy (Silver et al. 2004).

Hepatic Impairment In non-alcoholic fatty liver disease (NAFLD), supplementing with CoQ10 lowers biochemical markers and systemic inflammation. Nevertheless, bile is the excretion of CoQ10, as stated in pharmacokinetics. Therefore, it is not recommended to use in individuals who have biliary blockage (Campisi and La Motta 2022).

Renal Impairment Patients with nephrotic syndrome as a result of primary CoQ10 insufficiency may benefit from CoQ10 supplementation, according to a recent article on Improving Global Outcomes from Kidney Disease. CoQ10 supplements should only be used for recommended indications and should be avoided in individuals with renal impairment due to the limited evidence available.

Pregnancy Considerations The manufacturer advises against using CoQ10 during pregnancy. Nonetheless, a recent meta-analysis showed that clinical pregnancy rates

(CPR) in assisted reproductive technologies (ART) may benefit from CoQ10 supplementation.

Concerns with Breastfeeding The manufacturer advises against using CoQ10 during nursing. Naturally occurring in the body, Ubiquinone is a regular ingredient in human milk. Mothers with preterm infants had somewhat lower concentrations of CoQ10 in their breast milk. There is a paucity of safety evidence during nursing, and CoQ10 has no particular applications associated to lactation. As a result, usage is often discouraged.

Older Patients There are currently no precise dosage guidelines available for either paediatric or elderly patients.

Drug-Drug Interactions CoQ10 and vitamin K share a similar chemical makeup, and there have been reports of possible interactions between CoQ10 and warfarin in the literature. There is a chance that the warfarin therapy won't be effective if people use CoQ10 nutrients in addition to their medicine. This interaction is reversible.

Long-Term Safety Profile Long-term use of CoQ10 has not been associated with any notable adverse effects, and it is generally acceptable. Infrequently, people may experience minor stomach problems. Overdosing on CoQ10 is not conceivable, and there are no documented negative effects. More than 200 randomised controlled Research has shown that CoQ10 has health advantages; nevertheless, a clinical study that showed no effect from supplementing with CoQ10 at a daily dosage of 100 milligrams suggests that it may interfere with the clinical action of warfarin (Mantle 2017).

16.6 Future Directions and Research Opportunities

Diabetes patients can still develop DPN even with strict metabolic management of blood glucose, which lowers the prevalence of newly identified clinical neuropathy. Increased risk of hypoglycaemia episodes and other dangers are associated with intensified glucose-lowering medication. There is proof that antioxidant mechanisms are crucial in the development of diabetes, however a recent analysis found that using CoQ10 antioxidants for short-term (3 months) therapy of diabetic problems does not show any proven benefits (Golbidi et al. 2011). Ayaz et al. have demonstrated, however, that giving STZ rats CoQ10 for 5 weeks prevents the change towards slower CVs. Treatment with CoQ10 enhances blood flow and endothelial function; hence, it may be beneficial over the long run by increasing peripheral nerve oxygenation. Uncertainty surrounds the processes driving the outcomes following CoQ10 administration. To get more mechanistic insights, future studies should investigate the molecular pathways, such as nitric oxide generation, inflammatory cascades, and mitochondrial bioenergetics (Ayaz et al. 2008).

It is difficult to determine the true benefit of CoQ10 supplementation in clinical studies due to its limitations, which include problems with absorption, bioavailability, and the variability of research designs. Despite much study into new formulation techniques and the direct delivery of ubiquinol and ubiquinone, which lack of cellular degradation and are more easily absorbed by cell membranes, CoQ10's low bioavailability remains a major barrier to its use. Furthermore, the limited number of recent clinical trials on CoQ10 in diabetic neuropathy limits the applicability of the findings. To determine the effectiveness, ideal dosage, and long-term safety of CoQ10 supplementation in different phases of diabetic neuropathy, larger, randomised controlled studies are necessary. There needs to be further research on how long-term CoQ10 supplementation affects the course of the illness, nerve regeneration, and general quality of life in diabetic individuals with neuropathy. Therapy would be completely changed if CoQ10 could eventually slow down or even reverse neuropathy. Future research should focus on improving CoQ10 supplementation's bioavailability and conducting well-organised, consistent clinical trials to determine its efficacy in a variety of patient populations and medical conditions. Further research is needed to confirm the efficacy of CoQ10 and clarify the various ways in which it might be a good substitute in the management and prevention of diabetes and related heart disease.

16.7 Conclusion

CoQ10 is known to be a potent antioxidant with preventive qualities against a variety of illnesses, such as diabetes and its related cardiovascular problems. Recent research indicates that CoQ10 works by lowering inflammation, oxidative stress, and metabolic fat and carbohydrate metabolism. CoQ10 deficiency is usual among patients with type 2 diabetes for a number of reasons. Strong antioxidant CoQ10 is believed to be able to scavenge excess ROS and prevent cells from oxidative damage, particularly mitochondria. Thus, in order to enhance glycaemic management, restoring the CoQ10 level in T2DM patients with exogenous CoQ10 supplementation may help decreased oxidative stress and maintain mitochondrial function. A number of researches provided some evidence for this theory. Still, in order to prove its effectiveness and long-term safety, issues including low bioavailability and the requirement for more extensive, well planned clinical trials need to be resolved.

CoQ10 therapy has demonstrated encouraging outcomes in several clinical studies, reducing the percentage of cardiac-related deaths among patients with heart failure, despite its disadvantages. To clarify the precise effects of CoQ10 and its potential application in therapeutic settings, particularly in conditions of oxidative stress, the contradictory findings and the necessity for additional research are crucial.

References

Alam U, Sloan G, Tesfaye S. Treating pain in diabetic neuropathy: current and developmental drugs. Drugs. 2020;80(4):363–84.

Amini P, Sajedi F, Mirjalili M, Mohammadi Y, Mehrpooya M. Coenzyme Q10 as a potential add-on treatment for patients suffering from painful diabetic neuropathy: results of a placebo-controlled randomized trial. Eur J Clin Pharmacol. 2022;78(12):1899–910.

Ayaz M, Tuncer S, Okudan N, Gokbel H. Coenzyme Q (10) and alpha-lipoic acid supplementation in diabetic rats: conduction velocity distributions. Methods Find Exp Clin Pharmacol. 2008;30(5):367–74.

Campisi L, La Motta C. The use of the Coenzyme Q10 as a food supplement in the management of fibromyalgia: a critical review. Antioxidants. 2022;11(10):1969.

Casagrande D, Waib PH, Júnior AAJ. Mechanisms of action and effects of the administration of Coenzyme Q10 on metabolic syndrome. J Nutr Intermed Metabol. 2018;13:26–32.

Crane FL. Biochemical functions of coenzyme Q10. J Am Coll Nutr. 2001;20(6):591–8.

Fakhrabadi MA, Ghotrom AZ, Mozaffari-Khosravi H, Nodoushan HH, Nadjarzadeh A. Effect of coenzyme Q10 on oxidative stress, glycemic control and inflammation in diabetic neuropathy: a double blind randomized clinical trial. Int J Vitam Nutr Res. 2014;84(5–6):252–60.

Farsi F, Mohammadshahi M, Alavinejad P, Rezazadeh A, Zarei M, Engali KA. Functions of coenzyme Q10 supplementation on liver enzymes, markers of systemic inflammation, and adipokines in patients affected by nonalcoholic fatty liver disease: a double-blind, placebo-controlled, randomized clinical trial. J Am Coll Nutr. 2016;35(4):346–53.

Fuke C, Krikorian SA, Couris RR. FEATURES-alternative and complementary therapies-coenzyme Q10-clinical trials show CoQ10 may be valuable in preventing and treating conditions related to oxidative stress. US Pharm. 2000;25(10):28–41.

Galeshkalami NS, Abdollahi M, Najafi R, Baeeri M, Jamshidzade A, Falak R, et al. Alpha-lipoic acid and coenzyme Q10 combination ameliorates experimental diabetic neuropathy by modulating oxidative stress and apoptosis. Life Sci. 2019;216:101–10.

Gautam MK, Purohit V, Agarwal M, Singh A, Goel RK. In-vivo healing potential of Aegle marmelos in excision, incision and dead space wound models. Sci World J. 2014;2014:740107.

Ghirlanda G, Oradei A, Manto A, Lippa S, Uccioli L, Caputo S, et al. Evidence of plasma CoQ10-lowering effect by HMG-CoA reductase inhibitors: a double-blind, placebo-controlled study. J Clin Pharmacol. 1993;33(3):226–9.

Golbidi S, Alireza Ebadi S, Laher I. Antioxidants in the treatment of diabetes. Curr Diabetes Rev. 2011;7(2):106–25.

Heidari A, Hamidi G, Soleimani A, Aghadavod E, Asemi Z. Effects of coenzyme Q10 supplementation on gene expressions related to insulin, lipid, and inflammation pathways in patients with diabetic nephropathy. Iran J Kidney Dis. 2018;12(1):14.

Kagan V, Serbinova E, Packer L. Antioxidant effects of ubiquinones in microsomes and mitochondria are mediated by tocopherol recycling. Biochem Biophys Res Commun. 1990;169(3):851–7.

Kumar A, Kaur H, Devi P, Mohan V. Role of coenzyme Q10 (CoQ10) in cardiac disease, hypertension and Meniere-like syndrome. Pharmacol Ther. 2009;124(3):259–68.

Kumar M, Gautam MK, Singh A, Goel RK. Healing effects of Musa sapientum var. paradisiaca in diabetic rats with co-occurring gastric ulcer: cytokines and growth factor by PCR amplification. BMC Complement Altern Med. 2013;13:305. https://doi.org/10.1186/1472-6882-13-305.

Liang Y, Zhao D, Ji Q, Liu M, Dai S, Hou S, et al. Effects of coenzyme Q10 supplementation on glycemic control: a GRADE-assessed systematic review and dose-response meta-analysis of randomized controlled trials. EClinicalMedicine. 2022;52:101602.

Mantle D. Coenzyme Q10 supplementation for diabetes and its complications: an overview. Br J Diabet. 2017;17(4):145–8.

Mantle D, Heaton RA, Hargreaves IP. Coenzyme Q10, ageing and the nervous system: an overview. Antioxidants. 2021;11(1):2.

Mazidi M, Kengne AP, Banach M, Lipid and Blood Pressure Meta-analysis Collaboration Group. Effects of coenzyme Q10 supplementation on plasma C-reactive protein concentrations: a systematic review and meta-analysis of randomized controlled trials. Pharmacol Res. 2018;128:130–6.

Miles MV. The uptake and distribution of coenzyme Q (10). Mitochondrion. 2007;7:S72–7.

Mohammed-Jawad NK, Al-Sabbagh M, AL-Jezaeri KA. Role of L-carnitine and coenzyme Q10 as adjuvant therapy in patients with type 2 diabetes mellitus. Am J Pharmacol Sci. 2014;2(5):82–6.

Mondal NS, Islam A, Gautam MK, Mondal S, Ahmed M, Jamir SI, Khongwir LD, Mahajon B. Unraveling the mechanisms of diabetic wounds: insights into pathogenesis and advanced treatment strategies. Curr Diabetes Rev. 2025;22:113.

Murthy S, Gautam MK, Goel S, Purohit V, Sharma H, Goel RK. Evaluation of in-vivo wound healing activity of Bacopa monniera on different wound model in rats. Biomed Res Int. 2013;2013:972028.

Rauchova H, Drahota Z, Lenaz G. Function of coenzyme Q in the cell: some biochemical and physiological properties. Physiol Res. 1995;44:209.

Samimi F, Namiranian N, Sharifi-Rigi A, Siri M, Abazari O, Dastghaib S. Coenzyme Q10: a key antioxidant in the management of diabetes-induced cardiovascular complications—an Overview of mechanisms and clinical evidence. Int J Endocrinol. 2024;2024(1):2247748.

Schulz C, Obermüller-Jevic UC, Hasselwander O, Bernhardt J, Biesalski HK. Comparison of the relative bioavailability of different coenzyme Q10 formulations with a novel solubilizate (Solu™ Q10). Int J Food Sci Nutr. 2006;57(7–8):546–55.

Shen Q, Pierce JD. Supplementation of coenzyme Q10 among patients with type 2 diabetes mellitus. In: Healthcare, vol. 3, No. 2. MDPI; 2015. p. 296–309.

Shi TJS, Zhang MD, Zeberg H, Nilsson J, Grünler J, Liu SX, et al. Coenzyme Q10 prevents peripheral neuropathy and attenuates neuron loss in the db−/db− mouse, a type 2 diabetes model. Proc Natl Acad Sci. 2013;110(2):690–5.

Sifuentes-Franco S, Sánchez-Macías DC, Carrillo-Ibarra S, Rivera-Valdés JJ, Zuñiga LY, Sánchez-López VA. Antioxidant and anti-inflammatory effects of coenzyme Q10 supplementation on infectious diseases. In: Healthcare, vol. 10, No. 3. MDPI; 2022. p. 487.

Silver MA, Langsjoen PH, Szabo S, Patil H, Zelinger A. Effect of atorvastatin on left ventricular diastolic function and ability of coenzyme Q10 to reverse that dysfunction. Am J Cardiol. 2004;94(10):1306–10.

Sugiyama S, Kitazawa M, Ozawa T, Suzuki K, Izawa Y. Anti-oxidative effect of coenzyme Q 10. Experientia. 1980;36:1002–3.

Villalba JM, Navas P. Plasma membrane redox system in the control of stress-induced apoptosis. Antioxid Redox Signal. 2000;2(2):213–30.

Villalba JM, Parrado C, Santos-Gonzalez M, Alcain FJ. Therapeutic use of coenzyme Q10 and coenzyme Q10-related compounds and formulations. Expert Opin Investig Drugs. 2010;19(4):535–54.

Visnagri A, Kandhare AD, Kumar VS, Rajmane AR, Mohammad A, Ghosh P, et al. Elucidation of ameliorative effect of Co-enzyme Q10 in streptozotocin-induced diabetic neuropathic perturbation by modulation of electrophysiological, biochemical and behavioral markers. Biomed Aging Pathol. 2012;2(4):157–72.

Zhang YP, Eber A, Yuan Y, Yang Z, Rodriguez Y, Levitt RC, et al. Prophylactic and antinociceptive effects of coenzyme Q10 on diabetic neuropathic pain in a mouse model of type 1 diabetes. Anesthesiology. 2013;118(4):945–54.

Zhang SY, Yang KL, Zeng LT, Wu XH, Huang HY. Effectiveness of Coenzyme Q10 supplementation for Type 2 diabetes mellitus: a systematic review and meta-analysis. Int J Endocrinol. 2018;2018(1):6484839.

Part V

Integrative Approaches to Prevention and Cure of Diabetic Neuropathy

Dietary Strategies for Managing Diabetic Neuropathy

17

Sangita H. Shukla, Mahesh Kumar Gupta, Virendra Singh, Amit Lather, Tanuj Hooda, and Sangeeta Mathur

17.1 Introduction

As diabetes mellitus (DM) worsens and affects every system in the body, diabetic neuropathy is a frequent complication that can have serious repercussions. Whereas term "diabetic neuropathy" includes wide range of numerous nerve impairment conditions, diabetic peripheral neuropathy is the most common and well-studied. Diabetic neuropathy is a prevalent and debilitating complication of diabetes mellitus, affecting a significant proportion of individuals with the disease (Bunner et al. 2015a). It encompasses a range of nerve disorders that arise from prolonged hyperglycaemia, leading to nerve damage that manifests in various forms, including peripheral, autonomic, proximal and focal neuropathies (Nagpal et al. 2021).

The onset of diabetic neuropathy is often insidious, with signs varying from mild pain and insensibility in the extremities to serious complications that can hinder basic daily activities. As diabetes prevalence rises globally, so too does the incidence of diabetic neuropathy, making it a major public health concern. According to

S. H. Shukla
Indubhai Patel College of Pharmacy and Research Centre, Anand, Gujarat, India

M. K. Gupta
Career Point School of Pharmacy, Career Point University, Kota, Rajasthan, India

V. Singh
Pharmaceutical Chemistry, Sainik Pharmacy College, Prayagraj, Uttar Pradesh, India

A. Lather
Department of Pharmacology, Hindu College of Pharmacy, Sonipat, Haryana, India

T. Hooda (✉)
Department of Pharmaceutical Chemistry, Delhi Pharmaceutical Science and Research University, Delhi, New Delhi, India

S. Mathur
School of Arts and Humanities, Career Point University, Kota, Rajasthan, India

M. K. Gautam et al. (eds.), *Antioxidants and Functional Foods for Diabetic Neuropathy*, https://doi.org/10.1007/978-981-95-9326-2_17

estimates, up to 50% of diabetic people may develop neuropathy at some point in their lives, underscoring the critical need for effective management strategies. Understanding key phenotypes and risk factors is crucial for clinicians treating people with diabetes (Dobretsov et al. 2007; Calcutt and Backonja 2007).

17.2 Types of Diabetic Neuropathy

I. *Peripheral Neuropathy (Distal Symmetric Peripheral Neuropathy)*: It first affects the hands and arms, then the legs and feet. Tingling, reduced discomfort, numbness, muscular weakness, and intense sensitivity to touch are some of the symptoms. It is possible to develop severe foot issues including ulcers and infections (Azhary et al. 2010).

II. *Autonomic Neuropathy*: Involves the autonomic nervous system, controlling, heart rate, digestion, blood pressure, bladder function, and sexual response. Symptoms may include hypoglycaemia unawareness, decreased blood pressure upon standing, slow stomach emptying (gastroparesis), bladder or bowel problems and changes in vision adjustment (D'Souza et al. 2022).

III. *Proximal Neuropathy (Diabetic Polyradiculopathy)*: Affects nerves in the buttocks, hips, legs, thighs, and sometimes the abdominal and chest area. Usually unilateral, with pain, weakness, and sensory disturbances (Khdour 2020).

IV. *Focal Neuropathies*: These conditions involve damage to single nerves, often in specific areas (e.g., hand, head, torso, leg). Entrapment syndromes occur when nerves become compressed or trapped (e.g., carpal tunnel syndrome). Other focal neuropathies include cranial neuropathies affecting the nerves of the head (Huizinga and Peltier 2007).

17.3 Prevalence and Impact

Up to date there are no particular drugs available for the treatments of diabetic neuropathy. Clinical trials studies have concentrated on results of nerve conduction studies. According to several studies, tiny fibre damage from diabetic neuropathy occurs before big fibre dysfunction (Sumner et al. 2003). Diabetic neuropathy prevalence increased with disease timing. Heart disease risk factors such as smoking, obesity, high cholesterol and socioeconomic factors strongly predict Diabetic Neuropathy in type 1 diabetes. Recent studies have reported varying prevalence rates for diabetic neuropathy in type 1 diabetes. Using the Michigan Neuropathy Screening Instrument Questionnaire (MNSIQ), a reliable tool for assessing DPN symptoms, we found a prevalence of 11% among participants in the T1D Exchange Clinic Registry throughout the U.S. Although this incidence is lower than in some previous reports, it nevertheless represents the state of clinical care today (Mizokami-Stout et al. 2020). In addition, a research carried out in Scotland discovered that diabetic neuropathy was common in 22.7% of people with type 1 diabetes (Enders and Wright 2023).

17.4 Importance of Dietary Management in Diabetic Neuropathy

The Diabetes Prevention Program found that dietary changes and physical activity can lower the incidence of type 2 diabetes. There is encouraging evidence that dietary restrictions may help improve diabetic neuropathy in people, and recently published studies shown their efficacy in changing neuropathy in in-vivo models (Kumar et al. 2013). Dietary management is crucial in preventing the development of diabetic neuropathy. Consistent blood sugar control through a balanced diet helps maintain nerve health and overall well-being. Lifestyle changes, such as adopting a low-carbohydrate/high-fat diet and regular exercise, have shown promise in improving diabetic neuropathy symptoms (Pop-Busui et al. 2022; Strand et al. 2024).

Diabetic neuropathy is a complex condition arising from the extended exposure to elevated blood glucose levels, which significantly impacts nerve function. While medications play an important role in managing the symptoms, dietary strategies can complement treatment and enhance overall well-being. Here's a comprehensive overview in Fig. 17.1:

Balanced Diet with Macronutrients, Micronutrients, Vitamins (Vitamin B Complex, Vitamin D) and Minerals like Magnesium, Antioxidants and Polyphenols rich sources should be in your diet (Chowdhury et al. 2021; Young and Woodside 2001).

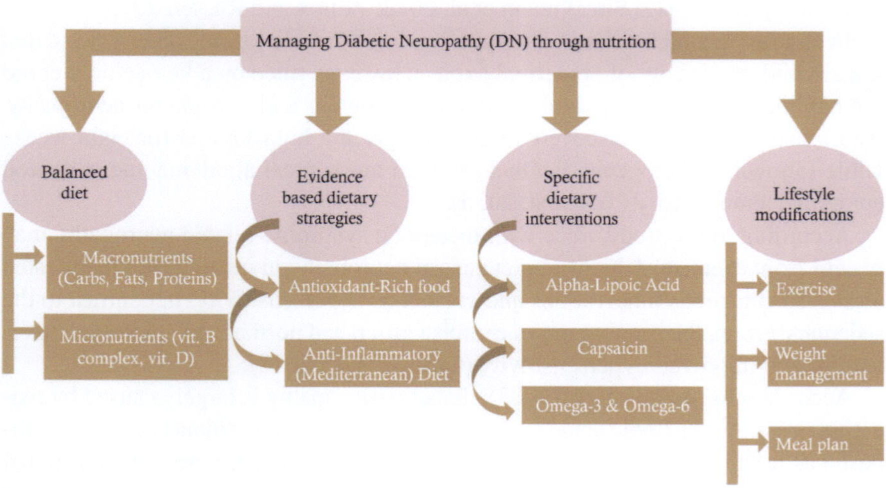

Fig. 17.1 Diabetic neuropathy management through nutrition

17.5 Evidence-Based Dietary Strategies

Increasing the use of antioxidant-rich foods in your diet provides numerous health benefits. These foods aid in the prevention of inflammatory and oxidative damage, two factors that affect the control of diabetes. Antioxidants are essential for protecting the body from damage caused by free radicals, which are outcomes of both internal cellular functions and external variables like pollution and UV rays. They work by neutralizing free radicals, thus preventing oxidative stress that results to chronic diseases such as heart disease, cancer, and cognitive decline (Hadi and Markad 2023; Muscolo et al. 2024). While antioxidants are abundant in fruits, vegetables, and whole grains, studies have shown mixed results regarding the effectiveness of antioxidant supplements. Research suggests that antioxidants work best in combination with other nutrients and plant chemicals, rather than as isolated supplements (Kozlov et al. 2024). Therefore, a diet rich in natural sources of antioxidants is recommended for optimal health benefits. Some of the foods include: Fruits and Vegetables, Green Tea and Dark Chocolate (Shapira 2019).

The Mediterranean diet, which is inspired by the traditional eating practice of Mediterranean countries, emphasizes whole, minimally processed foods and has been recognized as an effective anti-inflammatory eating pattern. While it's not a cure for diabetic neuropathy, adopting elements of the Mediterranean diet may support overall well-being and complement other management strategies. Here are the key principles of the Mediterranean diet: Plant-Based Healthy Fats Lean Proteins Herbs and Spices (Murthy et al. 2013). Whereas foods to be avoid includes limit processed, sugary, and fried foods. Incorporate whole grains such as quinoa, and brown rice should be used to regulate blood sugar levels because fibre slows digestion, preventing rapid spikes in blood glucose (Yan et al. 2024; Fattori et al. 2016). Specific dietary interventions and supplements included alpha-lipoic acid (ALA), capsaicin, essential fatty acids like omega-6 and omega-3 should also be included in diet (Shetty et al. 2022; Streckmann et al. 2022; Zilliox and Russell 2019).

In contrast to a high-fat diet, a research on mice fed a ketogenic diet revealed that it increased the density of epidermal axons. However, mice on a ketogenic diet did not exhibit mechanical allodynia, a frequent problem with peripheral neuropathy. Additionally, the study discovered that mice given a high-fat diet for eight weeks while following a ketogenic diet had reversed mechanical allodynia and enhanced epidermal axon density (Teli et al. 2025).

According to the study, mice with metabolic syndrome-related neuropathy may benefit from exercise. A high-fat diet caused neuropathy in mice, which manifested as altered cutaneous innervations and cutaneous hypersensitivity. In contrast to the sedentary group, the running wheel exercise group had normalized epidermal innervations and improved hypersensitivity (Groover et al. 2013).

According to Ametov et al. (2003), diabetic neuropathy is largely caused by oxidative stress. Alpha-lipoic acid (ALA) and other dietary antioxidants have been proposed as treatments for the illness. Daily intravenous ALA has been demonstrated in clinical studies to lessen neuropathic symptoms. It has been shown that taking three oral doses of ALA considerably lowers the overall diabetic neuropathy symptom score. The 600 mg OD dosage is the most beneficial. ALA may help diabetic neuropathy, according to a meta-analysis, even if a phase 3 research found no benefits (Ametov et al. 2003; Mondal et al. 2025).

Nicotinamide riboside produced nicotinamide adenine dinucleotide and present in diet may produce NAD^+, which might help prevent or treat diabetic autonomic neuropathy. This produced NAD^+ worked as critical metabolite in energy metabolism along with its usefulness in electron transfer in mitochondria. In diabetic neurons there is decrease in the levels of NAD^+ and elevated levels of NAD^+ can prevent oxidative injury (Trammell et al. 2016; Gautam et al. 2012a).

According to some study, high fat diets can cause diabetes neuropathy, whereas reducing the same high fat diet can minimize neuropathy. In order to improve essential fatty acid dysmetabolism, dietary intervention research was carried out to assess the utilization of omega-3 polyunsaturated fatty acids from seal oil in patients with diabetic neuropathy. After 1 year of diet supplementation they reported that treated patients have improved nerve fibre length (Hinder et al. 2017; Lewis et al. 2017).

17.6 Lifestyle Modifications and Their Role in Diabetic Neuropathy

Exercise and physical activities with low-impact cardiovascular activities like swimming, cycling, and walking improve blood flow, enhance nerve function, and promote overall health. Strength training (seated) exercises help in maintaining muscle strength and support joint stability. Balance and stability work like tai chi, yoga, and balance exercises reduce the risk of falls and improve proprioception. Regular physical activity supports glycaemic control, circulation, and overall well-being (Blaszkiewicz et al. 2024). Elevated body mass index (BMI) increases the chances of stroke, heart disease, and diabetes complications. Being underweight also poses health risks, such as osteoporosis and anaemia. Tailor weight management to your individual needs, aiming for a healthy weight range. Meal planning reduces stress, streamlines grocery shopping, and ensures balanced nutrition. Focus on nutrient-rich foods to support overall health. Select the recipes that align with your dietary goals. Prioritize whole foods, lean proteins, and fibre-rich options. Plan meals ahead to avoid last-minute unhealthy choices (Cell Death Discovery 2016).

17.7 Reported Clinical Studies

Crane and Sample in 1994 In a research, 21 individuals with painful systemic distal polyneuropathy and type 2 diabetes participated. A 25-day lifestyle improvement program consisting of (a) a vegan diet reduced in fat, calories, and fibre and devoid of added cholesterol and processed foods, and (b) 30 min of conditioning exercise each day was the subject of this uncontrolled intervention experiment. Group lectures and cooking demonstrations were provided to participants for avoiding alcohol, tea, and coffee. According to the research, glycaemia, blood pressure, and plasma lipids all improved, while 17 patients reported total pain relief, four reported partial reliefs, and 18 reported considerable weight reduction (Crane and Sample 1994).

Smith in 2006 Intra-epidermal Nerve Fibre Density (IENFD) was measured in 32 individuals with IGT who had neuropathy over the course of a year utilising standard care diet and exercise treatment as part of the Impaired Glucose Tolerance Neuropathy research (IGTN). Distal and proximal IENFD significantly improved, as did BMI, 2-h post-load glucose levels, and total blood cholesterol, according to the results. Pain ratings, however, did not significantly improve (Smith et al. 2006).

Kluding in 2012 The Michigan Neuropathy Screening Instrument (MNSI) questionnaire showed a substantial decrease in neuropathic features and pain in 17 diabetic neuropathy patients who participated in a 10-week aerobic and muscular exercise program. There were notable improvements in HbA1c levels, but no appreciable changes in BMI, waist circumference, NCS, or QST were noted. In managing diabetic neuropathy, the study emphasises the need of treating neuropathic symptoms (Kluding et al. 2012).

Fonseca et al. in 2013 In a double-blind, placebo-controlled, multisite research, 24 patients with diabetic peripheral neuropathy were given Pamlab LLC's Metanx, a wholesome meal that contains pyridoxal 5 phosphate, L-methylfolate, and methylcobalamin. The Neuropathy Total Symptom Score-6 was 2, and symptoms such as tingling, aching, burning, or lancinating pain greatly decreased at weeks 16 and 24, despite the study's lack of meaningful outcomes (Gautam et al. 2012b).

Singleton in 2014 The study assessed changes in IENFD at the proximal thighs and distal leg in 100 type 2 diabetic individuals without neuropathy in order to determine the effects of lifestyle counselling or a weekly exercise program and dietary counselling. The exercise group had significantly higher distal leg IENFD compared to the counselling cohort (1.5 ± 3.6 vs. -0.1 ± 3.2 fibres/mm, $P = 0.03$). After 1 year, the exercise group had substantially greater IENFD at the proximal thigh than the control group (12.1 ± 4.8 vs. 9.6 ± 4.7, $P = 0.02$). The exercise group showed a considerable increase in HDL levels but no discernible changes in NCS or clinical neuropathy scores (UENS) (Singleton et al. 2014).

Bunner et al. in 2015 Using a low-fat, plant-based diet and vitamin B12 supplements significantly improved pain ratings, quality of life, body weight, glycaemic management, and skin conduction in 34 participants with T2DM and painful distal polyneuropathy, ages 18–65, during the course of a 20-week research (Bunner et al. 2015b).

McGoey-Smith in 2019 After adopting a whole-food, plant-based diet that included nitrate-rich meals and six servings of leafy green vegetables per day, a

51-year-old Caucasian woman with type 2 diabetes and severe diabetic neuropathy was able to effectively manage her condition, improving her glycaemic control, losing weight, and requiring less medication (McGoey-Smith et al. 2019).

17.8 Emerging Therapies

Gene Therapy A ground-breaking avenue, gene therapy holds promise. Recent studies1 have explored adeno-associated virus (AAV)-based delivery of neurotrophic factors to adipose tissue, promoting nerve cell survival and regeneration (Ahmed and Al-Massri 2024).

Stem Cell-Based Interventions Mesenchymal stem cells (MSCs) emerge as regenerative warriors. Their paracrine effects secrete immunomodulatory substances, angiogenic agents, and neurotrophic factors, potentially ameliorating DN (Gemignani 2021; Zhang et al. 2024).

Immunomodulation Intravenous immunoglobulin (IVIG) therapy shows promise in managing painful DN4. By modulating the immune response, it may mitigate neuropathic symptoms (Pouwer et al. 2024).

Multimodal Approaches Revisited Combine these emerging therapies with lifestyle modifications, pharmacological agents, and patient education. The synergy may unlock new avenues for DN management. For neuropathic pain, pharmaceutical therapy is administered to as many as 69% of diabetic neuropathy patients. The FDA in the US has approved four drugs for the treatment of painful diabetic neuropathy (PDN): tapentadol, duloxetine, pregabalin, and an 8% capsaicin patch. Patients with diabetes mellitus may have less pain with non-pharmacological therapies including trans-cutaneous electrical nerve stimulation (TENS) and spinal cord stimulation (SCS) (Kim et al. 2021).

17.9 Conclusion and Future Directions

In this comprehensive overview of diabetic neuropathy (DN), we've traversed the intricate landscape of nerve damage, exploring epidemiology, diagnostic strategies, and therapeutic approaches. In summary, DN remains a difficult condition, but ongoing research provides hope for better result. As clinicians and researchers, we must continue our efforts to unravel its complexities and develop effective interventions as within patients' narratives lie seeds of progress, waiting to bloom (Davalos et al. 2022). While significant progress has been made, several areas warrant further investigation like biomarkers (identifying reliable biomarkers for early DN detection and progression), neuroprotection (developing strategies to prevent nerve damage), personalized medicine and patient education practices. In conclusion, diabetic neuropathy can be effectively treated with nutritional supplements because they change the metabolic process, have a major impact on glucose and lipid utilization, and promote axonal reconstruction through dietary and exercise treatments.

Conflict of Interest The study reported in this publication, according to the authors, could not have been impacted by any of their relationships or known financial conflicts.

References

Ahmed LA, Al-Massri KF. Exploring the role of mesenchymal stem cell–derived exosomes in diabetic and chemotherapy-induced peripheral neuropathy. Mol Neurobiol. 2024;61:5916–27. https://doi.org/10.1007/s12035-024-03916-z.

Ametov AS, Barinov A, Dyck PJ, Hermann R, Kozlova N, Litchy WJ, Low PA, Nehrdich D, Novosadova M, O'Brien PC, Reljanovic M, Samigullin R, Schuette K, Strokov I, Tritschler HJ, Wessel K, Yakhno N, Ziegler D. The sensory symptoms of diabetic polyneuropathy are improved with a-lipoic acid: the Sydney trial. Diabetes Care. 2003;26:770–6.

Azhary H, Farooq MU, Bhanushali M, Majid A, Kassab MY. Peripheral neuropathy: differential diagnosis and management. Am Fam Physician. 2010;81(7):887–92.

Blaszkiewicz M, Tao T, Mensah-Arhin K, Willows JW, Bates R, Huang W, Cao L, Smith RL, Townsend KL. Gene therapy approaches for obesity-induced adipose neuropathy: device-targeted AAV-mediated neurotrophic factor delivery to adipocytes in subcutaneous adipose. Mol Ther. 2024;32(4):104–7. https://doi.org/10.1016/j.ymthe.2024.03.010.

Bunner AE, Wells CL, Gonzales J, Agarwal U, Bayat E, Barnard ND. A dietary intervention for chronic diabetic neuropathy pain: a randomized controlled pilot study. Nutr Diabetes. 2015a;5(5):e158. https://doi.org/10.1038/nutd.2015.8.

Bunner AE, Wells CL, Gonzales J, Agarwal U, Bayat E, Barnard ND. A dietary intervention for chronic diabetic neuropathy pain: a randomized controlled pilot study. Nutr Diabetes. 2015b;5(5):e158. https://doi.org/10.1038/nutd.2015.8.

Calcutt NA, Backonja MM. Pathogenesis of pain in peripheral diabetic neuropathy. Curr Diab Rep. 2007;7:429–34.

Cell Death Discovery. Mesenchymal stem cells to treat diabetic neuropathy: a long and strenuous way from bench to the clinic (nature.com). Cell Death Discov. 2016;2:e16055. https://doi.org/10.1038/cddiscovery.2016.55.

Chowdhury SS, Tareq AM, Tareq SM, et al. Screening of antidiabetic and antioxidant potential along with phytochemicals of Annona genus: a review. Futur J Pharm Sci. 2021;7:144. https://doi.org/10.1186/s43094-021-00300-9.

Crane MG, Sample C. Regression of diabetic neuropathy with total vegetarian (vegan) diet. J Nutr Med. 1994;4(4):431–9. https://doi.org/10.3109/13590849409003592.

D'Souza RS, Barman R, Joseph A, et al. Evidence-based treatment of painful diabetic neuropathy: a systematic review. Curr Pain Headache Rep. 2022;26:583–94. https://doi.org/10.1007/s11916-022-01061-7.

Davalos L, Stino AM, Selvarajah D, Sakowski SA, Tesfaye S, Feldman EL. Precision medicine for diabetic neuropathy. In: Basu R, editor. Precision medicine in diabetes. Cham: Springer; 2022. https://doi.org/10.1007/978-3-030-98927-9_8.

Dobretsov M, Romanovsky D, Stimers JR. Early diabetic neuropathy: triggers and mechanisms. World J Gastroenterol. 2007;13:175–91.

Enders J, Wright DE. Lifestyle and dietary modifications: relevance in the management of diabetic neuropathy. In: Tesfaye S, Gibbons CH, Malik RA, Veves A, editors. Diabetic neuropathy. Contemporary diabetes. Cham: Humana; 2023. https://doi.org/10.1007/978-3-031-15613-7_22.

Fattori V, Hohmann MSN, Rossaneis AC, Pinho-Ribeiro FA, Verri WA. Capsaicin: current understanding of its mechanisms and therapy of pain and other pre-clinical and clinical uses. Molecules. 2016;21(7):844. https://doi.org/10.3390/molecules21070844.

Gautam MK, Gupta A, Vijaykumar M, Rao CV, Goel RK. Studies on the hypoglycemic effects of Murraya paniculata Linn. extract on alloxan-induced oxidative stress in diabetic and non-diabetic models. Asian Pacific J Trop Dis. 2012a;2(S1):S186–91.

Gautam MK, Gupta A, Rao CV, Goel RK. Antihyperglycemic and antioxidant potential of Murraya paniculata Linn. leaves: a preclinical study. J Pharm Res. 2012b;5(3):1334–7.

Gemignani F. Intravenous immunoglobulin therapy in painful diabetic neuropathy. Pain Med. 2021;22(11):2775–6. https://doi.org/10.1093/pm/pnab062.

Groover AL, Ryals JM, Guilford BL, Wilson NM, Christianson JA, Wright DE. Exercise-mediated improvements in painful neuropathy associated with prediabetes in mice. Pain. 2013;154:2658–67.

Hadi P, Markad P. Role of antioxidants in disease prevention: a review. Int J Health Sci. 2023:7023–31. https://doi.org/10.53730/ijhs.v6nS8.14124.

Hinder LM, O'Brien PD, Hayes JM, Backus C, Solway AP, Sims-Robinson C, Feldman EL. Dietary reversal of neuropathy in a murine model of prediabetes and metabolic syndrome. Dis Model Mech. 2017;10:717–25.

Huizinga MM, Peltier A. Painful diabetic neuropathy: a management-centered review. Clin Diabetes. 2007;25(1):6–15. https://doi.org/10.2337/diaclin.25.1.6.

Khdour MR. Treatment of diabetic peripheral neuropathy: a review. J Pharm Pharmacol. 2020;72(7):863–72. https://doi.org/10.1111/jphp.13241.

Kim K-S, Lee J-S, Park J-H, Lee E-Y, Moon J-S, Lee S-K, Lee J-S, Kim J-H, Kim H-S. Identification of novel biomarker for early detection of diabetic nephropathy. Biomedicine. 2021;9(5):457. https://doi.org/10.3390/biomedicines9050457.

Kluding PM, Pasnoor M, Singh R, Jernigan S, Farmer K, Rucker J, Sharma NK, Wright DE. The effect of exercise on neuropathic symptoms, nerve function, and cutaneous innervation in people with diabetic peripheral neuropathy. J Diabetes Complicat. 2012;26:424–9.

Kozlov AV, Javadov S, Sommer N. Cellular ROS and antioxidants: physiological and pathological role. Antioxidants. 2024;13(5):602. https://doi.org/10.3390/antiox13050602.

Kumar M, Gautam MK, Singh A, Goel RK. Healing effects of Musa sapientum var. paradisiaca in diabetic rats with co-occurring gastric ulcer: cytokines and growth factor by PCR amplification. BMC Complement Altern Med. 2013;13:305. https://doi.org/10.1186/1472-6882-13-305.

Lewis EJH, Perkins BA, Lovblom LE, Bazinet RP, Wolever TMS, Bril V. Effect of omega-3 supplementation on neuropathy in type 1 diabetes: a 12-month pilot trial. Neurology. 2017;88:2294–301.

McGoey-Smith KJ, Esselstyn CB, McGoey-Smith AD. Reversal of pulmonary hypertension, diabetes, and retinopathy after adoption of a whole food plant-based diet. IJDRP. 2019;1(2):10.

Mizokami-Stout KR, Li Z, Foster NC, Shah V, Aleppo G, McGill JB, Pratley R, Toschi E, Ang L, Pop-Busui R. T1D Exchange Clinic Network; the contemporary prevalence of diabetic neuropathy in type 1 diabetes: findings from the T1D exchange. Diabetes Care. 2020;43(4):806–12. https://doi.org/10.2337/dc19-1583.

Mondal NS, Islam A, Gautam MK, Mondal S, Ahmed M, Jamir SI, Khongwir LD, Mahajon B. Unraveling the mechanisms of diabetic wounds: insights into pathogenesis and advanced treatment strategies. Curr Diabetes Rev. 2025;22:113.

Murthy S, Gautam MK, Goel S, Purohit V, Sharma H, Goel RK. Evaluation of in-vivo wound healing activity of *Bacopa monniera* on different wound model in rats. Biomed Res Int. 2013;2013:972028. https://doi.org/10.1155/2013/972028.

Muscolo A, Mariateresa O, Giulio T, Mariateresa R. Oxidative stress: the role of antioxidant phytochemicals in the prevention and treatment of diseases. Int J Mol Sci. 2024;25(6):3264. https://doi.org/10.3390/ijms25063264.

Nagpal AS, Leet J, Egan K, et al. Diabetic neuropathy: a critical, narrative review of published data from 2019. Curr Pain Headache Rep. 2021;25:15. https://doi.org/10.1007/s11916-020-00928-x.

Pop-Busui R, Ang L, Boulton AJM, Feldman EL, Marcus RL, Kara Mizokami-Stout J, Singleton R, Ziegler D. Diagnosis and treatment of painful diabetic peripheral neuropathy. ADA Clin Compendia. 2022;2022(1):1–32. https://doi.org/10.2337/db2022-01.

Pouwer F, Mizokami-Stout K, Reeves ND, Pop-Busui R, Tesfaye S, Boulton AJM, Vileikyte L. Psychosocial care for people with diabetic neuropathy: time for action. Diabetes Care. 2024;47(1):17–25. https://doi.org/10.2337/dci23-0033.

Shapira N. The metabolic concept of meal sequence vs. satiety: glycemic and oxidative responses with reference to inflammation risk, protective principles and Mediterranean diet. Nutrients. 2019;11(10):2373. https://doi.org/10.3390/nu11102373.

Shetty SS, Suchetha Kumari N, Varadarajan R. The ratio of Omega-6/Omega-3 fatty acid: implications and application as a marker to diabetes. In: Patel VB, Preedy VR, editors. Biomarkers in diabetes. Biomarkers in disease: methods, discoveries and applications. Cham: Springer; 2022. https://doi.org/10.1007/978-3-030-81303-1_23-1.

Singleton JR, Marcus RL, Jackson JE, Lessard K, Graham TE, Smith AG. Exercise increases cutaneous nerve density in diabetic patients without neuropathy. Ann Clin Transl Neurol. 2014;1:844–9.

Smith AG, Russell JW, Feldman EL, Goldstein J, Peltier A, Smith S, Hamwi J, Pollari D, Bixby B, Howard J, Singleton JR. Lifestyle intervention for prediabetic neuropathy. Diabetes Care. 2006;29:1294–9.

Strand N, Anderson MA, Attanti S, et al. Diabetic neuropathy: pathophysiology review. Curr Pain Headache Rep. 2024;28:481–7. https://doi.org/10.1007/s11916-024-01243-5.

Streckmann F, Balke M, Cavaletti G, et al. Exercise and neuropathy: systematic review with meta-analysis. Sports Med. 2022;52:1043–65. https://doi.org/10.1007/s40279-021-01596-6.

Sumner CJ, Sheth S, Griffin JW, Cornblath DR, Polydefkis M. The spectrum of neuropathy in diabetes and impaired glucose tolerance. Neurology. 2003;60:108–11.

Teli A, Jena D, Saikia L, Mondal NS, Sen S, Mishra K, Gautam MK. Ethnobotanical use, phytochemistry and pharmacological activities of Clerodendrum colebrookianum (Lamiaceae). Curr Nutr Food Sci. 2025;21(2):148–55.

Trammell SA, Weidemann BJ, Chadda A, Yorek MS, Holmes A, Coppey LJ, Obrosov A, Kardon RH, Yorek MA, Brenner C. Nicotinamide riboside opposes Type 2 diabetes and neuropathy in mice. Sci Rep. 2016;6:26933. https://doi.org/10.1038/srep26933.:26933.

Yan S, Lu J, Chen B, Yuan L, Chen L, Ju L, Cai W, Wu J. The multifaceted role of alpha-lipoic acid in cancer prevention, occurrence, and treatment. Antioxidants. 2024;13(8):897. https://doi.org/10.3390/antiox13080897.

Young IS, Woodside JV. Antioxidants in health and disease. J Clin Pathol. 2001;54:176–86. Antioxidants in health and disease | Journal of Clinical Pathology (bmj.com)

Zhang EX, Yazdi C, Islam RK, et al. Diabetic neuropathy: a guide to pain management. Curr Pain Headache Rep. 2024; https://doi.org/10.1007/s11916-024-01293-9.

Zilliox LA, Russell JW. Physical activity and dietary interventions in diabetic neuropathy: a systematic review. Clin Auton Res. 2019;29:443–55. https://doi.org/10.1007/s10286-019-00607-x.

18 Nutritional Approaches to Preventing Diabetic Neuropathy: Guidelines and Strategies

Hemendra Gautam, Rajesh Shukla, Rita Mourya, Satyendra Singh, Ravi Kumar Gautam, and Bharat Kumar

18.1 Introduction to Diabetic Neuropathy and Its Nutritional Aspects

A higher quality of life, amputations, anxiety, depression, and gait issues are among the major consequences linked to diabetic peripheral neuropathy, which impacts up to 60% of individuals with insulin-dependent diabetes. Initially manifesting as injury to the terminal branches of peripheral nerves, the illness often affects little fibres that transmit temperature, light touch, and pain. Walking and balance problems might result from worsening neuropathy, which affects large fibres that regulate muscle tone and reflexes. Most people with diabetic peripheral neuropathy begin with odd, triggered, or spontaneous symptoms, such as numbness or pain in the lower extremities (Bunner et al. 2015). In clinical practice, pharmacologic drugs such as neuropathic painkillers like tricyclic antidepressants and serotonin-noradrenaline reuptake blockers, as well as anticonvulsants like gabapentin and pregabalin, have been effective first-line therapies. Nevertheless, pharmacologic treatments for this illness are linked to serious side effects and could not totally alleviate neuropathic pain. Additionally, it has been proposed that in order to address

H. Gautam (✉) · R. K. Gautam
Faculty of Pharmaceutical Sciences, Future University, Bareilly, UP, India

R. Shukla
Department of Pharmacy, Mangalayatan University-Jabalpur,
Jabalpur, Madhya Pradesh, India

R. Mourya
School of Pharmacy, SAM Global University, Bhopal, Madhya Pradesh, India

S. Singh
BPS Educational Institution of Pharmacy, Etah, UP, India

B. Kumar
School of Pharmaceutical Sciences, Bahra University, Shimla Hills, Himachal Pradesh, India

M. K. Gautam et al. (eds.), *Antioxidants and Functional Foods for Diabetic Neuropathy*, https://doi.org/10.1007/978-981-95-9326-2_18

pain and its impact on patients' lives, effective treatment techniques for chronic pain combine pharmaceutical and nonpharmaceutical therapy. Numerous nonpharmacologic treatments have been proposed to treat neuropathic pain, including nonsurgical approaches such integrated cognitive behavioural therapy, exercise therapy, and nutritional supplements (Abdelrahman and Hackshaw 2021).

18.1.1 Role of Nutrition in Disease Prevention

Nutrition, which is frequently seen as a lifestyle component, affects many chronic disorders, such as overweight, high blood pressure, heart disease, ischaemic stroke, diabetes, metabolic syndrome, several forms of cancer, and maybe some neurological problems. It is commonly known that illnesses and/or health issues can arise as a result of dietary deficits or inadequate nutritional condition. Many chronic diseases can be prevented with dietary changes, but once a problem has set in, it is sometimes necessary to alter a person's regular diet to help manage the illness or its symptoms. High blood sugar levels are a hallmark of diabetes, a chronic illness for which nutrition is essential to both prevention and treatment. Diabetes risk can be decreased, and blood sugar control can be enhanced by eating a diet high in whole, minimally processed foods and low in refined carbs, added sugars, and saturated and trans fats. Eating foods high in fibre, such as fruits, vegetables, and whole grains, can help improve insulin sensitivity and reduce the risk of diabetes (Gautam et al. 2012; Lateef et al. 2023).

According to the 2015 Dietary Guidelines Advisory Committee (DGAC), the United States under-consumed a number of nutrients, which it classified as shortfall nutrients; on the other hand, it overconsumed some nutrients. These nutrients were categorised as nutrients of public health concern when they were associated with common metabolic risk factors or unfavourable health outcomes.

- Iron is a deficiency nutrient among adolescents and premenopausal women; other deficiencies include folate, calcium, magnesium, fibre, potassium, and vitamins A, D, E, and C.
- Under-consumed nutrients of public health concern include potassium, fibre, vitamin D, and calcium.
- Overconsumption of saturated fat and salt is a public health hazard.

According to the nutritional intake, the majority of Americans consume insufficient amounts of vegetables, fruits, whole grains, and dairy products, essential dietary groups that are significant contributors of the deficiencies in nutrients. Their consumption of added sugars and processed carbohydrates is also excessive. In light of the role that food and associated lifestyle variables have in the development of noncommunicable diseases (NCDs), the Food and Agriculture Organisation and the World Health Organisation released expert dietary guidelines that are intended to reduce disease risk by encouraging good lifestyle choices throughout life (Kimokoti and Millen 2016). For many disorders, including peripheral neuropathy, a healthy diet is the first line of defence. Effective management of any risky medical condition is the best defence against peripheral neuropathy. Preventing diabetes, controlling diabetes that already exists, and preventing or at least postponing the onset of complications from diabetes are all made possible by medical nutrition treatment, or MNT. For this reason, it is crucial to avoid diabetes at every stage. Additionally, MNT is a crucial part of diabetes self-management training or education.

Dietary changes may be used to treat disease or its symptoms, but using dietary supplements may also be part of disease management. Two studies that examined the effectiveness of oral dietary supplements for individuals with diabetes or prediabetes were featured in this Special Issue. López-Gómez and colleagues demonstrated in their study how a hyperglycaemia-specific oral supplement might reduce the prevalence of malnutrition and sarcopaenia in people with diabetes and prediabetes (López-Gómez et al. 2022). Similarly, Basiri et al. compared the effects of usual wound care with dietary intake and information on inflammatory biomarkers in ulcerated diabetic feet patients. Education and supplements were shown to have a good effect on patients' inflammation (Basiri et al. 2022).

18.2 Pathophysiology of Diabetic Neuropathy and Nutritional Interventions

Diabetic neuropathy, or nerve damage in diabetes, is brought on by a confluence of inflammatory, vascular, and metabolic variables that gradually destroy the neurological system. Chronic hyperglycaemia, or elevated blood sugar, is one of the main causes, since it sets off detrimental metabolic reactions.

18.2.1 Mechanisms of Nerve Damage in Diabetes

The aetiology of diabetic DSP is well acknowledged to be multifaceted, including intricate relationships among age-related neuronal attrition, diabetes duration, blood pressure, cholesterol levels, weight, and the degree of glycaemic management. Numerous biochemical pathways that impact cellular metabolism are activated by hyperglycaemia, glucotoxicity, and altered insulin signalling, which work in tandem with other risk factors, according to experimental studies. These alterations result in neuronal death in the dorsal root ganglia and structural alterations such microangiopathy, segmental demyelination, and Wallerian degeneration, which subsequently harm and destroy myelinated and unmyelinated fibres. When paired with inadequate antioxidant defences, increased mitochondrial production of free radicals brought on by hyperglycaemia sets off additional detrimental pathways (Albers and Pop-Busui 2014; Mondal et al. 2025).

Additionally, nerve injury is significantly influenced by mitochondrial malfunction. Reactive oxygen species (ROS), which are produced when blood sugar levels are too high, overwhelm the body's antioxidant defences and harm the mitochondria, which are the cell's energy sources. This contributes to the decline of neural function by reducing the generation of energy, especially in long nerve fibres. Because high blood sugar damages the tiny blood arteries that provide oxygen and nourishment to neurones, vascular problems also play a role in nerve injury. Ischaemia, or decreased blood flow, causes nerve hypoxia and further harm.

Diabetes typically causes continuous low-grade inflammation, which is another crucial factor to consider. Pro-inflammatory cytokines like TNF-α and IL-6 may cause nerve damage, blood-nerve barrier disruption, and increased oxidative stress. Neural damage is made worse by inflammation, particularly when immune cells invade neural tissues. Diabetes also affects axonal transport, which is the movement of vital proteins and nutrients along nerve fibres. This impairs neurone survival and function, especially in longer peripheral nerves. Additionally, diabetes hinders the synthesis of neurotrophic factors, which are necessary for the upkeep and repair of neurones and include nerve growth factor (NGF). Nerves are less able to regenerate and are more vulnerable to harm when growth factors are not supported adequately. Finally, swelling and constriction brought on by the build-up of sorbitol and other metabolic by products might exacerbate nerve function by raising the osmotic pressure inside the nerves (Yagihashi et al. 2011; Gautam et al. 2015).

18.2.2 Key Nutrients Involved in Nerve Health

Nutrients from food are necessary to keep the body functioning properly. While micronutrients like vitamins and minerals are essential for metabolism and the maintenance of tissue function, macronutrients like proteins, lipids, and carbs provide us energy (El Soury et al. 2021).

Micronutrients can combat RNS and free radical ROS because of their antioxidant qualities. The passive scavenger's bilirubin, albumin, uric acid, and vitamins C and E have an impact on stratified defence. These are followed by enzymes that help convert radicals into H_2O, such as glutathione peroxidase (GSHPx) and superoxide dismutase (SOD), as well as their cofactors Cu, Zn, Mn, and Se. Diseases include central nervous system damage, peripheral nerve damage (demyelination or axonal damage), and a particular kind of myeloneuropathy (injury associated with the central and peripheral nervous systems) are brought on by micronutrient imbalance. Deficits in some micronutrients lead to pathological changes in the development of the nervous system and the genesis of "nutritional" neuropathies.

Antioxidant qualities of micronutrients enable them to combat free radical ROS and RNS. Uric acid, albumin, bilirubin, and vitamins C and E are examples of passive scavengers that have an impact on tiered defence. These are followed by cofactors Mn, Cu, Se, and Zn, as well as enzymes like GSHPx and SOD, which aid in converting radicals into H_2O. A unique kind of myeloneuropathy and peripheral nerve damage (demyelination or axonal damage) are caused by micronutrient imbalance. Pathological alterations in the nervous system's development and the emergence of "nutritional" neuropathies are caused by deficiencies in specific micronutrients. Normal pain sensations are caused by B1's protection of neurones from oxidative stress and aid in nerve regeneration. B6 aids in the production of neurotransmitters and the recovery of sensor nerve function. Improved or even full recovery of nerve function is the outcome of B12's promotion of nerve survival, remyelination, and myelin sheath preservation. Diabetic peripheral neuropathy is strongly linked to vitamin D insufficiency (Lahoda Brodska et al. 2023).

Nutraceuticals have been created to alleviate neuropathic pain in a number of conditions, such as diabetic neuropathy, fibromyalgia, and pain brought on by chemotherapy. Magnesium, vitamin B, vitamin D, and curcumin are all advised dietary supplements. Preclinical animal models of neuropathic pain, mostly in mice, have been used to investigate how dietary supplements affect neuropathic pain (Abdelrahman and Hackshaw 2021).

18.3 Macronutrient Considerations in Diabetic Neuropathy Prevention

Micronutrition Macronutrients The central nervous system's primary energy matrix is made up of carbohydrates. Carbohydrates, which are found in grains, tubers, roots, fruits, legumes, and sugars in general, are the most significant dietary sources. They account for 55–65% of your caloric intake. In the end, they are all transformed into glucose, the CNS's sole energy source. Since proteins make up almost every tissue in the human body, including the brain, they play a fundamentally changeable function. Along with its position as an intermediate in many metabolic events, it also plays a significant part in the immunological defences, hormones, mediators, and enzymes, as well as the plasma transport of many substances. They are converted into glucose, which gives you energy while you fast for a long time. Particularly in the formation of the cell membrane and as an energy store, lipids have a versatile function. They are precursors of a number of significant chemicals, including inflammatory mediators and some hormones (Nogueira-de-Almeida et al. 2023).

18.3.1 Carbohydrates: Glycaemic Control and Its Impact on Nerve Function

Glycaemic carbohydrates (GCs) are among the most important and cost-effective dietary energy sources for humans because they are readily absorbed and digested by the body and cause a glycaemic response. GCs, which are mostly sugars, oligosaccharides, and polysaccharides, can have a detrimental effect on cognition, particularly in diabetics, youths, and older adults. Their inclination may also lead to a rise in the number of youths with metabolic diseases like diabetes and obesity.

GCs have the ability to cause various glycaemic reactions. A major complication for type II diabetes (T2DM) and prediabetes, elevated postprandial blood glucose levels have become a global epidemic. Glycaemic swings can be significant due to the rapid rises and falls in blood glucose caused by the quickly hydrolysed carbs. On the other hand, postponing the digestion of starches lessens metabolic disorders by preserving glycaemic homeostasis (Yin et al. 2023).

Although chronic hyperglycaemia is still considered a major risk factor for these outcomes, it is being suggested that regular or large glucose fluctuations may be the source of complications associated with diabetes on their own. For instance, high GV (glycaemic variability) was found to be harmful to DN in a large HbA1c control cohort of patients with type II diabetes mellitus (T2DM), even when HbA1c levels were normal. Eating fat and protein before carbs has been shown to significantly improve glucose tolerance by delaying stomach emptying and raising insulin secretion. Additionally, it has been demonstrated that eating fibre lowers postprandial GV in diabetics. Researchers assessed the acute effects of fibre-enriched buckwheat and maize pasta on postprandial GV and found that patients with type 1 diabetes had more stable postprandial GV following fibre-enriched buckwheat pasta (Zhang et al. 2021).

A food's impact on blood glucose levels depends on both the kind and quantity of carbohydrates it contains (in grams). A food's impact on blood glucose is not necessarily predicted by the kind of carbohydrate it contains (such as starch vs sucrose). For instance, even though their chains are shorter, sugars like fructose and sucrose have a lower glycaemic index and glycaemic response. Actually, a number of inherent characteristics of a particular food might affect how it affects blood sugar levels. A meal high in carbohydrates can affect blood glucose levels depending on a number of parameters, including species, variety, physical form, ripeness, processing degree, starch type, and preparation technique. Insulin resistance, protein and fat intake, fasting or pre-prandial glucose levels, and past meal intake are examples of extrinsic variables that affect how a carbohydrate-containing diet affects blood glucose levels (Sheard et al. 2004; Murthy et al. 2013).

To sum up, carbohydrate-rich diets that promote steady but low blood sugar levels may aid in the management of diabetes' metabolic components and related effects. Slow-release carbohydrate diets reduce insulin and glucose movements throughout the day and improve fibrinolysis, which may be a promising therapy for type 2 diabetes. Moreover, several dietary elements may influence how meals regulate blood sugar levels. For example, a diet that includes fibre, certain proteins, or lipids may affect how quickly carbohydrates are absorbed and digested, which may help people with type 2 diabetes. T2D was closely linked to dietary fibres, which are mostly present in cereals, fruits, vegetables, or legumes. Glycaemic management in T2D patients was improved by increased fibre consumption, particularly soluble fibre. People can improve blood sugar control and lower their risk of developing diabetic neuropathy by consume a balanced diet with high fibre and complex carbs (Guo et al. 2020).

18.3.2 Fats: Omega-3 and Omega-6 Fatty Acids

Our body produces several lipid metabolites either intracellularly via the liver and adipose tissue through various routes or directly from the food we eat. Obesity is caused by a high-fat diet, which also causes lipid build-up in cells. A surplus of body fat raises hormones, proinflammatory cytokines, and other elements that contribute to insulin resistance. Free fatty acids damage the insulin signalling system by preventing Akt/PKB activation. Additionally, the mitochondria produce more reactive oxygen species, which impacts glucose regulation. Researchers looked at how persons with type 2 diabetes' diets affected their blood sugar and insulin levels. They discovered that substituting monounsaturated fatty acids (MUFAs) or polyunsaturated fatty acids for saturated fats may help them tolerate glucose or insulin better (Russell et al. 2016).

Over the last four decades, neuroprotective advantages have been shown in both humans and animals, with dietary consumption of polyunsaturated fatty acids (PUFAs) decreasing damage to neurones after traumatic or chemical nerve injury. These advantages, which come from meals like fish oil, avocado oil, and olive oil, can prevent neurones from dying.

Omega-3 The presence of ω-3 PUFAs, which are critical for both visual and cerebral development, has a substantial impact on the neurological system. These fatty acids are essential for maintaining the fluidity of cell membranes and lowering neuroinflammation and oxidative stress. High dietary intake of ω-3 PUFAs can enhance neuronal survival pathways, according to studies conducted on fat-1 mice. Systemic infusion of ω-3 PUFAs can prevent demyelination and nerve damage following low-dose radiation in rat foetuses. Indeed, rats with sciatic nerve CCI have been observed to heal and recover from neuropathic pain when fed a fish oil diet rich in ω-3 PUFAs.

Omega-6 Primarily, enzymes convert ω-6 PUFAs, such as linoleic acid, into ω-3 PUFAs, which are advantageous for cell adhesion, neurodevelopment, and brain plasticity. Oesophageal feeding of primrose oil, which is rich in the omega-6 fatty acid component, has been demonstrated to stimulate nerve regeneration in rats with sciatic nerve compression. As indicated by a greater toe-spreading trigger score, the rats recovered noticeably more quickly than the untreated ones. In addition, myelin and axon morphological characteristics were similar to those of healthy nerves 4 weeks following nerve injury (El Soury et al. 2021).

18.3.3 Proteins: Role in Maintaining Muscle and Nerve Health

The necessary amino acids that are needed to sustain protein synthesis and breakdown, as well as to support cellular functions including cell growth and development, make dietary proteins essential to life. Proteins in food increase the production of insulin, which lowers blood sugar levels in both healthy and diabetic people. Short-term, high-protein diets appear to improve glucose metabolism and weight reduction, dramatically improve insulin sensitivity, and reduce inflammation.

In humans, soy protein is beneficial due to its hypolipidaemic and hypocholesterolemic effects. Research has demonstrated that consumption of soy protein has a beneficial impact on glucose metabolism in addition to lowering blood cholesterol. Additionally, soy protein can boost the production of INSR mRNA, which will raise insulin signalling in the liver and fat and eventually increase these tissues' sensibility to insulin (Iritani et al. 1997; Gautam and Goel 2013).

Protein, the most abundant ingredient in lean fish, reduced triglyceride through low-density lipoprotein metabolism and improved the transit of cholesterol via high-density lipoprotein. Cod protein improved insulin-induced glucose transport by specifically increasing GLUT4 translocation to the T tubules and triggering the PI3K/AKT signalling pathway. Human studies have also shown that cod protein helps those with type 2 diabetes.

The creation or breakdown of proteins causes variations in the quantities of amino acids. Since the body cannot produce eight amino acids on its own, it must get them from nutrition. Amino acids are thought to be regulators of gene expression, including CHOP, which is crucial for the metabolism of glucose. The mTOR/S6K1 pathway is triggered by amino acids, and mTOR inhibition suppresses PI3K, which leads to insulin resistance (Kimball and Jefferson 2006).

18.4 Micronutrients Essential for Nerve Health

Micronutrients, which are essentially represented by vitamins and minerals, are chemicals that are required in trace amounts. Almost every micronutrient in the diet is used by the brain, although some are more crucial than others and are covered in the sections that follow.

18.4.1 Vitamin B12 (Cobalamin)

According to study, methylcobalamin, an enzyme involved in a number of digestion processes, can promote the production of myelin basic protein, which in turn can accelerate and improve myelination in nerve cells. In rats that have had acute sciatic nerve damage, vitamin B12 and dexamethasone together can enhance the diameter of regenerated axons, the number of SCs, and myelinated nerve fibres. This was followed by either improved outcome in motor end plate counts, axon and myelin investigations, and compound muscle action potential assessments, or an elevated sciatic functional index. Growth factors including brain-derived neurotrophic factor (BDNF) have been linked to elevated expression as part of the suggested mechanism for vitamin B12 activity (El Soury et al. 2021).

18.4.2 Vitamin C

The relationship between vitamin C intake and glycaemic control in diabetics has been the subject of several studies. Diabetes patients have been found to have lower amounts of antioxidants and vitamin C than healthy controls. Vitamin C is believed to improve insulin sensitivity, which might lead to improved glucose regulation. Because vitamin C may neutralise ROS, it may help lower the risk of having diabetes-related issues such eye disease, neuropathy, and nephropathy (Younes 2024).

Oral ascorbic acid therapy enhances the recovery of injured neurones, boosting motor and sensory capacities, electrophysiological results, and muscle growth. Additionally, it improves regenerated axons and myelin thickness. Ascorbic acid promotes the healing process by facilitating the elimination of myelin and axonal debris both in vitro and in vivo (Li et al. 2021).

18.4.3 Vitamin B6

The active form of vitamin B6, pyridoxal 50-phosphate (PLP), is engaged in 160 enzymatic processes that are connected to the metabolism of lipids, carbohydrates, and amino acids. Additionally, it functions as an antioxidant by limiting the production of advanced glycation end products (AGEs), which are connected to diabetes and ageing, and ROS. Supplementing with vitamin B6 has been demonstrated to lessen diabetic neuropathy (Mascolo and Vernì 2020).

18.4.4 Vitamin E

People with type 2 diabetes mellitus need vitamin E because it lowers their chance of developing complications from the disease. Human bodies only contain one of the eight forms of vitamin E, alpha-tocopherol. Vitamin E's antioxidant qualities have also been shown to have a significant role in blood sugar regulation in diabetics. This is because, according to several research, vitamin E administration reduces insulin resistance and the glycaemic index in people with type 2 diabetes (Udupa et al. 2012). Although vitamin E administration considerably decreases the HA1c in diabetic individuals, especially those with type 2 DM, the study showed no obvious effect on fasting blood sugar. Deficiency has a significant effect on diabetes since vitamin E is an antioxidant that reduces long-term problems like diabetic nerve damage and other effects including cardiovascular problems. Some studies have shown that people with vitamin E deficiencies might suffer from neurological issues, but testing were unable to confirm this since the deficit has been present for a long period (Blum et al. 2010).

Various investigations have been conducted to clarify how vitamin E affects the regeneration of peripheral nerves. After a sciatic nerve crush, differences in muscle reinnervation showed that more motor endplates were poly-innervated, and that this state persisted longer in rats fed a diet low in vitamin E than in animals on a regular diet. A lack of vitamin E can cause terminal sprouting and persistent poly-innervation of target muscles by decreasing the defence against free radicals, which can postpone nerve regeneration. According to a research on rat peripheral nerve regeneration, peripheral nerve regeneration was greatly enhanced and promoted by a chitosan conduit containing vitamin E, pyrroloquinoline quinone (PQQ), or both. In mice given a combination of vitamin E and PQQ, this improvement was more pronounced. According to the study, local vitamin E therapy may lessen oxidative stress, which may have an effect on neurone regeneration (Azizi et al. 2014).

18.4.5 Vitamin D (Calciferol)

Vitamin D is a precursor to steroid levels and a member of the lipid-soluble vitamin family. Vitamin D2 or ergocalciferol and cholecalciferol or vitamin D3, are the two main forms of it. Dietary ingestion of vitamin D has been shown to promote peripheral nerve regeneration. A in vivo model of a 10 mm autografted peroneal nerve transection was used to evaluate the therapeutic potential of vitamin D2 (Chabas et al. 2008). The study showed that ergocalciferol enhances axonal diameter and encourages axonal regeneration. Vitamin D's regulatory action on many neurotrophic factors involved in nerve regeneration, of which NGF showed an enhanced production, may be the cause of this improvement. Additionally, a decrease in axon degeneration at the damage site was noted in the vitamin D-treated group, indicating that vitamin D2 may have a neuroprotective function.

18.4.6 Magnesium

The fourth most common mineral, magnesium (Mg^{2+}), is vital to human health. Researchers have been trying to figure out how a diet high or deficient in magnesium affects the regeneration of peripheral nerves. Mg^{2+} supplementation following axotomy has been shown to enhance neurone regeneration by increasing the expression of S100 and neurofilament markers. Additionally, it helps motor units survive by rescuing muscle fibres from denervation-mediated atrophy and sciatic motor neurones from axotomy-induced cell death. In addition, improved sciatic nerve functional index, elevated compound muscle action potentials, and decreased nerve conduction delay were seen in mice receiving Mg^{2+} treatment (Chen et al. 2019). The beneficial effect of Mg^{2+} therapy on nerve regeneration may be connected to the inflammatory response. Inflammatory cytokines are released when there is a magnesium deficit, which results in macrophage deposits that impair neurone regeneration. However, by increasing the expression of BCl-2 and BCl-XL, removing cytochrome C, and inhibiting the formation of active caspase-3, Mg^{2+} supplementation suppresses the inflammatory response and stops nerve cell death (Pan et al. 2011).

18.4.7 Selenium

A trace element vital to human health, selenium is a dietary supplement present in a variety of foods. Because of its neuroprotective and antioxidant qualities, researchers have looked at how it affects peripheral nerve regeneration. After nerve compression, selenium therapy improved neurone regeneration, resulting in an increase in myelinated axon numbers, myelin thickness, and axon diameter. The protective effects might be exerted via lowering oxidative stress on nerve cells or by preventing or decreasing lipid peroxidation chain events. All things considered, selenium is an important trace element for general health (Kızılay et al. 2017).

18.4.8 Zinc

As a micronutrient that catalyses more than 100 enzymatic pathways in the body, zinc is essential for metabolic processes. Lower motor nerve conduction velocities, fewer myelinated axons, and smaller cross-sectional areas were seen in sciatic nerves of rats given a zinc-deficient diet. The growth, survival, and function of peripheral nerves depend on zinc, as shown by changes in neuronal structure, myelin and axon damage, and decreased Na and K-ATPase activity (Ünal et al. 2005).

18.5 Antioxidants and Anti-inflammatory Nutrients

Antioxidants are significant chemical substances that have a number of positive effects, such as scavenging or neutralising free radicals, lowering oxidative stress, preventing the body from degenerating oxidatively, regulating the formation of rancidity, avoiding the production of harmful oxidation products, preserving nutritional value, and prolonging the shelf life of food systems. Curcumin, carotenoids, flavonoids, melatonin, GSH, uric acid, polyamines (PAs), plant polyphenols, and alpha-lipoic acid (ALA) are examples of phytochemicals that are naturally occurring antioxidants. Strong evidence linking dietary antioxidant consumption to prevention against diabetes mellitus has been identified by epidemiological research. Previous studies show that antioxidant treatment retains beta-cell activity and shields it from oxidative stress-induced apoptosis. Antioxidants also lessen problems associated with diabetes (Shafras et al. 2024).

18.5.1 Alpha-Lipoic Acid

Numerous foods contain the antioxidant alpha-lipoic acid. Antioxidants can be used by the body to control or avoid oxidative stress, a process that causes tissue damage. One aspect of diabetic neuropathy illness is oxidative stress. Additional study is required to determine whether alpha-lipoic acid similarly lowers blood sugar levels. One kind of short-chain fatty acid that occurs naturally is lipoic acid (LA), also known as ALA. It is sometimes referred to as thioctic acid or 1,2-dithiolane-3-pentanoic acid. In plant and animal mitochondria, LA is produced enzymatically from fatty acid octanoic acid and sulphur-containing amino acid cysteine. It is recognised for its ability to oxidatively decarboxylate α-keto acids. Alpha-lipoic acid may be found in red meat, spinach, broccoli, potatoes, yams, and carrots. According to a recent multicentre, randomized, double-blinded study, people with neuropathy due to diabetes have shown improvements in their neurological capacities after taking ALA for 4 years (Ziegler et al. 2011).

According to the study, 30 days after rat sciatic nerve damage, oral ALA therapy enhanced fibre density, axon and myelin size parameters, and sciatic functional index values. ALA treatment has a protective effect on peripheral nerve damage, making it a viable adjuvant for treating peripheral injuries, according to a new study that also revealed enhanced connective and nerve tissue, fewer mast cells, and a quicker functional recovery (Oktay et al. 2021).

18.5.2 Carotenoids

Lipophilic substances called carotenoids are present in many types of bacteria, algae, and plants as colour pigments. Beta-carotene, xanthophyll, and lycopene are the three most common carotenoids. Carotenoids effectively scavenge ROS and are potent antioxidants. Specifically, carotenoids shield lipids from peroxidation and damage by scavenging peroxyl radicals. Carotenoids are more effective than tocopherols alone at enhancing antioxidant activity in the lipid part of the cellular membrane. Numerous investigations have demonstrated a robust association between DM and carotenoids (Roohbakhsh et al. 2017).

18.5.3 Docosahexaenoic Acid (DHA)

Deficits in DHA are linked to compromised neuronal function and the emergence of neurodegenerative disorders since it is an essential part of neuronal phospholipid membranes. By boosting the gene expression of Bcl-2 and Bcl-2-related protein A1 in the nerve cells and reducing the downstream effects of caspase-3 brought on by oxidative stress, DHA encouraged axonal regeneration. Furthermore, in a neuropathic pain model of median nerve CCI, DHA inhibited the c-Jun N-terminal kinase (JNK) signalling pathway, which, when administered intravenously, produced analgesia. Additionally, it was able to lessen systemic inflammatory responses (Huang and Tsai 2016).

18.5.4 Flavonoid

Fruits and vegetables include flavonoids, which are low molecular weight secondary metabolites that act as anti-inflammatory and antioxidant agents to aid in and control tissue repair. Many fruits, vegetables, and drinks including tea, coffee, and red wine contain them. By functioning as a nonenzymatic antioxidant defence and avoiding damage to nerves, the flavonoid naringenin, which is well-known for its antioxidant qualities, may also contribute to nerve regeneration and cellular redox equilibrium (Muratori et al. 2022).

18.5.5 Glutathione

The tri-peptic antioxidant GSH is soluble in water and composed of the amino acid glutamic acid, glycine, and cysteine. All animals and plants produce GSH, which is synthesised in the liver. By influencing the chain reactions of free radicals and functioning as a cofactor in the enzyme's glutathione reductase, glutathione transferase, and GPX, GSH shields healthy cells from oxidative attack. In vivo renal and neurological function were preserved and oxidative stress was decreased when GSH was added to animal models of diabetes mellitus (Hermes-Uliana et al. 2014).

18.5.6 Acetyl-L-Carnitine

The liver, kidneys, and brain all produce the chemical molecule acetyl-L-carnitine. It aids in converting food into energy. Additionally, it lessens oxidative stress, a process that damages tissue. Additionally, acetyl-L-carnitine contributes to the health of nerve cells. Acetyl-L-carnitine supplements have been shown to reduce pain in diabetic neuropathy patients in a few trials. Along with improvements in nerve function testing, they also showed improvements in their perception of vibrations. According to several of these trials, acetyl-L-carnitine reduces pain more effectively when taken shortly after the onset of diabetic neuropathy. However, additional study is required. When used as prescribed, acetyl-L-carnitine supplements are generally regarded as safe. Dry mouth, decreased appetite, difficulty sleeping, headaches, and irritability are possible side effects.

18.6 Patient Education and Practical Nutritional Guidelines

A multifaceted strategy involving medication, exercise, and dietary changes is needed to manage diabetic neuropathy. In order to effectively manage diabetes and associated consequences, including neuropathy, nutrition is essential. Patients' general quality of life may be enhanced, symptoms can be lessened, and the progression of nerve damage can be slowed down by teaching them the value of a balanced, anti-inflammatory diet. According to the American Diabetes Association's (ADA) standards, altering one's lifestyle is crucial to reaching the objectives of diabetes management. According to the American Diabetes Association's standards of care recommendations, lowering total carbohydrate consumption for diabetics has shown the strongest evidence of reducing glycaemia. This suggestion may be applied to a range of eating patterns that suit personal tastes and needs. For those with diabetes, the CDC recommends three dietary interventions: restricting portion sizes, using the plate method, and counting carbs (Sahyouni et al. 2024).

Initially, meal planning for individuals with type 1 diabetes was done using carbohydrate counting (CC). When calculating the appropriate bolus insulin dosage, the insulin-to-carbohydrate ratio (ICR) and the total amount of carbohydrates consumed at each meal are considered. After every meal, the exact amount of insulin needed to effectively control blood glucose levels is determined using these two factors. Research indicates that CC reduces the levels of glycosylated haemoglobin (HbA1c) and enhances metabolic regulation. Improving T1DM patients' glycaemic control may postpone long-term microvascular issues associated with the disease (Ibrahim et al. 2023).

It has been demonstrated that people with a BMI of 25 kg/m^2 can better control type 2 diabetes by following a low-carb diet of 20–50 g daily. According to a research with 26 T2D patients, a very low-carb ketogenic diet (20–50 g per day) worked better than a plate approach (Saslow et al. 2017). For the prevention and management of type 2 diabetes, research and clinical trials offer a variety of dietary recommendations that stress the significance of dietary practices and lifestyle decisions, including calorie restriction and exercise. When it comes to nutrition, quality matters more than quantity. Diets high in fruits, vegetables, legumes, and whole grains are advised for T2D patients in order to better manage glucose control. Low-GI (glycaemic index), low-carb, and high-protein dietary patterns will shield us from the occurrence of hyperglycaemia. Furthermore, it is advantageous to consume alcohol and nuts in moderation. Nutritional methods should be tailored to the cultures and genetic backgrounds of various communities or people since they have diverse dietary patterns, foods, and susceptibilities to illness (Guo et al. 2020).

18.6.1 Educating Patients on Balanced Meal Planning

Diabetes dietary recommendations have changed throughout time, becoming more patient-cantered and adaptable. The American Diabetes Association's (ADA) 2024 nutrition goals include the following (Care 2023):

(a) To enhance general health by encouraging and supporting a range of nutrient-dense meals in sensible portion sizes, as well as:
   - Reach and sustain body weight objectives.
   - Reach personalised cholesterol, blood pressure, and glycaemic targets.
   - Postpone or avoid diabetic consequences.

(b) The intention is to address each person's unique nutritional demands according to their cultural and personal preferences, health literacy, the accessibility of nutrient-dense foods, their desire to modify their behaviour, and any present barriers.
(c) To maintain the pleasure of eating by limiting food options only when backed by scientific evidence and promoting nonjudgmental attitudes towards them.
(d) To provide a diabetic with practical resources for developing a healthy diet rather than focussing on particular foods, macronutrients, or micronutrients.

18.6.2 Practical Tips for Integrating Nutritional Guidelines into Daily Life

For Carbohydrate Intake

- Focus a balanced diet on high-fibre foods including entire grains, legumes, vegetables, and fresh fruit.
- Brown rice, barley, quinoa, whole wheat, and whole oats are examples of all grain products.
- Make sure the first ingredient on the list is wholegrain and that less than 10% of the product's calories comes from sugars when you buy wholegrain products.
- Eat fruit, but go for whole fruit rather than fruit that has been dried, juiced, or processed further.
- Legumes are a great and affordable source of protein and fibre. Lentils or legumes can be used in lieu of ground meat in dishes like casseroles.
- Eat a variety of vegetables each day and avoid items that are very spicy or deep-fried.

For Fat Intake

- A person's diet should consist mostly of mono and polyunsaturated fats, which may be found in nuts, seeds, avocados, seafood, and oils such as canola, soybean, sunflower, and maize.
- Reduce consumption of red meat, processed and fast meals, and full-fat dairy products to reduce intake of saturated fats. When feasible, consider substituting skinless poultry, fish, beans, and nuts for red meat. You may also try reducing the fat content of dairy products like whole milk.
- Use liquid vegetable oils that are high in mono and polyunsaturated fats instead of butter or margarine for cooking and serving.
- Limit the amount of trans fat you consume. Limit fried meals and look for trans fats on food labels.

For Protein Intake

- Legumes, lentils, tofu, and tempeh are excellent plant protein sources (1/2c = 2 oz protein). Quorn and other plant-based meat substitutes may also be used, but avoid meat substitutes that have a lot of saturated fat and salt.
- It is crucial to promote plant-based protein sources such as Brazil nuts, cashews, hazelnuts, and almonds. Another choice is nut butters, but watch out for additional salt and sugar.
- Skinless poultry, fat of beef or pig, fish, or eggs, as well as reduced-fat dairy products (such as cheese and skim or low-fat milk and yoghurt) are all excellent sources of lean animal protein.
- For the majority of meals, protein sources should be used as an addition to fruits, vegetables, and whole grains rather than as the main course (Reynolds and Mitri 2024).

18.7 Conclusion and Future Directions in Nutritional Research for Diabetic Neuropathy Prevention

A diet that is well-balanced is vital because it gives you the energy and nutrients you need to live and remain healthy. The benefits of eating a nutritious diet and leading a healthy lifestyle are well-established, and they include a decreased risk of cardiovascular disease, cancer, and diabetes, and being obese, among other chronic conditions. Dietary components can improve peripheral nerve regeneration, according to recent research. According to studies, neuropathic pain—which is associated with damage to peripheral nerves—can be avoided or lessened by eating meals high in nutrients. Anti-inflammatory foods, avoiding processed and high-glycaemic meals, and getting adequate omega-3 fatty acids, antioxidants, vitamins B12 and D, and magnesium are important dietary measures for symptom reduction and nerve function protection. Since dietary decisions have a substantial influence on the development and course of neuropathy, it is essential to educate patients about these nutritional recommendations in order to maintain adherence. People are empowered to make significant lifestyle adjustments that promote nerve health and general metabolic well-being by following clear-cut, practical recommendations.

In future, a number of intriguing avenues for nutritional study may help us better understand how food affects diabetic neuropathy. Nutrigenomics, the study of how individual genetic differences impact food reactions, is one exciting field. More focused and efficient dietary treatments for the prevention of neuropathy may be provided by genetic profile-based personalised nutrition regimens. Another area of increasing study is the role of the gut bacteria in neuropathy and diabetes. Dietary treatments, such the use of probiotics and prebiotics, may be crucial in regulating the course of neuropathy since a balanced gut microbiota has been associated with decreased inflammation and enhanced insulin sensitivity. Furthermore, phytochemicals and polyphenols included in plant-based diets have the ability to lower oxidative stress and inflammation, which calls for more research into their potential as a neuropathy treatment. Whether getting these nutrients from food or supplements is better for preventing or treating neuropathy may also be clarified by ongoing study into nutritional supplementation, specifically for vitamin B12, co-enzyme Q10, vitamin D, and omega-3 fatty acids. Lastly, longer-term research on certain dietary regimens, such the low-carb or Mediterranean diets, may shed more light on how well they work to prevent and cure neuropathy.

References

Abdelrahman KM, Hackshaw KV. Nutritional supplements for the treatment of neuropathic pain. Biomedicine. 2021;9(6):674.

Albers JW, Pop-Busui R. Diabetic neuropathy: mechanisms, emerging treatments, and subtypes. Curr Neurol Neurosci Rep. 2014;14:1–11.

Azizi A, Azizi S, Heshmatian B, Amini K. Improvement of functional recovery of transected peripheral nerve by means of chitosan grafts filled with vitamin E, pyrroloquinoline quinone and their combination. Int J Surg. 2014;12(1):76–82.

Basiri R, Spicer M, Levenson C, Ledermann T, Akhavan N, Arjmandi B. Improving dietary intake of essential nutrients can ameliorate inflammation in patients with diabetic foot ulcers. Nutrients. 2022;14(12):2393.

Blum S, Vardi M, Brown JB, Russell A, Milman U, Shapira C, et al. Vitamin E reduces cardiovascular disease in individuals with diabetes mellitus and the haptoglobin 2-2 genotype. Pharmacogenomics. 2010;11(5):675–84.

Bunner AE, Wells CL, Gonzales J, Agarwal U, Bayat E, Barnard ND. A dietary intervention for chronic diabetic neuropathy pain: a randomized controlled pilot study. Nutr Diabetes. 2015;5(5):e158.

Care D. Standards of care in diabetes—2023. Diabetes Care. 2023;46:S1–S267.

Chabas JF, Alluin O, Rao G, Garcia S, Lavaut MN, Risso JJ, et al. Vitamin D2 potentiates axon regeneration. J Neurotrauma. 2008;25(10):1247–56.

Chen YJ, Cheng FC, Chen CJ, Su HL, Sheu ML, Sheehan J, Pan HC. Down-regulated expression of magnesium transporter genes following a high magnesium diet attenuates sciatic nerve crush injury. Neurosurgery. 2019;84(4):965–76.

El Soury M, Fornasari BE, Carta G, Zen F, Haastert-Talini K, Ronchi G. The role of dietary nutrients in peripheral nerve regeneration. Int J Mol Sci. 2021;22(14):7417.

Gautam MK, Goel RK. Wound healing effect of Ocimum sanctum leaves extract in diabetic rats. World Acad Sci Eng Technol. 2013;81:144–7.

Gautam MK, Gupta A, Rao CV, Goel RK. Antihyperglycemic and antioxidant potential of Murraya paniculata Linn. leaves: a preclinical study. J Pharm Res. 2012;5(3):1334–7.

Gautam MK, Gangwar M, Singh SK, Goel RK. Effects of Azadirachta indica on vascular endothelial growth factor and cytokines in diabetic deep wound. Planta Med. 2015;81(8):713–21.

Guo Y, Huang Z, Sang D, Gao Q, Li Q. The role of nutrition in the prevention and intervention of type 2 diabetes. Front Bioeng Biotechnol. 2020;8:575442.

Hermes-Uliana C, Panizzon CPDNB, Trevizan AR, Sehaber CC, Ramalho FV, Martins HA, Zanoni JN. Is L-glutathione more effective than L-glutamine in preventing enteric diabetic neuropathy? Dig Dis Sci. 2014;59:937–48.

Huang CT, Tsai YJ. Docosahexaenoic acid confers analgesic effects after median nerve injury via inhibition of c-Jun N-terminal kinase activation in microglia. J Nutr Biochem. 2016;29:97–106.

Ibrahim SMH, Shahat EA, Amer LA, Aljohani AK. The impact of using carbohydrate counting on managing diabetic patients: a review. Cureus. 2023;15(11):e48998.

Iritani N, Sugimoto T, Fukuda H, Komiya M, Ikeda H. Dietary soybean protein increases insulin receptor gene expression in Wistar fatty rats when dietary polyunsaturated fatty acid level is low. J Nutr. 1997;127(6):1077–83.

Kimball SR, Jefferson LS. New functions for amino acids: effects on gene transcription and translation. Am J Clin Nutr. 2006;83(2):500S–7S.

Kimokoti RW, Millen BE. Nutrition for the prevention of chronic diseases. Med Clin. 2016;100(6):1185–98.

Kızılay Z, Erken HA, Aktaş S, Ersoy N, Abas Bİ, Topçu A, et al. Effects of selenium on axon and myelin healing in an experimental sciatic nerve injury model. J Istanbul Facul Med. 2017;80(4):138–45.

Lahoda Brodska H, Klempir J, Zavora J, Kohout P. The role of micronutrients in neurological disorders. Nutrients. 2023;15(19):4129.

Lateef, Haseeb, Lucas M. The role of nutrition in disease prevention and management. Int J Med Dev Country. 2023;9:763.

Li L, Xu Y, Wang X, Liu J, Hu X, Tan D, et al. Ascorbic acid accelerates Wallerian degeneration after peripheral nerve injury. Neural Regen Res. 2021;16(6):1078–85.

López-Gómez JJ, Gutiérrez-Lora C, Izaola-Jauregui O, Primo-Martín D, Gómez-Hoyos E, Jiménez-Sahagún R, De Luis-Román DA. Real world practice study of the effect of a specific oral nutritional supplement for diabetes mellitus on the Morphofunctional assessment and protein energy requirements. Nutrients. 2022;14(22):4802.

Mascolo E, Vernì F. Vitamin B6 and diabetes: relationship and molecular mechanisms. Int J Mol Sci. 2020;21(10):3669.

Mondal NS, Islam A, Gautam MK, Mondal S, Ahmed M, Jamir SI, Khongwir LD, Mahajon B. Unraveling the mechanisms of diabetic wounds: insights into pathogenesis and advanced treatment strategies. Curr Diabetes Rev. 2025;22:113.

Muratori L, Fregnan F, Maurina M, Haastert-Talini K, Ronchi G. The potential benefits of dietary polyphenols for peripheral nerve regeneration. Int J Mol Sci. 2022;23(9):5177.

Murthy S, Gautam MK, Goel S, Purohit V, Sharma H, Goel RK. Evaluation of in-vivo wound healing activity of Bacopa monniera on different wound model in rats. Biomed Res Int. 2013;2013:972028.

Nogueira-de-Almeida CA, Zotarelli-Filho IJ, Nogueirade-Almeida ME, Souza CG, Kemp VL, Ramos WS. Neuronutrients and central nervous system: a systematic review. Cent Nerv Syst Agents Med Chem. 2023;23(1):1–12.

Oktay KNK, Akpinar P, Ozkan FU, Aktas I, Canillioglu YE, Senturk GE. The effect of alpha lipoic acid on the recovery of sciatic nerve injury in rats: a prospective randomized controlled trial. Am J Phys Med Rehabil. 2021;100(8):780–9.

Pan HC, Sheu ML, Su HL, Chen YJ, Chen CJ, Yang DY, et al. Magnesium supplement promotes sciatic nerve regeneration and down-regulates inflammatory response. Magnes Res. 2011;24(2):54–70.

Reynolds A, Mitri J (2024) Dietary advice for individuals with diabetes. Endotext; Feingold, KR, Anawalt, B., Blackman, MR, Boyce, A., Chrousos, G., Corpas, E., de Herder, WW, Dhatariya, K., Dungan, K., Hofland, J., et al., Eds.

Roohbakhsh A, Karimi G, Iranshahi M. Carotenoids in the treatment of diabetes mellitus and its complications: a mechanistic review. Biomed Pharmacother. 2017;91:31–42.

Russell WR, Baka A, Björck I, Delzenne N, Gao D, Griffiths HR, et al. Impact of diet composition on blood glucose regulation. Crit Rev Food Sci Nutr. 2016;56(4):541–90.

Sahyouni MJ, Acevedo LA, Rodriguez SC, Chiles JA, Pennings NJ. Neuropathy of diabetes following initiation of a low-carbohydrate diet: case report. Obes Pillars. 2024;11:100115.

Saslow LR, Mason AE, Kim S, Goldman V, Ploutz-Snyder R, Bayandorian H, et al. An online intervention comparing a very low-carbohydrate ketogenic diet and lifestyle recommendations versus a plate method diet in overweight individuals with type 2 diabetes: a randomized controlled trial. J Med Internet Res. 2017;19(2):e36.

Shafras M, Sabaragamuwa R, Suwair M. Role of dietary antioxidants in diabetes: an overview. Food Chem Adv. 2024;4:100666.

Sheard NF, Clark NG, Brand-Miller JC, Franz MJ, Pi-Sunyer FX, Mayer-Davis E, et al. Dietary carbohydrate (amount and type) in the prevention and management of diabetes: a statement by the American Diabetes Association. Diabetes Care. 2004;27(9):2266–71.

Udupa AS, Nahar PS, Shah SH, Kshirsagar MJ, Ghongane BB. Study of comparative effects of antioxidants on insulin sensitivity in type 2 diabetes mellitus. J Clin Diagn Res. 2012;6(9):1469.

Ünal B, Tan H, Orbak Z, Kiki İ, Bilici M, Bilici N, et al. Morphological alterations produced by zinc deficiency in rat sciatic nerve: a histological, electron microscopic, and stereological study. Brain Res. 2005;1048(1–2):228–34.

Yagihashi S, Mizukami H, Sugimoto K. Mechanism of diabetic neuropathy: where are we now and where to go? J Diabet Investig. 2011;2(1):18–32.

Yin J, Cheng L, Hong Y, Li Z, Li C, Ban X, et al. A comprehensive review of the effects of Glycemic carbohydrates on the neurocognitive functions based on gut microenvironment regulation and Glycemic fluctuation control. Nutrients. 2023;15(24):5080.

Younes S. The role of micronutrients on the treatment of diabetes. Human Nutr Metabol. 2024;35:200238.

Zhang X, Yang X, Sun B, Zhu C. Perspectives of glycemic variability in diabetic neuropathy: a comprehensive review. Commun Biol. 2021;4(1):1366.

Ziegler D, Low PA, Litchy WJ, Boulton AJ, Vinik AI, Freeman R, et al. Efficacy and safety of antioxidant treatment with α-lipoic acid over 4 years in diabetic polyneuropathy: the NATHAN 1 trial. Diabetes Care. 2011;34(9):2054–60.

The Role of Exercise and Physical Activity in Diabetic Neuropathy

19

Prasanth Damarasingu, Budumuru Padmasri, Mallikarjuna Rao Pichika, Madhu Katyayani Balijepalli, Kit-Kay Mak, and Prabhat Kumar Upadhyay

19.1 Introduction

The incidence of diabetes mellitus (DM) has increased at least fourfold over the last three decades, making it a significant public health and economic problem in the majority of countries. 90–95% of persons with diabetes have type 2 diabetes (T2D), which is growing more prevalent in kids, teens, and young adults. The main risk factor for T2D is the rising incidence of prediabetes (Tuell et al. 2023). The most common side effect of diabetes is neuropathy, which affects about half of those diagnosed with type 1 or type 2 diabetes mellitus. When diabetes is diagnosed, nerve conduction testing is used to demonstrate the presence of subclinical neuropathy (Albers et al. 2007; Gautam et al. 2012). Individuals diagnosed with diabetic neuropathy frequently experience decreased quality of life, insomnia, and

P. Damarasingu (✉)
Department of Pharmacology, School of Pharmacy and Life Sciences, Centurion University of Technology and Management, Balasore, Odisha, India

B. Padmasri
Department of Pharmaceutics, Sri Venkateswara College of Pharmacy, Srikakulam, Andhra Pradesh, India

M. R. Pichika
Newcastle University Medicine Malaysia, Iskandar Puteri, Johor, Malaysia

M. K. Balijepalli
Department of Pharmacology, Faculty of Medicine, Jenjarom, Selangor, Malaysia

K.-K. Mak
Department of Pharmaceutical Chemistry, School of Pharmacy, IMU University, Kuala Lumpur, Malaysia

P. K. Upadhyay
Institute of Pharmaceutical Research, GLA University, Mathura, Uttar Pradesh, India

M. K. Gautam et al. (eds.), *Antioxidants and Functional Foods for Diabetic Neuropathy*, https://doi.org/10.1007/978-981-95-9326-2_19

depression as a result of neuropathic pain, autonomic symptoms, and sensorimotor impairments.

As of right now, randomised, blinded (masked) clinical studies have not demonstrated the efficacy of any disease-modifying interventions for diabetic neuropathy. Previously conducted clinical trials that have been designed to test outcomes such as big fibre function based on length, such as nerve conduction investigations, have largely not been successful. Physical exercise and proper diet have been shown to lower the risk of T2D by the Diabetes Prevention Program. Adjusting one's lifestyle can also help counteract some of the bad consequences of diabetes, including weight reduction, improved blood pressure, glycaemic control, dyslipidaemia, and a lower chance of heart disease and the associated deaths (Wc 2002; Mondal et al. 2025).

When individuals with poor glucose tolerance take part in structured behavioural modification trials that involve weight reduction, dietary calorie restriction, and 150–175 min of physical exercise, their prevalence of T2D is decreased by 40% to 70%. Despite other cardiometabolic risk factors, this is the case. According to studies, when compared to traditional treatment, these programs also lower the incidence of body weight, fasting blood glucose, and T2D. There is a greater reduction in body weight when these instructions are followed (Colberg et al. 2016; Chandra et al. 2025).

Aerobic exercise has positive benefits on many of the mechanisms that diabetes negatively affects, in contrast to focused therapy. The body benefits from exercise in a variety of ways. Importantly, it can assist with insulin sensitivity, glucose management, obesity, and dyslipidaemia—three of the primary risk factors for DN. Additionally, exercise reduces humoral inflammation, improves end organ perfusion, lowers lipid and protein oxidation, improves hypertension, and stops adipocytes from producing free fatty acids and adipokines (Singleton et al. 2015).

19.2 Role of Physical Activity in Diabetes Management

It is commonly known that strict blood glucose control achieved via food, exercise, or medication can avoid diabetes and its related problems. More recently, a number of sizable randomised controlled studies proved that aerobic exercise helps diabetics with their insulin sensitivity, glycaemic management, and physical fitness. As a result, exercise is advised for diabetics in order to enhance glycaemic management and reduce problems related to their condition (Kluding et al. 2017; Gautam et al. 2015). Exercise for pain treatment, particularly low back pain, has gained popularity in recent years. Scholars have also supported studies on processes relating to exercise and pain, such as the consequences of exercise-induced hypoalgesia. In the meanwhile, studies have demonstrated the safety and efficacy of exercise for those with diabetes. The American Diabetes Association recommends 150 sessions of moderate-to-intense aerobic activity and two to three strength training sessions for diabetics. In the past, the American Diabetes Association has provided cautious advice on weight-bearing activities in cases with severe peripheral nerve damage (American Diabetes Association (2008); Luo et al. 2022). Since aerobic energy-producing systems are the main source of energy, aerobic activity involves using large muscle groups repeatedly. Resistance training exercises include body weight, free weights, weight machines, and elastic resistance bands. Exercises for flexibility enhance joint range of motion, whilst activities for balance strengthen gait and minimise the probability of falls. Resistance, versatility, and balance are all integrated into exercises like yoga and tai chi, which improve health outcomes for diabetics.

19.3 Exercise and Peripheral Nerve Health

In both prediabetic and diabetic animal models, exercise has been shown to have positive benefits on nerve function. Restored electrophysiological function, normalised epidermal innervation, enhanced neurone regeneration, and less pain are some of these outcomes. Notably, research on both humans and animals has demonstrated that substantial improvements in metabolic syndrome characteristics are not always associated with the preservation of nerve regeneration and the avoidance of excruciating early symptoms. These results provide credence to the theory that peripheral nerve plasticity is partially mediated by exercise via pathways unrelated to metabolic state (Kluding et al. 2017). Exercise has been demonstrated to be beneficial in restoring neurotrophin levels, lowering oxidative stress, and decreasing inflammation in animal models. The exact processes behind exercise-induced nerve protection or healing remain unclear. Exercise can also enhance electrophysiological performance by reducing $Ca2^{+}$ channel malfunction and preventing myelin damage (Shankarappa et al. 2011; Gautam et al. 2014). More and more evidence, that one of the main mediators of these advantages is the restoration of neuronal growth regulator support. Animal studies provide strong evidence that exercise influences several pathways to improve neurone regeneration and restore peripheral nerve milieus, which is a clear benefit over a single-factor strategy.

Physical activity has been shown to enhance many metabolic parameters that might impact nerve well-being and microvascular performance, thereby providing an indirect defence against injury to peripheral nerves. In a 4-year observational research, around 80 individuals with diabetes who did not exhibit any DPN symptoms or indications were followed up. Those who walked briskly under supervision for 4 h a week at the end of the experiment experienced fewer cases of motor or sensory neurological disorders than the control group (Balducci et al. 2006). Additionally, recent single-group trials that included an aerobic exercise intervention shown reduction in pain interference, physical and general fatigue, and both. Additionally, Dixit et al. found that aerobic activity increased nerve conduction velocity in a randomised controlled trial (Dixit et al. 2014).

19.3.1 Impact of Exercise on Nerve Function

The stimulation of afferent nerve cells from working muscles to the spinal cord is one recognized consequence of exercise. Exercise causes an increase in working muscle sensory fibre activity, which provides vital input on the cardiovascular and respiratory systems. The exercise pressor reflex (EPR), which regulates changes in heart rate (HR) and blood sugar during physical activity by activating sympathetic neurones, is an illustration of how exercise influences afferent nerve activity.

Regular exercise helps prevent muscle soreness because it reduces phosphorylation of the NR1 subunit of the brainstem's NMDA receptor, which affects nociception and subjective perceptions. Following resistance along with aerobic activity, there are improvements in the quantities of circulating nitrate in the cerebrospinal fluid and plasma. This suggests that the nitric oxide/cyclic GMP pathway may have an antinociceptive benefit during physical activity, as nitric oxide inhibitors no longer produce the analgesic effect of exercise (Cooper et al. 2016).

19.4 Mechanisms Through Which Exercise Improves Neuropathy

19.4.1 The Role of Exercise in Inflammatory Response

In order to coordinate immune responses and activate the pain pathway, immune cells, glia, and neurones all have an impact on chronic pain. These interactions include both neurotransmitters and inflammatory mediators. A significant contributing component to the pathogenesis of diabetic neuropathy (DNP) is the inflammatory response. An inflammatory response associated with diabetic neuropathy is triggered by elevated pro-inflammatory chemicals such as TNF-α, IL-10, and CRPs. Because exercise reduces pro-inflammatory factors, it can help DNP patients feel less pain and slow down the inflammatory response (Luo et al. 2022).

The decrease of pro-inflammatory cytokines and the rise of anti-inflammatory markers that result from endurance and resistance training regimens are two important benefits. It is suggested that once an exercise session is over, a sustained rise in anti-inflammatory cytokines occurs due to the activity's pro-inflammatory immediate effects. Both elderly and young people who regularly exercise have shown reduced inflammatory markers (Cooper et al. 2016).

19.4.2 The Role of Exercise in Micro-Ribonucleic Acids

Previous studies have shown that abnormal microRNA expression in the sural nerve, white blood cells, and the epidermis is associated with painful peripheral neuropathies. Aghdam et al., created a DNP rat model in which rats participated in a swimming exercise program for 10 weeks. In healthy rats, swimming activity and increased expression levels both reduced NaV1.3 expression, but swimming training restored the increased expression of NaV1.3 brought on by diabetes. These findings suggested that the beneficial effect of swimming against diabetic discomfort may be due to miR-96/NaV1.3 mRNA (Aghdam et al. 2018).

Another approach is the role of neurotrophin-3 (NT-3), which has been shown to encourage the development of new synapses and neurones as well as the longevity and division of existing neurones. It has been demonstrated that aerobic activity increases NT-3 expression in the skeletal muscles and spinal cord of diabetic and peripheral nerve injury healing rats. Additionally, it was just demonstrated to have relevant advantages for endogenous opiates like β-endorphin and met-enkephalin. These opiates are produced more often during exercise, which results in analgesia and feel-good feelings (like a "runner's high"). The last mechanism affected by exercise is heat shock protein 72 (HSP72), This lessens the impairment brought on by PN. HSP72 plays a crucial role in helping cells fight off stress. Aerobic exercise raises HSP72 expression throughout the body, especially in the brain's spinal cord and peripheral nerves, whereas diabetes and other diseases tend to decrease it (Dobson et al. 2014).

19.5 Types of Exercise Beneficial for Diabetic Neuropathy

19.5.1 Aerobic Exercise

An aerobic workout regimen improves the way the respiratory and cardiovascular systems interact and increases oxygen consumption. Aerobic exercises include swimming, cycling, running, rowing, walking, and jumping rope. Moderate aerobic exercise helps diabetic neuropathy patients keep their blood pressure stable (Mogensen 2002). Aerobic exercise improves lung and immune function, mitochondrial density, insulin sensitivity, insulin compliance and responsiveness, cardiovascular compliance and responsiveness, and cardiac output. In individuals with T1D & T2D, aerobic exercise dramatically lowers the risks of cardiovascular disease and overall mortality. In individuals with T1D, aerobic exercise increases endothelial function, lipid levels, insulin resistance, and cardiorespiratory fitness. Regular exercise reduces insulin resistance, lipids, blood pressure, and A1C in people with type 2 diabetes. Patients with type 1 diabetes can benefit from high-intensity interval training (HIIT), which promotes increases in skeletal muscle oxidative capacity, insulin sensitivity, and glycaemic management, without worsening their condition (Colberg et al. 2016).

19.5.2 Resistance Training

Resistance training is a crucial exercise for people with T2D, assisting in blood sugar regulation and reduce insulin resistance. It involves lifting weights against resistance, which depends on the equipment used. Resistance exercise improves heart rate, cholesterol levels, calcium levels, heart condition, muscle mass, body structure, strength, and mental as well as physical health. Combining resistance and aerobic exercise can reduce hypoglycaemia, especially when done earlier than aerobic exercise (Gordon et al. 2009).

19.5.3 Flexibility and Balance Training

In addition, passive and endurance exercise are complimentary treatment modalities for chronic conditions such as insulin-resistant diabetes mellitus. Utilising many big muscle groups during endurance exercise depends on the circulatory system supplying the muscles with oxygen. The patient's own body must be engaged in passive exercise, either by choice or by the involvement of another person or external force (Boulé et al. 2003).

Since advanced glycation end products frequently cause restricted joint mobility, balance and flexibility exercises are essential for older adults with diabetes. Although stretching increases flexibility, it has less impact on controlling blood sugar levels. Particularly for people with peripheral neuropathy, exercising balance can lower the risk of falls by enhancing gait and balance.

19.6 Exercise Prescription for Diabetic Neuropathy

Adults who engage in at least 150 min of moderate-intensity physical activity or 75 min of high-impact exercise each week, collected in 10-min or longer bursts, will see a considerable improvement in their metabolic health and cardiorespiratory fitness. Stricter rules are necessary for weight loss, with recommendations indicating that 200–300 min per week are necessary for sustained decreases (Jelleyman et al. 2015). According to the American Diabetes Association (ADA), people with diabetes should do moderate-intensity aerobic exercise 3–7 days a week for at least 30 min (American Diabetes Association (2008); Colberg et al. 2016). Exercise programs must include both aerobic and muscle-building exercises. Moderate to intense aerobic exercise (65–90% of maximum heart rate) raises VO2max cardiac output in people with T2D, and this is strongly associated with a decreased risk of cardiovascular disease and mortality. Before starting an exercise program, those who have been sedentary should get permission from their healthcare physician. Therefore, muscle strength may be used as a result indicator for resistance training exercises aimed at improving blood glucose control; the more muscle power is affected by resistance training, the more successfully HbA1c is lowered. A diabetic's safe and appropriate exercise regimen depends on a number of criteria being carefully considered and good clinical judgement being used. The optimal physical activity and risk level may be determined with the help of an individual's medical record and physical examination (Colberg et al. 2016). The following are the ADA's general guidelines:

Aerobic Exercise moderate-to-intense activity for at least 150 min each week, spaced out across 3–7 days, with no more than 2 days between sessions. It is advised that younger, more physically fit individuals participate in vigorous-intensity or interval exercise for at least 75 min per week in order to maximise insulin activation. Either as a continuous workout or as high-intensity interval training.

Resistance Exercise Twice or at least three times a week on days that aren't consecutive, a progressive, moderate to vigorous resistance training program should include at least eight to ten exercises, each with one to three sets of ten to fifteen repetitions. It is advised that older adults with T2D engage in flexibility and balancing exercises two to three times per week. It is also recommended to participate in supervised training programs in order to maximise the health benefits of exercise.

19.6.1 Challenges and Barriers to Exercise in Diabetic Neuropathy

Scientists have examined and identified many obstacles that people with diabetes face while attempting to participate in physical activity (PA). Since these barriers are impacted by a multitude of factors, such as cultural, ecological, psychological, and economic traits, their shapes differed significantly between backgrounds. One study assessed the amount of PA in adult T2DM patients in Argentina. The results showed that 52.3% of the patients who were tested performed poorly. This low level of PA is associated with female sex, diabetic foot history, higher HbA1c values, and advanced age. The most often cited barriers to completing PA were a lack of time, energy, and willpower. Out of four participants, one said that social considerations were a barrier (Martin et al. 2021).

Since they have been demonstrated to be consistent with most previous studies and are associated with Relevant variables include height, sex, age, body mass index (BMI), and low PA levels. The relationship between PA level and age and BMI was inverse, with women experiencing greater levels of low PA. Being physically unable to exercise and developing bad behaviours as a result of being overweight have been demonstrated to be barriers to obesity. In terms of metabolic control, the study's findings showed a robust association between low levels of PA and higher HbA1c readings (Heiss and Petosa 2014). The inability to regulate one's own behaviour, emotions, or determination is known as a lack of willpower. Additional obstacles to doing PA include a lack of duration, inspiration, and energy, which is the sensation of being exhausted or drowsy. Furthermore, a study has explored that particular traits are rather common (Lidegaard et al. 2016). It may be necessary to develop self-efficacy, defined as "people's beliefs about their capabilities to exercise control over their own level of functioning and over events that affect their lives," in order to overcome these obstacles to movement.

The investigations revealed that social effects accounted for one in every four patient obstacles with respect to other barriers. Alghafri et al. discovered that a common obstacle was the absence of social support (Alghafri et al. 2017). In a similar vein, physical challenges and time constraints were more common hindrances than environmental or financial resource issues among low-income persons in the USA with type 2 diabetes. Furthermore, in the cohort, the biggest obstacles to exercise for people with T1D who had insulin resistance (IR) were diabetes-specific factors, particularly fear of hypoglycaemia; in contrast, the biggest obstacles to exercise for people with T1D who did not have IR were fitness and generic physical health issues. When people with T1D exercise, hypoglycaemia frequently occurs and can be challenging to anticipate, prevent, and control (Alobaid et al. 2023).

19.7 Safety Considerations for Exercise in Diabetic Neuropathy

It is frequently recommended that diabetic patients participate in regular physical activity. Patients with more severe neuropathy and insensate feet should exercise caution when doing weight-bearing activities since they run the risk of developing diabetic foot ulcers. Proliferative nephropathy patients, however, should refrain from any activities that might raise their blood pressure. Patients with diabetes may often tolerate supervised exercise regimens successfully. A pharmacological therapy would not only be more expensive than a successful lifestyle intervention with a multisystemic impact, but it would also be less likely to have side effects. Apart from the adverse effects per se, patients may have reduced physical activity due to common drug side effects such as weariness, unbalance, or gastrointestinal discomfort. Because lethargy has a detrimental effect on overall health and quality lifestyles in diabetes, this is especially troublesome for people with chronic conditions and low levels of physical exercise.

Diabetic neuropathy caused by therapy should be recognised. This is a painful, reversible autonomic neuropathy with an early onset that can manifest acutely in people with previously poorly managed diabetes, soon after they begin rigorous glucose management. Neuropathy brought on by therapy may arise from severe caloric restriction alone. Because of their heightened vulnerability to hypoglycaemia, poor thermoregulation, orthostatic hypotension, and diminished cardiac reactivity to exercise, individuals with autonomic neuropathy in particular may be more vulnerable to exercise-induced damage. Additionally, as cardiac autonomic neuropathy is an independent risk indicator for the occurrence of silent myocardial infarction and cardiovascular mortality, individuals with this illness should be screened and maybe put through an exercise stress test prior to starting an exercise program. Once a person is approved to exercise, the best method for recommending exercise intensity is the heart rate reserve approach, which monitors the maximal heart rate directly (Zilliox and Russell 2019).

19.7.1 Risk of Falls and Injuries

One of the primary issues linked to impaired autonomic function is an increase in falls. People with type 2 diabetes are more likely to fall or be at increased risk of falling, especially if they are over 65. This disorder may lead to increased morbidity and death. Therefore, it's crucial to assess these people's fall risk before beginning an exercise program (Hansen et al. 2013).

According to a research, those who experience orthostatic hypotension and fall frequently are more likely to fall. Those with a history of falls had a median symptom score of 5 on the Survey of Autonomic Symptoms, whereas those without a history had a median score of 3.5. The initial group's median Total Impact Score for fallers was 14, whereas the subsequent groups were 9, suggesting a correlation between dysautonomia symptoms and fall rates (Zilliox and Russell 2019).

19.7.2 Monitoring Blood Glucose Levels

Hypoglycaemia is a more common adverse effect after exercise training in individuals with type 2 diabetes. There are several risk factors and circumstances to consider while attempting to optimise blood glucose levels during exercise in people with type 2 diabetes. Numerous factors, such as the kind of medicine used, the amount of exercise and muscle mass, the degree of physical fitness, the blood glucose level before to exercise, nutritional condition, and the presence of counter regulating hormones, might affect the risk of hypoglycaemia. The risk is increased by sulfonylurea use and exogenous insulin treatment, as well as by increased exercise volume or total muscle activation. The danger may also be raised by training under fasting conditions, nutritional state, or glycogen depletion. Monosaccharides should be kept on hand for these individuals, who should eat them when hypoglycaemia symptoms appear. To prevent hypoglycaemia, patients with type 2 diabetes on exogenous insulin treatment should modify their insulin dosage prior to physical activity. If recurrent hypoglycaemia episodes do occur during training or in normal life, the exercise training plan, glucose intake, and medication should be modified in collaboration with a physician (Hansen et al. 2013).

Hydration is crucial when blood glucose levels rise beyond 16.7 mml/L when exercising. Before beginning their normal exercise routine again, people should speak with their healthcare physician or a diabetes expert if their blood glucose values are continuously elevated. A person with type 1 diabetes should be screened for ketones if their levels of blood sugar are greater than 300 mg/dl (16.7 mml/L). If so, the patient should follow the increased ketones care plan, drink, and put off activities. However, there is no need to postpone physical activity because of hyperglycaemia alone if the patient is feeling well and there are either no or very few blood or urine ketones (Colberg 2013).

19.7.3 Foot Care and Injury Prevention

Neuropathic ulcers and consequent skin disintegration have been clearly linked to too much physical pressure on the insensitive feet of DPN patients. High amounts of localised stress, which are frequently experienced when walking, have been linked to skin disintegration. Since plantar pressure (stress) is highest behind the metatarsal heads, this typically happens there. The understanding that too much stress may cause skin degradation was accompanied by the insight that protecting the skin from stress might speed up the healing process. When the ulcers are unloaded using protective walking boots or full contact casting, the majority of neuropathic plantar ulcers can heal in 6–8 weeks, per certain randomised controlled trials (Kluding et al. 2017). When it comes to foot injuries, peripheral neuropathy is an issue. Moderate weight-bearing activity is safe for people without acute ulcers. It is important to pay close attention to proper footwear selection and any indications of foot harm. A balance assessment is necessary because of the elevated risk of falls and potential compromising of balance. To ascertain the patient's risk, think about consulting a podiatrist.

19.8 The Role of Multidisciplinary Care in Exercise Management

Today's healthcare system is primarily built to address individual illnesses rather than provide complete, patient-centred care, which can result in inconsistent and often fragmented treatment. Treatment dissatisfaction, worse HRQOL, and poor treatment results are caused by incoherent patient pathways. Consequently, there is a dearth of contact between patients and healthcare professionals (HCP). Optimising prevention, management, and healthcare reconfiguration is required to achieve a good balance between disease-oriented specialist treatment and patient-centred generalist care. Maintaining an exercise routine in people with type 2 diabetes is crucial since it is associated with long-term cardiovascular health and decreased mortality. Patients receiving care from primary care doctors, physiotherapists, and nursing professionals who specialise in diabetes education are crucial in helping patients understand the value of an exercise program as a therapeutic choice for managing their condition. Simple behavioural counselling during normal clinic visits has been shown in studies to have positive effects on improving compliance.

The rigorous nature of the programs necessitating additional visits for specific sessions makes it tough to stick to exercise routines for longer than 3 months. The rate of compliance decreased from 80% for 6 weeks to less than 50% for 3 months in a 10-year trial including 255 diabetic individuals participating in a diabetes education program that prioritised exercise. One year later, the compliance rate had fallen even further, to less than 20%. Maximising compliance will be facilitated by a coordinated interprofessional strategy involving educators and clinicians (Borhade and Singh 2022).

19.8.1 Role of Physical Therapists and Rehabilitation Specialists

Physical therapists ought to be the primary healthcare practitioners in the multidisciplinary team when it comes to managing and preventing diabetes. When it comes to helping patients meet their physical activity goals, first preference should go to physical therapists whether they are at chance for developing diabetes or have been tested for it. A crucial element of diabetes therapy should be a physical therapist-led intervention that includes recommendations for safe physical exercise. Merely 2% of physical therapy referrals are related to managing diabetes. This could indicate that medical providers have not acknowledged diabetes as a contributing factor to numerous diagnosed conditions that physical therapists frequently treat and examine, such as frozen shoulder, myocardial infarction, stroke, and carpal tunnel syndrome (Harris-Hayes et al. 2020). Physical therapists can help patients put into practice a successful approach for managing their lifestyle. During the interview to get the pertinent medical history, find out if the diabetic patient is following a comprehensive lifestyle management program. As part of lifestyle management, physical therapists evaluate their patients' current level of exercise to create the most appropriate and secure regimen. Within the current healthcare system, there are several chances to collaborate and form connections with physicians, such as endocrinologists, diabetes educators, and clinicians. Physical therapists' abilities and knowledge may also be advantageous to outreach initiatives and community partners. They are qualified to safely and effectively oversee exercise and suggest physical activity to patients with complex medical conditions.

19.9 Case Studies and Clinical Evidence

The body benefits from exercise in a variety of ways. Importantly, it can assist with insulin sensitivity, glucose management, obesity, and dyslipidaemia—three of the primary causes for diabetic neuropathy. Exercise also lowers oxidation of fat and protein, enhances end organ perfusion, improves hypertension, prevents adipocyte generation of adipokines and free fatty acids, and lessens humoral inflammation. The following describes the specifics of exercise trials on diabetic neuropathy:

In a study by Smith et al. (2006), thirty-two patients with impaired glucose tolerance (IGT) and neuropathy received a one-year intervention consisting of standard care, including diet and exercise advice. Changes in intraepidermal nerve fibre density (IENFD) in the proximal thigh and distal leg were assessed in this research. The distal IENFD exhibited a little improvement, while the proximal IENFD showed a more notable improvement. The subjects also had substantial reductions in their total serum cholesterol, BMI, and 2-h post-load hyperglycaemia; however, their pain levels only indicated a non-significant improvement (Smith et al. 2006).

In a 2012 research by Kluding et al., 17 individuals with diabetic neuropathy participated in a 10-week aerobic and strengthening exercise program. The Michigan Neuropathy Screening Instrument (MNSI) revealed notable reductions in neuropathic symptoms and discomfort, whereas proximal skin biopsies revealed an increase in intraepidermal nerve fibre branching. Despite the significant improvement in HbA1c, no appreciable changes were seen in the NCS, QST, BMI, or waist circumference (Kluding et al. 2012).

In a research conducted in 2014 by Singleton et al., 100 patients with type 2 diabetes were split into two groups: 60 got weekly exercise and nutritional therapy, and 40 received lifestyle counselling. In comparison to the control group, the exercise group had substantially greater IENFD (1.5 ± 3.6 fibres/mm) and higher IENFD in the proximal thigh. HDL levels significantly improved in the exercise group as well, although neither the NCS nor the clinical neuropathy ratings (UENS) changed significantly (Singleton et al. 2014).

Another research examined the potential and initial effects of a 12-week foot-ankle fitness regimen on clinical, practical, and biomechanical effects in people with DPN. Thirty DPN patients were randomly selected to either regular therapy (control) or normal care plus supervised exercise (test sample). While the intervention group showed notable increases in toe strength, walking contact time, and DPN symptoms, there were noticeably higher heel maximum forces and strength in the control group. The workout program was feasible given a motivated and contented populace and a fair recruiting rate. When compared to usual treatment, the intervention showed a number of positive initial advantages over time (Monteiro et al. 2020).

In a follow-up study, the effects of Tai Chi exercise on glucose control, neuropathy scores, balance, and quality of life in people with type 2 diabetes and neuropathy were examined. The participants had been diagnosed with type 2 diabetes for over 12 years, with a mean age of 64. The Tai Chi group's (n = 20) participants' status was considerably better than that of their control group's (n = 19) in terms of overall symptom ratings, glucose management, balance, and quality of life (Ahn and Song 2012).

Whole Body Vibration (WBV) is a novel technique that has been tested by another group of researchers and may be useful in relieving chronic pain related to diabetic peripheral neuropathy (DPN). According to preliminary studies, WBV may be useful in managing DPN's chronic pain symptoms. The treatment plan consisted of four 12-min sessions (four bouts of 3 min each) spread out over 4 weeks, with at least 1 day off in between. The members of the control group employed a fake vibration software and were blinded to the therapy for 2 weeks. WBV successfully reduces DPN-related pain during a 2–4 week timeframe. It became an initial study to demonstrate this using a simulated vibration control. Additionally, they noted less discomfort even after the day of treatment, which may indicate that WBV therapy has long-term effects (Kessler et al. 2020).

19.10 Conclusion

The area of biological research on exercise as a therapeutic intervention is expanding quickly. But more study is desperately needed to understand how exercise affects sensory nerve problems. Our limited understanding of the routes and molecular alterations that lead to the advantages of exercise impedes the medical community's attempts to employ it as a tool for patient treatment. Exercise is crucial for managing diabetic neuropathy since it improves nerve function, mobility, balance, and quality of life while lowering pain. Resistance and aerobic exercise enhance nerve conduction velocity, reduce symptoms of neuropathy, and improve glucose control. For the management of pain, high-intensity interval training, or HIIT, is very helpful. Balance and proprioception exercises increase stability and lower the risk of falls. As a result, future study should examine the therapeutic context of exercise and academics and doctors should keep investigating its advantages. Along with regular exercise, people should be encouraged to cut back on their daily idleness and switch between sitting and inactivity often.

References

Aghdam AM, Shahabi P, Karimi-Sales E, Ghiasi R, Sadigh-Eteghad S, Mahmoudi J, Alipour MR. Swimming exercise induced reversed expression of miR-96 and its target gene NaV1. 3 in diabetic peripheral neuropathy in rats. Chin J Physiol. 2018;61(2):124–9.

Ahn S, Song R. Effects of tai chi exercise on glucose control, neuropathy scores, balance, and quality of life in patients with type 2 diabetes and neuropathy. J Altern Complement Med. 2012;18(12):1172–8.

Albers JW, Herman WH, Pop-Busui R, Martin CL, Cleary P, Waberski B, Diabetes Control and Complications Trial (DCCT)/Epidemiology of Diabetes Intervention and Complications (EDIC) Research Group. Subclinical neuropathy among Diabetes Control and Complications Trial participants without diagnosable neuropathy at trial completion: possible predictors of incident neuropathy? Diabetes Care. 2007;30(10):2613–8.

Alghafri T, Alharthi SM, Al Farsi YM, Bannerman E, Craigie AM, Anderson AS. Perceived barriers to leisure time physical activity in adults with type 2 diabetes attending primary healthcare in Oman: a cross-sectional survey. BMJ Open. 2017;7(11):e016946.

Alobaid AM, Zulyniak MA, Ajjan RA, Brož J, Hopkins M, Campbell MD. Barriers to exercise in adults with type 1 diabetes and insulin resistance. Can J Diabetes. 2023;47(6):503–8.

American Diabetes Association. Standards of medical care in diabetes—2008. Diabetes Care. 2008;31(Supplement_1):S12–54.

Balducci S, Iacobellis G, Parisi L, Di Biase N, Calandriello E, Leonetti F, Fallucca F. Exercise training can modify the natural history of diabetic peripheral neuropathy. J Diabetes Complicat. 2006;20(4):216–23.

Borhade MB, Singh S. Diabetes and exercise. In: StatPearls [Internet]. StatPearls Publishing; 2022.

Boulé NG, Kenny GP, Haddad E, Wells GA, Sigal RJ. Meta-analysis of the effect of structured exercise training on cardiorespiratory fitness in Type 2 diabetes mellitus. Diabetologia. 2003;46:1071–81.

Chandra P, Shahjad P, Rastogi V, Mittal P, Verma A, Sachan N, Prabahar EA, Gautam MK. Progresses in wound healing: integrating nutrition, physical therapy, traditional and alternative medicine, and novel technologies. Curr Diabetes Rev. 2025;22:68.

Colberg SR. Exercise and diabetes: a clinician's guide to prescribing physical activity. American Diabetes Association; 2013.

Colberg SR, Sigal RJ, Yardley JE, Riddell MC, Dunstan DW, Dempsey PC, et al. Physical activity/exercise and diabetes: a position statement of the American Diabetes Association. Diabetes Care. 2016;39(11):2065.

Cooper MA, Kluding PM, Wright DE. Emerging relationships between exercise, sensory nerves, and neuropathic pain. Front Neurosci. 2016;10:372.

Dixit S, Maiya AG, Shastry BA. Effect of aerobic exercise on peripheral nerve functions of population with diabetic peripheral neuropathy in type 2 diabetes: a single blind, parallel group randomized controlled trial. J Diabetes Complicat. 2014;28(3):332–9.

Dobson JL, McMillan J, Li L. Benefits of exercise intervention in reducing neuropathic pain. Front Cell Neurosci. 2014;8:102.

Gautam MK, Gupta A, Vijaykumar M, Rao CV, Goel RK. Studies on the hypoglycemic effects of Murraya paniculata Linn. extract on alloxan-induced oxidative stress in diabetic and non-diabetic models. Asian Pacific J Trop Dis. 2012;2(S1):S186–91.

Gautam MK, Purohit V, Agarwal M, Singh A, Goel RK. In-vivo healing potential of Aegle marmelos in excision, incision and dead space wound models. Sci World J. 2014;2014:740107.
Gautam MK, Gangwar M, Singh SK, Goel RK. Effects of Azadirachta indica on vascular endothelial growth factor and cytokines in diabetic deep wound. Planta Med. 2015;81(8):713–21.
Gordon BA, Benson AC, Bird SR, Fraser SF. Resistance training improves metabolic health in type 2 diabetes: a systematic review. Diabetes Res Clin Pract. 2009;83(2):157–75.
Hansen D, Peeters S, Zwaenepoel B, Verleyen D, Wittebrood C, Timmerman N, Schotte M. Exercise assessment and prescription in patients with type 2 diabetes in the private and home care setting: clinical recommendations from AXXON (Belgian Physical Therapy Association). Phys Ther. 2013;93(5):597–610.
Harris-Hayes M, Schootman M, Schootman JC, Hastings MK. The role of physical therapists in fighting the type 2 diabetes epidemic. J Orthop Sports Phys Ther. 2020;50(1):5–16.
Heiss V, Petosa R. Correlates of physical activity among adults with type 2 diabetes: a systematic literature review. Am J Health Educ. 2014;45(5):278–87.
Jelleyman C, Yates T, O'Donovan G, Gray LJ, King JA, Khunti K, Davies MJ. The effects of high-intensity interval training on glucose regulation and insulin resistance: a meta-analysis. Obes Rev. 2015;16(11):942–61.
Kessler NJ, Lockard MM, Fischer J. Whole body vibration improves symptoms of diabetic peripheral neuropathy. J Bodyw Mov Ther. 2020;24(2):1–3.
Kluding PM, Pasnoor M, Singh R, Jernigan S, Farmer K, Rucker J, et al. The effect of exercise on neuropathic symptoms, nerve function, and cutaneous innervation in people with diabetic peripheral neuropathy. J Diabetes Complicat. 2012;26(5):424–9.
Kluding PM, Bareiss SK, Hastings M, Marcus RL, Sinacore DR, Mueller MJ. Physical training and activity in people with diabetic peripheral neuropathy: paradigm shift. Phys Ther. 2017;97(1):31–43.
Lidegaard LP, Schwennesen N, Willaing I, Færch K. Barriers to and motivators for physical activity among people with Type 2 diabetes: patients' perspectives. Diabet Med. 2016;33(12):1677–85.
Luo J, Zhu HQ, Gou B, Zheng YL. Mechanisms of exercise for diabetic neuropathic pain. Front Aging Neurosci. 2022;14:975453.
Martin CG, Pomares ML, Muratore CM, Avila PJ, Apoloni SB, Rodríguez M, Gonzalez CD. Level of physical activity and barriers to exercise in adults with type 2 diabetes. AIMS Public Health. 2021;8(2):229.
Mogensen CE. Nephropathy: early in handbook of exercise in diabetes. Alexandria: ADA; 2002.
Mondal NS, Islam A, Gautam MK, Mondal S, Ahmed M, Jamir SI, Khongwir LD, Mahajon B. Unraveling the mechanisms of diabetic wounds: insights into pathogenesis and advanced treatment strategies. Curr Diabetes Rev. 2025;22:113.
Monteiro RL, Ferreira JS, Silva ÉQ, Donini A, Cruvinel-Júnior RH, Verissímo JL, et al. Feasibility and preliminary efficacy of a foot-ankle exercise program aiming to improve foot-ankle functionality and gait biomechanics in people with diabetic neuropathy: a randomized controlled trial. Sensors. 2020;20(18):5129.
Shankarappa SA, Piedras-Rentería ES, Stubbs EB Jr. Forced-exercise delays neuropathic pain in experimental diabetes: effects on voltage-activated calcium channels. J Neurochem. 2011;118(2):224–36.
Singleton JR, Marcus RL, Jackson JE, Lessard K, Graham TE, Smith AG. Exercise increases cutaneous nerve density in diabetic patients without neuropathy. Ann Clin Transl Neurol. 2014;1(10):844–9.
Singleton JR, Marcus RL, Lessard MK, Jackson JE, Smith AG. Supervised exercise improves cutaneous reinnervation capacity in metabolic syndrome patients. Ann Neurol. 2015;77(1):146–53.
Smith AG, Russell J, Feldman EL, Goldstein J, Peltier A, Smith S, et al. Lifestyle intervention for pre-diabetic neuropathy. Diabetes Care. 2006;29(6):1294–9.
Tuell DS, Los EA, Ford GA, Stone WL. The role of natural antioxidant products that optimize redox status in the prevention and management of type 2 diabetes. Antioxidants. 2023;12(6):1139.
Wc K. Reduction in the incidence of type 2 diabetes with lifestyle intervention or metformin. N Engl J Med. 2002;346:393–403.
Zilliox LA, Russell JW. Physical activity and dietary interventions in diabetic neuropathy: a systematic review. Clin Auton Res. 2019;29:443–55.

Mental Health Strategies for Managing Diabetic Neuropathy

20

Subhashree Pramanik, Manoj Kumar Yadav, Divyang Patel, Abhishek Shrivastava, Nidhi Saxena, and Shalini Singh Negi

20.1 Introduction

Diabetes has a detrimental impact on several organ systems, increases the risk of chronic complications, and raises the death rate. Diabetic neuropathies (DNs) are among the most prevalent chronic consequences, affecting up to 50% of individuals with diabetes. The most researched disorders related to DN include autonomic neuropathies and distal symmetric polyneuropathy (DSPN). The most prevalent type of DN is called DSPN, and it is brought on by injury to both major and tiny peripheral nerve fibres. The symptoms and indications of DN, whether they occur alone or in combination, have been linked to increased mortality rates, a higher risk of catastrophic cardiovascular events, a worse quality of life (QoL), and reduced psychological well-being. The biological risk factors associated with diabetic neuropathy (DN) are well-established and include of height, cardiometabolic variables (obesity, dyslipidaemia, hypertension, and smoking), and glycaemic exposure (HbA1c, age,

S. Pramanik (✉)
Centurion University of Technology and Management, Gopalpur, Odisha, India

M. K. Yadav
Institute of Pharmacy & Paramedical Sciences, Dr Bhimrao Ambedkar University, Agra, UP, India

D. Patel
Institute of Pharmaceutical Sciences, Faculty of Pharmacy, Parul University, Vadodara, Gujarat, India

A. Shrivastava · N. Saxena
Sagar Institute of Research and Technology Pharmacy (SIRT PHARMACY), Sanjeev Agrawal Global Educational University, Bhopal, MP, India

S. S. Negi
Seth Vishambhar Nath Institute of Engineering and Technology (SVNIET), Barabanki, UP, India

M. K. Gautam et al. (eds.), *Antioxidants and Functional Foods for Diabetic Neuropathy*, https://doi.org/10.1007/978-981-95-9326-2_20

and duration of diabetes). The contribution of socioeconomic determinants of health to the development of DN is, however, significantly less understood. Factors pertaining to mental health are also significant (Pouwer et al. 2024).

20.1.1 The Connection Between Diabetic Neuropathy and Mental Health

Mental illnesses associated with diabetes patients include depression, which is the most frequent mental condition characterised by poor mood. Depression and poor mood are characterised by symptoms of melancholy, anxiety, worry, low self-esteem, exhaustion, rage, frustration, thoughts of suicide, changes in activity and sleep patterns that last longer than 2 weeks, and low self-esteem. Four or more of the previously listed indicators and symptoms are present in depressed people. Depression is categorised as a significant and extremely prevalent medical disorder. Depression is more likely to worsen the symptoms of type 2 diabetes mellitus (T2D), and diabetes can exacerbate the symptoms of depression. This is because there is a strong correlation between depression and unhealthy lifestyle choices, such as smoking, eating a lot of food, and not exercising, which all lead to obesity and impaired glucose regulation (Alghafri et al. 2020). Research has indicated that individuals suffering from diabetic neuropathy have an increased probability of experiencing sadness, anxiety, and cognitive impairment. Damage to nerves alters the body in ways that are related to these problems. Important impulses to the brain are interfered with when the nerves that carry messages throughout the body deteriorate, resulting in an imbalance of neurotransmitters that impacts mood.

Additionally, the chronic pain and suffering associated with the illness can cause mental anguish. According to research, neuropathy harms the autonomic nervous system and activates the sympathetic nervous system, which can lead to anxiety and depression symptoms in diabetics. Furthermore, the psychological anguish associated with neuropathy might be attributed to worse quality of life and social isolation, both of which are typical in people with this illness. In addition, having difficulty sleeping due to pain might lead to increased emotional turmoil (Vileikyte et al. 2005).

20.1.2 Importance of Mental Health in Managing Chronic Pain Conditions

According to meta-analytic evaluations, people with diabetes and depression are 38% more likely to die from cardiovascular disease, all-cause mortality, microvascular illnesses, and macrovascular complications (Beran et al. 2022). In those with diabetic foot issues, the idea that depression is linked to poor future health outcomes has also been validated. Depression was linked to a 68% increased risk for foot ulcers according to the Hospital Anxiety and Depression Scale (HADS) standard unit in people without a history of a foot ulcer diagnosis, according to a systematic evaluation of 15 studies including more than 12,000 participants (Westby et al.

2020). Not only is depression linked to a much lower quality of life, but it is also a major factor in the poor health outcomes that occur with diabetes.

Based on research, pain intensity and related symptoms are significantly reduced when mental health issues are treated in addition to pain. There is a strong correlation between pain and its severity and despair and anxiety. Pain also seems to emerge more frequently in those with moderate-to-severe depression symptoms. As evidenced by earlier research in PDN, anxiety disorders were found to be equally prevalent as depression, Collins et al.'s suggestion that anxiety disorders (32%) were more common than depression (22%) (Twiddy et al. 2021). Furthermore, research on particular pain situations indicates that poorer sleep quality (insomnia, waking in the middle of the night, and restless sleep) is positively correlated with increased pain intensity. Individuals who experience both depression and chronic pain together also frequently report higher pain levels, more pain-related interference, lower activity levels, and higher degrees of impairment. Thus, before beginning psychological therapies for pain, it is imperative that basic depression or anxiety disorders be addressed (Gore et al. 2005).

20.2 Psychological Impact of Diabetic Neuropathy

Painful DN (PDN), which affects up to 30% of all DN patients, is perhaps the most incapacitating neuropathic symptom. Since of its detrimental impacts on quality of life, PDN is still difficult to treat since doctors mostly rely on pharmaceutical interventions, which have significant dose-related adverse effects that restrict their efficacy. As a result, the proportion of PDN patients who get clinically meaningful pain reduction is still too low. As a result, insufficient pain management in individuals with post-stroke neuropathy (PDN) leads to extreme sleep loss, anxiety, depressive symptoms, and compromised cognitive and motor abilities. Furthermore, insufficient pain management results in lower productivity at work (Pouwer et al. 2024).

People who engage in pain catastrophising use a variety of negative cognitive processes, such as rumination, amplification, and helplessness, to process their pain experiences. Patients with PDN may experience mental discomfort as a result of pain catastrophising, which has been connected to increased disability and a decreased quality of life because of a perceived decrease in physical activity. Selvarajah and colleagues found that a key contributing factor to elevated anxiety and depression levels was catastrophic thinking (Selvarajah et al. 2014). However, Kioskli et al. discovered that a cross-sectional sample of PDN patients showed a correlation between increased pain acceptance and a decrease in pain intensity, pain distress, functional impairment, and the degree and impact of depression (Kioskli et al. 2019).

Researchers who conducted a comprehensive analysis concluded that among people with PDN, the prevalence of depression varied from 13.6% to 50.6% and the prevalence of anxiety from 7.8% to 60%. It is commonly recognised that anxiety and depression frequently coexist: 31% to 26% of PDN patients reported having both anxiety and depression, while 42% to 44% of PDN patients reported having sleep difficulties. The results of a Dutch study revealed a number of particular worries that individuals with PDN who are in pain face, including fear of pain worsening, physical harm, falling, losing their identity, and receiving unfavourable feedback from others (Pouwer et al. 2024).

The most studied variables and those that are consistently correlated with the degree of pain include quality of life, sadness, anxiety, and sleep quality, according to recent systematic investigations of psychosocial factors associated with PDN. The idea that psychological problems are common and that many of these traits may make pain perception worse is therefore well supported by the available data.

20.3 The Biopsychosocial Model in Diabetic Neuropathy

For clinical practice and research, psychologists, nurses, doctors, and social workers can use Engel's biopsychosocial model of health and sickness. Official endorsement of Engel's concept has come from the American Psychiatric Association and the American Board of Psychiatry and Neurology. Every illness, including depression, that is brought on by psychological (lifestyle), biological (genetic or physiological predispositions), or social (socioeconomic status, and social support) factors is covered by Engel's model. The relationship between these factors and how the patient's body and mind are affected is shown by the model (Habtewold et al. 2016).

According to the biopsychosocial concept of pain, biological variables have a significant role in how aberrant sensations caused by injured nerves are interpreted and experienced. Social and environmental factors also have an impact on these assessments and behavioural reactions. Because chronic pain is an illness rather than a symptom, managing it necessitates using a biopsychosocial approach and acknowledging that all contributing variables need to be taken into consideration. Personalised patient care is best achieved through a multimodal/multidisciplinary team approach that includes psychologists, physical therapists, nurses, podiatrists, and pain specialists. Internal medicine specialists, endocrinologists, neurologists, pain specialists and general practitioners are frequently involved in initial consultations (Young-Hyman et al. 2017).

20.3.1 Interaction Between Biological, Psychological, and Social Factors

Several important aspects are taken into consideration in order to offer complete treatment in the biopsychosocial assessment of patients with painful diabetic neuropathy. Comprehending the type and cause of the pain, evaluating the physical prognosis, and figuring out if the pain is medically appropriate or makes biological sense are all considered biomedical components. It also entails assessing the necessity for additional medical or surgical procedures and taking any drug adverse effects into account.

The patient's degree of physical deconditioning and the recommended amount of physical exercise are the main topics of physiotherapy. It's critical to evaluate the patient's pacing of activities and look for signs of excessively limited mobility brought on by anxiety avoidance.

Aspects of occupational therapy look at how pain affects every area of the workplace, including daily schedules and organisational systems. Determine whatever changes and assistance might help the person perform better. You should also consider any unresolved concerns that could need a follow-up medical examination.

The psychological components of pain management include how pain affects mood and activity, pain beliefs and actions such avoidance and anxiety, and iatrogenic discomfort. In addition, it is critical to evaluate the person's locus of control, motivation for change, and past mental health history. Together with treatment adherence for other comorbidities, goals and value-led activities are essential for controlling the psychological component (Twiddy et al. 2021).

By incorporating these areas into the assessment, healthcare providers may provide a comprehensive approach to addressing the physical, emotional, and functional effects of painful diabetic neuropathy.

20.3.2 Application of the Model to Diabetic Neuropathy Management

It may be feasible to incorporate the knowledge acquired in other chronic pain populations into the field of PDN, as people with PDN often share the comorbidities of despair and fear (consequently and disability) with these other groups. The Fear-Avoidance paradigm is a well-known and widely utilised biopsychosocial paradigm for explaining chronic pain. According to this concept, scared patients predict that particular actions, activities, or events would injure their bodies by producing (re) injury. Negative illogical (catastrophising) ideas regarding the pain emerge. The vicious cycle of persistent, incapacitating pain is exacerbated by dread, which encourages avoidance behaviour. Avoidance behaviour can have serious health effects, including depression, physical deconditioning brought on by inactivity, and disability. It's interesting to note that our cross-sectional study also identified this mechanism: PDN was linked to catastrophic thinking, and we hypothesise that catastrophic thinking caused low QOL, increased disability, and a reported drop in physical activity. Strangely, catastrophising about pain was linked to the subjective perception of losing physical activity because of the pain, but not to the assessed degree of real activity.

As a result, it is suggested that PDN patients' overall quality of life (QOL) may be optimally improved when concomitant anxiety and negative emotions are appropriately identified, diagnosed, and treated. Therefore, rather than concentrating just on reducing physical pain, a multidisciplinary therapy strategy that addresses these psychological components is advised for the treatment of PDN (Geelen et al. 2017).

20.4 Cognitive Behavioural Therapy (CBT) for Neuropathic Pain

Psychotherapy called cognitive-behavioural therapy (CBT) has lately been used with chronic pain sufferers. Numerous studies have shown that CBT reduced pain and associated issues, whether it was used in isolation or in conjunction with medical care. Both individual cognitions and tissue damage may cause pain, and the core idea of cognitive behavioural therapy (CBT) is that unhelpful thinking contributes to the persistence of behavioural issues and emotional suffering (Lim et al. 2018). In order to address depression, Beck created CBT initially. This form of treatment is time-limited, problem-focused, and regimented. A third wave of cognitive behavioural therapy has emerged as a result of scientific advancements. It contains a variety of approaches, including metacognitive therapy (MCT), acceptance and commitment therapy (ACT), and mindfulness-based cognitive therapy (MBCT). A group of therapies known as cognitive behavioural therapy (CBT) alters maladaptive thoughts in order to alter maladaptive behaviours and emotional suffering. Pain management-related grey matter alterations in the brain and/or functional connections within these areas may be linked to the effects of cognitive behavioural therapy (CBT) on chronic pain (Li et al. 2023).

20.4.1 Techniques to Manage Pain-Related Thoughts and Behaviours

The fundamental tenet of cognitive behavioural therapy (CBT) is that your mental processes influence your feelings, which can then influence your actions. CBT, for example, emphasises how unfavourable ideas can result in unfavourable emotions and behaviours. Rethinking your ideas, however, might result in happier emotions and constructive actions.

Among the methods that are most frequently employed in conjunction with CBT are the following approaches:

Cognitive restructuring or reframing: Examining negative thinking patterns closely is necessary for this. Individuals undergoing treatment are assisted in identifying automatic negative thinking patterns, including "This pain will never go away" or "I can't do anything because of my pain". In order to lessen emotional suffering and enhance coping, the therapist works with patients to reframe these ideas into more realistic, balanced viewpoints.

Guided discovery: The therapist will become familiar with your perspective via guided discovery. Subsequently, they will pose queries intended to refute your convictions and extend your perspective. During the process, you'll develop the ability to view things from several angles, particularly ones you might not have previously thought about.

Activity scheduling and behaviour activation: This is done in an effort to promote meaningful and enjoyable activities even in the face of pain. Patients are urged to progressively resume enjoyable activities that they may have postponed because of discomfort. This method assists in interrupting the pattern of avoidance and inactivity associated with pain, which can exacerbate symptoms on the physical and psychological levels.

Relaxation and stress reduction techniques: This method aids in lessening the psychological and physiological impacts of pain. Patients are taught methods such as progressive muscle relaxation, deep breathing, and guided visualisation to help them cope with the tension and worry that comes with pain. By soothing the neurological system, regular practice helps reduce the overall impression of pain.

Pacing and Activity Management: This method assists patients in controlling their degree of physical activity without making their discomfort worse. Patients are instructed to take regular pauses and to divide chores into smaller, more manageable chunks. Pacing gives patients a sense of control over their daily routines and helps prevent overexertion, which can lead to flare-ups in pain.

Distraction Techniques: This method tries to divert attention from pain perception. In this, patients are urged to think or do things that take their minds off of their pain. This might involve doing creative work, listening to music, or visualising happy memories in your mind (Eccleston et al. 2013; Ehde et al. 2014).

20.4.2 Evidence of Effectiveness in Diabetic Neuropathy Patients

According to a chronic pain study, cognitive behavioural therapy (CBT) significantly improved patients' clinical outcomes, including pain catastrophising, the chronic pain self-efficacy scale, and the pain survey. It also decreased the intrinsic functional connectivity (iFC) between the amygdala/periaqueductal grey and the anterior default mode network, while increasing it between the right secondary somatosensory cortex and the basal ganglia network in comparison to pain education. Furthermore, a research discovered that patients increased grey matter volumes in the dorsolateral prefrontal and sensorimotor cortexes following CBT intervention were linked to a reduction in pain catastrophising. Neuropathic pain brought on by diabetic peripheral neuropathy, HIV-related peripheral neuropathy, and spinal cord injuries has been effectively treated using cognitive behavioural therapy. These RCTs have demonstrated that Cognitive Behavioural Therapy (CBT) is practical and helpful for those suffering from neuropathic pain (Li et al. 2023).

20.5 Mindfulness-Based Stress Reduction (MBSR)

20.5.1 Concept of Mindfulness and Its Benefits in Chronic Pain

In order to help patients with chronic diseases who were not responding to recommended medical therapies, Kabat-Zinn developed mindfulness-based stress reduction (MBSR), a group psychosocial intervention. By practicing mindfulness, individuals can learn to adopt a metacognitive perspective on their experiences, which enhances their capacity to perceive stressors (such pain) objectively and improves their self-regulation. Instead of catastrophic ruminations or instinctive reactions, this makes it easier to choose more adaptive answers. It is possible to reduce the threat and perception of damage that come with pain, as well as its intensity, by adopting a more welcoming and open mindset towards this difficult experience (Nathan et al. 2017).

The fundamental mechanism by which MBSR is thought to impact health is via its impact on stress. Additionally, a number of studies have shown that mindfulness has a beneficial impact on stress indicators such serum cortisol. These structural and functional alterations may be facilitated by cortisol and brain-derived neurotrophic factor. The hippocampus, amygdala, prefrontal cortex, and other brain regions are susceptible to structural and functional changes brought on by stress. Reduced parasympathetic tone, decreased stress hormone release, and decreased sympathoadrenal activity are all linked to MBSR. Additionally, there's a decrease in the release of inflammatory indicators including interleukin-6, C-reactive protein, and tumour necrosis factor α (Priya and Kalra 2018).

20.5.2 Techniques for Integrating Mindfulness Practices

By including mindfulness-based stress reduction (MBSR) practices into diabetic neuropathy treatment, a comprehensive strategy for managing the illness's psychological and physical symptoms may be implemented. The following are a few methods for managing pain that use mindfulness practices:

Mindful Breathing Exercises: This method aids in lowering tension and raising body awareness. Through this, the patients learnt how to focus their attention on their breathing while deep diaphragmatic breathing. This can assist focus attention on the breath instead of the pain during neuropathic pain flare-ups.

Mindful Movement (Yoga or Tai Chi): This method encourages moderate exercise, which can enhance mental clarity, muscular elasticity, and circulation while reducing discomfort and increasing range of motion. It combines low-impact exercises like Tai Chi and yoga with conscious breathing and body awareness to improve the mind-body connection. Patients can begin with mild, adaptable motions and advance according to their own schedule. These procedures have to be modified in light of their physical constraints.

Body Scan Meditation: It eases tight muscles and raises awareness of bodily sensations, which may lessen the perception of pain. Using a mental scan from head to toe, the patient is guided through the method, identifying any pain, tension, or discomfort without passing judgement. Using relaxation techniques, the goal is to become aware of these feelings and relieve tension. This may be performed while seated or laying down, and it may aid patients in comprehending the type and location of their neuropathic pain.

CBT conjunction with Mindfulness: By addressing unfavourable thinking patterns that intensify pain perception, this therapy assists patients in managing the psychological components of neuropathy. Patients can become aware of habitual negative ideas about pain and replace them with more adaptable, positive thought patterns by combining mindfulness practices with cognitive behavioural therapy (CBT) (Bawa et al. 2015; Zeidan et al. 2012).

20.6 Clinical Evidence Supporting MBSR for Diabetic Neuropathy

One research found that MBSR helps people develop meditative awareness via practicing self-regulation of attention, being aware of the present moment, and responding to stress rather than reacting to it. Therefore, with attitudes of nonjudgment and acceptance, people with diabetes may feel more in control of their worried mood and more conscious of their current feeling. This may help to promote mindfulness and lower levels of anxiety and sadness. According to the results of a meta-analysis, MBSR had a substantial and clinically meaningful impact on lowering anxiety and depressive symptoms at the post-intervention stage. These reductions persisted over the follow-up phase, which lasted an average of 4.3 months but may range from one to twelve months (Fisher et al. 2023). Another study found that when MBSR was used instead of usual care for PDPN patients whose pharmacotherapy had been optimised in a chronic pain clinic, the patients' function improved, their pain intensity, pain catastrophising, depression, and perceived stress decreased, and their health-related quality of life improved (Nathan et al. 2017).

Rosenzweig et al. conducted a brief, prospective, observational pilot study of 14 patients to compare the effects of MBSR on measures of A1C, blood pressure, body weight, and psychological symptoms, including anxiety, depression, somatization, and general psychological distress. Patients who completed the program saw decreases in depression, anxiety, and general psychological discomfort in addition to a 0.5% drop in A1C and a 6 mmHg drop in mean arterial pressure, according to this uncontrolled experiment (Rosenzweig et al. 2007).

20.7 The Role of Social Support in Mental Health

Social support refers to the resources that others offer to an individual in order to assist them in managing their difficulties in a more efficient manner. According to the Stress-Buffering Model, SS shields people from the negative impacts of stress on their health and wellbeing. According to a number of studies, patients who get greater social support adjust better and experience less mental anguish. Numerous studies have demonstrated that social support can mitigate the negative consequences of diabetes-related anxiety on a range of health outcomes. Social support, for example, has been shown to reduce the association between depression symptoms, glycaemic control, and diabetic distress (Parviniannasab et al. 2024). People's subjective well-being increases and they feel cared for, accepted, and valued when they get social support from others. Consequently, they take a more optimistic view of their chronic illnesses. Decisions about one's health are influenced by friends, family, and spouses in general. This is especially true for those with type 2 diabetes. The psychological comfort and ability to manage health difficulties will increase for those who get help. The ultimate aim of managing diabetes is to maintain adherence to self-care, and integrating social support into therapy for depression may help eliminate psychological barriers to this.

Additionally, a study on the impact of depression among Type 2 Diabetes patients on perceived social support revealed that a supportive environment is a source of social support that aids in the improvement of patients' mental health. Patients' support network members offer basic disease knowledge, self-care or management skills, and problem-solving tactics. They converse with their peers and exchange best practices and experiences in diabetes management. In order to manage stressful events, adhere to their treatment plan during difficult times, and lower their risk of developing stress-related diseases, patients can benefit from social support by learning coping mechanisms and creating structure in their daily routines (Azmiardi et al. 2022).

Social support plays a vital role in the lives of diabetics and other individuals with chronic illnesses in general. The social structure of a person's life as well as the distinct roles that different types of interpersonal interactions play in that person's life are both included in the broad concept of social support. While the functional component of support is divided into two categories: perceived support (people's subjective interpretation of the support they believe to be available to them) and received support (assistance provided by others), the structural component of support is determined by evaluating social integration, which indicates the degree to which the individual is a part of social networks. Three factors that influence health include social support's ability to control thoughts, feelings, and behaviour; to encourage healthy lifestyles that nurture a person's sense of purpose in life; and to support both excellent physical and mental health. The degree to which a patient adjusts psychosocially to a chronic illness is largely determined by the availability of social support. When a person needs help coping with the demanding demands of their illness, they typically go to their larger social network. Long-term therapy can benefit from a patient's view of the availability of social support for his issues, as it gives him a sense of genuine support from those closest to him, even during depressive episodes (Wasserman and Trifonova 2006).

A better work environment may be achieved via the significance of workplace evaluations and informing coworkers and bosses about the situation. Several individuals in the current study reported feeling lonely and socially isolated. The results of this study and those of other researchers, who observed that patients began to fear growing weaker or crippled and being more reliant on friends and family, were shown to have similarities. Similar findings were found by another researcher who proposed that a decrease in the quantity and/or quality of personal interactions a patient had was the primary factor influencing health outcomes. According to our participants, peripheral neuropathy has both physical and psychological side effects. These include anxiety, mood swings, and sleep deprivation. These findings align with the long-term effects of chemotherapy-induced peripheral neuropathy as well as the mood disorders associated with neuropathic pain (Alkandari and Hollywood 2023).

20.7.1 Benefits of Support Networks in Managing Diabetic Neuropathy

Support networks are essential for managing diabetic neuropathy because they provide both practical and emotional assistance. In terms of emotions, they ease feelings of loneliness by enabling people to talk about their struggles and experiences. This helps to lower stress levels. The drive to follow medication regimens, go for routine examinations, and participate in activities that help reduce neuropathic symptoms can all be enhanced by this sense of community. Furthermore, support groups facilitate a knowledge and resource sharing among members who discuss how to manage symptoms, take medicine, and alter their way of life. By increasing health literacy, people are better able to make knowledgeable decisions regarding their treatment (Peyrot and Rubin 2007).

Another advantage of support network is social responsibility, as members may motivate one another to adhere to treatment plans and take up healthy routines, including working out or abstaining from actions that exacerbate neuropathy. These networks also help people create better coping strategies by allowing them to talk about problems and work together to create workable solutions for dealing with pain and emotional difficulties. This cultivates resilience and an optimistic mindset, both of which are critical for managing a chronic illness (Tang et al. 2006).

Furthermore, support groups frequently provide a direct line of communication between patients and medical experts, such as therapists or diabetes educators, providing knowledgeable advice on available treatments and cutting-edge therapies. Lastly, people with chronic diseases like diabetic neuropathy sometimes experience emotions of loneliness and isolation; however, the sense of community found in support networks helps to lessen these symptoms. These networks greatly improve a person's capacity to manage their disease, which improves overall health outcomes and quality of life, by fostering mental well-being and emotional support (Fukunishi et al. 1998).

20.8 Sleep Management Strategies

A higher incidence of sleep problems, particularly obstructive sleep apnoea, poor sleep efficiency, sleep fragmentation, and recurrent nocturnal hypoxia, is seen in patients with type 2 diabetes mellitus who also have difficulties from peripheral neuropathy. Moreover, there is a correlation between sleep disturbances and the risk of death as well as diabetic complications. The risk of death for individuals with type 2 diabetes mellitus rose in a large adult cohort from the United States from 1.84 for a 7-h sleep duration to 2.78 and 3.67 for 5 and 10 h, respectively, when compared to those without the disease. According to similar findings, people with recently diagnosed type 2 diabetes mellitus who have sleep disruptions are more likely to develop cardiovascular disease and die from all causes, according to data from the Korean National Health Insurance Program (Raizada and Madhu 2021).

20.8.1 Relationship Between Neuropathy, Pain, and Sleep Disturbances

One of the most frequent concerns made by people with chronic pain is lack of sleep, and there is a complicated correlation between the two. Pain can make it harder and more disruptive to go asleep and remain asleep, and not getting enough sleep can make pain worse the next day. Moreover, persistent discomfort may make it harder to go back asleep after awakenings and "lighten" sleep. There was a substantial positive correlation between the level of neuropathic pain and the quality of sleep. As a result, the prognosis for pain may be predicted by the quality of one's sleep. It has been established that rats in neuropathic pain paradigm exhibit less slow-wave sleep, or less time spent in the NREM phase of sleep. In mice, pain interference during NREM raises somatosensory and cortico-autonomic arousal, which reduces the effectiveness of sleep. Additionally, Chloe et al. found that chronic sleep deprivation enhanced pain sensitivity more than chronic sleep fragmentation in an animal sleep deprivation model. Chen et al. discovered in their clinical investigation that sleep deprivation can lower a healthy person's pain threshold, which can lead to oxidative stress and hyperalgesia. Thus, in both animal and human research, neuropathic pain has been shown to interfere with sleep, and poor sleep can further modify pain sensitivity (Zhu and Huang 2023).

20.8.2 Cognitive and Behavioural Strategies for Improving Sleep Quality

Improved sleep quality can be achieved with nonpharmacological, cognitive and behavioural techniques, which are commonly combined under the name Cognitive Behavioural Therapy for Insomnia (CBT-I). These techniques assist people in recognising and altering ideas, actions, and routines that keep them from getting enough sleep. Identifying and disputing unfavourable ideas about sleep is a key component of cognitive restructuring, one of several tactics for lowering anxiety associated with sleep. Through advice to go to bed only when tired and get out of bed if awakened throughout the night, stimulus control treatment promotes the idea that the bed is exclusively associated with sleep. Sleep restriction treatment gradually increases the amount of time spent in bed as sleep quality improves, but it is limited to the actual amount of time spent sleeping. Deep breathing, progressive muscle relaxation, and mindfulness are a few relaxation strategies that can assist lower physical tension and stress that prevents you from falling asleep. By motivating people to give up on attempting to fall asleep and instead concentrate on remaining awake, paradoxical intention helps people with their sleep anxiety. Education on good sleep hygiene supports these methods by encouraging sound sleeping practices, and Cognitive Behavioural Therapy for Pain (CBT-P) assists in controlling pain-related emotions and thoughts that interfere with sleep. When combined, these techniques can greatly improve the quality of sleep, particularly for people who are experiencing chronic pain (Murawski et al. 2018).

20.8.3 Sleep Hygiene Tips for Diabetic Neuropathy Patients

A set of guidelines for encouraging sleep and avoiding insomnia has been produced by experts in the area, and these guidelines are frequently applied as a behavioural insomnia remedy (Markwald et al. 2018).

The following are some behavioural suggestions for getting good-quality sleep:

- Adhere to a regular sleep pattern. Even on weekends and holidays, rise at the same hour each day.
- Establish a bedtime that will allow you to sleep for a minimum of seven hours each night.
- Save your bedtime for when you're very tired.
- Create a calming evening ritual.
- Create a calm and serene sleeping environment. Maintain a cool, pleasant temperature in the space.
- Minimise your nighttime exposure to bright light.
- Switch off electronics at least half an hour before going to bed.
- Avoid consuming a heavy meal just before bed. Snack on something light and nutritious if you're hungry at night.
- Work out frequently. Excessive activity right before bedtime might cause sleep disturbances in certain people. Find out what suits you the most.
- Stay away from alcohol right before bed.
- Don't drink caffeine in the late afternoon or evening.

20.9 The Role of Healthcare Professionals in Managing Mental Health for Diabetic Neuropathy Patients

In the event that psychosocial issues are found, a thorough assessment, diagnosis, and treatment should be sought from a licensed behavioural and/or mental health specialist, preferably one with expertise in diabetes. Referral criteria might include positive test results for general work-life stress, diabetic distress, trouble managing diabetes, depression, anxiety, disordered eating, and cognitive impairment. Preventing the onset of a specific issue or decline in metabolic or psychological condition by integrating psychosocial assessment and therapy into routine care is preferred. In order to effectively treat patients with diabetes and address the psychosocial components of the disease, health care providers need identify behavioural and mental health specialists. A list of mental health practitioners with specific knowledge or training in psychosocial and behavioural aspects of diabetes is offered by the American Diabetes Association. Professionals in mental health have to be integrated into diabetes treatment environments. This specialised psychosocial treatment may also be offered by other health care providers who have received training in behavioural and mental health interventions due to the restricted availability of behavioural health resources. Optimising the connection between a health care practitioner and a person with diabetes may enhance the possibility that the individual may accept a referral for additional therapies, even if some medical experts may not feel prepared to manage psychological issues. The treatment of diabetes, the effects of depression, and psychosocial functioning have all been shown to benefit with collaborative care interventions and a team approach (ElSayed et al. 2023).

20.10 Conclusion

In order to improve overall quality of life, controlling diabetic neuropathy requires attending to both physical and mental health. Pharmacological therapies such as anxiolytics, antidepressants, and anticonvulsants, which help reduce the feelings of pain, anxiety, and depression, are important mental health measures. The psychological toll of neuropathy can also be effectively managed by cognitive and behavioural treatments, which include cognitive restructuring, relaxation methods, and teaching on good sleep hygiene. These methods encourage more restful sleep, lessen stress, and strengthen emotional fortitude. Better coping strategies, happier moods, and more effective pain management are some of the long-term advantages of mental health care in diabetic neuropathy. Patients may gradually become less sensitive to neuropathic pain, have less anxiety and sadness, and feel more in control of their illness. Good mental health treatment can help increase diabetes management adherence, which can lower complications and enhance results. Physical, emotional, and psychological assistance must all be integrated into a holistic approach to care. Patients suffering from diabetic neuropathy might experience more thorough and long-lasting relief when their medical care is combined with mental health techniques. In order to address the whole range of issues provided by the illness, this strategy promotes collaboration among healthcare practitioners, patients, and carers.

References

Alghafri RM, Gatt A, Formosa C. Depression symptoms in patients with diabetic peripheral neuropathy. Rev Diab Stud. 2020;16(1):35–40.

Alkandari M, Hollywood A. People's experiences living with peripheral neuropathy: a qualitative study. Front Pain Res. 2023;4:1162405.

Azmiardi A, Murti B, Febrinasari RP, Tamtomo DG. Low social support and risk for depression in people with type 2 diabetes mellitus: a systematic review and meta-analysis. J Prev Med Public Health. 2022;55(1):37.

Bawa FLM, Mercer SW, Atherton RJ, Clague F, Keen A, Scott NW, Bond CM. Does mindfulness improve outcomes in patients with chronic pain? Systematic review and meta-analysis. Br J Gen Pract. 2015;65(635):e387–400.

Beran M, Muzambi R, Geraets A, Albertorio-Diaz JR, Adriaanse MC, Iversen MM, et al. European Depression in Diabetes (EDID) Research Consortium the bidirectional longitudinal association between depressive symptoms and HbA1c: a systematic review and meta-analysis. Diabet Med. 2022;39:e14671.

Eccleston C, Morley SJ, Williams ADC. Psychological approaches to chronic pain management: evidence and challenges. Br J Anaesth. 2013;111(1):59–63.

Ehde DM, Dillworth TM, Turner JA. Cognitive-behavioral therapy for individuals with chronic pain: efficacy, innovations, and directions for research. Am Psychol. 2014;69(2):153.

ElSayed NA, Aleppo G, Aroda VR, Bannuru RR, Brown FM, Bruemmer D, et al. 5. Facilitating positive health behaviors and well-being to improve health outcomes: standards of care in diabetes—2023. Diabetes Care. 2023;46(Supplement_1):S68–96.

Fisher V, Li WW, Malabu U. The effectiveness of mindfulness-based stress reduction (MBSR) on the mental health, HbA1C, and mindfulness of diabetes patients: a systematic review and meta-analysis of randomised controlled trials. Appl Psychol Health Well Being. 2023;15(4):1733–49.

Fukunishi I, Horikawa N, Yamazaki T, Shirasaka K, Kanno K, Akimoto M. Perception and utilization of social support in diabetic control. Diabetes Res Clin Pract. 1998;41(3):207–11.

Geelen CC, Kindermans HP, van den Bergh JP, Verbunt JA. Perceived physical activity decline as a mediator in the relationship between pain catastrophizing, disability, and quality of life in patients with painful diabetic neuropathy. Pain Pract. 2017;17(3):320–8.

Gore M, Brandenburg NA, Dukes E, Hoffman DL, Tai KS, Stacey B. Pain severity in diabetic peripheral neuropathy is associated with patient functioning, symptom levels of anxiety and depression, and sleep. J Pain Symptom Manag. 2005;30(4):374–85.

Habtewold TD, Islam MA, Radie YT, Tegegne BS. Comorbidity of depression and diabetes: an application of biopsychosocial model. Int J Ment Heal Syst. 2016;10:1–9.

Kioskli K, Winkley K, McCracken LM. Might psychological flexibility processes and Acceptance and Commitment Therapy (ACT) apply in adults with painful diabetic neuropathy? A cross-sectional survey. J Contextual Behav Sci. 2019;13:66–73.

Li C, Hou W, Ding D, Yang Y, Gu S, Zhu Y. Evidence mapping based on systematic reviews of cognitive behavioral therapy for neuropathic pain. Neural Plast. 2023;2023(1):2680620.

Lim JA, Choi SH, Lee WJ, Jang JH, Moon JY, Kim YC, Kang DH. Cognitive-behavioral therapy for patients with chronic pain: implications of gender differences in empathy. Medicine. 2018;97(23):e10867.

Markwald RR, Iftikhar I, Youngstedt SD. Behavioral strategies, including exercise, for addressing insomnia. ACSMs Health Fit J. 2018;22(2):23–9.

Murawski B, Wade L, Plotnikoff RC, Lubans DR, Duncan MJ. A systematic review and meta-analysis of cognitive and behavioral interventions to improve sleep health in adults without sleep disorders. Sleep Med Rev. 2018;40:160–9.

Nathan HJ, Poulin P, Wozny D, Taljaard M, Smyth C, Gilron I, et al. Randomized trial of the effect of mindfulness-based stress reduction on pain-related disability, pain intensity, health-related quality of life, and A1C in patients with painful diabetic peripheral neuropathy. Clin Diabetes. 2017;35(5):294–304.

Parviniannasab AM, Faramarzian Z, Hosseini SA, Hamidizadeh S, Bijani M. The effect of social support, diabetes management self-efficacy, and diabetes distress on resilience among patients with type 2 diabetes: a moderated mediation analysis. BMC Public Health. 2024;24(1):477.

Peyrot M, Rubin RR. Behavioral and psychosocial interventions in diabetes: a conceptual review. Diabetes Care. 2007;30(10):2433–40.

Pouwer F, Mizokami-Stout K, Reeves ND, Pop-Busui R, Tesfaye S, Boulton AJ, Vileikyte L. Psychosocial care for people with diabetic neuropathy: time for action. Diabetes Care. 2024;47(1):17–25.

Priya G, Kalra S. Mind–body interactions and mindfulness meditation in diabetes. Eur Endocrinol. 2018;14(1):35.

Raizada N, Madhu SV. Sleep: an emerging therapeutic target in diabetes care. Int J Diabetes Dev Ctries. 2021;41:1–3.

Rosenzweig S, Reibel DK, Greeson JM, Edman JS. Mindfulness-based stress reduction is associated with improved glycemic control in type 2 diabetes mellitus: a pilot study. Altern Ther Health Med. 2007;13(5):36.

Selvarajah D, Cash T, Sankar A, Thomas L, Davies J, Cachia E, et al. The contributors of emotional distress in painful diabetic neuropathy. Diab Vasc Dis Res. 2014;11(4):218–25.

Tang TS, Funnell MM, Anderson RM. Group education strategies for diabetes self-management. Diab Spectr. 2006;19(2):99–105.

Twiddy H, Frank B, Alam U. A consideration of the psychological aspects to managing patients with painful diabetic neuropathy: an insight into pain management services at a tertiary centre in the UK. Diab Ther. 2021;12(2):487–98.

Vileikyte L, Leventhal H, Gonzalez JS, Peyrot M, Rubin RR, Ulbrecht JS, et al. Diabetic peripheral neuropathy and depressive symptoms: the association revisited. Diabetes Care. 2005;28(10):2378–83.

Wasserman LI, Trifonova EA. Diabetes mellitus as a model of psychosomatic and somatopsychic interrelationships. Span J Psychol. 2006;9(1):75–85.

Westby M, Norman G, Vedhara K, Game F, Cullum N. Psychosocial and behavioural prognostic factors for diabetic foot ulcer development and healing: a systematic review. Diabet Med. 2020;37(8):1244–55.

Young-Hyman D, de Groot M, Hill-Briggs F, Gonzalez JS, Hood K, Peyrot M. Erratum. Psychosocial care for people with diabetes: a position statement of the American Diabetes Association. Diabetes Care 2016; 39: 2126–2140. Diabetes Care. 2017;40(2):287.

Zeidan F, Grant JA, Brown CA, McHaffie JG, Coghill RC. Mindfulness meditation-related pain relief: evidence for unique brain mechanisms in the regulation of pain. Neurosci Lett. 2012;520(2):165–73.

Zhu M, Huang H. The underlying mechanisms of sleep deprivation exacerbating neuropathic pain. Nat Sci Sleep. 2023;15:579–91.

Pharmacological Interventions in Diabetic Neuropathy

21

Shweta Singh, Amarjeet Yadav, Priyanka Sonker, Rajeshwari Chandrakar, Nirbhay Narayan Tiwari, and Vandita Vishwakarma

21.1 Introduction

Diabetic neuropathy is a kind of nerve injury that can occur in patients with diabetes, generally as a result of chronic high blood glucose levels. About 50% of diabetic patients may have it at some point throughout their condition, making it one of the most prevalent consequences of the disease. The primary symptoms of the disease are tingling, pain, and paralysis in the extremities, particularly the hands and feet. In addition, it may have an impact on the cardiovascular, urinary, and digestive systems, which may result in a range of symptoms and consequences.

The underlying cause of diabetic neuropathy, chronic hyperglycaemia, destroys the circulatory capillaries that supply the nerves, known as the vasa nervorum, and induces inflammation and oxidative stress. Over time, this leads to degeneration and compromised nerve function. The mechanisms that cause oxidative stress in neurones and microvessels, namely polyol, glycation, protein kinase C, poly (ADP-ribose) polymerase (PARP), and hexosamine, are all overactivated by hyperglycaemia. These pathogenic pathways affect the structure and function of nerves. The myelin sheath is disrupted, Schwann cells separate from both myelinated and unmyelinated nerve fibres, produce less neurotrophic factor, and eventually die. Disruptions to axonal transport and signalling, perhaps at the axo-glial junction, then result in axonal loss. This happens to large, myelinated axons following

S. Singh (✉)
School of Pharmacy, Madhyanchal Professional University, Bhopal, MP, India

A. Yadav
Department of Pharmacy, Guru Ghasidas Vishwavidyalaya, Bilaspur, Chhattisgarh, India

P. Sonker · R. Chandrakar · V. Vishwakarma
Department of Pharmacy, Institute of Technology and Management, GIDA, Gorakhpur, India

N. N. Tiwari
Department of Pharmacy, Tahira Institute of Medical Sciences, GIDA, Gorakhpur, UP, India

M. K. Gautam et al. (eds.), *Antioxidants and Functional Foods for Diabetic Neuropathy*, https://doi.org/10.1007/978-981-95-9326-2_21

unmyelinated and small, sparsely myelinated fibres (Zochodne 2014; Gautam et al. 2014; Smith et al. 2022a, b).

Diabetic neuropathy can take many different forms, the most prevalent of which is peripheral neuropathy, which affects the limbs. Other forms include focal neuropathy, which damages certain body nerves, and autonomic neuropathy, which affects internal organs. Serious side effects from diabetic neuropathy might include infections, ulcers on the feet, and amputations because of lack of feeling in the feet. Furthermore, nerve loss can cause balance and coordination problems, which raises the possibility of falls. Autonomic neuropathy, meanwhile, can result in gastrointestinal difficulties, cardiovascular irregularities, and sexual dysfunction.

A multimodal strategy that includes medication management, lifestyle changes, and blood sugar control is necessary to manage diabetic neuropathy. Rather, DN preventive and management measures are recommended by national and international clinical guidelines, which base their recommendations on the best available data. Prevention measures aim at dealing with DN before indications arise or to stop DN from worsening, whereas therapy techniques target DN symptoms that patients currently have. For those with painful DN, pharmacotherapies are recommended; however, because they fail to assist the 70% of DN patients who do not experience pain, these are frequently not the best alternatives (Pop-Busui et al. 2022). Excruciating DPN can be treated with both pharmacological and nonpharmacologic therapy. Few superior, in tandem clinical study comparing these therapy techniques exist, making it difficult to identify which treatment plan may be most effective. Furthermore, the methods used in the available studies vary.

21.2 Classes of Pharmacological Interventions

Current guidelines concentrate on solutions for painful DN because there aren't many options for managing insensate or painless DN patients. Among these are pharmacotherapies such opioids, topical analgesics, anticonvulsants, tricyclic antidepressants (TCAs), serotonin and norepinephrine reuptake inhibitors (SNRIs), and intravenous (IV) drugs.

Anticonvulsants For diabetic neuropathy, anticonvulsants such pregabalin and gabapentin are frequently recommended. By blocking calcium channels, these medications calm overactive neurones and lessen the release of neurotransmitters that are involved in pain perception. They are especially helpful for those who don't react well to conventional forms of pain relief.

Antidepressants Neuropathic pain is frequently treated with antidepressants, especially tricyclic antidepressants (TCAs) and serotonin–norepinephrine reuptake inhibitors (SNRIs). Amitriptyline and other TCAs block serotonin and norepinephrine from being reabsorbed, which alters pain signalling pathways. Similar methods of action allow SNRIs like duloxetine to effectively reduce pain without as many adverse effects as TCAs.

Opioid Analgesics When other therapies fail to relieve severe pain, opioids like tramadol and oxycodone are prescribed. Opioids are useful in treating pain, but they also come with dangers of addiction, tolerance, and unfavourable side effects such diarrhoea and sleepiness. As a result, they are usually recommended for brief periods of time and with careful medical monitoring.

Topical Agents Topical drugs, such as capsaicin cream and lidocaine patches, relieve pain locally without having a harmful impact on the body. Substance P, a neuropeptide involved in pain transmission, is depleted by capsaicin, and lidocaine functions as a local anaesthetic to numb the affected region.

Novel and Emerging Therapies Research is being done on more recent treatments, such as nerve growth factor modulators that target pain pathways, sodium channel blockers, aldose reductase inhibitors, and TRPV1 antagonists. These medications seek to reduce adverse effects while increasing the effectiveness of pain management (Snyder et al. 2016).

21.3 Antidepressants in Neuropathic Pain

The original purpose of antidepressants, one of the earliest drugs for neuropathic pain, was to help people with chronic pain feel less depressed and less discomfort. According to a preliminary theory, these medications may enhance the activity of downward inhibitory systems that connect the brain stem to the spinal cord's dorsal horn. This was accomplished by blocking the reabsorbed serotonin and noradrenaline, which flow into spinal synapses between spinothalamic and nociceptors neurones. Through their attachment to membrane receptors on their surface, neurotransmitters can prevent synaptic communication between these neurones. For example, noradrenaline can bind to alpha-2 adrenergic receptors. Alternatively, they can cause interneurons to produce inhibitory substances, such as gamma-aminobutyric acid (GABA) or endogenous opioids, such as serotonin at its metabotropic receptors and noradrenaline at alpha-1 adrenergic receptors (Fornasari 2017).

21.3.1 Use of Serotonin–Norepinephrine Reuptake Inhibitors (SNRIs)

First-line medication for PDN is thought to use antidepressants like duloxetine. Duloxetine is an SNRI that was first administered for depression but is now extensively researched in the treatment of post-depression illness. It is taken at a 60–120 mg/day. Duloxetine helps PDN patients with their overall quality of life, depressive symptoms, and long-term pain results (Wernicke et al. 2006). In the UK and the USA, it is only authorised for DN; other neuropathic pains are not covered. An SNRI called duloxetine inhibits pain pathways and decreases dopamine reuptake. Its side effects are similar to those of anticonvulsants, but it can also lead to sleep problems and sexual dysfunction (Wernicke et al. 2006). The dose of the selective serotonin reuptake inhibitor venlafaxine is 37.5–225 mg once day. Research indicates that it greatly lowers pain when compared to a placebo, even though it is not FDA-approved for postmenopausal depression (PDN). It should only be taken into consideration, though, if duloxetine is not working or is having serious adverse effects (Bansal 2013).

Inhibitors of serotonin and noradrenaline reuptake, such as venlafaxine, milnacipran, and duloxetine, induce a balanced inhibition of these neurotransmitters. These medications are also known as balanced noradrenaline and serotonin inhibitors. Though the features of the blockage differ from those of tricyclic antidepressants, ventlafaxine inhibits sodium channels without producing postsynaptic effects. A variety of postsynaptic receptors and sodium channels are not significantly impacted by duloxetine, which is a strong, well-balanced inhibitor of serotonin and noradrenaline reuptake (Sindrup et al. 2005; Kumar et al. 2013).

21.3.2 Role of Tricyclic Antidepressants (TCAs)

Alternatively, amitriptyline is recommended as a first-generation or second-generation treatment for discomfort-causing DN. Although it is not allowed for this usage in the USA, it is the only TCA that is authorised in the UK to treat neuropathic pain. Major depressive disorder is the main condition it is used to treat. There is not much data in favour of amitriptyline. Although amitriptyline has been shown in many RCTs to reduce DN pain (Boyle et al. 2012).

Although amitriptyline's precise mode of action is unknown, it is known to have a variety of impacts on the body and neurological system. This drug works by blocking the central nervous system's sodium, potassium, and N-methyl-D-aspartate (NMDA) channels, which are all linked to neuropathic pain, as well as the neuronal terminals' absorption of noradrenaline and serotonin. It is advised to begin taking 10–25 mg daily, and this amount can be progressively raised to 75 mg daily. Diarrhoea, mouth dryness, pain in the head, arrhythmias, sleep problems, sexual dysfunction, and postural hypotension are common side effects. The latter is especially problematic in elderly people and requires close monitoring (Smith et al. 2022a, b).

Studies have also examined the efficacy of nortriptyline and desipramine in the treatment of painful DN. Desipramine dramatically decreased the symptoms of painful DN, as demonstrated by many RCTs. It may have a comparable therapeutic effect to amitriptyline (Max et al. 1991).

21.4 Anticonvulsants for Neuropathic Pain

Anticonvulsants such as pregabalin and gabapentin, together with lamotrigine, topiramate, oxcarbazepine, valproate, and lacosamide, are used to treat painful deep neural discharge (DPN). Although a recent meta-analysis indicates gabapentin is the best choice, the American Academy of Neurology and Toronto recommendations support pregabalin as the first-line therapy (Ziegler and Fonseca 2015).

There is some evidence to support the use of drugs to treat painful DN; gabapentin and pregabalin 300 mg have been shown to significantly reduce pain intensity. In comparison to a placebo, pregabalin decreased pain intensity by at least 30% and 50%, respectively, according to a recent Cochrane systematic review. At doses of 1200 mg or more, gabapentin was proven to have moderate to significant benefits (Gautam et al. 2012a, b; Cooper et al. 2017). Studies have been conducted on several anticonvulsants, including phenytoin, carbamazepine, valproic acid, lamotrigine, and lamotrigine; however, the evidence supporting their usage is weak.

21.4.1 Gabapentin and Pregabalin: Mechanism of Action in Modulating Pain

Despite being produced from GABA, both pregabalin and gabapentin have no impact on the GABAergic system. When medications bind to voltage-gated calcium channels in the spinal cord and central nervous system, they have specific pharmacological effects. These channels control the release of neurotransmitters and open in response to peripheral action potentials. They are found on presynaptic terminals. The fusion of synaptic vesicles and the release of neurotransmitters into the synaptic cleft are made easier by calcium ions entering these channels. Research suggests that aberrant the transmission of signals in the spinal cord may result from chronic pain issues because of an increase in voltage-gated calcium channels. By binding to the alpha-2/delta-1 subunit, gabapentinoids may help these channels internalise. This might alter the number of open calcium channels in the plasma membrane by upsetting the macromolecular structure that keeps the calcium channel on the surface of the presynaptic terminal (Stahl et al. 2013).

21.4.2 Efficacy and Safety

Based on a dose-dependent response, pregabalin consistently reduces pain when compared to a placebo. In seven RCTs (n = 1596), pregabalin and placebo were compared. The results demonstrated that 39% and 47% of patients, respectively, saw a significant decrease in pain when receiving 300 mg and 600 mg daily at BD, compared to 22% with a placebo. After 5 days for 300 mg and 4 days for 600 mg, the effects were apparent. A daily dose of 150 mg of medicine did not work (Spallone et al. 2012; Gautam et al. 2012a, b, 2014).

At doses of 1200 mg daily or more, gabapentin is more effective than a placebo and has an effect comparable to that of pregabalin, according to nine Randomised Controlled Trials (RCTs), with at least five of these RCTs reporting comparable efficacy.

More superior data on pregabalin's effects, more certain pharmacokinetics, shorter calibration periods, and the option to administer twice daily make it a better choice for first-line treatments than gabapentin. However, gabapentin is the less expensive generic alternative and has fewer medication interactions. The only aspects of quality of life that are enhanced by gabapentin in comparison to pregabalin, which is better at many other aspects of life, are mental health and vigour (Snyder et al. 2016).

The usual side effects of pregabalin and gabapentin, which are normally well-tolerated drugs, include drowsiness, dry mouth, oedema, and impaired vision. In addition to peripheral oedema and disorientation, more than 20% of patients report feeling sleepy and lightheaded. Both drugs have a 4% to 4% discontinuation rate because of somnolence, and they are both dose-dependent and transient (Bockbrader et al. 2010).

21.5 Opioid Analgesics in Diabetic Neuropathy

Both light opioids like tramadol and powerful opioids like oxycodone, hydromorphone, and morphine are beneficial in treating neuropathic pain compared to other drugs. The quantity required to treat these medications is comparable to that of antidepressants. However, because of adverse drug responses and worries about misuse, diversion, and addiction, they have long been regarded as second-line, and more recently, third-line, medications. In the USA and the UK, Tramadol is approved for medium to extreme pain rather than specifically for neuropathic pain. With the greatest affinity for the latter, it interacts directly to the δ, κ, and μ receptors. Furthermore, it prevents noradrenaline and serotonin from being reabsorbed, hence it shouldn't be used with SNRIs or SSRIs. It's possible that tramadol has a lower misuse risk than other opioids (Attal et al. 2010; Dunn et al. 2019).

The FDA has authorised tapentadol, an opioid, as a third-line treatment for PDN management. The recommended dosage for tapentadol is between 100 and 250 mg twice day. Numerous studies show that when compared to a placebo, extended-release tapentadol greatly lessens the degree of pain. Both methadone and levorphanol, synthetic opioids, have demonstrated promise in the treatment of neuropathic pain. Levorphanol, like gabapentin and TCAs, works well in high-strength formulations. Even in individuals who have not reacted well to traditional opioids, methadone has demonstrated promise in the treatment of neuropathic pain. While individuals on other opioids must gradually acclimatise, patients who have never taken opioids previously are frequently put on 2.5–5 mg every 8–12 h. To determine if they are suitable for PDN, more research is required (Staudt et al. 2022).

Analgesia is caused by opioids' effects on the brain, vertebral column, and peripheral terminals of primary afferent neurones. Mu, delta, and kappa transmembrane G protein-coupled receptors are the three kinds to which they bind. Activation of opioid receptors prevents intracellular cAMP production and adenylate cyclase. Analgesia is primarily caused by opioid receptors coupling to potassium and calcium channels (Fornasari 2017).

21.5.1 Indications and Guidelines for Opioid Use

Opioid analgesics should only be used in cases where gabapentinoids and SNRIs, two first-line medications, are ineffective either alone or in combination to reduce pain. Therefore, before contemplating opioids, it is crucial to ensure adherence to treatment plan employing these medicines. Checking and adjusting the dosage and timing of administration to the day-to-day fluctuations in symptoms is recommended. For instance, give pregabalin or duloxetine 1–2 h before the anticipated start of symptoms. The next stage before employing opioids as an adjuvant treatment is to make sure that these medications have been taken to the highest level of tolerance. Consider switching between medications (e.g., duloxetine and pregabalin, or vice versa) or augmenting treatment plans (e.g., monotherapy and combination therapy) (Balhara et al. 2020).

21.5.1.1 Using Screening Tools for Risk Stratification of Patients

These instruments may improve the accuracy of risk prediction for outcomes like opioid misuse and opioid use disorder when used in conjunction with clinical evaluation. The Opioid Safety instrument is one such validated instrument. To predict the occurrence of recurring aberrant drug use behaviours, the revised Screener and Opioid Screening for Individuals with Pain is also used. These instruments can be used proactively to identify patients who require higher levels of clinical monitoring (Butler et al. 2015).

21.5.1.2 Medication Counselling

A patient's participation in therapy is maximised by effective patient education and counselling. Both the patient and the doctor should participate in the decision-making process before beginning opioid analgesics. It is necessary to do a reasonable cost-benefit analysis before beginning opioid analgesics for PDN and to communicate the findings with the patient and carers. The patient has to be made aware of the possibility of opioid use disorder, overdose, and the necessity to refrain from co-administering benzodiazepines and alcohol.

21.5.1.3 Pain Medicine Agreements and Urine Drug Testing

These agreements are basically treatment contracts between patients and clinicians that outline the dangers and benefits of opioid usage as well as a promise not to change prescriptions, take more medicine than is recommended, exceed daily dosages, exceed recommended therapy durations, or share pharmaceutical supplies with others. These agreements often include periodic urine drug testing for adherence monitoring and to identify opioid use disorder or diversion (Fishman et al. 1999).

21.5.1.4 Drug Prescription

The course of PDN symptomatology might vary greatly. In order to relieve symptoms, doctors should only administer the minimal amount of opioids. Even though there are no discernible differences among long-acting and short-acting opioids in terms of potency or side effect outcomes, the lowest dose of short-acting opioids is advised. Short-acting medicine must be used in an emergency if there is severe pain or if unpleasant feelings get worse at night. Prescriptions ought to have a set duration, and they shouldn't be renewed automatically. Careful dosage titration is necessary if a long-acting opioid is recommended since the medication accumulates and can cause unwanted effects (Albrecht et al. 2015).

21.5.1.5 Periodic Screening and Follow-up

Regular screening for the onset of opioid-use disorder is necessary for patients with PDN receiving opioid treatment. Frequent evaluation for mental health issues including anxiety and depression could also be necessary. There are validated screening instruments for opioid use disorder. A 17-item survey called the Current Opioid Misuse Measure was created and approved in order to determine whether individuals are actively abusing their prescription opioid medication. The Addiction Behaviour Checklist is a 20-item tool that helps patients with chronic pain detect hazardous behaviours linked to long-term opioid usage. A follow-up visit should take place within 1–4 weeks and can be customised to the patient's needs (Butler et al. 2007).

21.5.2 Risks and Controversies Surrounding Opioid Treatments

Chronic opioid treatment may be helpful for certain PDN patients, but there are unique hazards associated with its use. For instance, the issue of abuse and misuse with long-term opioid treatment is now well acknowledged. Abuse is defined as when a substance is used when it becomes harmful or illegal, while misuse is defined as using a drug against medical advice. Further, when it comes to long-term opioid usage, it's critical to distinguish between tolerance a physiological adaptation that lowers the effectiveness of the medication and physical dependence a physiological adaptation that may be triggered by abruptly stopping the drug or reducing dosage. It is important to get a history of previous and present drug usage and addiction when evaluating a person with PDN for the possible use of chronic opioid treatment. This will allow the physician to address any pertinent concerns prior to writing an opioid prescription (Gautam & Goel 2014).

The development of tolerance may lead to an increase in dosage, thus the physician must be aware of the dosage-dependent risks involved in treating opioids. The lowest effective dosage should always be the limit while using opioids. Regular evaluation of side effects and atypical behaviours is required of the physician, and opioid-sparing measures like interventional treatment must be frequently considered. An additional problem with long-term therapy is the onset of opioid-induced hyperalgesia, which, despite its lack of understanding, makes some patients genuinely feel more pain. If a dosage reduction is identified, it can be quite effective (Rivat and Ballantyne 2016).

21.6 Topical Therapies for Neuropathic Pain

A kind of NP that affects superficial tissues, localised neuropathic pain (LNP) is typified by regular, bounded regions of extreme pain that are linked to either positive or negative sensory indicators or sporadic symptoms. In this kind of NP, topical agents are warranted (Casale et al. 2017). Topical medicines are a more effective option than first-line systemic medications due to their low systemic absorption and localised activity. This reduces the likelihood of drug-drug interactions and prevents issues like gastrointestinal disturbances and fluctuating blood concentrations. The European Medicines Agency (EMA) has granted licenses for two topical therapies for peripheral neuropathy (NP): capsaicin 179 mg cutaneous patch for all types and lidocaine 5% medicated plaster for post-herpetic neuralgia alone. The Food and Drug Administration has authorised both for treating NP linked to postherpetic neural discomfort (Sommer and Cruccu 2017).

21.6.1 Capsaicin and Lidocaine: Efficacy in Localised Pain

Capsaicin is a local analgesic used in formulations with low (0.075%) and high (8%) dosages for the treatment of PDN. Although the results are not always consistent, there have been reports of significant reductions in the intensity of neuropathic pain. According to additional studies, 8% capsaicin is not inferior to pregabalin, duloxetine, and gabapentin and has a comparatively more tolerable undesirable side effect profile. The soreness of the skin is one of the adverse effects (Simpson et al. 2017).

Capsaicin, the active component of fruits in the genus Capsicum, functions as an agonist on the transient receptor potential vanilloid-I receptor (TRPVI). TRPV1-expressing cutaneous nociceptors have an initial increase in sensitivity, which is followed by a long-lasting desensitisation that produces a long-lasting analgesic effect. In healthy volunteers, capsaicin significantly reduces epidermal nerve fibre density morphologically; however, this loss is followed by recovery over a 24-week period (Kennedy et al. 2010). Designed to administer a single therapeutic dosage of capsaicin over 30 min (for feet) or 60 min (for other sites), each 14 × 20 cm patch is meant to be removed after that period. A doctor should identify and designate the treatment region, and only a doctor or a healthcare practitioner working under a doctor's supervision should administer the 8% capsaicin patch. If necessary, a therapy consisting of no more than four patches may be repeated every 90 days.

Even though the precise mechanism of action is unknown, topical lidocaine is known to inhibit neuronal action potentials by binding 1:1 to voltage-gated sodium channels. The lidocaine 5% medicated plaster may provide long-term pain relief by counteracting central sensitisation due to a combination of reduced peripheral nerve input and decreased nerve fibre density in the epidermis (Bhaskar and Mittal 2011). Each 10 × 14 cm medicated plaster contains 700 mg (5% w/w) of lidocaine in a white hydrogel plaster. In a 24-h period, up to three patches may be applied once for a maximum of 12 h. Although patients are able to use the patch on their own, it is advised that they undergo another evaluation 2–4 weeks later to make sure the analgesic effect is realised. Although only about 3% of the lidocaine is absorbed systemically, patients with severe liver or kidney impairment are advised to exercise caution when using lidocaine 5% medicated plasters over the long term due to the potential risks associated with secondary metabolites (Sommer and Cruccu 2017). Several studies have shown that topical lidocaine (5%) can effectively treat localised neuropathic pain by significantly lowering discomfort, enhancing quality of life, and raising patient satisfaction in comparison to a placebo. Amitriptyline, gabapentin, pregabalin, and capsaicin have demonstrated sustained efficacy with somewhat improved tolerability in noninferiority tests. Another example of a side effect is dermal irritation (Staudt et al. 2022).

21.6.2 Benefits and Limitations of Topical Treatments

Lidocaine medicated plasters are especially advised for senior patients. Capsaicin patches have been suggested as a second or third-line therapy for postherpetic neuralgia and peripheral neuropathic pain. Due to low-quality data, lidocaine patches were not suggested as the first-line treatment in a recent revision of the NeuPSIG guidelines. Capsaicin patches were suggested as a second-line treatment instead because of their strong evidence, modest effect size, training requirements, and possible long-term safety issues with feeling. It was more likely that high-dose 8% capsaicin patches would result in side effects such dryness, erythema, pruritus, papules, oedema, burning, and pain. Olvanil is a novel and potentially superior topical analgesic medication, according to recent study, since it immediately desensitises TRPV1 channels and has less side effects when applied topically (Casale et al. 2017). However, a significant drawback of the topical cream is that, in order to provide the most pain relief, it needs to be administered three or four times a day for up to 2 months. Furthermore, a lot of people are intolerant to topical capsaicin's side effects, which mostly include searing discomfort while in hot weather or upon contact with warm or hot water.

21.7 Novel and Emerging Pharmacological Agents

These are the key novel pharmacotherapies for neuropathic pain in brief: Though new and cutting-edge medicines are being developed to address underlying processes and enhance neurone regeneration, the main focus of current pharmacological treatments is on treating symptoms such as pain and discomfort. With the use of novel compounds, repurposed medications, and biologics, DN may be managed more successfully. Some of the most promising agents under investigation right now are listed here.

Ambroxol A recent study indicated that ambroxol, an active metabolite of bromhexine, may have several applications in the treatment of neuropathic pain because of its various effects. When it comes to the creation and transmission of pain signals, ambroxol affects the activity of voltage-gated sodium channels, particularly Nav1.8. Many animal models of neuropathic pain were used in pre-clinical research to examine ambroxol's analgesic effects. Studies on ambroxol's analgesic properties have been conducted, despite the paucity of clinical data regarding its other uses. Using topical ambroxol cream (20%), which provides individual pain decreases over many hours, was utilised to treat seven patients with severe neuropathic pain who were not responding to traditional therapy such as lidocaine or capsaicin patches (Russo et al. 2023).

Cannabidiol The cannabis plant (Cannabis sativa L.) naturally contains the non-psychoactive cannabinoid chemical cannabidiol, or CBD. Because of its possible analgesic and anti-inflammatory properties, CBD has gained attention as a potential therapy option for neuropathic pain. CBD interacts with the body's endocannabinoid system (ECS), which regulates a number of physiological functions, including the experience of pain. CBD modulates pain signalling and reduces inflammation via binding to cannabinoid receptors, namely the CB1 and CB2 receptors. Additionally, transient receptor potential (TRP) channels and other receptors and ion channels implicated in pain transmission may be impacted by CBD. Numerous pre-clinical and clinical investigations have shown that CBD has the ability to reduce the symptoms of neuropathic pain (Petzke et al. 2022).

Nerve Growth Factor (NGF) Modulators The survival and functionality of peripheral neurones are critically dependent on NGF. Novel treatments that alter NGF signalling seek to improve nerve regeneration and provide injury prevention. Tanezumab, for example, is a monoclonal antibody that inhibits NGF, and its potential to treat neuropathic pain by decreasing NGF pathway hyperactivity is being investigated. Pfizer created the recombinant humanised monoclonal antibody (IgG) known as tanezumab to target NGF in the management of various pain disorders. Despite encouraging first results, long-term safety problems are still being investigated, especially with regard to joint degeneration (Patel et al. 2018).

Aldose Reductase Inhibitors (ARIs) Aldose reductase inhibitors are a family of drugs that obstruct the polyol pathway, a particular metabolic process that breaks down glucose, and may be able to prevent or delay the course of neuropathy. By blocking the enzyme aldose reductase, aldose reductase inhibitors—like epalrestat—aim to block this route and lessen the harmful consequences of sorbitol accumulation. In Japan, it is authorised for the treatment of abnormal changes in heart rate, abnormal vibration sensation, and subjective neuropathy symptoms related to diabetic peripheral neuropathy. After analysing data from six clinical studies, it was shown that, when compared to baseline and placebo, epalrestat 50 mg three times per day may reduce subjective neuropathy symptoms and enhance motor and sensory nerve conduction velocity (Ramirez and Borja 2008).

Sodium Channel Blockers The most apparent strategy for treating peripheral neuropathic pain would be to inhibit or lessen excitability in both injured and undamaged sensory neurones, with sodium channels serving as the primary target. Nav1.7, Nav1.8, and Nav1.9 in particular are voltage-gated sodium channels that are important in the production of neuropathic pain signals. A potential medication that selectively blocks the Nav1.8 channel, VX-150, helps to relieve neuropathic pain without having the negative effects on the body that come with other sodium channel blockers. It is a pro-drug that is readily absorbed via the mouth that quickly transforms into its active component, a highly selective (>400-fold) inhibitor of NaV1.8 in comparison to other sodium channel subtypes (Hijma et al. 2021).

TRPV1 Antagonists The nonspecific cation channel TRPV1 is involved in the sense of pain and temperature. TRPV1 antagonists are useful in preventing TRPV1 activation, preventing pain signals from peripheral neurones from reaching the central nervous system, and relieving pain brought on by different types of nerve injury. In preclinical studies, TRPV1 antagonists like SB-705498 have demonstrated promise by lowering hyperalgesia. Creating novel TRPV1 channel modulators for diabetes-related pain syndrome is a potential alternative to traditional therapy, since TRPV1 has demonstrated the special benefits of concurrently controlling blood sugar and pain. Although dysregulated body temperature was the reason why early TRPV1 antagonists did not work as intended, more recent iterations of these medications are being developed to avoid this negative effect (Liu et al. 2023; Zhu et al. 2023).

Gene Therapy A novel method of treating DN is gene therapy, which directly modifies the genetic processes responsible for nerve damage. Gene therapy technology has been investigated as a potential means of managing pain in order to get around the drawbacks of the currently available medications, which are all based on tiny molecules. For the treatment of painful DPN, VM202 (Engensis), a nonviral plasmid DNA product, is the first gene therapy to reach advanced clinical studies. It provides hepatocyte growth factor (HGF), a substance whose neuroprotective and regenerative property have showed promise in preliminary clinical studies. HGF has the ability to reverse the harm produced by DN by promoting angiogenesis and nerve regeneration (Kessler et al. 2021).

SIRT1 Activators A protein called sirtuin 1 (SIRT1) is involved in ageing and cellular metabolism. Because SIRT1 regulates the synaptic plasticity of spinal dorsal horn neurones, it is known to lessen diabetic neuropathic pain. SRT2104 is the most selective small molecule activator of SIRT1 among its modulators and is the one that is moving into the clinical translation stage at this time. In both human studies and animal models, its tolerability has demonstrated promise (Chang et al. 2024).

21.8 Multimodal Pharmacological Approaches

Since monotherapy frequently fails to provide a significant enough impact, pharmacotherapeutic treatments are frequently coupled with additional procedures in clinical practice. Drug therapy should be started as soon as possible. The effectiveness of complementary therapies to pharmacotherapeutic interventions is frequently evaluated just as monomodal treatment. For example, acupuncture is utilised to boost endorphin production in the central nervous system. Acupuncture research demonstrates beneficial benefits on several forms of peripheral polyneuropathy, including those brought on by chemotherapy and diabetes mellitus (Iravani et al. 2020). A unique kind is acupuncture on the scalp. It is predicated on the ancient Chinese medical notion of meridians in conjunction with the functional arrangement of stimulation throughout the cerebral cortex. Research indicates that there is a favourable impact on cognitive capacities as well as an enhancement of sensory and motor abilities. Additionally, commonly employed are methods from the fields of mind-body medicine and biofeedback. They can help to improve quality of life by lowering pain intensity, pain catastrophisation, sadness, and perceived stress (Nathan et al. 2017). Another crucial type of therapy for the management of peripheral neuropathy is exercise therapy. It may also have an impact on blood circulation and the metabolic state of diabetes. Foot massages and massages in general have a beneficial effect on polyneuropathy symptoms. Transcutaneous electrostimulation has been the subject of studies for many years, with positive therapeutic outcomes (Souza et al. 2021).

The management and prevention of peripheral neuropathy are significantly influenced by nutrition. Consuming foods rich in unsaturated fatty acids, phytochemical-containing meals, extracts from medicinal plants, and vitamins B and E is recommended. It is said that they enhance the sirtuin enzymes' activity and boost patients' quality of life when they suffer from peripheral polyneuropathy brought on by chemotherapy. Multimodal, complicated treatments can be employed in severe cases of peripheral neuropathy, in addition to specific operations or combination therapies. The term "complex therapies" refers to the tight temporal coordination of high-frequency and indication-related therapeutic techniques. In order to develop a patient-specific therapy plan with goals at the start of treatment, a multiprofessional team must be present (Romeyke and Stummer 2024).

21.9 Adverse Effects and Considerations

21.9.1 Common Side Effects of Pharmacological Agents

Patients with epilepsy, cardiac incompensation, or cardiac conduction abnormalities cannot use tricyclic antidepressants. Tricyclic antidepressant side effects, which can be irritating and cause many patients to stop taking the medication, include dry mouth, sweating, dizziness, weariness, constipation, and issues with micturition. Although SNRIs and SSRIs are better tolerated, both classes of medications have adverse effects. Similar adverse effects are observed with SNRIs and SSRIs, including nausea, vomiting, and dyspepsia. Venlafaxine may also cause blood pressure to rise (Sindrup et al. 2005).

Treatment with extended-release (ER) duloxetine (Cymbalta) has been linked to the following side effects: hyponatraemia, seizures, abnormal bleeding, activation of mania or hypomania, falls, syncope, hepatotoxicity, serotonin syndrome, and urinary hesitation and retention. Furthermore, gabapentin has been linked to dizziness, somnolence, sleepiness, driving impairment, multiorgan hypersensitivity (drug responses accompanied by eosinophilia and systemic symptoms), suicidal behaviour and thoughts, and sedation when taken for its authorised use of postherpetic neuralgia in adults. Furthermore, the main adverse events linked to the medicated patch containing 5% lidocaine are application-site responses. These responses may manifest as blisters, bruises, burning, dermatitis, depigmentation, erythema, oedema, exfoliation, irritation, papules, petechiae, pruritus, and vesicles (Cohen et al. 2015).

21.9.2 Long-Term Safety Concerns and Patient Monitoring

It's important to pay attention to a few adverse effects. Tricyclic antidepressants may produce orthostatic hypotension, which has been linked to hip fractures and falls, according to population-based research. An increased risk of hip fractures has been linked to both SSRIs and tricyclic antidepressants, according to a recent study. Suicidal ideation linked to cardiac arrhythmia is the most concerning adverse effect of tricyclic antidepressants. According to a retrospective cohort research, SSRIs and tricyclic antidepressants at dosages less than 100 mg/day did not raise the risk of SC mortality, but greater doses of these medications were linked to a relative risk that was two to three times higher (Sindrup et al. 2005). When starting antidepressant treatment for whatever reason, all patients should have proper monitoring and attentive observation for signs of clinical deterioration, suicidality, and unexpected behavioural changes. This is especially important in the early months of treatment or when there are dosage adjustments or reductions. Patients above the age of 60 shouldn't use amitriptyline due to the possibility of comorbidities. TCAs should also not be used by those who have taken antidepressants in the 14 days prior or who are still recovering from a myocardial infarction (Cohen et al. 2015).

Opioids' analgesic impact makes them useful for managing pain over the long term, but their efficacy is disputed because of concerns about acceptability, possible tolerance, and addiction risks. Over 20% of patients reported opioid-related side effects, and over 50% of patients quit their treatment as a result. Dry mouth, drowsiness, diarrhoea, nausea, constipation, itching, respiratory depression, and urine retention are typical adverse medication reactions. These problems might result in addiction and usually go away with tolerance. The most severe opioid adverse medication responses should be taken into consideration, along with constipation, as they can have a major negative impact on the patients' quality of life. These side effects include drowsiness and cognitive impairment. In response to the ongoing rise in accidental overdose deaths, a global conversation about addiction, diversion, and abuse has emerged in the previous few years, beginning in the USA (Labianca et al. 2012).

21.10 Conclusion

A major consequence of diabetes that significantly lowers a patient's quality of life is diabetic neuropathy (DN), which causes chronic pain, sensory loss, and motor impairment. Pharmacological therapies continue to be a mainstay in the care of diabetic neuropathy (DN), with the goals of reducing pain, stopping more nerve damage, and treating related mental health issues. Antidepressants like amitriptyline and duloxetine, as well as anticonvulsants like pregabalin and gabapentin, are the mainstays of contemporary pharmacotherapy due to their effectiveness in decreasing neuropathic pain via altering pain pathways. Opioids and tramadol are used in more severe or treatment-resistant situations; however, because of the potential for dependency and adverse effects, their usage is frequently limited. Additional, pain management options include the use of topical medications such as capsaicin and lidocaine. Innovative methods in medication therapy encompass the creation of new drugs that target underlying pathophysiological systems, such as gene therapy, sodium channel blockers, NGF modulators, and aldose reductase inhibitors. It is anticipated that the therapeutic landscape will change in the future due to a stronger focus on personalised therapy, targeting certain biochemical pathways, and combining pharmaceutical and nonpharmacological methods. A better knowledge of the pathophysiology of diabetic neuropathy and advancements in medication research might lead to more specialised and successful therapies in the future.

References

Albrecht JS, Khokhar B, Pradel F, Campbell M, Palmer J, Harris I, Palumbo F. Perceptions of patient provider agreements. J Pharm Health Serv Res. 2015;6(3):139–44.

Attal N, Cruccu G, Baron RA, Haanpää M, Hansson P, Jensen TS, Nurmikko T. EFNS guidelines on the pharmacological treatment of neuropathic pain: 2010 revision. Eur J Neurol. 2010;17(9):1113–e88.

Balhara YPS, Singh S, Kalra S. Pragmatic opioid use in painful diabetic neuropathy. Eur Endocrinol. 2020;16(1):21.

Bansal D. Comparative efficacy and safety of six antidepressants and anticonvulsants in painful diabetic neuropathy: a network meta-analysis. Pain Physician. 2013;16:E705–14.

Bhaskar A, Mittal R. Local therapies for localised neuropathic pain. Rev Pain. 2011;5(2):12–20.

Bockbrader HN, Wesche D, Miller R, Chapel S, Janiczek N, Burger P. A comparison of the pharmacokinetics and pharmacodynamics of pregabalin and gabapentin. Clin Pharmacokinet. 2010;49:661–9.

Boyle J, Eriksson ME, Gribble L, Gouni R, Johnsen S, Coppini DV, Kerr D. Randomized, placebo-controlled comparison of amitriptyline, duloxetine, and pregabalin in patients with chronic diabetic peripheral neuropathic pain: impact on pain, polysomnographic sleep, daytime functioning, and quality of life. Diabetes Care. 2012;35(12):2451–8.

Butler SF, Budman SH, Fernandez KC, Houle B, Benoit C, Katz N, Jamison RN. Development and validation of the current opioid misuse measure. Pain. 2007;130(1–2):144–56.

Butler SF, McNaughton EC, Black RA. Tapentadol abuse potential: a postmarketing evaluation using a sample of individuals evaluated for substance abuse treatment. Pain Med. 2015;16(1):119–30.

Casale R, Symeonidou Z, Bartolo M. Topical treatments for localized neuropathic pain. Curr Pain Headache Rep. 2017;21:1–9.

Chang N, Li J, Lin S, Zhang J, Zeng W, Ma G, Wang Y. Emerging roles of SIRT1 activator, SRT2104, in disease treatment. Sci Rep. 2024;14(1):5521.

Cohen K, Shinkazh N, Frank J, Israel I, Fellner C. Pharmacological treatment of diabetic peripheral neuropathy. Pharm Therap. 2015;40(6):372.

Cooper TE, Derry S, Wiffen PJ, Moore RA. Gabapentin for fibromyalgia pain in adults. Cochrane Database Syst Rev. 2017;2017(1):CD012188.

Dunn KE, Bergeria CL, Huhn AS, Strain EC. A systematic review of laboratory evidence for the abuse potential of tramadol in humans. Front Psych. 2019;10:704.

Fishman SM, Bandman TB, Edwards A, Borsook D. The opioid contract in the management of chronic pain. J Pain Symptom Manag. 1999;18(1):27–37.

Fornasari D. Pharmacotherapy for neuropathic pain: a review. Pain Ther. 2017;6:25–33.

Gautam MK, Goel RK. Toxicological study of Ocimum sanctum Linn. leaves: Hematological, biochemical and histopathological studies. J Toxicol. 2014;2014:135654.

Gautam MK, Gupta A, Vijaykumar M, Rao CV, Goel RK. Studies on the hypoglycemic effects of Murraya paniculata Linn. extract on alloxan-induced oxidative stress in diabetic and nondiabetic models. Asian Pacific J Trop Dis. 2012a;2(S1):S186–91.

Gautam MK, Singh A, Rao CV, Goel RK. Toxicological evaluation of Murraya paniculata (L.) leaves extract on rodents. Am J Pharmacol Toxicol. 2012b;7(2):62–7.

Gautam MK, Purohit V, Agarwal M, Singh A, Goel RK. In-vivo healing potential of Aegle marmelos in excision, incision and dead space wound models. Sci World J. 2014;2014:740107.

Hijma HJ, Siebenga PS, de Kam ML, Groeneveld GJ. A phase 1, randomized, double-blind, placebo-controlled, crossover study to evaluate the pharmacodynamic effects of VX-150, a highly selective NaV1. 8 inhibitor, in healthy male adults. Pain Med. 2021;22(8):1814–26.

Iravani S, Kazemi Motlagh AH, Emami Razavi SZ, Shahi F, Wang J, Hou L, et al. Effectiveness of acupuncture treatment on chemotherapy-induced peripheral neuropathy: a pilot, randomized, assessor-blinded, controlled trial. Pain Res Manag. 2020;2020(1):2504674.

Kennedy WR, Vanhove GF, Lu SP, Tobias J, Bley KR, Walk D, et al. A randomized, controlled, open-label study of the long-term effects of NGX-4010, a high-concentration capsaicin patch, on epidermal nerve fiber density and sensory function in healthy volunteers. J Pain. 2010;11(6):579–87.

Kessler JA, Shaibani A, Sang CN, Christiansen M, Kudrow D, Vinik A, et al. Gene therapy for diabetic peripheral neuropathy: a randomized, placebo-controlled phase III study of VM202, a plasmid DNA encoding human hepatocyte growth factor. Clin Transl Sci. 2021;14(3):1176–84.

Kumar M, Gautam MK, Singh A, Goel RK. Healing effects of Musa sapientum var. paradisiaca in diabetic rats with co-occurring gastric ulcer: cytokines and growth factor by PCR amplification. BMC Complement Altern Med. 2013;13:305.

Labianca R, Sarzi-Puttini P, Zuccaro SM, Cherubino P, Vellucci R, Fornasari D. Adverse effects associated with non-opioid and opioid treatment in patients with chronic pain. Clin Drug Investig. 2012;32:53–63.

Liu C, Miao R, Raza F, Qian H, Tian X. Research progress and challenges of TRPV1 channel modulators as a prospective therapy for diabetic neuropathic pain. Eur J Med Chem. 2023;245:114893.

Max MB, Kishore-Kumar R, Schafer SC, Meister B, Gracely RH, Smoller B, Dubner R. Efficacy of desipramine in painful diabetic neuropathy: a placebo-controlled trial. Pain. 1991;45(1):3–9.

Nathan HJ, Poulin P, Wozny D, Taljaard M, Smyth C, Gilron I, et al. Randomized trial of the effect of mindfulness-based stress reduction on pain-related disability, pain intensity, health-related quality of life, and A1C in patients with painful diabetic peripheral neuropathy. Clin Diabetes. 2017;35(5):294–304.

Patel MK, Kaye AD, Urman RD. Tanezumab: therapy targeting nerve growth factor in pain pathogenesis. J Anaesthesiol Clin Pharmacol. 2018;34(1):111–6.
Petzke F, Tölle T, Fitzcharles MA, Häuser W. Cannabis-based medicines and medical cannabis for chronic neuropathic pain. CNS Drugs. 2022;36(1):31–44.
Pop-Busui R, Ang L, Boulton AJ, Feldman EL, Marcus RL, Mizokami-Stout K, et al. Diagnosis and treatment of painful diabetic peripheral neuropathy. Arlington: American Diabetes Association; 2022.
Ramirez MA, Borja NL. Epalrestat: an aldose reductase inhibitor for the treatment of diabetic neuropathy. Pharmacotherapy. 2008;28(5):646–55.
Rivat C, Ballantyne J. The dark side of opioids in pain management: basic science explains clinical observation. Pain reports. 2016;1(2):e570.
Romeyke T, Stummer H. Multimodal approaches in the treatment of chronic peripheral neuropathy—evidence from Germany. Int J Environ Res Public Health. 2024;21(1):66.
Russo MA, Baron R, Dickenson AH, Kern KU, Santarelli DM. Ambroxol for neuropathic pain: hiding in plain sight? Pain. 2023;164(1):3–13.
Simpson DM, Robinson-Papp J, Van J, Stoker M, Jacobs H, Snijder RJ, et al. Capsaicin 8% patch in painful diabetic peripheral neuropathy: a randomized, double-blind, placebo-controlled study. J Pain. 2017;18(1):42–53.
Sindrup SH, Otto M, Finnerup NB, Jensen TS. Antidepressants in the treatment of neuropathic pain. Basic Clin Pharmacol Toxicol. 2005;96(6):399–409.
Smith S, Normahani P, Lane T, Hohenschurz-Schmidt D, Oliver N, Davies AH. Pathogenesis of distal symmetrical polyneuropathy in diabetes. Life. 2022a;12(7):1074.
Smith S, Normahani P, Lane T, Hohenschurz-Schmidt D, Oliver N, Davies AH. Prevention and management strategies for diabetic neuropathy. Life. 2022b;12(8):1185.
Snyder MJ, Gibbs LM, Lindsay TJ. Treating painful diabetic peripheral neuropathy: an update. Am Fam Physician. 2016;94(3):227–34.
Sommer C, Cruccu G. Topical treatment of peripheral neuropathic pain: applying the evidence. J Pain Symptom Manag. 2017;53(3):614–29.
Souza D, Pessoni A, de Lira CAB, Vancini RL, Ramírez-Vélez R, Gentil P. Supervised home-based resistance training for managing idiopathic peripheral polyneuropathy–a case report. J Bodyw Mov Ther. 2021;28:126–30.
Spallone V, Lacerenza M, Rossi A, Sicuteri R, Marchettini P. Painful diabetic polyneuropathy: approach to diagnosis and management. Clin J Pain. 2012;28(8):726–43.
Stahl SM, Porreca F, Taylor CP, Cheung R, Thorpe AJ, Clair A. The diverse therapeutic actions of pregabalin: is a single mechanism responsible for several pharmacological activities? Trends Pharmacol Sci. 2013;34(6):332–9.
Staudt MD, Prabhala T, Sheldon BL, Quaranta N, Zakher M, Bhullar R, et al. Current strategies for the management of painful diabetic neuropathy. J Diabetes Sci Technol. 2022;16(2):341–52.
Wernicke JF, Pritchett YL, D'souza DN, Waninger A, Tran P, Iyengar S, Raskin J. A randomized controlled trial of duloxetine in diabetic peripheral neuropathic pain. Neurology. 2006;67(8):1411–20.
Zhu K, Wang L, Liao T, Li W, Zhou J, You Y, Shi J. Progress in the development of TRPV1 small-molecule antagonists: novel strategies for pain management. Eur J Med Chem. 2023;261:115806.
Ziegler D, Fonseca V. From guideline to patient: a review of recent recommendations for pharmacotherapy of painful diabetic neuropathy. J Diabetes Complicat. 2015;29(1):146–56.
Zochodne DW. Mechanisms of diabetic neuron damage: molecular pathways. In: Handbook of clinical neurology, vol. 126; 2014. p. 379–99.

Integrating Antioxidant Therapy and Medications: Advancing the Management of Diabetic Neuropathy

22

Prity Das, Anuradha Gupta, Ranjit Das, Priyanka Verma, Sandip Mondal, and Tanay Samanta

22.1 Introduction

Diabetic neuropathy (DN) is one of the common conditions of diabetic patients where the nerves (peripheral nerves are most common) are damaged due to the elevated blood glucose level, the presence of free radicals (oxidative stress), and inflammation, etc. It can be both sympathetic and parasympathetic in nature, and the classification of the neuropathy can be broad based on its site of prevalence. DN can occur in any type of diabetic patient at any point of the diseased condition, even in secondary diabetic patients. It takes quite a toll on the patient's body, and the foremost symptom is pain: it can be sometimes sharp or stabbing, even; numbness with burning sensations; fatigue; coordination problems; irregularity in digestive conditions: nausea, vomiting, diarrhoea; vision problems, etc. The frequently used pharmaceuticals for the DN are gabapentin, pregabalin, and duloxetine (Feldman et al. 2019).

P. Das · T. Samanta
Department of Pharmaceutical Technology, School of Health and Medical Sciences, Adamas University, Kolkata, West Bengal, India

A. Gupta
School of Materials Science and Technology, Indian Institute of Technology, Banaras Hindu University, Varanasi, Uttar Pradesh, India

R. Das
Department of Microbiology, All India Institute of Medical Sciences (AIIMS), Kalyani, Nadia, West Bengal, India

P. Verma
Kamla Raja Girls Government Post Graduate College, Gwalior, Madhya Pradesh, India

S. Mondal (✉)
Amity Institute of Pharmacy, Amity University Kolkata, Kolkata, West Bengal, India

M. K. Gautam et al. (eds.), *Antioxidants and Functional Foods for Diabetic Neuropathy*, https://doi.org/10.1007/978-981-95-9326-2_22

In relation to the "antioxidant therapy," it indicates the use of antioxidants such as vitamins, coenzymes Q10, polyphenols, as free radical neutralisers, which reduces oxidative stress and decreases the inflammation. The neutralisation of unstable free radicals or molecules helps in the prevention of many diseases and health conditions, and diabetic neuropathy is one of them.

As mentioned earlier, DN is mostly problematic to people for its excruciating pain, and the pain is mostly driven by the oxidative stress and inflammation, where the body is unable to neutralise the excessive free radical. This imbalance of the neutralisation and production of free radicals can be diminished by the presence of the antioxidant. The reduced inflammation due to reduced oxidative stress leads to the protection of nerves and minimises the nerve damage (Bansal et al. 2006).

Antioxidants alone cannot treat the DN completely, as they can only be considered as a complementary way to control the problem. The blood glucose level control and lifestyle management are also required for the betterment of the patient, and for the comprehensive approach, the combination of medicine is preferred. The usage of medications along with the antioxidant therapy has a promising beneficiary site where the potency of relieving the diabetic neuropathy is comparatively more than what antioxidant therapy can achieve alone. Therefore, the need of combining medication is quite undeniable (Kukkar et al. 2013).

The aim of the book chapter is to elaborate on the requirement of the combination of medication along with the antioxidant therapy for the treatment of the diabetic neuropathy condition.

22.2 History and Background

22.2.1 Rise of Diabetic Neuropathy Throughout the Century

In the eighteenth century, the complication of diabetic neuropathy was studied for the first time, and by the Western physicians Thomas Willis and Thomas Sydenham, the symptoms of diabetic neuropathy, such as numbness, pain, and nerve-related damage due to diabetes, were noted down (Karamanou et al. 2016).

In 1798, the symptoms of DN were seen in the patient and were informed for the first time (Nkonge et al. 2023).

After that, in the early twentieth century, along with the discovery of insulin, the interest in diabetic complications inflicted interest among people. Then DN was acknowledged as one of the major complications among the patients (March et al. 2022).

Around the mid-twentieth century, the diagnosis of diabetic neuropathy became more prominent, and the classification of DN, such as peripheral and autonomic, was done in this period of time, which left a significant mark on the history of diabetes complications. The research on the mechanism for the action of nerve damage was also initiated in this era (Tesfaye et al. 2010).

In the late twentieth century, the methods of diagnosis have improved a lot along with the research background. The increased understanding and awareness increased significantly. The use of gabapentin to relieve DN pain was taking over the market. It leaves a milestone for the medical treatment of DN in the research world (Backonja et al. 1998).

Now, in the twenty-first century, the priority of early detection & prevention has taken over. The development of the new medication has also been seen. The research is also enabling the greatest possibilities for the people and the health-care system.

22.2.1.1 Mechanism of Diabetic Neuropathy

As we know, diabetic neuropathy depends on many factors, such as oxidative stress, ROS, AGEs, and inflammation, and all of them have a significant role in the mechanism of diabetic neuropathy. A notable point is that they are all interconnected to each other.

Firstly, hyperglycaemia-induced oxidative stress

The excess glucose present in the blood under chronic hyperglycaemic conditions participates in the polyol pathway, where NADPH consumption occurs. Due to the lack of NADPH, the antioxidant level (glutathione) decreases in the body, resulting in an increased number of AGEs and PKC, as well as mitochondrial dysfunction. Three of these factors further generate ROS, damaging nerve cells and causing inflammation, which inflicts pain (Pang et al. 2020).

Secondly: advanced glycation end-products (AGEs)

Due to chronic hyperglycaemia (not hypoglycaemia), the excess glucose present in the body undergoes nonenzymatic glycation, producing advanced glycation end-products (AGEs). These AGEs are irreversible in nature and cause alterations in the structure and function of neurones, blood vessels, and extracellular matrix (ECM) proteins. They also bind to the specific surface receptor RAGE (receptor of advanced glycation end-product) and activate NF-κB (nuclear factor kappa-light-chain-enhancer of activated B cells) and MAPK (mitogen-activated protein kinase) signalling pathways. This activation leads to the secretion of pro-inflammatory cytokines and ROS elements, causing inflammation and nerve damage that results in neuropathic symptoms (Yue et al. 2022).

Thirdly, dysfunctional mitochondria

Chronic hyperglycaemic conditions (not hypoglycaemic) allow excess glucose to enter nerve cells, causing irregularities in bodily functions. This includes the formation of excess metabolites due to overloaded glycolysis, leading to the generation of excess electrons that overload the electron transport chain (ETC). The excess electrons tend to leak and form reactive oxygen species (ROS) as a byproduct. ROS has the potency to disrupt mitochondrial function, decrease ATP generation, increase mitochondrial ROS production, and result in energy deficiency in nerve cells. This energy deficit condition leads to gradual neuronal degeneration and inflammation (Feldman et al. 2019).

Fourthly, inflammation and nerve damage
Due to the excessive presence of glucose, pro-inflammatory cytokines such as TNF-alpha, IL-1β, and IL-6 are activated. These cytokines further trigger an immune response, activating immune cells (macrophages, T cells, etc.). Consequently, these immune cells secrete oxidative stress mediators in the body, which then damage neuronal components and cause neuronal damage (apoptosis) (Zhu et al. 2024).

22.2.2 Analysis of the Drawback Other Methods

For the treatment of diabetic neuropathy, a number of noncombination methods have been investigated, such as lifestyle changes, occupational therapy, physiotherapy-based therapies, device-based stimulation, and surgical procedures. Nevertheless, these approaches have a number of drawbacks, including the inability to sustain lifestyle modifications, sluggish clinical progress, poor desire, high expense, restricted accessibility, and insufficient availability of qualified experts. Massage, physical therapy, and occupational therapy are examples of nonpharmacological treatments that frequently exhibit inconsistent effectiveness and have scant long-term supporting data. In a similar vein, device-based therapies (TENS, SCS, and PNS) need to be standardised and thoroughly validated in order to avoid discomfort or consequences, uneven results, and other issues. Regular use of surgical procedures is less practical due to the hazards of nerve injury, infection, poor wound healing, and the need for trained surgeons. Therefore, improving accessibility, standardising protocols, enhancing follow-up tracking, and strengthening evidence through research are essential to overcome these drawbacks (Table 22.1) (Shukla; Somers and Somers 1999; Singleton et al. 2015).

Table 22.1 All the other methods for treatment for diabetic neuropathy other than combination therapy

Name	Methodology	Drawback	Possible way to rectify	Reference
1. Lifestyle modification	Exercise: Aerobic strength training in the morning decreases the glucose level in the body by using it in the generation of energy.	1. Not being able to maintain the lifestyle changes, 2. Cost-effectiveness, 3. Slow progression of the changes, 4. Complexity in maintaining the change, 5. Lack of motivation to continue.	Personalisation of the activities, tracking of the changes, realistic milestone set-up, and having a peer support group to help each other out	Obrosova et al. (2007), Singleton et al. (2015), Ghavami et al. (2018), Callaghan et al. (2020)
	Weight management: reduces visceral fat, increases the sensitivity of the insulin, and improves nerve function by decreasing its pressure.			
	Healthy diet (low sugar, high fibre): it decreases the presence of glucose in the body by not intaking it in large amounts.			
	Stress reduction (meditation, yoga): improves neuronal circulation, reduces high cortisol and pro-inflammatory cytokines in the body			
2. Massage and physical therapy	Massage therapy: deep tissue massage, reflexology, myofascial release	1. Limited trained therapists in the society 2. Cost-effectiveness 3. Limited accessibility 4. Limited data on the long-term benefits	Enhanced accessibility to the general population, more research to get the data, home-based exercise program	Akyuz and Kenis (2014), Bayat et al. (2020)
	Physical therapy: aerobics; joint mobilisation; heat, cold & electrical stimulation; balance exercise, etc.			

(continued)

Table 22.1 (continued)

Name	Methodology	Drawback	Possible way to rectify	Reference
3. Occupational therapy	It is basically performing daily activity without focusing on the DN symptoms. Here therapeutic exercises, sensory assessments, cognitive retraining, etc., are performed as programs for better functional ability and pain & discomfort management.	1. Variable efficacy 2. Keeping track of the changes 3. Limited data on long-term benefits in the research area. 4. Limited specialist	Promotion via online platforms, standardisation of the tracking, outcome tracking, and more research initiatives	Shukla (2021)
4. Device-based treatments:	Transcutaneous electrical nerve stimulation (TENS): Here, the delivery of electrical impulses is done through electrodes to stimulate nerves, improving function and pain relief.	1. Variability in efficacy 2. Limited understanding 3. Skin irritation 4. Limited accessibility 5. Limited research	Standardising the protocols, more research required, modifying the development of the device	Somers and Somers (1999)
	Spinal cord stimulation (SCS): To reduce the neuropathic symptoms, electrical impulses are passed to the spinal cord to disrupt the pain signal.	1. Surgical risk 2. Cost-effectiveness 3. Tendency to have complications 4. Limited data	More research, increased insurance coverage, modification of the devices, and more improved surgical techniques	Henson et al. (2021)
	Peripheral nerve stimulation (PNS): the delivery of electrical impulses to the specific peripheral nerves, such as the sural nerve, saphenous nerve, etc., to reduce the chronic pain.	1. Surgical risk, 2. Surgical complication 3. Limited efficacy 4. Tendency to damage nerve 5. Cost-effectiveness	More research Improvised devices and surgical techniques	Wang et al. (2022)

(continued)

Table 22.1 (continued)

Name	Methodology	Drawback	Possible way to rectify	Reference
5. Surgical interventions:	Decompression surgery (nerve decompression): Here it can be done in 3 ways: 1. Neurolysis: compressed nerves are released from the surrounding tissues. 2. Neurectomy: damage nerve or its segments are removed permanently. 3. Epineurolysis: removal of the fibrotic tissues around the nerve to relieve it from the pain.	1. Surgical risk 2. Nerve damage 3. Variable response rate 4. Limited data 5. Cost-effectiveness 6. Limited skilled surgeon	Improved surgical technique required with more research on it, insurance coverage increment, standardisation of the protocol	Melenhorst et al. (2009)
	Nerve grafting: surgical reconstruction of previously damaged nerve by inserting a nerve graft (obtained from the donor) between the damaged nerve's end and connecting it to the surrounding nerves for betterment.	1. Infection, 2. Poor wound healing and graft rejection, 3. Adverse reaction to anaesthesia 4. Lack of donor	Use of bioengineered nerves, uses of biocompatible material to reduce the risk of the infection, stem cell therapy for the regeneration.	Navarro et al. (1997)

22.3 Medicine Therapy

- *What is medicine therapy*

From the context, it's clear that we're discussing a combination of medications, specifically "Medicine Combination Therapy." This approach involves using two or more medications to treat a particular disease, typically complex conditions or health issues, where a single medication cannot meet all the body's needs. To enhance the effectiveness of the active pharmaceutical ingredient (API), reduce symptoms, and minimise side effects, combination therapy is preferred. The complementary mechanisms of action of the drugs reduce the risk of resistance, promoting more effective treatment outcomes for patients (Sun et al. 2016).

For an example we can talk about, antidepressants such as Duloxetine & Amitriptyline: Duloxetine is a serotonin and norepinephrine reuptake inhibitor (SNRI) class of antidepressant (Smith and Nicholson 2007), whereas amitriptyline is a tricyclic class of antidepressant (TCA) (Max et al. 1987). Combinations of these drugs are used for chronic pain, diabetic neuropathy, or for major depressive disorder. Though the mechanisms of action of the drugs are different, they are quite complementary in nature with each other, so these are preferred combinations (Tripathi et al. 2016).

Anticonvulsants (e.g., pregabalin and gabapentin): Gabapentin (Kukkar et al. 2013) and pregabalin (Azmi et al. 2019) are two different anticonvulsants that can be used for neuropathic pain or analgesia, respectively. However, combining these two drugs is rare due to their overlapping mechanisms of action, which increases the risk of side effects. Typically, gabapentin or pregabalin is combined separately with opioids, NSAIDs, or antidepressants for the treatment of chronic pain (Senderovich and Jeyapragasan 2018).

Opioids (e.g., Tapentadol, Tramadol): When other methods to control pain are ineffective, opioids such as Tapentadol or Tramadol come into play. A combination of these two drugs is sometimes used in the treatment of neuropathic pain and complex pain conditions (Roulet et al. 2021).

22.3.1 Reason Behind the Combination Therapy

The need for combination therapy in the medicinal field arises from its holistic nature. Complex diseases often involve multiple biological pathways that cannot be effectively controlled or prevented through a single type of medication. Furthermore, side effects and adverse reactions are frequent concerns. To provide better healthcare, combination therapy targets several aspects of complex diseases. The primary objectives of combination therapy are:

1. Multifaceted actions (Yorek 2024)
2. Symptom relief
3. Decreased drug resistance
4. Reduced disease progression (e.g., in HIV)
5. Minimised side effects (Guerrero-García and Rubio-Guerra 2018)
6. Lower dose requirements
7. Treatment of multiple conditions simultaneously

These benefits positively impact healthcare systems, leading to improved outcomes among patients with complicated diseases. Consequently, managing such conditions becomes more manageable (Vorobeychik et al. 2011; Tani et al. 2020; Serrano Afonso et al. 2021; Saul et al. 2023).

22.3.2 Mechanism of Action of Combination Therapy

In combination therapy for diabetic neuropathy, many medication classes operate on several pathogenic pathways at once using a multitarget strategy. While vascular protective medications enhance neuronal blood flow and lessen oxidative vascular damage, anti-inflammatory medicines reduce pain and inflammation by blocking

COX-2, NF-κB signalling, and pro-inflammatory cytokines. Antidepressants and anticonvulsants primarily alleviate symptoms by altering ion channels and neurotransmitters, which lessens neuropathic pain and enhances movement and sleep. Through central μ-receptor effects, opioids provide strong analgesia, but their use is restricted because of tolerance and dependence hazards. While ACE inhibitors, vitamin B complex, and mitochondrial enhancers promote nerve protection and lower oxidative stress, topical medications offer local pain relief with less systemic effects. Overall, combination therapy improves efficacy by synergistic action and reduced individual drug doses, although side effects, variable patient responses, and limited nerve regeneration remain challenges (Table 22.2) (Collins et al. 2000; Tajti et al. 2016).

- *Significance of medicine combination in current scenario:*

At present, it is evident that managing diabetic neuropathy (DN) requires a multifaceted approach, incorporating lifestyle modifications alongside medication. As previously discussed, relying solely on antioxidants is insufficient for treating this chronic pain condition; thus, medicinal intervention is necessary. The combination of lifestyle management and medication can address multiple conditions simultaneously (Smith et al. 2022).

Medication can target the origin of pain, while antioxidants provide stability and enhance onset activity, leading to prolonged effects in some cases. Diabetic patients are prone to DN, and proactive medication can serve as a preventative measure, even before symptoms arise, ensuring more effective management (Bril 2012).

From an economic perspective, this preventative approach has made combination therapy a popular choice. Long-term benefits prioritise improved quality of life. Patients can move without debilitating pain, making daily activities like walking and manual work significantly easier. Disease progression slows, and customisation of medication is possible according to individual preferences and needs (Medeiros Dantas et al. 2023).

22.4 Antioxidant Therapy

- *What is antioxidant therapy*

Antioxidant therapy involves using antioxidants to treat diseases or conditions where free radicals and oxidative stress are primary contributing factors. By neutralising excess free radicals, antioxidant therapy reduces oxidative stress, prevents cellular damage, and promotes overall well-being through a holistic approach (Tan et al. 2018).

Table 22.2 Mechanism of action of different type of drugs that are used in the combination therapy

Class of drug	Mechanism of action	Example	Benefits	Limitation	Reference
Anti-inflammatory	It can act by many pathways, such as: 1. Cox-2 inhibition 2. Inhibition of NF-κB pathway 3. Reducing pro-inflammatory cytokine 4. Complementing antioxidant activity	Naproxen, Ibuprofen	Reduction of pain, diminishing of the inflammation, dual protection of nerves due to the presence of both anti-inflammatory as well as antioxidant, synergy effect, decreased dosage requirement.	Ulcer, GIT irritation, hypertension, arrhythmia, limited effectivity, potential drug dependency (e.g., corticosteroids), and nephron activity are also affected.	Joussen et al. (2002)
Vascular protective agents	There are different types of pathways based on the type of drug that is used: 1. Increasing flow of blood to the neurones (e.g., ARBs) 2. Reduction of the vascular inflammation 3. Reduction of oxidative stress	Amlodipine (calcium channel blocker), Metoprolol (beta-blocker)	Neuronal protection, lessening of the progression of the disease, reduction of blood pressure,	Indirect action as a pain reliever, increasing the risk of hypotension, affects the activity of the kidney.	Tajti et al. (2016)
Antidepressant	The 2 main MOAs are: 1. Neurotransmitter modulation, as TCA & SNRI, tends to increase 5-HT & norepinephrine (NE). 2. Sodium channel blocking mechanism	Amitriptyline, Duloxetine	Effective pain management, anxiety reduction, and a multifaceted approach to the disease.	TCA class of drug to cause drowsiness, weight gain, and irregular bowel movement; similarly, SNRI tends to increase BP and dizziness; aside from this, it does not have onset action & has the potency to cause drug tolerance.	Goodnick (2001)

Anticonvulsant	The anticonvulsant can work in the following way: 1. Modulation of calcium channel blocker 2. Reducing the release of excitatory neurotransmitters such as gabapentin, NE, and substance P in the body 3. Reduction of the hyperactivity of signalling pathways	Gabapentin, Pregabalin	Pain-relieving activity, nerve protection, improved mobility of the body and the sleep cycles, decreased the dosage, and synergic activity	Dizziness, coordination issues, increased drowsiness, very little effect on the nerve regeneration, drug tolerance, and withdrawal effect, no onset actions	Collins et al. (2000)
Opioids	1. Reduction of the pain sensation by binding at mu-receptor 2. Reduction of pain by inhibiting pain signal transmission in CNS 3. By inducing antidepressant activity	Tapentadol, tramadol	Nerve-protecting activity, potent analgesic activity, reduces anxiety or depression if associated with DN	Addiction, drug abuse, drug tolerance, drug dependency (tend to show withdrawal effect) gradual onset activity does not have an impact on the nerve regeneration, respiratory depression, nausea, sedation, etc.	Wong et al. (2007)
Topical agents	Capsaicin: reduction of substance P Lidocaine: blocking of sodium channel	Lidocaine patches, capsaicin cream	Application site as target, reduced systemic effect, comparatively fewer side effects, long-term usability, zero addiction tendency	Do not have any effect on the widespread pain; absorption rate is not constant, and there is a limited effect on the pain originating from the depth of the nervous system.	Tandan et al. (1992), Barbano et al. (2004)

(continued)

Table 22.2 (continued)

Class of drug	Mechanism of action	Example	Benefits	Limitation	Reference
ACE inhibitors	Inhibiting ACE in the body.	Lisinopril, Captopril	Reduction of high BP decreases oxidative damage and provides a comprehensive approach for nerve health and renal homeostasis management.	Hypotension, hypokalaemia, variable response due to the patient's individuality, indirect pain relief	Reja et al. (1995)
Vitamin B	1. Neural health protection 2. Decreasing homocysteine 3. Neutralisation of free radicals & reduction of oxidative stress	B6 (pyridoxine), B1 (thiamine), B12 (cobalamin)	Regeneration & repairment of nerve, increased energy production, analgesic action, comparatively fewer side effects	Requirement other pain-managing strategies, different responses in patients, can increase neuropathy if an excess amount is used	Khalil et al. (2021)
Mitochondrial function enhancer	Increasing the activity of mitochondria, as mitochondrial dysfunction has a significant role as the diabetic neuropathy precursor.	Metformin, Pioglitazone	Decreased ROS production, reduction of apoptosis, synergistic effect, slower progression of disease due to the effect of the drug, nerve repair	Variance in response, emerging topic, hypersensitivity induction, costly	Lós et al. (2019)

As mentioned earlier, there are different types of antioxidants used in therapy. The most commonly used antioxidants, along with examples represent in Fig. 22.1.

The method of administration varies based on the requirement and dosage form. While antioxidants are commonly used as dietary supplements, they are also increasingly used in pharmaceutical formulations, such as topical preparations and parenteral formulations (intravenous therapy) (Oyenihi et al. 2015).

The growing popularity of antioxidants stems from their overwhelmingly positive response in treatment. Benefits include (Hennekens et al. 1994; Yoshihara et al. 2010):

- Improved cellular activity and health
- Enhanced immunity
- Reduced disease risk factors
- Enhanced cognitive functions
- Comprehensive well-being

- *The necessity of antioxidant therapy:*

- Antioxidant therapy is necessary for specific populations due to various reasons:
  1. Smokers: Smoking has become a prevalent habit in modern lifestyles, leading to increased free radicals and oxidative stress in the body. Antioxidants help neutralise these effects (Kinnula 2005).
  2. Individuals with chronic diseases: Patients with cardiovascular, cancer, or neurodegenerative diseases often develop conditions linked to oxidative stress, such as:
     - Hypertension (Sinha and Kumar Dabla 2015)
     - High cholesterol (Yang et al. 2009)
     - Heart failure (Korantzopoulos et al. 2003)

     DNA mutations due to excessive oxidative stress (Wei and Lee 2002)

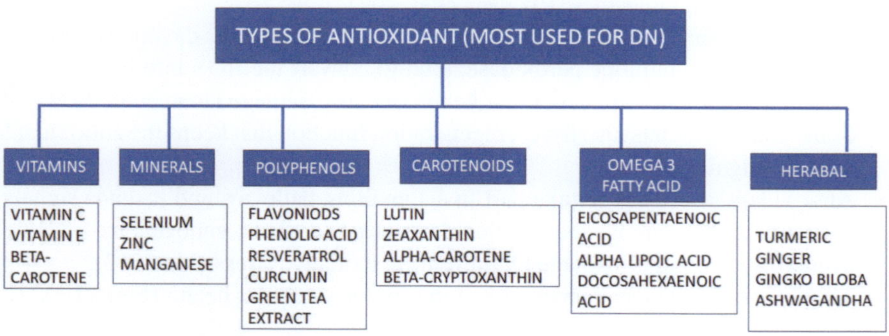

Fig. 22.1 Types of antioxidants for the treatment of diabetic neuropathy. (Based on the nature)

- Increased cancer risk factors (Wei and Lee 2002)
- Alzheimer's disease (AD) (Grundman and Delaney 2002)
- Parkinson's disease (PD) (Prasad et al. 1999)

These conditions highlight the need for antioxidants in treating chronic diseases.

- Athletes: Due to excessive physical activity, athletes experience increased free radicals and oxidative stress, damaging cells and impairing performance. Antioxidant therapy helps reduce oxidative stress (Antonioni et al. 2019).
- Pregnant women: Antioxidants promote cellular health, supporting healthy foetal development (Obeagu et al. 2023).
- Ageing individuals: Antioxidants promote cellular health, longevity, and healthy ageing, slowing down the ageing process (Fletcher et al. 2003).
- Individuals exposed to environmental toxins: Modern lifestyles expose people to UV radiation, environmental toxins, stress, and other external sources of free radicals, emphasising the need for antioxidants (Limón-Pacheco and Gonsebatt 2009).

- *How the antioxidant therapy works:*

Due to several reasons, such as UV radiation, smoking, and stress, metabolic processes increase the oxidative stress in the body. It gives rise to the ROS production in the body; this ROS factor tends to take electrons from the protein, lipid, and DNA molecules. This electron primarily gives stability to the ROS but causes protein alteration in functional and structural proteins, cell damage, and gives rise to the cancer (Fig. 22.2).

When the antioxidant is given, it works on the damage in different methods:

- Antioxidants tend to decrease the oxidation stress in the cell, which provides cellular stability and prevents apoptosis (Forman and Zhang 2021).
- Antioxidants themselves provide electrons to the ROS, and ROS stabilises without affecting biomolecules (Rizwana et al. 2021).
- In some scenarios, some antioxidants, such as superoxide dismutase (SOD), catalase, and glutathione peroxidase, tend to convert the ROS into less reactive molecules that don't affect cellular health or cause damage (Juan et al. 2021).
- Some active oxidants also have a regeneration function that keeps the antioxidant activity ongoing.
- Antioxidant also tends to take part in a signalling pathway, and it sends signals to increase the endogenous antioxidant enzyme. This endogenous enzyme decreases the cellular damage by preventing oxidative stress (Ma 2013).
  Thus, antioxidants mechanism of action promotes cellular health (Fig. 22.3).

- *Evidence of the effective antioxidant therapy as neuroprotective agent:*

In 2010, a study conducted by the Neuroscience Research Center at Shahid Beheshti University in Iran investigated the neuroprotective effects of curcumin, a poly-

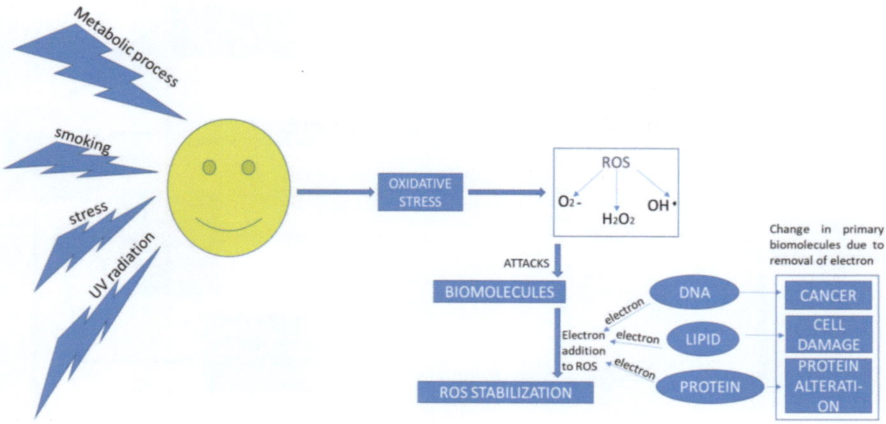

Fig. 22.2 How the oxidative stress produces ROS in the body causes cell damage

phenolic antioxidant compound. The study involved male rats and utilised D-L-Homocysteine (Hcy) as a neurotoxin.

The results showed that curcumin demonstrated significant neuroprotective activity against Hcy-induced toxicity. Notably, curcumin exhibited this protective effect at both low and high doses, primarily by:

- Reducing oxidative stress through decreased lipid peroxidation.
- Neutralising reactive oxygen species (ROS), evidenced by reduced superoxide anion (SOA) and malondialdehyde (MDA) formation

The study concluded that curcumin has potent neuroprotective properties (Fig. 22.4) (Ataie et al. 2010)

- *Classification of antioxidant for the diabetic neuropathy:*

22.5 Rationale for Combination Therapy

- *Synergistic effect of drug and antioxidant:*

Antioxidants don't target a specific organ; instead, they work systemically throughout the body. When combined with pharmaceuticals, therapeutic outcomes improve significantly. The synergistic effect enhances anti-inflammatory actions.

Diabetic neuropathy is a complex condition involving multiple mechanisms. Synergistic action addresses the root cause in several ways:

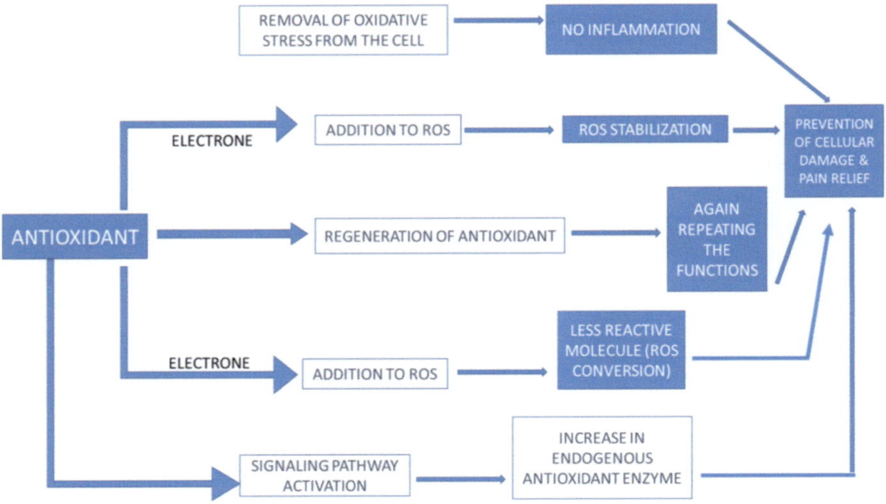

Fig. 22.3 Different mechanism of actions of antioxidant for the diabetic neuropathy

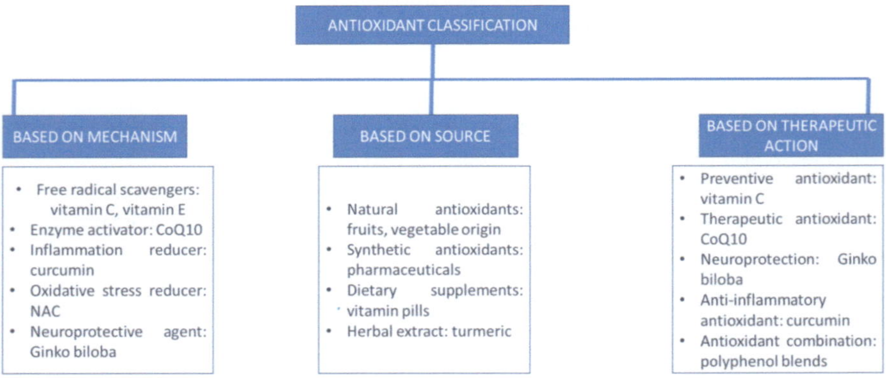

Fig. 22.4 Classification of the antioxidant based on mechanism, source of antioxidants & therapeutic action of the antioxidants for diabetic neuropathy

1. Reduced oxidative stress: Combining ACE inhibitors with antioxidants decreases ROS production and improves cellular health (Elfowiris and Banigesh 2022)
2. Combined anti-inflammatory action: NSAIDs and antioxidants reduce pain and swelling due to nerve damage (Oyenihi et al. 2015)
3. Improved vascular health: ACE inhibitors (Malik 2000) and calcium channel blockers (Murray 1991), along with antioxidants, enhance blood flow, ensuring nutrient supply to neurons.
4. Enhanced neuroprotection: Antidepressants or anticonvulsants with antioxidants stabilise nerve cells and reduce pain (Kośmider et al. 2023).
5. Enhanced pain relief: Opioids with antioxidants modulate CNS signals, reducing oxidative stress (Thorsen et al. 2020)

Specific antioxidants, such as alpha-lipoic Lipoic Acid (ALA), manage blood glucose levels, preventing neuropathic worsening (Viana et al. 2022).
The synergistic effect offers additional benefits:

- Reduced required dosage (Elfowiris and Banigesh 2022)
- Lower risk of side effects

- *Efficacy enhancement of drug by antioxidant:*
  - The activity of the drug enhances to varying degrees depending on the combination of medication and antioxidant. Primarily, this enhancement increases the bioavailability of the active pharmaceutical ingredient (API) while reducing rapid metabolism (Ribaudo et al. 2020).
    Additionally, antioxidants protect nerves, allowing the drug to target the root cause of the condition without harming neurons. As previously discussed, the synergistic activity amplifies the effects of both the antioxidant and medication, leading to better outcomes (Viana et al. 2022).
    Furthermore, antioxidants promote targeted drug delivery, increasing drug availability at the site of action and enhancing efficacy comprehensively (Chen et al. 2023)
    This improved API action, facilitated by antioxidants, not only manages symptoms of diabetic neuropathy but also provides neuroprotection and improves patients' quality of life (Fig. 22.5) (Viana et al. 2022).
- *Potential of improved outcomes for the patient:*

- *Mitigation of pharmacological adverse effects:*

Due to the synergy effect, the required dose decreases, significantly reducing the tendency for dose-dependent adverse side effects and opioid dependence (Thorsen et al. 2020).

This combination therapy also provides protection against organ damage, safeguarding the liver, kidney, gastrointestinal tract (GIT), and other organs from harm. Additionally, it improves vascular tissue health, enhancing blood and fluid circulation, which can be compromised by long-term use of other therapies (Senderovich and Jeyapragasan 2018).

Enhanced neuroprotection reduces cognitive and motor impairment, mitigating the risk of long-term cognitive side effects. Notably, sole NSAID use can compromise gastrointestinal health, but antioxidant coadministration protects the lumen and tissues, preventing:
- Nausea
- Vomiting
- Acid reflux
- Ulcer formation (Joussen et al. 2002; Oyenihi et al. 2015)

By promoting mitochondrial function, this combination decreases oxidative stress, counteracting metabolic side effects of medications (Lós et al. 2019).

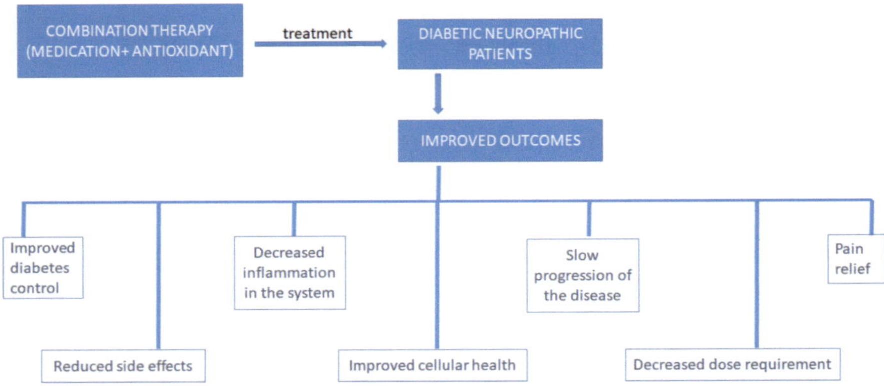

Fig. 22.5 Positive impact on the diabetic neuropathic patients (Vorobeychik et al. 2011)

In summary, combining medications with antioxidants mitigates pharmacological adverse events and lessens side effects.

- *Biological processes that elucidate the advantages of combination treatment:*
  As previously discussed, reducing oxidative stress protects cells from damage, maintaining cellular health. Similarly, this approach decreases the inflammatory response in affected nerves, reducing neuronal inflammation. Combination therapy also supports neuroprotection by stabilising neurons and preventing degeneration (Joussen et al. 2002; Oyenihi et al. 2015).
  Moreover, this therapy improves mitochondrial function, boosting cellular energy production, which is essential since mitochondrial dysfunction is a significant factor in diabetic neuropathy (Lós et al. 2019).
- Additional benefits include:
  - Enhanced vascular health, increasing blood flow and nutrient delivery to tissues (Tajti et al. 2016)
  - Modulation of pain signalling pathways, providing comprehensive pain relief (Kirkpatrick et al. 2015)
  - Antioxidant properties that improve glucose control, minimising hyperglycaemic damage (Bajaj and Khan 2012).
  - Synergistic effects that reduce required dosage (Elfowiris and Banigesh 2022), lowering drug-related side effects and dependency (particularly with opioid analgesics) (Karajibani et al. 2017)

  This combination therapy provides more tolerable treatment options for patients.
- *The function of inflammatory pathways and neurotrophic factors:*
  In diabetic neuropathy, the pathway of inflammation has a very significant role in the progression of this particular disease, where the key factors are nuclear factor Kappa b (Cheng et al. 2024), tumour necrosis factor-alpha (Baum et al. 2021), interleukin-1 beta (Xue et al. 2021), interleukin-6 (Baum et al. 2021),

cyclooxygenase-2 (Xue et al. 2021). And results in the production of pro-inflammatory cytokines, nerve damage, cellular apoptosis, axonal degeneration, & Schwann dysfunction in neurons, disruption in the blood-nerve barrier, i.e., it's affecting the homeostasis of the system and negatively impacting the body.

Whereas neurotrophic factors are those that support the growth, maintenance, and survival of the nerves. Nerve growth factor, brain-derived neurotrophic factors, neurotrophic-3, neurotrophic-4, glial cell line-derived neurotrophic factors, etc., are a few significant ones. Primarily, it helps in the neuronal survival and growth of the neuron, regeneration of the axons, synaptic plasticity, modulation of the pain, and neuro-protection (Fornaro et al. 2020; Cheng et al. 2024).

And in the study, it has been found there is a complex relationship between the inflammatory pathways and neurotrophic factors. For example, inflammation (due to the inflammatory pathway) tends to reduce neurotrophic factor expressions; similarly, neurotrophic factors also inhibit the inflammatory pathway so that inflammation does not occur in the body. And because of this interconnection, there is always a crosstalk between neurotrophic factor signals and nuclear factor Kappa B & an imbalance between pro-inflammatory and anti-inflammatory signals (Lima Giacobbo et al. 2019).

- *Case studies demonstrating effective combo therapy:*

  In the study, the synergic effect of levetiracetam with gabapentin along with antioxidants (alpha-lipoic acid, coenzyme Q10) is used for the diabetic neuropathy treatment. Here mouses were used, and DN was induced in them to use as a model. Via the tail flick test, the combination of medication as well as individual medications was tested. Later, with coadministration in a specific dose fraction, the effects were tested.

  For the study, it can be concluded that the combination therapy of medicine with antioxidants in the treatment of diabetic neuropathy had a positive response in the model, and it discloses the path for more exploration in this area (Stepanović-Petrović et al. 2017).

- *Drugs for Combination Medications:*

  The frequently suggested combination therapies for diabetic neuropathy, which combine typical neuropathic treatments with antioxidants to improve treatment results. Antioxidants such alpha-lipoic acid, omega-3 fatty acids, acetyl-L-carnitine, vitamins C/E, coenzyme Q10, and NAC are mixed with drugs like gabapentin, pregabalin, duloxetine, amitriptyline, and anticonvulsants/opioids. By lowering oxidative stress, enhancing mitochondrial activity, decreasing ROS generation, and promoting nerve repair, these combinations primarily seek to improve pain relief and neuroprotection. Overall, the table indicates that through synergistic and disease-modifying effects, antioxidant cotherapy may increase the effectiveness of conventional medications (Table 22.3) (Akorede et al. 2023; Pingali et al. 2024).

Table 22.3 The combination of medication and antioxidant (which are used)

Name	General uses	Combining action with antioxidant	Mechanism of action	References
Gabapentin	Anticonvulsant	Alpha-lipoic acid	Nerve damage prevention and neuroprotection	Pingali et al. (2024)
Duloxetine	Antidepressant	Omega-3-fatty acid	Nerve repairment, anti-inflammatory action	Rowin (2020)
Amitriptyline	Antidepressant	Acetyl-L-carnitine	Reduction of the oxidative stress	Herzmann et al. (2005)
Pregabalin	Anticonvulsant	Vitamin C	Reduction of the oxidative stress, neuroprotection	Pingali et al. (2024)
Carbamazepine	Anticonvulsant	Vitamin C, Vitamin E	Pain relief, Reduction of the oxidative stress	Akorede et al. (2023)
Valproate	Anticonvulsant	Alpha-lipoic acid, Coenzyme Q10	Targets the different pain signalling pathway, neuroprotection	Mohammed et al. (2021)
Venlafaxine	Antidepressant	Omega-3 fatty acids, N-Acetylcysteine	Reduction of oxidative stress, neuro-regeneration	Kelkar and Kelkar (2020)
Desipramine	Antidepressant	Acetyl-L-carnitine, Polyphenols (curcumin, resveratrol)	Anti-inflammatory action, neuro-protection	Hu et al. (2024)
Tramadol	Anti-inflammatory	Vitamin C, Magnesium	Reduced oxidative stress, neuro-regeneration	Carr and McCall (2017)
Tapentadol	Anti-inflammatory	Alpha-lipoic acid, Vitamin B6	Reduced oxidative stress, neuro-regeneration	Jang and Oh (2023)
Oxycodone	Analgesic	Coenzyme Q10, N-Acetylcysteine	Mitochondrial function enhancement, reduced oxidative stress	Singh et al. (2024)
Fentanyl	Analgesic	Omega-3 fatty acids, Antioxidant cocktail (Vitamins C, E, and beta-carotene)	Protection from the damage	Chong et al.
Epalrestat	Aldose-reductase inhibitor	Vitamin C, Vitamin E	Reduced oxidative stress, reduced ROS production	Ziegler (2023)
Tolrestat	Aldose-reductase inhibitor	Alpha-lipoic acid, Magnesium	Reduced oxidative stress, reduced ROS production	Rendell (2021)
Sorbinil	Aldose-reductase inhibitor	Acetyl-L-carnitine, Coenzyme Q10	Mitochondrial function enhancement, reduced oxidative stress	Malone et al. (1996)

22.6 Clinical Trials with Evidence

- *Overview:*
  In the clinical trials, a screening phase was held for the diabetic neuropathic patients, and the people who were fulfilling the required criteria for the specific study were selected. And among the selected ones who consented to be participants in the study were under the clinical trial.
  In most of the scenarios, the usual hypoglycaemic drugs were given; additionally, the combination of the medication and antioxidant was given to only one group; the rest were given different medications (only antioxidants, only medication, or placebo). The trial was conducted with proper care and checkups for the betterment of the patients.
- *Study on combination therapy: standard drugs with antioxidants:*
- Alpha-lipoic acid and pregabalin, carbamazepine (anticonvulsant): 150 DN patients participated in the study where they were divided into 3 groups. One group was administered pregabalin, another was administered carbamazepine, and the 3rd group was administered alpha-lipoic acid. From the study, it can be concluded that the use of the combination can help in pain reduction in the patient's body, proving its effectiveness as they have fewer side effects in the body (Patel et al. 2014).
- Alpha-lipoic acid and gabapentin (anticonvulsant): It was a controlled randomised trial of 82 people where they were divided into four groups. One group was given gabapentin, another was given alpha-lipoic acid, and the third group was given both of them in combination. The fourth one was given a placebo. The study was conducted for 12 weeks (84 days). At the end of the study, it was concluded that alpha-lipoic acid can reduce oxidative stress, but in the presence of gabapentin, there is a synergistic effect, and it works more effectively (Komurcu et al. 2020).
- Combined metabolic medication with antioxidants: in this study, the combined metabolic medication (CMM) contained both API and antioxidants: riboflavin, inosine, nicotinamide, and succinic acid. Here riboflavin and succinic acid were used as antioxidants. 269 diabetic neuropathic patients were under the clinical trial for 24 weeks. The trial was a multicentred, randomised, double-blind, placebo-controlled, parallel-group trial. In this study it could be concluded that there was significant improvement in the diabetic neuropathy symptoms along with improved quality of life & proper safety with a tolerability profile (Kharitonova et al. 2022).

22.7 Future Prospects

- Research and development: As neither all the possible combinations of drugs nor the mechanism of action nor all the possible outcomes are known to humankind, there are always broader prospects in the research area. Due to the demand, the change in need in the market leaves the open spot for the development. The modification in the existing formulation provides much more benefit to the company and the economy of the region (Tavakoli et al. 2023)

- Public awareness and education: The awareness among the people about the need for antioxidant therapy along with the medications can lead to a healthier next generation, where more options are available for the treatment & comprehensive approach towards a better lifestyle (Vincent et al. 2004).
- Emerging antioxidant & novel therapy: As we can see, antioxidant therapy alone can't cure any condition or disease but has a great role when medications are added alongside it, which emerges the opening of novel therapies in the health-care world. With the current data in the market, there's a huge possibility of a positive outcome in the near future (Feldman 2003).
- Customisation of medication: every individual has individual needs, and the body has different outcomes due to the same medication. To fulfil the requirement of the individual, there is a huge demand in the market for the customised medicines. If the personalisation becomes possible, then it can be used for the greater good in the forthcoming (Klonoff 2008).
- Integrative and holistic approach: The antioxidant does not have a single use in the body; neither do the medications. If the study on the synergic effect of the can is done further, then there is a huge possibility of preparing formulations that have the capability to cure multiple conditions in one go. This will not only boost the market but also tend to become revolutionary research for mankind (Edwards et al. 2008).
- Genetic and molecular-based therapies combined with antioxidants: The genetic and molecular-based therapies have a promising future. If the antioxidant therapy is added along with them, then the targeted delivery with synergic action can have better prospects with treatment specification (Hajdú et al. 2024).

22.8 Research Gaps

- Pharmacokinetics of antioxidants: There is very limited data on the absorption of antioxidants from food and supplements, and also the targeted delivery with the distribution is hectic. The metabolic pathway and the possible impact of metabolism are still unidentified. The different nature of the antioxidants has different therapeutic windows, leading to different excretion rates, which results in one of the major research gaps in the present scenario (Rapisarda et al. 2022)
- Bioavailability of antioxidants: The different kinds of antioxidants have different pharmacokinetic and pharmacodynamic properties, and it has not been studied on a long-term basis. The bioavailability of the antioxidant doesn't have concrete evidence. And due to that, the dose amount, duration of action, onset action timing, and half-life data are still not known properly (Briançon et al. 2011).
- Potential effect of long-term use of combination therapy: We already know the use of antioxidants in a high dose already has a toxic effect on the body. But on a long-term basis, in a combination therapy, it has the potential to cause accumu-

lation in the system, leading to more toxic effects and ultimately causing damage and inflammation in the cell (nonneuroprotective action) (Briançon et al. 2011).

- Cost-effectiveness: The cost-effectiveness in diabetic neuropathy for the combination of medication with antioxidant therapy is always essential but under research. As it is a still ongoing research study, health-care providers cannot recommend every other patient with full confidence. The variability in effects and dosing, limited clinical study data, along with the financial condition of the patient, leaves a proper gap for being used on a large amount in the market (Gordois et al. 2003).
- New pharmaceutical candidates under investigation: There are more medications that are still under trial to combine with antioxidant therapy for the treatment of diabetic neuropathy. All the present meditation more or less has side effects and adverse effects in different individuals. Therefore, the upcoming new pharmaceutical formulations are wanted in the market with better outcomes and which can be used on a long-term basis without affecting the patient's health (Syed et al. 2023).
- Progress in genetic understanding and personalised treatments: As discussed earlier in the future prospect, the customisation or personalisation of medications is dearly needed. As every other individual has different ways of responding to antioxidant therapy and the medications are combined with antioxidants, therefore the synergic effect is also different in individuals. So the genetic understanding (for the difference) in this scenario is very much required so that the customisation of the treatment can be progressed (Nair 2010).

22.9 Conclusion

The objective of our topic was to discuss the use of medicine combined with antioxidant therapy for diabetic neuropathy. And during the study, it can be concluded that it is definitely one of the more comprehensive ways to treat and control the diabetic neuropathic pain and symptoms, and it has a better prospect in the future too. Though there is limited data, still in the current scenario, it is one of the better alterations to treat the patients, and hereby we can state that we can expect better results in this method of treatment.

References

Akorede GJ, Ambali SF, Olatunji AO, Aremu A, Ahmed AO, Basiru A, Azeez MO, Sanusi F, Kadir RA, Abdulmajeed I. Vitamin C protects against blood and thyroid toxicities induced by the chronic use of Carbamazepine in rats. Iranian J Toxicol. 2023;17:147–56.

Akyuz G, Kenis O. Physical therapy modalities and rehabilitation techniques in the management of neuropathic pain. Am J Phys Med Rehabil. 2014;93:253–9.

Antonioni A, Fantini C, Dimauro I, Caporossi D. Redox homeostasis in sport: do athletes really need antioxidant support? Res Sports Med. 2019;27:147–65.

Ataie A, Sabetkasaei M, Haghparast A, Moghaddam AH, Kazeminejad B. Neuroprotective effects of the polyphenolic antioxidant agent, Curcumin, against homocysteine-induced cognitive impairment and oxidative stress in the rat. Pharmacol Biochem Behav. 2010;96:378–85.

Azmi S, ElHadd KT, Nelson A, Chapman A, Bowling FL, Perumbalath A, Lim J, Marshall A, Malik RA, Alam U. Pregabalin in the management of painful diabetic neuropathy: a narrative review. Diab Ther. 2019;10:35–56.

Backonja M, Beydoun A, Edwards KR, Schwartz SL, Fonseca V, Hes M, LaMoreaux L, Garofalo E, Group, G.D.N.S., Group, G.D.N.S. Gabapentin for the symptomatic treatment of painful neuropathy in patients with diabetes mellitus: a randomized controlled trial. JAMA. 1998;280:1831–6.

Bajaj S, Khan A. Antioxidants and diabetes. Indian J Endocrinol Metabol. 2012;16:S267–71.

Bansal V, Kalita J, Misra U. Diabetic neuropathy. Postgrad Med J. 2006;82:95–100.

Barbano RL, Herrmann DN, Hart-Gouleau S, Pennella-Vaughan J, Lodewick PA, Dworkin RH. Effectiveness, tolerability, and impact on quality of life of the 5% lidocaine patch in diabetic polyneuropathy. Arch Neurol. 2004;61:914–8.

Baum P, Toyka KV, Blüher M, Kosacka J, Nowicki M. Inflammatory mechanisms in the pathophysiology of diabetic peripheral neuropathy (DN)—new aspects. Int J Mol Sci. 2021;22:10835.

Bayat D, Mohammadbeigi A, Parham M, Mehrandasht A, Hashemi M, Mahlooji K, Asghari M. The effect of massage on diabetes and its complications: a systematic review. Crescent J Med Biol Sci. 2020;7:22.

Briançon S, Boini S, Bertrais S, Guillemin F, Galan P, Hercberg S. Long-term antioxidant supplementation has no effect on health-related quality of life: the randomized, double-blind, placebo-controlled, primary prevention SU. VI. MAX trial. Int J Epidemiol. 2011;40:1605–16.

Bril V. Treatments for diabetic neuropathy. J Peripher Nerv Syst. 2012;17:22–7.

Callaghan BC, Gallagher G, Fridman V, Feldman EL. Diabetic neuropathy: what does the future hold? Diabetologia. 2020;63:891–7.

Carr AC, McCall C. The role of vitamin C in the treatment of pain: new insights. J Transl Med. 2017;15:77.

Chen S, Wang J, Feng J, Xuan R. Research progress of Astaxanthin nano-based drug delivery system: applications, prospects and challenges? Front Pharmacol. 2023;14:1102888.

Cheng Y, Chen Y, Li K, Liu S, Pang C, Gao L, Xie J, Wenjing L, Yu H, Deng B. How inflammation dictates diabetic peripheral neuropathy: an enlightening review. CNS Neurosci Ther. 2024;30:e14477.

Chong CR, Phuong H, Choi B, Raman B, Horowitz JD, Omega-3 fatty acids modulate Weibel-Palade body degranulation and actin cytoskeleton rearrangement in PMA-stimulated human umbilical vein endothelial cells .

Collins SL, Moore RA, McQuay HJ, Wiffen P. Antidepressants and anticonvulsants for diabetic neuropathy and postherpetic neuralgia: a quantitative systematic review. J Pain Symptom Manag. 2000;20:449–58.

Edwards JL, Vincent AM, Cheng HT, Feldman EL. Diabetic neuropathy: mechanisms to management. Pharmacol Ther. 2008;120:1–34.

Elfowiris A, Banigesh A. Evaluation of antioxidant therapeutic value of ACE inhibitor as adjunct therapy on type 2 diabetes mellitus patients with cardiovascular disease. ACS Pharmacol Transl Sci. 2022;5:413–8.

Feldman EL. Oxidative stress and diabetic neuropathy: a new understanding of an old problem. J Clin Invest. 2003;111:431–3.

Feldman EL, Callaghan BC, Pop-Busui R, Zochodne DW, Wright DE, Bennett DL, Bril V, Russell JW, Viswanathan V. Diabetic neuropathy. Nat Rev Dis Prim. 2019;5:1–18.

Fletcher AE, Breeze E, Shetty PS. Antioxidant vitamins and mortality in older persons: findings from the nutrition add-on study to the Medical Research Council Trial of Assessment and Management of Older People in the Community. Am J Clin Nutr. 2003;78:999–1010.

Forman HJ, Zhang H. Targeting oxidative stress in disease: promise and limitations of antioxidant therapy. Nat Rev Drug Discov. 2021;20:689–709.

Fornaro M, Giovannelli A, Foggetti A, Muratori L, Geuna S, Novajra G, Perroteau I. Role of neurotrophic factors in enhancing linear axonal growth of ganglionic sensory neurons in vitro. Neural Regen Res. 2020;15:1732–9.
Ghavami H, Radfar M, Soheily S, Shamsi SA, Khalkhali HR. Effect of lifestyle interventions on diabetic peripheral neuropathy in patients with type 2 diabetes, result of a randomized clinical trial. Agri. 2018;30:165–70.
Goodnick PJ. Use of antidepressants in treatment of comorbid diabetes mellitus and depression as well as in diabetic neuropathy. Ann Clin Psychiatry. 2001;13:31–41.
Gordois A, Scuffham P, Shearer A, Oglesby A, Tobian JA. The health care costs of diabetic peripheral neuropathy in the US. Diabetes Care. 2003;26:1790–5.
Grundman M, Delaney P. Antioxidant strategies for Alzheimer's disease. Proc Nutr Soc. 2002;61:191–202.
Guerrero-García C, Rubio-Guerra AF. Combination therapy in the treatment of hypertension. Drugs in context. 2018;7:212531.
Hajdú N, Rácz R, Tordai DZ, Békeffy M, Vági OE, Istenes I, Körei AE, Kempler P, Putz Z. Genetic variants influence the development of diabetic neuropathy. Int J Mol Sci. 2024;25:6429.
Hennekens CH, Buring JE, Peto R. Antioxidant vitamins—benefits not yet proved. Mass Medical Soc. 1994:1080–1.
Henson JV, Varhabhatla NC, Bebic Z, Kaye AD, Yong RJ, Urman RD, Merkow JS. Spinal cord stimulation for painful diabetic peripheral neuropathy: a systematic review. Pain Ther. 2021;10:1–14.
Herzmann C, Johnson MA, Youle M. Long-term effect of acetyl-L-carnitine for antiretroviral toxic neuropathy. HIV Clin Trials. 2005;6:344–50.
Hu M, Cai J-Y, He Y, Chen K, Hao F, Kang J-S, Pan Y, Tie L, Li X-J. Protective effects of curcumin on desipramine-induced islet β-cell damage via AKAP150/PKA/PP2B complex. Acta Pharmacol Sin. 2024;45:327–38.
Jang HN, Oh TJ. Pharmacological and nonpharmacological treatments for painful diabetic peripheral neuropathy. Diabetes Metab J. 2023;47:743.
Joussen M, Poulaki V, Mitsiades N, Kirchhof B, Koizumi K, Döhmen S, Adamis P. Nonsteroidal anti-inflammatory drugs prevent early diabetic retinopathy via TNF-α suppression. FASEB J. 2002;16:438–40.
Juan CA, Pérez de la Lastra JM, Plou FJ, Pérez-Lebeña E. The chemistry of reactive oxygen species (ROS) revisited: outlining their role in biological macromolecules (DNA, lipids and proteins) and induced pathologies. Int J Mol Sci. 2021;22:4642.
Karajibani M, Montazerifar F, Feizabad AK. Study of oxidants and antioxidants in addicts. Int J High Risk Behav Addict. 2017;6
Karamanou M, Protogerou A, Tsoucalas G, Androutsos G, Poulakou-Rebelakou E. Milestones in the history of diabetes mellitus: the main contributors. World J Diabetes. 2016;7:1.
Kelkar S, Kelkar S. Treatment of painful diabetic neuropathy. In: Diabetic neuropathy and clinical practice; 2020. p. 241–63.
Khalil H, Ang CD, Khalil V. Vitamin B for treating diabetic peripheral neuropathy–a systematic review. Diabetes Metab Syndr Clin Res Rev. 2021;15:102213.
Kharitonova T, Shvarts YG, Verbovoy AF, Orlova NS, Puzyreva VP, Strokov IA. Efficacy and safety of the combined metabolic medication, containing inosine, nicotinamide, riboflavin and succinic acid, for the treatment of diabetic neuropathy: a multicenter randomized, double-blind, placebo-controlled parallel group clinical trial (CYLINDER). BMJ Open Diab Res Care. 2022;10:e002785.
Kinnula V. Focus on antioxidant enzymes and antioxidant strategies in smoking related airway diseases. Thorax. 2005;60:693–700.
Kirkpatrick DR, McEntire DM, Hambsch ZJ, Kerfeld MJ, Smith TA, Reisbig MD, Youngblood CF, Agrawal DK. Therapeutic basis of clinical pain modulation. Clin Transl Sci. 2015;8:848–56.
Klonoff DC. Personalized medicine for diabetes. SAGE Publications; 2008. p. 335–41.

Komurcu HF, Baran P, Erel O, Anlar O. The efficacy of alpha-lipoic acid and/or gabapentin on the oxidant-antioxidant system in patients with diabetic polyneuropathy. Int J Res Med Sci. 2020;8:2777.

Korantzopoulos P, Galaris D, Papaioannides D, Siogas K. The possible role of oxidative stress in heart failure and the potential of antioxidant intervention. Med Sci Monit. 2003;9:RA120-RA125.

Kośmider K, Kamieniak M, Czuczwar SJ, Miziak B. Second generation of antiepileptic drugs and oxidative stress. Int J Mol Sci. 2023;24:3873.

Kukkar A, Bali A, Singh N, Jaggi AS. Implications and mechanism of action of gabapentin in neuropathic pain. Arch Pharm Res. 2013;36:237–51.

Lima Giacobbo B, Doorduin J, Klein HC, Dierckx RA, Bromberg E, de Vries EF. Brain-derived neurotrophic factor in brain disorders: focus on neuroinflammation. Mol Neurobiol. 2019;56:3295–312.

Limón-Pacheco J, Gonsebatt ME. The role of antioxidants and antioxidant-related enzymes in protective responses to environmentally induced oxidative stress. Mutation Res Genet Toxicol Environ Mutagen. 2009;674:137–47.

Lós DB, de Oliveira WH, Duarte-Silva E, Sougey WWD, de Freitas EDSR, de Oliveira AGV, Braga CF, de França MER, da Rocha Araújo SM, Rodrigues GB. Preventive role of metformin on peripheral neuropathy induced by diabetes. Int Immunopharmacol. 2019;74:105672.

Ma Q. Role of nrf2 in oxidative stress and toxicity. Annu Rev Pharmacol Toxicol. 2013;53:401–26.

Malik RA. Can diabetic neuropathy be prevented by angiotensin-converting enzyme inhibitors? Ann Med. 2000;32:1–5.

Malone JI, Lowitt S, Salem AF, Miranda C, Korthals JK, Carver J. The effects of acetyl-L-carnitine and sorbinil on peripheral nerve structure, chemistry, and function in experimental diabetes. Metabolism. 1996;45:902–7.

March CA, Libman IM, Becker DJ, Levitsky LL. From antiquity to modern times: a history of diabetes mellitus and its treatments. Horm Res Paediatr. 2022;95:593–607.

Max M, Culnane M, Schafer S, Gracely R, Walther D, Smoller B, Dubner R. Amitriptyline relieves diabetic neuropathy pain in patients with normal or depressed mood. Neurology. 1987;37:589.

Medeiros Dantas J, de Jesus Oliveira M, Silva LAO, Batista S, Dagostin CS, Schachter DC. Monotherapy versus combination therapy in the treatment of painful diabetic neuropathy: a systematic review and meta-analysis. Clin Drug Investig. 2023;43:905–14.

Melenhorst WB, Overgoor ML, Gonera EG, Tellier MA, Houpt P. Nerve decompression surgery as treatment for peripheral diabetic neuropathy: literature overview and awareness among medical professionals. Ann Plast Surg. 2009;63:217–21.

Mohammed MA, Gharib DM, Reyad HR, Mohamed AA, Elroby FA, Mahmoud HS. Antioxidant and anti-inflammatory properties of alpha-lipoic acid protect against valproic acid–induced liver injury. Can J Physiol Pharmacol. 2021;99:499–505.

Murray K. Calcium-channel blockers for treatment of diabetic nephropathy. Clin Pharm. 1991;10:862–5.

Nair SR. Personalized medicine: striding from genes to medicines. Perspect Clin Res. 2010;1:146–50.

Navarro X, Sutherland DE, Kennedy WR. Long-term effects of pancreatic transplantation on diabetic neuropathy. Ann Neurol. 1997;42:727–36.

Nkonge KM, Nkonge DK, Nkonge TN. Screening for diabetic peripheral neuropathy in resource-limited settings. Diabetol Metab Syndr. 2023;15:55.

Obeagu EI, Ubosi NI, Uzoma G. Antioxidant supplementation in pregnancy: effects on maternal and infant health. Int J Adv Multidiscip Res. 2023;10:60–70.

Obrosova IG, Ilnytska O, Lyzogubov VV, Pavlov IA, Mashtalir N, Nadler JL, Drel VR. High-fat diet–induced neuropathy of pre-diabetes and obesity: effects of "healthy" diet and aldose reductase inhibition. Diabetes. 2007;56:2598–608.

Oyenihi AB, Ayeleso AO, Mukwevho E, Masola B. Antioxidant strategies in the management of diabetic neuropathy. Biomed Res Int. 2015;2015:515042.

Pang L, Lian X, Liu H, Zhang Y, Li Q, Cai Y, Ma H, Yu X. Understanding diabetic neuropathy: focus on oxidative stress. Oxidative Med Cell Longev. 2020;2020:9524635.

Patel N, Mishra V, Patel P, Dikshit RK. A study of the use of carbamazepine, pregabalin and alpha lipoic acid in patients of diabetic neuropathy. J Diabetes Metab Disord. 2014;13:1–7.

Pingali U, Kammila S, Mekala P, Yareeda S, Penugonda S. A study to evaluate the effect of alpha-lipoic acid on neuropathic symptoms in diabetic neuropathy patients on gabapentin or pregabalin. Cureus. 2024;16:e70299.

Prasad KN, Cole WC, Kumar B. Multiple antioxidants in the prevention and treatment of Parkinson's disease. J Am Coll Nutr. 1999;18:413–23.

Rapisarda P, Amenta M, Ballistreri G, Fabroni S, Timpanaro N. Distribution, antioxidant capacity, bioavailability and biological properties of anthocyanin pigments in blood oranges and other citrus species. Molecules. 2022;27:8675.

Reja A, Tesfaye S, Harris N, Ward J. Is ACE inhibition with lisinopril helpful in diabetic neuropathy? Diabet Med. 1995;12:307–9.

Rendell MS. The time to develop treatments for diabetic neuropathy. Expert Opin Investig Drugs. 2021;30:119–30.

Ribaudo G, Bortoli M, Pavan C, Zagotto G, Orian L. Antioxidant potential of psychotropic drugs: from clinical evidence to in vitro and in vivo assessment and toward a new challenge for in silico molecular design. Antioxidants. 2020;9:714.

Rizwana N, Agarwal V, Nune M. Antioxidant for neurological diseases and neurotrauma and bioengineering approaches. Antioxidants. 2021;11:72.

Roulet L, Rollason V, Desmeules J, Piguet V. Tapentadol versus tramadol: a narrative and comparative review of their pharmacological, efficacy and safety profiles in adult patients. Drugs. 2021;81:1257–72.

Rowin J. Neuropathy and neuropathic pain. In: Integrative neurology, vol. 243; 2020.

Saul H, Deeney B, Imison C, Tesfaye S. Combination therapy for painful diabetic neuropathy is safe and effective. BMJ. 2023;381:866.

Senderovich H, Jeyapragasan G. Is there a role for combined use of gabapentin and pregabalin in pain control? Too good to be true? Curr Med Res Opin. 2018;34:677–82.

Serrano Afonso A, Carnaval T, Videla Cés S. Combination therapy for neuropathic pain: a review of recent evidence. J Clin Med. 2021;10:3533.

Shukla A. The effect of enhanced occupational therapy interventions on the balance performance and fall risk among patients suffering from diabetic neuropathy. Int J Res Anal Rev. 2021;8:786.

Singh M, Kim A, Young A, Nguyen D, Monroe CL, Ding T, Gray D, Venketaraman V. The mechanism and inflammatory markers involved in the potential use of N-acetylcysteine in chronic pain management. Life. 2024;14:1361.

Singleton JR, Smith AG, Marcus RL. Exercise as therapy for diabetic and prediabetic neuropathy. Curr Diab Rep. 2015;15:1–8.

Sinha N, Kumar Dabla P. Oxidative stress and antioxidants in hypertension–a current review. Curr Hypertens Rev. 2015;11:132–42.

Smith T, Nicholson RA. Review of duloxetine in the management of diabetic peripheral neuropathic pain. Vasc Health Risk Manag. 2007;3:833–44.

Smith S, Normahani P, Lane T, Hohenschurz-Schmidt D, Oliver N, Davies AH. Prevention and management strategies for diabetic neuropathy. Life. 2022;12:1185.

Somers DL, Somers MF. Treatment of neuropathic pain in a patient with diabetic neuropathy using transcutaneous electrical nerve stimulation applied to the skin of the lumbar region. Phys Ther. 1999;79:767–75.

Stepanović-Petrović R, Micov A, Tomić M, Pecikoza U. Levetiracetam synergizes with gabapentin, pregabalin, duloxetine and selected antioxidants in a mouse diabetic painful neuropathy model. Psychopharmacology. 2017;234:1781–94.

Sun W, Sanderson PE, Zheng W. Drug combination therapy increases successful drug repositioning. Drug Discov Today. 2016;21:1189–95.

Syed O, Jancic P, Knezevic NN. A review of recent pharmacological advances in the management of diabetes-associated peripheral neuropathy. Pharmaceuticals. 2023;16:801.

Tajti J, Szok D, Majláth Z, Csáti A, Petrovics-Balog A, Vécsei L. Alleviation of pain in painful diabetic neuropathy. Expert Opin Drug Metab Toxicol. 2016;12:753–64.

Tan BL, Norhaizan ME, Liew W-P-P, Sulaiman Rahman H. Antioxidant and oxidative stress: a mutual interplay in age-related diseases. Front Pharmacol. 2018;9:1162.

Tandan R, Lewis GA, Krusinski PB, Badger GB, Fries TJ. Topical capsaicin in painful diabetic neuropathy: controlled study with long-term follow-up. Diabetes Care. 1992;15:8–14.

Tani H, Takasu S, Uchida H, Suzuki T, Mimura M, Takeuchi H. Factors associated with successful antipsychotic dose reduction in schizophrenia: a systematic review of prospective clinical trials and meta-analysis of randomized controlled trials. Neuropsychopharmacology. 2020;45:887–901.

Tavakoli M, Klingelhöfer D, Fadavi H, Groneberg DA. The landscape of global research on diabetic neuropathy. Front Endocrinol. 2023;14:1220896.

Tesfaye S, Boulton AJ, Dyck PJ, Freeman R, Horowitz M, Kempler P, Lauria G, Malik RA, Spallone V, Vinik A. Diabetic neuropathies: update on definitions, diagnostic criteria, estimation of severity, and treatments. Diabetes Care. 2010;33:2285.

Thorsen R, Dalla Costa D, Beghi E, Ferrarin M. Myoelectrically controlled FES to enhance tenodesis grip in people with cervical spinal cord lesion: a usability study. Front Neurosci. 2020;14:412.

Tripathi CD, Mehta AK, Yadav AM. Drug combinations in diabetic neuropathic pain: an experimental validation. J Basic Clin Physiol Pharmacol. 2016;27:617–24.

Viana MDM, Lauria PSS, Lima AAD, Opretzka LCF, Marcelino HR, Villarreal CF. Alpha-lipoic acid as an antioxidant strategy for managing neuropathic pain. Antioxidants. 2022;11:2420.

Vincent AM, Russell JW, Low P, Feldman EL. Oxidative stress in the pathogenesis of diabetic neuropathy. Endocr Rev. 2004;25:612–28.

Vorobeychik Y, Gordin V, Mao J, Chen L. Combination therapy for neuropathic pain: a review of current evidence. CNS Drugs. 2011;25:1023–34.

Wang EJ, Berninger LE, Komargodski O, Smith TJ. Painful diabetic neuropathy-spinal cord stimulation, peripheral nerve stimulation, transcutaneous electrical nerve stimulation, and scrambler therapy: a narrative review. Pain Physician. 2022;25:E1161.

Wei Y-H, Lee H-C. Oxidative stress, mitochondrial DNA mutation, and impairment of antioxidant enzymes in aging. Exp Biol Med. 2002;227:671–82.

Wong M-C, Chung JW, Wong TK. Effects of treatments for symptoms of painful diabetic neuropathy: systematic review. BMJ. 2007;335:87.

Xue T, Zhang X, Xing Y, Liu S, Zhang L, Wang X, Yu M. Advances about immunoinflammatory pathogenesis and treatment in diabetic peripheral neuropathy. Front Pharmacol. 2021;12:748193.

Yang ZW, Ji BP, Zhou F, Li B, Luo Y, Yang L, Li T. Hypocholesterolaemic and antioxidant effects of kombucha tea in high-cholesterol fed mice. J Sci Food Agric. 2009;89:150–6.

Yorek M. Combination therapy is it in the future for successfully treating peripheral diabetic neuropathy? Front Endocrinol. 2024;15:1357859.

Yoshihara D, Fujiwara N, Suzuki K. Antioxidants: benefits and risks for long-term health. Maturitas. 2010;67:103–7.

Yue Q, Song Y, Liu Z, Zhang L, Yang L, Li J. Receptor for advanced glycation end products (RAGE): a pivotal hub in immune diseases. Molecules. 2022;27:4922.

Zhu J, Hu Z, Luo Y, Liu Y, Luo W, Du X, ShengLiang P, Luo Z, Hu J. Diabetic peripheral neuropathy: pathogenetic mechanisms and treatment. Front Endocrinol. 2024;14:1265372.

Ziegler D. Pathogenetic treatments for diabetic peripheral neuropathy. Diabetes Res Clin Pract. 2023;206:110764.

Part VI

Case Studies and Future Perspectives

Clinical Trials and Outcomes in Antioxidant Therapy for Diabetic Neuropathy

23

Hemendra Gautam, Manish Kumar Mishra, Ajit Kumar Yadav, Vikas Pandey, Ramakant, and Vandana Sharma

23.1 Introduction to Antioxidant Therapy in Diabetic Neuropathy

Diabetic neuropathy (DN), a microvascular consequence of diabetes, refers to peripheral nerve problems in persons suffer with diabetes that causes have been ruled out. Asadi et al. have connected the aetiology of diabetic neuropathy (DN) to systemic oxidative stress caused by chronic hyperglycaemia. When reactive oxygen species (ROS) and antioxidant systems are out of equilibrium, oxidative stress results (Asadi et al. 2019). It affects nerve cell activity through a number of different molecular signalling pathways. Diabetes-related impaired glucose breakdown is a major mechanism for oxidative stress induction because it redirects unused glucose to other metabolic or nonmetabolic pathways (Pang et al. 2020).

Diabetes must be managed to avoid and control DN, as well as nerve damage from worsening. There are currently no therapies to prevent the beginning or progression of DN. Pain management in these individuals with traditional drugs is unsatisfactory. The severe side effects of neuropathy, such as ulcers and a decreased quality of life, make alternative treatments for diabetic foot pain (DSPN)

H. Gautam (✉) · Ramakant
Faculty of Pharmaceutical Sciences, Future University, Bareilly, UP, India

M. K. Mishra
Department of Pharmaceutical Chemistry, SGRR University, Dehradun, Uttarakhand, India

A. K. Yadav
R K Institute of Pharmacy, Bareilly, UP, India

V. Pandey
Department of Pharmacy, Mangalayatan University-Jabalpur, Jabalpur, Madhya Pradesh, India

V. Sharma
School of Pharmaceutical Sciences, Bahra University, Shimla Hills, Himachal Pradesh, India

M. K. Gautam et al. (eds.), *Antioxidants and Functional Foods for Diabetic Neuropathy*, https://doi.org/10.1007/978-981-95-9326-2_23

imperative. Since there isn't yet a good treatment for DN, antioxidants or drugs that block free radicals may be useful (Asadi et al. 2019).

Antioxidants enriched foods like anthocyanins, phenols, vitamins, minerals and other phytonutrients that scavenge free radicals and reduce oxidative damage are abundant in fruits, vegetables, and berries. These antioxidants have been discovered to be therapeutic agents for neurological illnesses and also slow down the ageing process. Eating nuts, for example, can help avoid brain damage while obese (Nuzzo 2021).

The most researched dietary antioxidant supplements include active ingredients such polyphenols/flavonoids, vitamin C & E, and carotenoids Gautam & Goel (2014). The Antioxidants work in two ways: First, they limit the production of ROS and inhibit and grab the produced radicals. This is done by antioxidant defence mechanisms, which are activated in response to oxidative damage. Aqueous compartments and cell membranes contain these systems, which can be either enzymatic or nonenzymatic. It has been established that an enzymatic antioxidant system including glutathione peroxidase, catalase, and superoxide dismutase is the primary defence system. The water- and lipid-soluble, as well as lesser thiol and nonenzymatic antioxidants, constitute the second line of defence. Second, degraded biomolecules are eliminated as part of the antioxidant healing procedure before they congregate and change cell metabolism. The repair systems' involvement includes the use of certain enzymes to repair oxidatively damaged nucleic acids, proteolytic systems to remove oxidised proteins, and phospholipases, peroxidases, or acryl transferases to repair oxidised lipids (Gautam et al. 2012; Gangwar et al. 2014; Lee et al. 2020).

23.2 Key Antioxidants Studied in Diabetic Neuropathy

23.2.1 Vitamin E: Mechanisms and Trial Outcomes

Tocopherols and tocotrienols, which are potent lipophilic antioxidants, are part of the class of compounds known as vitamin E, which has also been found to mitigate the unpleasant neuropathic repercussions of diabetes mellitus. A higher incidence of diabetes is linked to low vitamin E levels, and some evidence indicates that antioxidant levels are lower in diabetics. Because hyperglycaemia increases the generation of free radicals, people with diabetes may also need more antioxidants (Ogbera et al. 2014).

23.2.1.1 Mechanism of Action

A strong lipophilic antioxidant is vitamin E. It prevents oxidative damage to cellular membranes and lipoproteins, especially LDL, by neutralising free radical species generated during regular cellular metabolism. Additionally, it interacts with antioxidants that are soluble in water, such glutathione. By reducing protein glycation, lipid oxidation, and suppression of platelet adhesion and aggregation, it may help prevent and cure major consequences of diabetes, including cardiovascular disease, nephropathy, and neuropathy. In clinical trials involving people with diabetes, vitamin E's effects on microvascular consequences of diabetes, insulin sensitivity, glycaemic management, protein glycation, cardiovascular disease, and its risk factors have been investigated (Gautam et al. 2015; O'Connell 2001).

In one study, electrophysiological parameters were assessed in relation to vitamin E, the main regulator of free radical activity, in patients with insulin-dependent peripheral sensorimotor polyneuropathy that were selected for length of illness and metabolic therapy. This double-blind, randomized, placebo study involved twenty-one individuals with type II diabetes; ten patients were given a placebo, and eleven patients were given vitamin E. Over the course of 6 months, individuals were randomised to get a control drug or 900 mg of vitamin E. Over the course of the investigation, the subjects' usual nutritional vitamin E intake remained consistent. The electrophysiological tests that evaluated nerve conduction served as the primary outcome measure. According to this study, therapeutic dosages of vitamin E-containing supplements may help diabetics with mild-to-moderate chronic peripheral neuropathy by improving impaired nerve transmission (Tütüncü et al. 1998).

The safety and efficacy of treating painful diabetic neuropathy with vitamin E, an antioxidant, and Eve Primrose combined were assessed in a study initiated at the Diabetic Centre of Lagos State University Teaching Hospital. The one-year experiment involved 80 individuals with type II diabetes. Clinical and biochemical assessments were performed on participants who took vitamin E (400 mg) and 500–1000 mg of eve primrose each day. The decrease in neuropathy symptoms was the main outcome measure. The results demonstrated that eve primrose and vitamin E are effective treatments for mild-to-severe diabetic neuropathy (Ogbera et al. 2014).

Strong anti-inflammatory and antioxidant qualities of tocotrienol-rich vitamin E (Tocovid) are thought to target these pathogeneses in order to improve DPN. According to a research comparing Tocovid's impact on blood biomarkers and nerve conduction measurements in diabetics, Tocovid considerably raised nerve conduction velocities when compared to a placebo. According to the study, the Tocovid group's serum NGF levels were noticeably greater than the placebo groups at 8 weeks. After 8 weeks, the nerve conduction velocities of those who received Tocovid showed a considerable improvement, suggesting that Tocovid had a major effect on the health of diabetes patients (Ng et al. 2020).

23.2.2 Alpha-Lipoic Acid (ALA) in Clinical Research

In vitro and in vivo, ALA treatment raises decreased glutathione (GSH), a significant endogenous antioxidant. Together, GSH and lipoic acid facilitate essential cellular processes. Oxidative stress results in reduced artery circulation and endoneurial damage from oxidative stress because diabetes is linked to increased production and inadequate elimination of ROS. An effective lipid-based free radical scavenger of peripheral neurones is ALA (Vallianou et al. 2010).

A major endogenous antioxidant, reduced glutathione (GSH), is increased both in vitro and in vivo by alpha-lipoic acid therapy. It works well as a lipophilic free radical scavenger of peripheral neurones when coupled with lipoic acid. Diabetes may enhance production and inadequate elimination of ROS, which leads to oxidative stress and endoneurial oxidative damage. Six of fifteen nerve conduction electrophysiological parameters are significantly optimised by taking α-lipoic acid nutrients. Oral α-lipoic acid supplementation helps people with diabetic peripheral neuropathy via enhancing the conduction of motor nerves in their whole body (Vijayakumar et al. 2014).

Various methods, dosages, and treatment durations of alpha-lipoic acid (ALA) were employed to treat diabetic peripheral neuropathy (DPN). A further research evaluated the effectiveness of oral 600 mg ALA twice a day for 6 months in treating DPN patients. In a randomised, single, double-blind, placebo trial, 200 DPN patients were randomised to receive oral 600 mg twice-day ALA or a placebo for 6 months at Mansoura Specialised Hospital, Mansoura University. Visual analogue pain, neurological symptom, neurological disability, and vibration perception threshold ratings were used to assess the efficacy of the therapy. After comparing the research groups to the baseline data, no discernible differences were found. ALA-treated patients, however, performed noticeably better on next visits in terms of nearly every outcome indicator (NSS, NDS, VAS, and VPT). For DPN, it was determined that taking 600 mg of ALA orally twice a day for 6 months is safe, efficacious, and acceptable (El-Nahas et al. 2020).

In comparison to controls, α-lipoic acid medication dramatically improved TSS and decreased diabetic neuropathic pain in a trial with 110 patients. Patients were divided into the same control and treatment groups as part of a randomised control experiment (Siddique et al. 2021).

23.2.3 Co-Enzyme Q10: Clinical Evidence and Trial Results

COQ10 is a well-known supplement that is used to treat a number of illnesses, such as diabetes, neurological disorders, cardiovascular diseases, metabolic abnormalities including hypertension and hyperlipidaemia, and diseases of the mitochondria. Its antioxidant, anti-inflammatory, antihypertensive, hypoglycaemic, and anti-hyperlipidaemia qualities are allegedly responsible for these benefits, which are intended to alleviate diabetes and its cardiovascular consequences (Samimi et al. 2024).

Clinical studies suggest that using anti-inflammatory and antioxidant supplements, such as CoQ10, may aid diabetics with PDN. By mixing pregabalin with coenzyme Q10 (CoQ10), the research sought to assist patients with painful diabetic neuropathy (PDN) in managing their symptoms. A total of 121 participants were randomised to receive either a placebo plus pregabalin or CoQ10 plus pregabalin. The intent-to-treat population analysis revealed that the CoQ10 plus pregabalin regimen reduced pain more than the placebo + pregabalin regimen. Furthermore, those who received CoQ10 supplements saw a greater decrease in pain-associated SIS than those are took the control sample (Amini et al. 2022).

A further investigation was conducted to assess the effects of oral ubiquinone in diabetic polyneuropathy, as well as the function of nerve growth factor (NGF-β) and lipid peroxidation (LPO). For 12 weeks, participants in a double-blind, placebo clinical trial were randomised to receive either 400 mg of ubiquinone or a placebo every day. The main outcomes were safety, LPO, NGF-β, clinical ratings, and nerve conduction tests. Ubiquinone therapy for 12 weeks improves diabetic polyneuropathy's clinical results and nerve conduction metrics. It also lowers oxidative stress without causing serious side effects (Hernández-Ojeda et al. 2012).

Another study looked at how CoQ10 affected rats with sciatic nerve chronic constriction injury (CCI) in terms of pain reduction. Four groups of Wistar rat (male) were randomly arrange. The sciatic nerve was damaged by four loose ligatures. CoQ10 at dose 200 mg per kg or vehicle intraperitoneal injections were administered daily for 10 days. Behaviour tests were performed before nerve damage and 5 and 10 days after the lesion. The pain threshold was assessed using radiant heat, Von Frey filaments, and the Randall Selitto test. In CCI rats, CoQ10 was demonstrated to eradicate neuropathic pain. One possible explanation for the prevention of cell loss is the pain relief effect of CoQ10. Therefore, it might be utilized as a neuropathic pain reliever (Janzade et al. 2014).

23.2.4 Other Antioxidants of Interest: N-Acetylcysteine, Flavonoids, etc.

The earliest usage of the sulfhydryl-containing compound N-acetylcysteine (NAC) in medicine dates back to 1967. Its anti-inflammatory, mucolytic, and antioxidant qualities include scavenging free radicals, stabilising protein structures, boosting glutathione S-transferase activity, and replenishing glutathione. It has been demonstrated that NAC helps animals with diabetic retinopathy, diabetic nephropathy, and diabetic peripheral neuropathy. When administered as an adjuvant therapy to pregabalin, NAC (600 mg daily) has demonstrated considerable improvement in painful peripheral neuropathy in human studies, lowering pain ratings by more than 50% when compared to a placebo (Schwalfenberg 2021).

The study looked at how N-acetylcysteine (NAC) can improve peripheral nerve function in cirrhosis caused by oxidative stress by inhibiting maternally expressed 3 (MEG3)/protease activated receptor 2 (PAR2). Five groups of thirty Wistar rats were created: thiacetamide (TAA), TAA + NAC (50, 100, or 200 mg/kg/day), and naïve. PCR and the von Frey test were used to measure the expression of MEG3, NF-PB, and neuronal PAR2 in the liver and brain. According to the study's findings, NAC may help cirrhotic rats with peripheral neuropathy (Mohamed et al. 2020).

In a research, the effects of naringenin on streptozotocin-induced diabetic neuropathic pain in Wistar rats were investigated. The results revealed considerable allodynia and hyperalgesia, which were linked to reduced motor nerve conduction velocity, inflammatory mediator generation, MMP-9 activation, and increased oxidative-nitrosative stress. Neuropathy emerged during an 8-week diabetic induction period. In a dose-dependent way, naringenin treatment (25, 50, and 100 mg/kg) for 4 weeks dramatically reduced behavioural, biochemical, and molecular alterations as well as motor nerve conduction velocity. In diabetic rats, the combination insulin-naringenin therapy was more effective than either drug by itself. Insulin treated hyperglycaemia and partially restored pain sensitivity in diabetic rats. In diabetic rats, neuropathic pain was alleviated by combining insulin and naringenin (Singh et al. 2020).

The antioxidant, anti-inflammatory, and antinociceptive qualities of the flavonoid's quercetin and rutin have led to their classification as cell-protective agents. The antioxidant properties of these medications may help avoid the unpleasant peripheral neuropathy caused by oxaliplatin, according to our theory. Nine Swiss male mice were administration via intravenous of oxaliplatin at 1 mg/kg twice a week. The onset of sensory changes, including mechanical and thermal allodynia, was assessed using the von Frey test and the tail immersion test in cold water (10 °C). An intraperitoneal dose of rutin and quercetin at 25–100 mg/kg was given half an hour prior to each oxaliplatin injection. Quercetin and rutin significantly improved the mechanical and thermal nociceptive response by blocking the effects of oxaliplatin at all doses. The results of the oxidative stress tests mostly shown that quercetin and rutin lessened the effects of spinal cord peroxidation brought on by oxaliplatin. Additionally, the flavonoids reduced nitrotyrosine immunostaining and inducible nitric oxide synthase in the spinal cord's dorsal horn. It has been demonstrated that rutin and quercetin reduce the painful peripheral neuropathy that oxaliplatin causes in rats (Azevedo et al. 2013).

23.3 Major Clinical Trials and Their Outcomes

An important clinical investigation called the SYDNEY 2 experiment sought to determine how α-lipoic acid (ALA) affected neuropathic impairments and pleasurable sensory experiences in diabetic individuals with distal symmetric polyneuropathy. For 5 weeks, 181 patients from Israel and Russia participated in the trial and were given either ALA or a placebo every day. Including searing pain, paraesthesia, stabbing pain, and sleep-induced foot numbness, the main end endpoint was the Total Symptom Score (TSS) compared with initial state. The Neuropathy Symptoms and Change (NSC) score, Neuropathy Impairment Score (NIS), individual TSS symptoms, and the patients' overall efficacy assessment were secondary objectives. 600 mg taken orally once daily produced the highest risk-benefit ratio (Ziegler et al. 2006).

Over the course of 4 years, individuals with mild-to-moderate diabetic distal symmetric sensorimotor polyneuropathy (DSPN) participated in the NATHAN 1 experiment, which assessed the safety and efficacy of ALA. For 4 years, 460 individuals were randomised to receive oral medication with 600 mg of ALA or a placebo. Nerve conduction, quantitative sensory tests, NIS, and NIS-LL were examples of secondary outcome measures. The trial was a noteworthy investigation in this field. Though it had no impact on the primary composite end objective, 4 years of ALA treatment was demonstrated to be well tolerated, lead to a clinically significant enhancement and stop the development of neuropathic deficits in mild-to-moderate DSPN patients. ALA had a greater chance of serious adverse reactions (38.1%) compared to a blank (28.0%) (Ziegler et al. 2011).

Alpha-lipoic acid intravenous infusions or a placebo were administered to 328 diabetic patients with symptomatic peripheral neuropathy who did not have insulin-dependent diabetes for 3 weeks as part of the multicentre, double-blind study and controlled ALADIN Study. According to the results, ALA at a dose of 600 mg per day for 3 weeks was more efficient than a placebo at reducing diabetic peripheral neuropathy symptoms without having any notable side effects. The study assessed neuropathic pain at baseline and throughout each visit (Ziegler et al. 1995).

Alpha-lipoic acid administered intravenously and then orally was tested for safety and effectiveness in the ALADIN III Study in people with type II diabetes who also had symptomatic polyneuropathy. Alpha-lipoic acid 600 mg, a placebo, and a placebo were the three therapies assessed in a multicentre study with 509 outpatients. Alpha-lipoic acid therapy administered intravenously for 3 weeks and then orally for 6 months did not appreciably alter neuropathic symptoms, according to the results. This could result from a greater degree of intercenter variation in symptom assessment. The drug did, however, have a beneficial effect on neuropathic deficits and no significant adverse effects (Ziegler et al. 1999).

23.3.1 Significant Trials with Positive Outcomes

In a tertiary care hospital study of 92 patients with diabetic neuropathy, vitamin E administration significantly improved physical health, reduced random blood sugar levels, and decreased the overall pain score when compared to the usual care group. The intervention group was given their regular meds plus vitamin E, oral hypoglycaemic drugs, and pregabalin. Using the Neuropathy Pain Score and the RAND 36 questionnaire, the study assessed participants' quality of life and pain levels. Blood samples from baseline and 12 weeks after the intervention were analysed. According to the study's findings, vitamin E is a naturally occurring antioxidant that helps diabetic neuropathy sufferers with their pain levels (Rajanandh et al. 2014).

In a double-blind, randomised, placebo clinical study, 80 people with diabetes participated. The placebo group and the treatment group were both given to the participants. They received either a placebo or 80 mg of Nano-curcumin over the course of 8 weeks. The severity of DSPN, food consume, physical workout, glycaemic indices, and anthropometric measures were assessed during the course. In comparison with the control group, the addition of Nano-curcumin dramatically decreased temperature, glycated haemoglobin, total neuropathy score, total reflex score, and rapid blood sugar. Curcumin pills for 2 months improved and lessened the severity of DSPN, according to the study's findings (Asadi et al. 2019).

Another clinical trial comprised 23 individuals with diabetic neuropathy of the nerves, 16 of whom were men and 7 of whom were women. They received a single film tablet containing 600 mg of ALA every day for 60 days after receiving an IV dose of 600 mg for 10 days. By day ten, there was a noticeable change in the clinical signs of diabetic mononeuropathy, such as mydriasis, ocular movement and position, upper eyelid droops, and dual vision. In 17 individuals, or 74%, peripheral neuropathy was found. By the conclusion of the second month, foot discomfort had decreased by 65.5%, with a 33% drop seen on the tenth day. According to the study's findings, α-lipoic acid seems to be a potent medication that can help patients fully recover from diabetic mononeuropathy of the cranial nerves as well as peripheral and autonomic diabetic neuropathy (Tankova et al. 2005).

In individuals with diabetic neuropathy, the combination of two potent antioxidant medicines improves both subjective and objective metrics. In another study, it was shown that a new combination of alpha lipoic acid and superoxide dismutase was useful in the treatment of diabetic neuropathy. This study was conducted in an ambulatory (outpatient) care environment. Fifty individuals with diabetes mellitus who also showed a loss in motor and sensory nerve conduction participated in a prospective, nonrandomized, open-label trial. For 4 months, the patient was treated with ALA600SOD®, a novel combination of SOD and ALA. Before and after therapy, electroneurogram variables and pain perception were measured. Following 4 months of therapy, patients' electroneurographic markers and pain perception considerably improved. The biggest improvements were seen in sensory nerve conduction (Bertolotto and Massone 2012).

Type 2 diabetes and symptomatic DPN in men and women aged 45–74 years were the subjects of a double-blind, placebo-controlled clinical investigation. To assess the effectiveness and safety of treating DPN, the study employed the antioxidants succinic acid, inosine, nicotinamide, and riboflavin (SINR). After receiving it intravenously for 10 days, patients were given either the trial medication or a placebo orally for 75 days. Regardless of haemoglobin A1c levels, the SINR group displayed less symptoms; however, results were better for those with a baseline TSS <7.5. On the basis of numbness, burning intensity, and paraesthesia intensity, the study discovered statistically significant differences between groups (Bertolotto and Massone 2012).

NAC's effectiveness in reducing PDN symptoms is most likely due to its antioxidant properties. In a research, the effects of oral N-acetylcysteine (NAC) in PDN were examined in comparison to pregabalin. Over the course of 8 weeks, 102 people with type II diabetes and PDN participated in the study. They were randomly selected to receive either N-acetylcysteine or pregabalin. In addition to other metrics including pain score, sleep interference score, and patient overall impression of change, overall antioxidant potential levels in the blood, total thiol groups, malondialdehyde, catalase enzyme activity, and NO were assessed at the start and finish of the trial. Every patient tolerated NAC nicely. Mean pain ratings decreased, and SIS increased similarly in both groups. The PGIC and CGIC of the NAC group improved more than those of the baseline. Notably, NAC decreased blood MDA and NO levels while increasing TAC, and CAT levels. Pregabalin significantly lesser serum MDA and NO levels while increasing TAC levels (Sajedi et al. 2024).

23.3.2 Trials with Inconclusive or Negative Outcomes

Since it had no beneficial effects on DPN or heart perfusion and did not halt the progression of the illness, a combined antioxidant treatment regimen may have been detrimental in the study population of type 1 diabetics with mild-to-moderate CAN. This study assessed how a combination triple antioxidant drug affected the cardiovascular autonomic neuropathy (CAN) and myocardial blood flow (MBF) in individuals with type I diabetes. A similar, placebo-controlled, randomised study was carried out. To allocate participants to therapy, the research chemist was given opaque, sealed, sequentially numbered envelopes. Both examiners and participants were not informed of the treatment allocation. The subjects were evaluated by adenosine stress testing, positron emission tomography employing [11C] HED and [13 N] ammonia, and cardiovascular autonomic sensitivity analysis. After a day, F2-isoprostanes in the urine served as markers of oxidative stress. Diabetic peripheral neuropathy was evaluated using intra-epidermal nerve fibre density, electrophysiology, symptoms, and signs. The participants received antioxidant therapy consisting of nicotinamide, α-lipoic acid, and allopurinol, or a placebo, for a duration of 24 months. The primary discovery was the difference in the global [11C] HED retention index (RI) between patients receiving the active medication and those receiving a placebo after 24 months (Pop-Busui et al. 2013).

The effectiveness of dietary mixture tocotrienols for people with diabetic peripheral neuropathy was assessed in follow-up research. A similar double-blind, placebo-controlled trial, the Vitamin E in Neuroprotection trial (VENUS), recruited volunteers between January 2011, and December 2014, with a 12-month monitoring period. 14,289 diabetic patients from five local hospitals, six medical facilities, and outpatient centres in Malaysia were examined in this study. The Neuropathy Impairment Score (NIS) and Total Symptom Score (TSS) were used to further evaluate 391 individuals who had reported neuropathic symptoms. Patients with a TSS of 3 or higher and a NIS of 2 or higher who were 20 years of age or older were recruited for the study. Participants were given a matched placebo or 200 mg of tocotrienols BD for a duration of 12 months. According to the study, using 400 mg OD of oral mixed tocotrienol for a year did not lessen overall neuropathic symptoms (Hor et al. 2018).

23.4 Challenges and Limitations in Current Clinical Trials

Although observational research and experimental model studies with antioxidants clearly imply that antioxidants should have positive benefits in lowering neuropathic complications in diabetes, there is insufficient clinical data to support their usage. It should be noted that the majority of clinical research of antioxidants in diabetes emphasised the usage of vitamin C as well as vitamin E, and more recently, α-lipoic acid. Numerous limitations affect the reliability and effectiveness of current clinical trials investigating antioxidants as a therapy for diabetic neuropathy. Key difficulties consist of:

Patient Populations: A lot of trials use very specific patient groups or tiny sample numbers, which might not be representative of the general diabetes community. It is difficult to extrapolate results from trials to early-stage or preventative settings since they are frequently carried out on individuals who have severe diabetic neuropathy. Additionally, trial outcomes may be impacted by variables including age, the duration of diabetes, and concomitant illnesses like cardiovascular disease; however, study designs may not always account for or stratify these variables. In the absence of an accurate understanding of the underlying mechanistic approach, the specific antioxidants under investigation were chosen based on observational and epidemiological data. The bulk of the research involved older patients (average age 65.4 years) and lasted little more than 5 years, despite the fact that observational studies are based on entire populations and demonstrate how dietary habits impact individuals throughout their lives. People whose medical illnesses had progressed to the point where antioxidant intervention was no longer beneficial may have been included in the research populations.

Heterogeneity of Study Designs: The methods used in clinical trials can differ greatly, which makes it challenging to compare the findings of different research. Inconsistent results are caused by variations in the kinds of antioxidants employed, doses, length of treatment, and techniques for evaluating neuropathy. Lack of standardisation makes it challenging to combine data from several research for meta-analyses.

Short Duration of Trials: A lot of antioxidant studies are carried out over a little time frame, usually a few weeks to a few months. Because diabetic neuropathy is a chronic illness, short-term research may not fully reflect the long-term safety or effectiveness of antioxidant therapies. This makes it more challenging to evaluate the long-term benefits or dangers of these therapies.

Diverse Antioxidant Compounds: It is difficult to determine the overall effectiveness of the antioxidant compounds studied because of their diversity, which includes coenzyme Q10, vitamin E, and alpha-lipoic acid, among others. Despite having different pharmacokinetics and modes of action, antioxidants are sometimes clubbed together in talks, even if their results may vary. It is difficult to make firm judgements on the general effectiveness of antioxidant treatment in DN because of the variety of therapeutic substances.

Variability in Outcome Measures: Various clinical and neurophysiological endpoints, including nerve conduction investigations, pain scales, and quality of life evaluations, are often used in trials. Comparisons between research are made more difficult by the lack of agreement on the outcome measures to employ. Additionally, subjective metrics such as pain scales are less accurate for determining the actual benefit of treatment since they are susceptible to placebo effects and patient variability.

Lack of biomarkers: The endpoints employed in the human studies up to this point were general indicators of health, such the impact on mortality, rather than having a direct connection to oxidative stress. The length of the diabetes disease states, which is undoubtedly a significant confounding element, was not evaluated in the trials. Certain oxidative stress indicators should be monitored in order to evaluate oxidative stress and the impact of antioxidants on it.

Inconsistent dosing and Bioavailability: The bioavailability of various antioxidants varies greatly, and there is a great deal of heterogeneity in the dosage schedules employed for antioxidant therapy. For instance, compared to other antioxidants like alpha-lipoic acid, oral vitamin E may have a reduced bioavailability, which might have an impact on clinical trial results. Another difficulty is achieving therapeutic levels of antioxidants in target tissues, such as peripheral nerves, because bioavailability varies depending on the amount given (Oyenihi et al. 2015; Johansen et al. 2005; Vincent and Feldman 2004).

23.5 Future Directions in Antioxidant Therapy for Diabetic Neuropathy

23.5.1 Emerging Antioxidant Therapies in Development

Anti-AGE agents stop AGEs from forming and building up. Antioxidants including aspirin, aminoguanidine, and benfotiamine are well-known for their capacity to stop AGE–RAGE interactions, which might worsen the oxidative stress damage caused by DN (Haupt et al. 2005).

To treat DPN, a combined therapy that includes ACE inhibition and either neprilysin or NEP might be employed. Numerous studies have demonstrated that ACE inhibition ameliorated diabetic peripheral neuropathy in both human and animal species. Suppression of the angiotensin system via pharmacologically successfully avoids diastolic pressure, according to a clinical retrospective cohort examination of people with type II diabetes. Fascinatingly, natriuretic peptides like endothelin, bradykinin, adrenomedullin, CGRP, and CNP are all broken down by NEP. AVE7688, a vasopeptidase inhibitor that would subsequently be known as Ilepatril, was the first ACE and NEP inhibitor combo that we looked at. Along with lowering nitrotyrosine and superoxide levels, AVE7688 reduced hindpaw hypoalgesia, reversed diabetes-induced endoneurial blood flow, improved vascular relaxation by epineurial arterioles, and significantly increased the conduction velocity of motor and sensory neurones. However, testing on human subjects revealed that ilepatril increased the risk of angioedema, which led to the termination of its clinical development.

A novel medication combination called LCZ696 (ENTRESTO®) lowers the risk of angioedema and promotes the use of angiotensin II receptor blockers rather than ACE inhibitors. Sacubitril, the first medication of its sort, is combined with valsartan. Studies indicate that valsartan by itself is not as efficient as sacubitril/valsartan in enhancing vascular and neuronal function. During the initial treatment regimen, the intervention restored vascular reactivity, motor and sensory nerve conduction velocities, and cutaneous and corneal sensory nerve sensitivity/regeneration, therefore slowing the course of diabetic peripheral neuropathy (DPN).

A different research looked at how menhaden (fish) oil, enalapril, and α-lipoic acid affected peripheral neuropathy. Almost all of the vascular and nerve impairments in diabetic rats restored to levels that were close to control when they were given enalapril, menhaden oil, or ALA. The sole exception was the motor nerve conduction velocity, which was much lower than that of normal rats but also significantly better than that of diabetic rats. Despite the fact that healthy volunteers have not been used to evaluate a combination of these substances, it is expected to be safe and perhaps improve DPN.

The study looked at how ω-3 fatty acids and salsalate affected the diabetic peripheral neuropathy (DPN) in rats with type II diabetes. It has been discovered that salsalate works better than menhaden oil to stop the deterioration of vascular relaxation, heat nociception, corneal nerve density, and motor and sensory nerve conduction velocity brought on by high-fat diets (Yorek 2024).

The fact that each medication or ingredient in these formulations has been proven to be safe for human use is heartening. In addition to a mix of human-safe chemicals and medications, as well as appropriate dietary and activity changes, future clinical studies should think about using updated main endpoints, including nerve conduction impairments, that don't require years to develop.

23.6 Conclusion

A primary feature of diabetes is hyperglycaemia, which is the cause of multiple mechanisms that generate oxidative stress, which ultimately results in diabetic neuropathy (DN). The two main causes of diabetes, decreased insulin production and insulin resistance, have been related to oxidative stress. The most well-understood mechanism by which diabetes develops into diabetic neuropathy is still hyperglycaemia-induced oxidative stress. Therefore, therapies that target both hyperglycaemia and oxidative stress may be safe and economical ways to prevent or treat diabetes and diabetic neuropathy.

A comprehensive study of traditional antioxidants as a possible remedy is necessary to address the ongoing problem of diabetic neuropathy. ALA, vit A, vit E, and vit C are examples of antioxidants that have been shown in human and animal models to rapidly scavenge free radicals. It is essential to combine these antioxidants with medicinal plant products that have antihyperglycaemic and antioxidant properties. These combinations have a minimal risk for harm, a range of effects that target distinct phases of diabetic neuropathy development, and natural dietary constituents. To evaluate a wider range of end-points, clinical trials should be centred around this approach.

In summary, ALA therapy for 4 years was well tolerated and linked to improvements in neuropathic deficits but not nerve conduction characteristics in patients with mild-to-moderate DSPN. As a result, there was little change in the principal composite end point. The way DSPN developed clinically in the control group was somewhat slow in the context of this randomised clinical investigation, which had an outstanding patient retention rate over time. Consequently, given epidemiologic data showing notable progress, it might not be suitable to forecast a therapeutic response meant to halt the progression of DSPN. Long-term prediction of a chronic neuropathic state should be the goal of future research in similar diabetic populations.

References

Amini P, Sajedi F, Mirjalili M, Mohammadi Y, Mehrpooya M. Coenzyme Q10 as a potential add-on treatment for patients suffering from painful diabetic neuropathy: results of a placebo-controlled randomized trial. Eur J Clin Pharmacol. 2022;78(12):1899–910.

Asadi S, Gholami MS, Siassi F, Qorbani M, Khamoshian K, Sotoudeh G. Nano curcumin supplementation reduced the severity of diabetic sensorimotor polyneuropathy in patients with type 2 diabetes mellitus: a randomized double-blind placebo-controlled clinical trial. Complement Ther Med. 2019;43:253–60.

Azevedo MI, Pereira AF, Nogueira RB, Rolim FE, Brito GA, Wong DVT, et al. The antioxidant effects of the flavonoids rutin and quercetin inhibit oxaliplatin-induced chronic painful peripheral neuropathy. Mol Pain. 2013;9:1744–8069.

Bertolotto F, Massone A. Combination of alpha lipoic acid and superoxide dismutase leads to physiological and symptomatic improvements in diabetic neuropathy. Drugs R&D. 2012;12:29–34.

El-Nahas MR, Elkannishy G, Abdelhafez H, Elkhamisy ET, El-Sehrawy AA. Oral alpha lipoic acid treatment for symptomatic diabetic peripheral neuropathy: a randomized double-blinded placebo-controlled study. Endocr Metab Immune Disord Drug Targets. 2020;20(9):1531–4.

Gangwar M, Gautam MK, Sharma A, Tripathi YB, Goel RK, Nath G. Antioxidant capacity and radical scavenging effect of Mallotus philippinensis fruit extract on human erythrocytes: an in vitro study. Sci World J. 2014;2014:279451.

Gautam MK, Goel RK. Toxicological study of Ocimum sanctum Linn. leaves: hematological, biochemical and histopathological studies. J Toxicol. 2014;2014:135654.

Gautam MK, Gupta A, Rao CV, Goel RK. Antihyperglycemic and antioxidant potential of Murraya paniculata Linn. leaves: a preclinical study. J Pharm Res. 2012;5(3):1334–7.

Gautam MK, Gangwar M, Singh SK, Goel RK. Effects of Azadirachta indica on vascular endothelial growth factor and cytokines in diabetic deep wound. Planta Med. 2015;81(8):713–21.

Haupt E, Ledermann H, Köpcke W. Benfotiamine in the treatment of diabetic. Int J Clin Pharmacol Ther. 2005;43(2):71–7.

Hernández-Ojeda J, Cardona-Muñoz EG, Román-Pintos LM, Troyo-Sanromán R, Ortiz-Lazareno PC, Cárdenas-Meza MA, et al. The effect of ubiquinone in diabetic polyneuropathy: a randomized double-blind placebo-controlled study. J Diabetes Complicat. 2012;26(4):352–8.

Hor CP, Fung WY, Ang HA, Lim SC, Kam LY, Sim SW, et al. Efficacy of oral mixed tocotrienols in diabetic peripheral neuropathy: a randomized clinical trial. JAMA Neurol. 2018;75(4):444–52.

Janzade A, Jameie SB, Choobchian S, Nasirinezhad F. Neuroprotective effect of coenzyme Q10 in chronic constriction injury-induced neuropathic pain in rat. Thrita. 2014;3(1):e16607.

Johansen JS, Harris AK, Rychly DJ, Ergul A. Oxidative stress and the use of antioxidants in diabetes: linking basic science to clinical practice. Cardiovasc Diabetol. 2005;4:1–11.

Lee KH, Cha M, Lee BH. Neuroprotective effect of antioxidants in the brain. Int J Mol Sci. 2020;21(19):7152.

Mohamed DI, Khairy E, Khedr SA, Habib EK, Elayat WM, El-Kharashi OA. N-acetylcysteine (NAC) alleviates the peripheral neuropathy associated with liver cirrhosis via modulation of neural MEG3/PAR2/NF-κB axis. Neurochem Int. 2020;132:104602.

Ng YT, Phang SCW, Tan GCJ, Ng EY, Botross Henien NP, Palanisamy M, UD, et al. The effects of tocotrienol-rich vitamin E (Tocovid) on diabetic neuropathy: a phase II randomized controlled trial. Nutrients. 2020;12(5):1522.

Nuzzo D. Role of natural antioxidants on neuroprotection and neuroinflammation. Antioxidants. 2021;10(4):608.

O'Connell BS. Select vitamins and minerals in the management of diabetes. Diab Spectr. 2001;14(3):133–48.

Ogbera AO, Ezeobi E, Unachukwu C, Oshinaike O. Treatment of diabetes mellitus-associated neuropathy with vitamin E and Eve primrose. Indian J Endocrinol Metabol. 2014;18(6):846–9.

Oyenihi AB, Ayeleso AO, Mukwevho E, Masola B. Antioxidant strategies in the management of diabetic neuropathy. Biomed Res Int. 2015;2015(1):515042.

Pang L, Lian X, Liu H, Zhang Y, Li Q, Cai Y, et al. Understanding diabetic neuropathy: focus on oxidative stress. Oxidative Med Cell Longev. 2020;2020(1):9524635.

Pop-Busui R, Stevens MJ, Raffel DM, White EA, Mehta M, Plunkett CD, et al. Effects of triple antioxidant therapy on measures of cardiovascular autonomic neuropathy and on myocardial blood flow in type 1 diabetes: a randomised controlled trial. Diabetologia. 2013;56:1835–44.

Rajanandh MG, Kosey S, Prathiksha G. Assessment of antioxidant supplementation on the neuropathic pain score and quality of life in diabetic neuropathy patients–a randomized controlled study. Pharmacol Rep. 2014;66(1):44–8.

Sajedi F, Abdi A, Mehrpooya M, Faramarzi V, Mohammadi Y, Sheida F. Comparison of therapeutic effects of N-Acetylcysteine with pregabalin in improving the clinical symptoms of painful diabetic neuropathy: a randomized, double-blind clinical trial. Clin Diab Endocrinol. 2024;10(1):15.

Samimi F, Namiranian N, Sharifi-Rigi A, Siri M, Abazari O, Dastghaib S. Coenzyme Q10: a key antioxidant in the management of diabetes-induced cardiovascular complications—an overview of mechanisms and clinical evidence. Int J Endocrinol. 2024;2024(1):2247748.

Schwalfenberg GK. N-acetylcysteine: a review of clinical usefulness (an old drug with new tricks). J Nutr Metabol. 2021;2021(1):9949453.

Siddique N, Awais FF, Shakil M, Sarwar S, Kakar NH, Ullah MI. Effect of anti-oxidant (α-lipoic acid) treatment on improvement of diabetic neuropathic pain. Rawal Med J. 2021;46:33.

Singh P, Bansal S, Kuhad A, Kumar A, Chopra K. Naringenin ameliorates diabetic neuropathic pain by modulation of oxidative-nitrosative stress, cytokines and MMP-9 levels. Food Funct. 2020;11(5):4548–60.

Tankova T, Cherninkova S, Koev D. Treatment for diabetic mononeuropathy with α-lipoic acid. Int J Clin Pract. 2005;59(6):645–50.

Tütüncü NB, Bayraktar M, Varli K. Reversal of defective nerve conduction with vitamin E supplementation in type 2 diabetes: a preliminary study. Diabetes Care. 1998;21(11):1915–8.

Vallianou N, Evangelopoulos A, Koutalas P. Alpha-lipoic acid and diabetic neuropathy. Rev Diab Stud. 2010;6(4):230.

Vijayakumar PA, Kalshetti S, Bhatt J. Supplementation of α-lipoic acid in diabetic peripheral neuropathy: a prospective open label randomized controlled trial. Int J Pharm Pharm Sci. 2014;6:90–3.

Vincent AM, Feldman EL. New insights into the mechanisms of diabetic neuropathy. Rev Endocrine Metabol Disorders. 2004;5:227–36.

Yorek M. Combination therapy is it in the future for successfully treating peripheral diabetic neuropathy? Front Endocrinol. 2024;15:1357859.

Ziegler D, Hanefeld M, Ruhnau KJ, Meiner HP, Lobisch M, Schütte K, et al. Treatment of symptomatic diabetic peripheral neuropathy with the anti-oxidant α-lipoic acid: a 3-week multicentre randomized controlled trial (ALADIN study). Diabetologia. 1995;38:1425–33.

Ziegler D, Hanefeld M, Ruhnau KJ, Hasche H, Lobisch M, Schütte KLEMENS, et al. Treatment of symptomatic diabetic polyneuropathy with the antioxidant alpha-lipoic acid: a 7-month multicenter randomized controlled trial (ALADIN III study). ALADIN III Study Group. Alpha-Lipoic acid in diabetic neuropathy. Diabetes Care. 1999;22(8):1296–301.

Ziegler D, Ametov A, Barinov A, Dyck PJ, Gurieva I, Low PA, et al. Oral treatment with α-lipoic acid improves symptomatic diabetic polyneuropathy: the SYDNEY 2 trial. Diabetes Care. 2006;29(11):2365–70.

Ziegler D, Low PA, Litchy WJ, Boulton AJ, Vinik AI, Freeman R, et al. Efficacy and safety of antioxidant treatment with α-lipoic acid over 4 years in diabetic polyneuropathy: the NATHAN 1 trial. Diabetes Care. 2011;34(9):2054–60.

Novel Biomarkers for Early Diagnosis of Diabetic Neuropathy: Current Research and Prospects

24

Rajesh Kumar Sharma, Prithu Pathak, Mallikarjuna Rao Pichika, Madhu Katyayani Balijepalli, Mohd Tabish, and Manikandan

24.1 Introduction

The most common and early of the three main problems linked to diabetes is diabetic neuropathy (DN), yet it is frequently recognised only after the illness has advanced significantly (Feldman et al. 2019). For this reason, early diagnosis is crucial to care and treatment. According to reports, asymptomatic instances of diabetic peripheral neuropathy (DPN) might account for as much as 50% of cases. For the early identification of DN, therefore, it is necessary to develop new biological markers that are both practical and precise in order to determine the intensity or stage progression in addition to clinical symptoms and neurological abnormalities. Several different kinds of variables have been found as possible DN biomarkers because biomarkers show the intricacy of these aetiologies and the development of the pathological disease. These include elements connected to the aetiology,

R. K. Sharma (✉)
Department of Pharmacy, Faculty of Health and Allied Science, KAAF University, Buduburam, Ghana

P. Pathak
Sagar Institute of Pharmacy and Technology, Bhopal, MP, India

M. R. Pichika
Newcastle University Medicine Malaysia, Iskandar Puteri, Johor, Malaysia

M. K. Balijepalli
Department of Pharmacology, Faculty of Medicine, Jenjarom, Selangor, Malaysia

M. Tabish
Department of Pharmacy, Institute of Biomedical Education and Research, Mangalayatan University, Aligarh, Uttar Pradesh, India

Manikandan
Department of Pharmaceutical Analysis, SRM College of Pharmacy, SRM Institute of Science and Technology, Chennai, Tamilnadu, India

M. K. Gautam et al. (eds.), *Antioxidants and Functional Foods for Diabetic Neuropathy*, https://doi.org/10.1007/978-981-95-9326-2_24

molecules involved in the cellular response to damage to vascular endothelial cells and neurodegeneration, and variables linked to the evolution of pathological diseases such inflammation (Galer et al. 2000).

24.2 Newly Developed Biomarkers for Early Identification and Forecasting

24.2.1 Genetic Predisposition

It is still highly fascinating to look for genetic factors because high blood sugar levels and other established risk indicators cannot completely predict the onset and progression of DN in any given patient. Aside from the uncommon exceptions mentioned below, additional studies have examined potential DN-sensitive genes; however, most of these investigations had small sample numbers, which made it more challenging to assess the role of common alleles (Spallone 2017).

In order to evaluate the connection with ACE gene variations and DPN, a systematic review and meta-analysis examined five study designs that were carried out in five different nations. The ACE gene's insertion or depletion polymorphism was shown to have a legitimate correlation with the development of DPN by the meta-analysis. In this meta-analysis, four case–control studies on the gene methylene-tetrahydrofolate reductase (MTHFR), which changes homocysteine into methionine, were examined. For nucleotide 677, an MTHFR polymorphism was shown to be associated with DPN (Wu et al. 2017). Research on hereditary risks for chronic illnesses has increased as a result of advancements in gene sequencing technology; a meta-analysis found that genetic alterations are linked to neuropathic pain such as HLA, GCH1, TNFA, IL-6, and OPRM1 (Veluchamy et al. 2018). Furthermore, DPN and neuropathic pain in diabetes have been linked to genetic variations. Numerous gene polymorphisms linked to painful-DPN have been discovered through genome-wide association studies. Additionally, it has been shown that small fibre neuropathy and painful DPN are associated with rare genetic polymorphisms of the ion-gated sodium channel Nav 1.7 (Li et al. 2015). Although confirmation is still required, the ion-gated sodium channel Nav 1.7 study has discovered more mutation sites linked to painful DPN. Ten out of 111 people with painful DPN have an uncommon Nav 1.7 variation, which causes increased intensity of pain and higher sensitivity to pressure triggers. These results suggest a relationship between genetic variations and clinical phenotype that might be used to forecast treatment outcome. Additionally, a recent study discovered that those receiving treatment with the anticonvulsant lacosamide for small fibre neuropathy with Nav 1.7 mutations experienced noticeably less pain than those receiving a placebo (de Greef et al. 2019). In general, DPN and genome-wide association studies provide some evidence for genetic susceptibility to DPN; nonetheless, large-scale research are the only sources of conclusive data about genetic biomarkers for DPN.

24.2.2 Clinical Phenotype

Somatosensory phenotyping finds patient subgroups with nerve structural abnormalities using surveys like the QST or the Neuropathic Pain Symptom Inquiry (NPSI). By analysing 13 criteria of sensory loss and gain aberrations, the German Research Network on Neuropathic Pain (DFNS) approach finds three clusters of somatosensory profiles in neuropathies with various aetiologies. In contrast to pleasant DPN, heat hyposensitivity is a sensory loss in painful DPN, according to extensive studies employing this QST approach (Yang et al. 2020).

24.2.3 Advanced Glycation End Products (AGEs) and Their Precursors

Protein glycation that is non-enzymatic is the primary cause of diabetes-related complications such nephropathy and macroangiopathy. It has been demonstrated that DN peripheral nerve tissue exhibits an excessive build-up of AGEs, which is correlated with a reduction in nerve fibre count. On the other hand, it appears that AGEs are not a good biomarker and that measuring them in tissues presents challenges with regard to sample collection and quantification. Reactive dicarbonyl, on the other hand, which are precursors of AGEs and include methylglyoxal and α-oxoaldehyde, are gaining a lot of interest as biomarkers for DN progression and onset prediction (Fujita et al. 2021).

Mice treated with methylglyoxal showed reduced nerve conduction velocity, increased production of peptides linked to the calcitonin gene from cutaneous nerve terminals, and caused hyperalgesia in response to mechanical and heat stimulation. Similarly, thermal hyperalgesia and chemical pain perceptions are also induced when methylglyoxal microinjections are made into the skin of healthy human volunteers (Düll et al. 2019). These findings suggest that the main reason for the hyperalgesia seen in DN patients may be methylglyoxal. It is known that reactive dicarbonyl, such as methylglyoxal, can be broken down in vivo by the rate-determining enzyme glyoxalase I (GLO I). A recent study found a substantial correlation with painful DN and reduced GLO I activity in blood samples. Consequently, GLO I activity could be a helpful biomarker (Skapare et al. 2013).

24.3 Molecules Involved in Inflammation Development

24.3.1 Toll-Like Receptor (TLR)

TLR are one of several pattern recognition receptors (PRRs) that the innate immune system uses to recognise danger-associated molecular patterns (DAMPs) and initiate subsequent inflammatory cascades. According to recently discovered evidence, DN is linked to TLR-mediated inflammatory cascades. The endoneurial capillaries of the sciatic nerves of leptin-deficient ob/ob mice express more TLR4. An experimental DN model's mechanical allodynia and hyperalgesia were reduced by TLR4 downregulation. In sciatic nerves from mice with type 2 diabetes, TLR2 inhibitors have been demonstrated to decrease inflammation and promote M2 phenotype macrophages. One important modulator that promotes the polarisation of inflammatory macrophages is TLR2 (Cheng et al. 2024).

In their investigation of TLR4 and its genes in type II diabetes individuals, Zhu et al. discovered that TLR4 expression levels were greater in those with neuropathy than in healthy people and diabetics without neuropathy. TLR4 may be a helpful DN marker, as diabetics who develop neuropathy also have considerably higher levels of TNF-α and IL-6 (Zhu et al. 2015).

24.3.2 Adiponectin

In the human body, adiponectin regulates fat, metabolism of glucose, and insulin sensitivity, among other processes. Serum adiponectin levels in 487 type II diabetic patients were measured in a cross-sectional research conducted in India. Compared to those without neuropathy, diabetics with neuropathy had noticeably greater blood adiponectin levels. According to a study by Pradeepa et al., Adiponectin levels above normal were associated with an increased risk of neuropathy (Pradeepa et al. 2015).

24.3.3 MicroRNAs (miRNAs)

miRNAs, which are non-coding, single-stranded RNA molecules that range in length from 20 to 22 nucleotides, control genetic activity by complementary attach to the target mRNA's 3′ untranslated region and preventing translation or destruction. Since the patterns of expression of miRNAs are associated with pathophysiological processes, miRNAs are garnering interest as novel biomarkers. In dorsal root ganglia (DRG) neurones, Wang and colleagues found that hyperglycaemia increased levels of IL-1 receptor activated kinase (IRAK1) and tumour necrosis factor receptor-associated factor 6 (TRAF6). In a study of individuals with type 1 diabetes, DPN patients had greater blood levels of miR-518d-3p along with miR-618 than people without microvascular problems (Fujita et al. 2021).

24.3.4 Monocyte Chemoattractant Protein-1 (MCP-1)

In their chemokine family, monocyte binding is an essential step whereby dendritic cells, monocytes, and macrophages release MCP-1 at the site of injury, damage, and infection. During inflammation, pro-inflammatory mediators for example TNF-α,

IL-1β, and PDGF trigger the production of MCP-1. In reaction to inflammation and hypoxia, MCP-1 promotes the migration of mononuclear phagocytes. Through C-C motif ligand receptor 2 (CCR2) receptors, monocyte binding is associated with an upregulation of MCP-1 production, which helps to regulate and treat diabetic neuropathic pain. In addition to improving excitatory synaptic transmission, the inhibition of MCP-1 is associated with the management of discomfort and neuropathic pain (Laddha and Kulkarni 2020).

24.3.5 Vascular Endothelial Growth Factor (VEGF)

The angiogenic factor VEGF is released by a range of cells, such as keratinocytes, renal mesangial cells, platelets, and macrophages. VEGF is essential for biological processes such development, wound healing, haematopoiesis, and bone formation. Both the protein and mRNA levels of the VEGF pathway in Spinal ganglion and Schwann cells are impacted by hyperglycaemia. Researchers found that severe diabetic neuropathic symptoms might be cured using VEGF gene transfer trials. During inflammation, DM also affects neuropeptide Y and VEGF, two molecules that are critical to the pathophysiology of DN. Additionally, it has been noted that VEGF levels in the blood are downregulated in individuals with DPN who are symptomatic. As a result, it may be applied as a biomarker to identify DN (Darvishi 2023).

24.4 The Molecules Linked to Nerve Damage

24.4.1 Neuron-Specific Enolase (NSE)

Li et al. explored the relationship among DN and blood NSE concentrations in lung cancer and neuroblastoma, emphasising that long-term hyperglycaemia can change the production of NSE during nerve regeneration and degeneration (Li et al. 2013). Furthermore, it has been recently revealed that the value declined as the neuropathy improved in response to therapy, as well as with its initiation and development.

24.4.2 Semaphorins

A vast family of proteins known as semaphorins was first discovered to be involved in the guiding of growing nervous system axons. These are multifunctional proteins that are essential to several biological processes, including angiogenesis, organogenesis, and immunological responses. As a result, neurotrophic factors are thought to be potential biomarkers for DN, yet there aren't many reports of these investigations in people (Sun et al. 2018).

24.4.3 Heat Shock Protein (HSP27)

Cell protection from stress is significantly aided by the little heat shock protein HSP27. The spinal cord DRG cells of diabetic mice have been shown to have elevated levels of HSP27. This protein also contributes to the defence of Schwann cells against programmed cell death. Conflicting findings have been found despite several investigations looking into the connection between patients with DN and blood HSP27 levels (Fujita et al. 2021).

It has been revealed that HSP27 has a role in both neurone survival and inherited neuropathy types. Not all target tissues of diabetic problems have been shown to overproduce HSP27. Two hypothesised pathways that provide a neuroprotective effect include apoptosis suppression and actin cytoskeleton stabilisation (Pourhamidi et al. 2011).

There are contradicting findings about HSP27 expression in DRG neurones. While one research found no change, another report found that the only stress/plasticity indication with increased gene expression was HSP27. A variety of neuropathic defects, such as loss of epidermal innervation, mechanical allodynia, loss of footpad temperature sensitivity, and reduced sensory conduction velocity, were prevented in diabetic mice by overexpressing HSP27 in the neurones. Therefore, HSP27 could be advantageous to DPN (Bellini et al. 2017).

24.5 Local Biomarkers

24.5.1 Mitochondrial SOD2

Human cells depend on the antioxidant enzyme manganese SOD2. Apart from possessing antiapoptotic qualities in favour of inflammatory cytokines and oxidative stress, which are linked to diabetic neuropathy, it detoxifies superoxide in mitochondria. In individuals with newly diagnosed type 2 diabetes from the initial cohort of the German Diabetes Study, overexpression of subepidermal SOD2 was associated with a longer duration of diabetes and a more balanced sympathovagal surroundings (Ziegler et al. 2015).

24.5.2 Langerhans Cells

The development of DN may be influenced by skin inflammation. It is possible to detect Langerhans cells (LCs) by using antibodies against CD207 (Langerin) on LCs. The forearm's cutaneous LC number dropped right away when diabetes initially manifested in T1D individuals. However, this did not occur in those who had diabetes for 6 months. In a different research, the quantity of LCs in the foot ulcer epidermis was associated with healing in those with T1D and T2D who had previously had neuropathy. The possibility that a cutaneous immunogenic imbalance towards inflammation might be facilitated by the early loss of cutaneous LCs in T2D patients should be studied (Bönhof et al. 2019).

24.5.3 Transient Receptor Potential Vanilloid Type 1 (TRPV1)

The capsaicin receptor, also known as TRPV1, is distributed throughout the PNS's nociceptive pathway, including epidermal nerve fibres. When triggered by heat, capsaicin, or acidic environments, it mediates burning or painful feelings. The possible involvement of TRPV1 in painful neuropathies is of great interest, since it is implicated in nociception and may be involved in many forms of hyperalgesia. Since such, it may be a viable target for analgesic therapy in the future (Lauria et al. 2006).

The two major symptoms of DPN are hyperalgesia and allodynia. The overexpression of TRPV1 in peripheral nerves is the main cause of the excruciating pain in DN patients. TRPV1 levels were found to be higher in diabetics because their reflex arcs functioned better. The findings of the research demonstrated that TRPV1 is essential to heat hyperalgesia and chemical pain perception generated by methylglyoxal through selective pharmacological inhibition. According to the study's findings, methylglyoxal may mediate diabetes-induced neuropathic pain by sensitising ligand-gated sodium channel subtype 1.8 and activating TRPV1 (Agarwal et al. 2020).

24.5.4 p75 Neurotrophin Receptor

Immune cells such as macrophages and microglia, like Schwann cells and axons, express the neurotrophin receptor p75 (p75). It is essential for synaptic plasticity. High expression of p75NTR has been found in the myelin sheaths surrounding axons that are prone to degeneration in diabetic patients, sparking curiosity in a possible role in the onset of DPN. The marker for Schwann cells, p75NTR, was employed in two investigations. In an intracutaneous axotomy model, the study discovered unique patterns of Schwann cell migration and p75NTR expression in individuals with persistent neuropathies after nerve transection (Bönhof et al. 2019).

Using a urine sample and a home-developed ELISA method, evaluated the levels of the extracellular domain (ECD) of p75NTR in relation to creatinine. Diabetic patients and those with diabetic neuropathy provided urine samples for collection. Compared to controls or corresponding sex-matched diabetic patients without neuropathy, diabetic patients with neuropathy had substantially higher p75NTR ECD levels. It is possible that this is due to the increased concentration of p75NTR on Schwann cells in diabetic neuropathy patients, which causes an increased turnover or an increase in the number of accessible receptors for cleavage. Both throughout development and following nerve damage, substantial concentrations of the soluble form of ECD are created. Uncertainty surrounds its exact biological function. It might scavenge neurotrophins and proneurotrophins in the extracellular environment by functioning as a decoy receptor, or it could help p75NTR-mediated death signalling (Jourdi et al. 2023).

24.6 Enzyme Biomarkers

24.6.1 Nicotinamide Adenine Dinucleotide Phosphate Oxidase (NOX1)

The electron transport route from NADPH to molecular oxygen is mediated by the enzyme NOX. They are often found in the main brain cells, including astrocytes, microglia, vascular endothelial cells, and phagocytic cells. By causing phagocytic cells to produce ROS, such superoxide, NOXs have an impact on neuropathic pain. NOX-ROS are crucial to the oxidative stress-related inflammatory pathways that damage vascular tissue, including neurones. These pathways are essentially connected to inflammation, angiogenesis, and nerve cell hypertrophy. These pathways cause the rise of NOX in DN (Gray et al. 2014).

24.6.2 Haem Oxygenase-1 (HO-1)

HO-1 is extensively distributed in systemic organs and is linked to the stress response. Numerous actions of HO-1 have been documented, including immunomodulatory, antiproliferative, anti-inflammatory, antioxidant, and antiapoptotic properties. Leng et al. investigated the role of HO-1 in DN. After 6 weeks of induction, diosgenin (50 and 100 mg/kg) was administered for 14 days to diabetic animals. The study found that diosgenin significantly increased mechanical excessive pain and tail pull-out delay (Leng et al. 2020). In a western blot examination, diabetic animals have HO-1 down-regulated in comparison to the normal control group. In contrast to diabetic rats, diosgenin therapy increased the expression of HO-1. The study found that via lowering oxidative stress in the Nrf2/HO-1 pathway, diosgenin had a neuroprotective impact in DPN. In this sense, HO-1 may be viewed as a biomarker for DN early detection (Laddha and Kulkarni 2020).

24.6.3 Ceruloplasmin

A necessary component for maintaining homeostasis is copper. The liver cells that generate ceruloplasmin, a 2-glycoprotein or serum ferroxidase, store over 95% of the copper present in plasma. According to scientific research, neuropathy and neurodegenerative illnesses are among the neuronal diseases linked to aberrations in the ceruloplasmin gene. Additionally, researchers calculated that DN had considerably higher quantities of ceruloplasmin. To comprehend ceruloplasmin's mode of action in DN, much research is necessary (Adki and Kulkarni 2023).

24.6.4 Dipeptidyl Peptidase-4 (DPP4)

DPP4 is an immune-regulated signal transduction enzyme that plays a role in glucose metabolism. It facilitates the breakdown of glucose in conjunction with growth factors, cytokines, and GLP-1. DPP4 inhibitors raise levels of incretin hormones, which regulate blood sugar, and insulin sensitivity. Obesity, diabetes, and non-alcoholic fatty liver disease have all been associated with elevated DPP4 blood levels. Neuroinflammation and changes in DPP4 regulation are linked to hyperthermia (Nargis and Chakrabarti 2018).

24.6.5 Other Biomarkers for DPN

New biomarkers for DPN have been discovered as a result of several pathophysiological causes. The presence of DPN may also favourably correlate with other growth factors, such as neuron-specific enolase, cystatin C, EGF, platelet-derived growth factors AB/BB, and VEGF. These results were published by cross-sectional studies.

24.6.6 Sirtuin 1 (SIRT1)

One of the histone deacetylases that depends on nicotinamide adenine dinucleotide is SIRT, also known as the metabolic sensor. Through the control of oxidative stress, inflammation, and mitochondrial activity, it is essential for determining the body's energy state. The development of different types of DN may be caused by an imbalance between the supply and demand for energy (Yerra et al. 2017). One of the main findings in type 2 DN is the downregulation of SIRT1. DN may proceed as a result of hypoxia, oxidative damage, and metabolic stress caused by SIRT1 downregulation.

24.6.7 Metastasis-Associated Lung Adenocarcinoma Transcript 1 (MALAT1)

MALAT1, a conserved lncRNA, is the first reported example of a non-protein coding gene. The p53 gene controls it during haematopoiesis, and it is found in the brain, liver, lung, kidney, and heart. MALAT1 has a role in maintaining haematopoietic cell proliferation. MALAT1 is activated by inflammatory mediators when blood sugar levels are out of equilibrium, which might increase serum levels of amyloid antigen 3. As DN developed, overexpression of MALAT1 was seen. MALAT1 has also been connected to an anti-inflammatory role during nerve damage, according to research data. A potential biomarker for DN might be MALAT1, the study found (Adki and Kulkarni 2023).

24.6.8 Caveolin 1

The membrane-integral proteins known as caveolins are essential for receptor-independent endocytosis and are involved in signal transduction, cholesterol homeostasis, vesicular transport, and cancer prevention (Williams and Lisanti 2004). According to studies, caveolin 1 phosphorylates the Nmethyl-D-aspartate receptor 2B subunit in DN rat models and regulates the expression of toll-like receptor 4. The pathophysiological movement of DPN is augmented by the damage to caveolin 1 (Pascariu et al. 2004).

24.7 Conclusion

Patients with diabetic neuropathy have a worse quality of life. Thus, it is necessary to establish diagnostic techniques for the early diagnosis of DN. All a plethora of novel biomarkers representing various pathogenic pathways, including inflammation and oxidative stress, have emerged in recent years. These discoveries should ultimately lead to the improvement of our treatment arsenal against this prevalent, crippling, or even fatal illness. Nevertheless, none of the biomarker candidates that have been identified can be fully implemented in clinical practice, as this review explains. One of the reasons for this appears to be the complex nature of the disease's aetiology. Moreover, the pathological status varies according to how the disease progresses from the beginning, making therapeutic management challenging. Biomarkers that are important to the pathophysiology of diabetic neuropathic pain (DN) include PARP a, MCP-1, Ceruloplasmin, HO-1, SIRT1, Carolin 1, NOX 1 & 4, VEGF, MALAT1, DPP4, and miRNA. These biomarkers can also be utilised to diagnose DN and find novel treatments. In order to find a biomarker for DN, the diseased condition must be selected according to the proper diagnostic criteria, and it must be assessed at each stage using the same methodology.

References

Adki KM, Kulkarni YA. Biomarkers in diabetic neuropathy. Arch Physiol Biochem. 2023;129(2):460–75.

Agarwal N, Taberner FJ, Rojas DR, Moroni M, Omberbasic D, Njoo C, et al. SUMOylation of enzymes and ion channels in sensory neurons protects against metabolic dysfunction, neuropathy, and sensory loss in diabetes. Neuron. 2020;107(6):1141–59.

Bellini S, Barutta F, Mastrocola R, Imperatore L, Bruno G, Gruden G. Heat shock proteins in vascular diabetic complications: review and future perspective. Int J Mol Sci. 2017;18(12):2709.

Bönhof GJ, Herder C, Strom A, Papanas N, Roden M, Ziegler D. Emerging biomarkers, tools, and treatments for diabetic polyneuropathy. Endocr Rev. 2019;40(1):153–92.

Cheng Y, Chen Y, Li K, Liu S, Pang C, Gao L, et al. How inflammation dictates diabetic peripheral neuropathy: an enlightening review. CNS Neurosci Ther. 2024;30(4):e14477.

Darvishi M. Diagnostic biomarkers of diabetic neuropathy. J Clin Res Rep. 2023;13(1):01–10.

de Greef BT, Hoeijmakers JG, Geerts M, Oakes M, Church TJ, Waxman SG, et al. Lacosamide in patients with Nav1. 7 mutations-related small fibre neuropathy: a randomized controlled trial. Brain. 2019;142(2):263–75.
Düll MM, Riegel K, Tappenbeck J, Ries V, Strupf M, Fleming T, et al. Methylglyoxal causes pain and hyperalgesia in human through C-fiber activation. Pain. 2019;160(11):2497–507.
Feldman EL, Callaghan BC, Pop-Busui R, Zochodne DW, Wright DE, Bennett DL, et al. Diabetic neuropathy. Nat Rev Dis Prim. 2019;5(1):1–18.
Fujita Y, Murakami T, Nakamura A. Recent advances in biomarkers and regenerative medicine for diabetic neuropathy. Int J Mol Sci. 2021;22(5):2301.
Galer BS, Gianas A, Jensen MP. Painful diabetic polyneuropathy: epidemiology, pain description, and quality of life. Diabetes Res Clin Pract. 2000;47(2):123–8.
Gray SP, Jha JC, Di Marco E, Jandeleit-Dahm KA. NAD (P) H oxidase isoforms as therapeutic targets for diabetic complications. Exp Rev Endocrinol Metabol. 2014;9(2):111–22.
Jourdi G, Fleury S, Boukhatem I, Lordkipanidzé M. Soluble p75 neurotrophic receptor as a reliable biomarker in neurodegenerative diseases: what is the evidence? Neural Regen Res. 2023;19:536.
Laddha AP, Kulkarni YA. NADPH oxidase: a membrane-bound enzyme and its inhibitors in diabetic complications. Eur J Pharmacol. 2020;881:173206.
Lauria G, Morbin M, Lombardi R, Capobianco R, Camozzi F, Pareyson D, et al. Expression of capsaicin receptor immunoreactivity in human peripheral nervous system and in painful neuropathies. J Peripher Nerv Syst. 2006;11(3):262–71.
Leng J, Li X, Tian H, Liu C, Guo Y, Zhang S, et al. Neuroprotective effect of diosgenin in a mouse model of diabetic peripheral neuropathy involves the Nrf2/HO-1 pathway. BMC Complementary Med Ther. 2020;20:1–9.
Li J, Zhang H, Xie M, Yan L, Chen J, Wang H. NSE, a potential biomarker, is closely connected to diabetic peripheral neuropathy. Diabetes Care. 2013;36(11):3405–10.
Li QS, Cheng P, Favis R, Wickenden A, Romano G, Wang H. SCN9A variants may be implicated in neuropathic pain associated with diabetic peripheral neuropathy and pain severity. Clin J Pain. 2015;31(11):976–82.
Nargis T, Chakrabarti P. Significance of circulatory DPP4 activity in metabolic diseases. IUBMB Life. 2018;70(2):112–9.
Pascariu M, Bendayan M, Ghitescu L. Correlated endothelial caveolin overexpression and increased transcytosis in experimental diabetes. J Histochem Cytochem. 2004;52(1):65–76.
Pourhamidi K, Dahlin LB, Boman K, Rolandsson O. Heat shock protein 27 is associated with better nerve function and fewer signs of neuropathy. Diabetologia. 2011;54:3143–9.
Pradeepa R, Surendar J, Indulekha K, Chella S, Anjana RM, Mohan V. Association of serum adiponectin with diabetic microvascular complications among south Indian type 2 diabetic subjects—(CURES-133). Clin Biochem. 2015;48(1–2):33–8.
Skapare E, Konrade I, Liepinsh E, Strele I, Makrecka M, Bierhaus A, et al. Association of reduced glyoxalase 1 activity and painful peripheral diabetic neuropathy in type 1 and 2 diabetes mellitus patients. J Diabetes Complicat. 2013;27(3):262–7.
Spallone V. Might genetics play a role in understanding and treating diabetic polyneuropathy? Diabetes Metab Res Rev. 2017;33(4):e2882.
Sun Q, Tang DD, Yin EG, Wei LL, Chen P, Deng SP, Tu LL. Diagnostic significance of serum levels of nerve growth factor and brain derived neurotrophic factor in diabetic peripheral neuropathy. Med Sci Monit. 2018;24:5943.
Veluchamy A, Hebert HL, Meng W, Palmer CN, Smith BH. Systematic review and meta-analysis of genetic risk factors for neuropathic pain. Pain. 2018;159(5):825–48.
Williams TM, Lisanti MP. The caveolin proteins. Genome Biol. 2004;5:1–8.
Wu S, Han Y, Hu Q, Zhang X, Cui G, Li Z, Guan Y. Effects of common polymorphisms in the MTHFR and ACE genes on diabetic peripheral neuropathy progression: a meta-analysis. Mol Neurobiol. 2017;54:2435–44.

Yang H, Sloan G, Ye Y, Wang S, Duan B, Tesfaye S, Gao L. New perspective in diabetic neuropathy: from the periphery to the brain, a call for early detection, and precision medicine. Front Endocrinol. 2020;10:929.

Yerra VG, Kalvala AK, Kumar A. Isoliquiritigenin reduces oxidative damage and alleviates mitochondrial impairment by SIRT1 activation in experimental diabetic neuropathy. J Nutr Biochem. 2017;47:41–52.

Zhu T, Meng Q, Ji J, Lou X, Zhang L. Toll-like receptor 4 and tumor necrosis factor-alpha as diagnostic biomarkers for diabetic peripheral neuropathy. Neurosci Lett. 2015;585:28–32.

Ziegler D, Strom A, Brüggemann J, Ziegler I, Ringel B, Püttgen S, et al. Overexpression of cutaneous mitochondrial superoxide dismutase in recent-onset type 2 diabetes. Diabetologia. 2015;58:1621–5.

Pioneering Technologies in the Management of Diabetic Neuropathy

25

Saroj Yadav, Angesh Kumar, Brijesh Shivhare, Shivli Nomani, Ashish Jorasia, and Sejal ben Gunvantbhai Patel

25.1 Introduction

Diabetes is one of the conditions that is growing at the fastest pace in the modern era, which has quadrupled in the last 20 years. According to the International Diabetes Federation, 700 million people worldwide—or 10.9% of the total population—will have diabetes by 2045 (Neeland and Patel 2019). Even though the majority of instances are avoidable, diabetic foot ulcers and amputations are among the most significant and detrimental consequences of diabetes (Kim et al. 2014). Peripheral nerve dysfunction, characterised by a distal-to-proximal progression with symptoms and indications of pain and/or sensitisation, muscular weakness, and numbness or desensitisation, is known as diabetic neuropathy. Although the precise origin of diabetic neuropathy is unknown, elevated blood sugar levels are thought to play a role. Some individuals still do not receive an appropriate diagnosis

S. Yadav (✉)
School of Medical and Allied Sciences, K.R.Mangalam University, Sohna Gurugram, Haryana, India

A. Kumar
Department of Pharmaceutical Technology, Meerut Institute of Engineering and Technology (MIET), Meerut, Uttar Pradesh, India

B. Shivhare
Department of Botany, Faculty of Sciences, Baba Mastnath University, Rohtak, Haryana, India

S. Nomani
SNS College of Pharmacy, Motihari, Bihar, India

A. Jorasia
School of Arts and Humanities, Career Point University, Kota, Rajasthan, India

S. b. G. Patel
Nootan Pharmacy College, Sankalchand Patel University, Mahesana, Gujarat, India

M. K. Gautam et al. (eds.), *Antioxidants and Functional Foods for Diabetic Neuropathy*, https://doi.org/10.1007/978-981-95-9326-2_25

or treatment for DPN, despite advancements in diagnostic and therapeutic techniques. Furthermore, because definitive diagnostic standards are lacking, it is impossible to determine the frequency of DPN in various groups with any degree of reliability. Regretfully, due to their time-consuming nature, specialised equipment needed, and need for trained specialists, Clinical settings do not lend itself to the use of gold conventional techniques from clinical trials (Tesfaye et al. 2010). Additionally, there is variation in the threshold values and sensitivity and specificity of various DPN diagnostic techniques.

25.2 Tools and Technologies in the Diagnosis and Treatment of DN

Quality of life (QOL), which encompasses sleep, mood, energy, and mobility, is significantly impacted by pain. A favourable prognosis for DPN patients depends on an early diagnosis, and prompt, thorough therapy can help avoid falls and minimised the negative impact on patients' QOL. Regretfully, about 75% of elderly participants in a population-based survey were not aware that they had the illness. For individuals with DPN or at risk for it, there is currently no clinical test that can be used to detect or forecast the onset or exacerbation of symptoms. Furthermore, there is no agreement on the exact drug regimen to treat DPN, and improved glucose management is the only recognised disease-modifying therapy for DPN that works. As a result, identifying individuals early on will offer a window of opportunity to determine targeted therapy to alter the course of DPN (Callaghan et al. 2012). According to the American Diabetes Association's 2018 clinical practice guidelines, a 10-g Semmes Weinstein monofilament should be used to evaluate protective feeling once a year. This is because the tendency of diabetic foot ulcers and potentially eventual amputation is significantly increased when one is unable to experience protective feeling.

25.3 The Most Commonly Used Clinical Diagnostic Modalities Involved

25.3.1 Symptoms and Signs

In order to measure general neuropathic impairments, composite grading techniques based on indicators, complaints, or each has been devised. The Neuropathy Disability Score (NDS), the Neuropathy Condition Profile, and the Neuropathy Symptom Score of Dyck are a few examples of these systems. A good diagnostic tool for neuropathic pain, which includes needles, numbness, electrical shocks, tingling, pins and itching, is the Douleur Neuropathique en 4 (DN4). Additionally, tactile dynamic allodynia and touch and pin hypoesthesia are tested (Petropoulos et al. 2018). These scoring schemes improve diagnostic precision by aggregating the findings from several tests. Using a visual analogue scale or reliable measures and questionnaires, clinicians treating patients with painful DPN can evaluate the degree of neuropathic pain and the patients' response to therapy.

25.3.2 Quantitative Sensory Testing (QST)

The function of both large and small nerve fibres is evaluated using the NeuroSensory Analyser, a popular quantitative sensory testing (QST) method. The TSA-II NeuroSensory Analyser from Medoc Advanced Medical Systems is often utilised. NerveCheck is a newer, more portable, and less expensive equipment. When compared to conventional QST equipment, this instrument has shown remarkable consistency and diagnostic accuracy for identifying DPN (Carmichael et al. 2021).

25.3.3 Neurophysiology

The elements of the peripheral nervous system are included in neurophysiology, and the gold standard for diagnosing DPN is nerve conduction testing (NCS). By measuring the peripheral nerve's electrical impulse transmission capacity, NCS evaluates the beginning and course of DPN. Nevertheless, NCS is costly, time-consuming, unskilled, and necessitates medical specialists. Additionally, it can only evaluate large nerve fibres, which are affected first in DPN patients, and screening huge data samples is difficult. Tiny fibre neuropathies cannot be assessed by standard electrophysiological testing. Only in situations of unclear diagnosis or atypical clinical presentation may electrophysiology be employed. As a result, early intervention and treatment require straightforward yet effective strategies (Yu 2021).

25.3.4 Skin Biopsy

In neuropathies, such as DN, small-fibre damage is diagnosed by immunohistochemistry or immunofluorescent nerve fibre staining with PGP9.5 and a 3-mm punch biopsy skin sample. Within a year of being diagnosed with type 2 diabetes, intraepidermal nerve fibre density (IENF) falls, and IENF loss is a typical feature of manifest small fibre neuropathy (SFN) (Bönhof et al. 2019).

Diabetes peripheral neuropathy (DPN) may be diagnosed with a sensitivity of 61–90% and a specificity of 64–82.8% using IENFD, a measurement of nociceptor density in the epidermis. Because of its rapid natural rate of depletion of epidermal innervation, it corresponds with other measures of neural function and might be an early indicator of DPN. Neuropathic pain is unaffected by it, though. In addition to being utilised as a clinical outcome, studies indicate that diet and exercise recommendations may help IENFD (Yang et al. 2020).

There are very few researches that have looked at the temporal sequence of DSPN and SFN cutaneous nerve fibre loss. IENFD and dermal nerve fibre immunoreactivity decreased in a 6-month surveillance of a sample with T1D or T2D and

DSPN, while clinical symptoms, scores, and a number of sensory tests remained unchanged. These results suggest that dermal nerve fibres and IENFs may develop into useful indicators of the course of DSPN (Narayanaswamy et al. 2012). According to another study, individuals with T2D who had just been diagnosed had an IENFD reduction rate that was many times higher than that of controls. This study also proposed that the rate of IENF loss in these patients was relative rather than linear, given the baseline IENFD (Divisova et al. 2016).

25.3.5 Corneal and Retinal Innervation

Numerous eye tests, including pupil responsiveness, retinal nerve fibre layer thickness, and corneal confocal microscopy (CCM), been put out as alternative markers for DPN and other neurological disorders (Yang et al. 2020). CCM is a relatively new method for morphometrically diagnosing corneal SFN. Deficits in the CCM parameters are evident at the beginning of DSPN, and such deficiencies worsen as DSPN severity rises. Rapid and non-invasive, CCM has become a popular method for studying corneal innervation and for DPN diagnosis. For the diagnosis of DPN, it has a high sensitivity (68–92%) and a specificity of 40–64%. Moreover, CCM measurements on skin biopsy correlate with IENFD. A shorter corneal nerve fibre length was shown by Pritchard et al. to be indicative with incident DPN (Pritchard et al. 2015). Furthermore, researchers are discovered that ocular nerve characteristics dropped off quickly before foot problems appeared (Yang et al. 2020). Furthermore, diminished pain/temperature perception and other SFN measurements have been connected to deficits in CCM parameters. When considered collectively, these findings imply that the severity of DSPN might be categorised using CCM. Diabetic patients can have nerve problems diagnosed utilising procedures such as CCM and OCT. Before DSPN appears, CCM can assess the specific anatomical characteristics of nerve anomalies, such as reduced CNFL in T1D (Stem et al. 2014). OCT has been used to identify retinal nerve fibre loss in diabetic retinopathy, which is linked to diabetics losing their retinal nerve fibre layer (RNFL). However, because these methods need expensive equipment and specialised knowledge, they are not yet broadly accessible (Iqbal et al. 2018).

25.3.6 High-Frequency Ultrasound

A vital diagnostic technique for DPN patients, high-frequency ultrasonography improves clinical diagnosis and lowers missed diagnoses. By measuring the nerve's size, blood vessels, echo, and motility, it can identify nerve tissue injury. Blood flow, echogenicity, boundary ambiguity, longitudinal section, and cross-sectional area (CSA) are all markedly increased in DPN patients. When it comes to subclinical neuropathy, high-resolution ultrasonography is very helpful in identifying anomalies and peripheral nerve activity in individuals with satisfactory electrical reports.

The benefits of high-frequency ultrasound include its non-invasiveness, strong repeatability, and wide range of potential applications in DPN auxiliary diagnostic and prognosis determination. The tested result of DPN can be increased, and absent or incorrect diagnoses can be prevented, by combining nerve conduction testing with ultrasonography. It is helpful in determining the incidence and outcome of DPN with unclear border, elevated CSA, and hypoechoic area, although certain limitations for the examination of the inner neuromuscular plexus and lumbosacral plexus (Yu 2021).

25.3.7 Unique Primary Screening and Predictive Tools

Precise identification of various DPN characteristics, such as tiny fibre neuropathy, is essential for the planning of clinical trials and to enable focused treatment approaches. For research purposes, the 2020 ACTTION criteria are used as standard diagnostic criteria for idiopathic large, small, and mixed fibre neuropathies. For small, large, and mixed fibre polyneuropathies, respectively, at least one small or large fibre symptom and sign, and abnormal IENFD or sensory NCS (or both) 18. It is anticipated that DPN's exact aligned criteria would be published in 2022. In order to prevent false-positive findings that may arise from inadequate handling, the biopsy material must be processed and analysed in highly skilled laboratories. Small fibre neuropathy can still be confirmed pathologically with IENFD, despite the test being less intrusive. We need to discover less intrusive and more user-friendly biomarkers to help with diagnosis and the design of clinical trials for early intervention or disease prevention. Here are several new instruments that might be utilised for DN early detection.

25.3.8 Nerve Excitability Testing (NET)

One promising new experimental neurophysiological indicator of early axonal deterioration may be nerve excitability testing (NET). In order to detect axonal damage before it manifests itself as NCS results, NET analyses the axonal firing thresholds in response to submaximal and supramaximal current given by non-invasive electrodes. It is hoped that NET can identify early alterations in DPN axonal function prior to irreversible axonal degeneration, as seen in oxaliplatin-induced neurotoxicity. In contrast to healthy controls, patients with minor DPN (distal sensory nerve conduction slowness or missing H-reflexes) had aberrant axonal excitability profiles, including lower threshold changes to 50% depolarising current, a longer relative refractory time, and a less noticeable shift in super excitability. There are certain restrictions on the technology: (1) it is not commonly accessible and necessitates specialised training and equipment; (2) it is more accurate and repeatable in motor nerves as opposed to sensory nerves; (3) it excludes information regarding the condition of small fibre nerves; (4) no clinically meaningful normative ranges have been established; and (5) NET still requires validation as a substitute biomarker for diabetic peripheral neuropathy (Fan and Smith 2022).

25.3.9 Evoked Responses

To assess the activity of Aδ fibres in people with neuropathic pain, the European Federation of the Neurological Societies (EFNS) created evoked potentials, a reliable and non-invasive technique (Burgess et al. 2021). Pain or heat are examples of external small-fibre stimuli that can overcome the invasiveness and psychophysical variability of the equipment currently employed to measure SFNs. Synchronous electroencephalography is used to detect and record evoked responses, such as laser-evoked potentials (LEPs), contact heat-evoked potentials (CHEPs), and pain-related evoked potentials (PREPs). Many studies that looked at CHEPs or LEPs in people with SFN or DSPN who also had diabetes discovered a strong relationship between evoked potentials and SFN characteristics (IENFD, thermal thresholds) (Bönhof et al. 2019).

25.3.10 Laser Doppler Flare

Elicited responses can be utilised to noninvasively monitor small-fibre function utilising the laser Doppler imaging (LDI) flare method. Microvascular blood flow is increased by neurogenic vasodilation, which happens when C-fibres are activated. Cholinergic C-fibres are activated by heat, isotylcholine, or histamine iontophoresis, which results in local vasodilation. By evaluating the neurogenic flare, the vasomotor axon response may be measured. In identifying diabetic peripheral neuropathy (DPN), the particularity and sensitivity of LDIFl are range from medium to high, and there is a noteworthy connection among the left and right feet. Nevertheless, LDI is devoid of normative data and a systematic approach (Kubasch et al. 2017).

25.3.11 Neurometer

The non-invasive QST Neurometer may be used to measure current perception threshold (CPT). The CPT measures the functional state of three different types of nerves: giant myelinated (Aβ) fibres, medium-size myelinated (Aδ) fibres, and unmyelinated (C) fibres, at 2000, 250, and 5 Hz, respectively. It is a quick, painless device that can tell the difference between hyperaesthesia and hypoaesthesia. Studies have shown that CPT evaluation using the Neurometer is more sensitive than vibration perception threshold (VPT) and monofilament testing in identifying milder DPN. It was also noted that while the CPT showed a correlation with neuropathic symptoms and indications, it adds less information in comparison to traditional testing methods (Cheng et al. 1999).

25.3.12 Microneurography

One technique for recording the intricate action potentials of peripheral nerves, including heat-sensitive C-fibres, is microneurography. In order to stimulate the skin to identify C-fibres with slower conduction velocities, A nerve fascicle is punctured with an electrode made of tungsten metal. With a spike raster plot, this data is shown in real time. According to microneurography, people with painful neuropathy or neuropathy symptoms have aberrant sensitivity to mechanical and thermal stimuli, which causes C-fibre action potentials to double and spike (Burgess et al. 2021).

25.3.13 Imaging Techniques

Recent research has shown that improvements in imaging technology may potentially be utilised to evaluate peripheral nerve injury and alterations in the composition and function of the brain in people with diabetes (Selvarajah et al. 2014).

25.3.14 Brain Magnetic Resonance Imaging

It has been discovered that neuropathy, which is typically thought of as a PNS disease, affects other parts of the CNS, underscoring the value of brain imaging techniques in science. Since there are relationships between brain volume and pain-processing areas such the anterior cingulate cortex and IENFD, MRI has been utilised extensively to investigate individuals with neuropathic pain. According to Selvarajah et al. (2014), T1D patients' brain MRI scans showed reduced grey matter volume in relation to severe DSPN (Selvarajah et al. 2014). In patients with painful DN, recent studies employing single-photon emission CT showed increased cerebral blood flow in the anterior cingulate cortex, which returned to normal once pain was relieved (Watanabe et al. 2018).

25.3.15 Magnetic Resonance Neurography

Using high-resolution magnetic resonance neurography (MRN), Peripheral nerve morphology can be evaluated, and lesions that are clinically important can be found. Diabetes patients with diabetic peripheral neuropathy (DSPN) had a gradient of sciatic nerve lesions from proximal to distal, with a higher number of lesions at the thigh level than in more distal nerve sections, according to a recent magnetic resonance imaging study of the condition that called into question the conventional distal-to-proximal pattern of nerve pathology. These researchers also observed variations in sciatic nerve lesions across those with Type 1 and Type 2 diabetes, as well as associations between these lesions and the severity of clinical DSPN. They postulated that these variations might be due to distinct pathogenic variables in T1D as opposed to T2D that are implicated in the development of DSPN (Jende et al. 2018).

25.3.16 Diffusion Tensor Imaging (DTI)

A quantitative assessment of nerve integrity is made possible by a magnetic resonance technique known as DTI, which is sensitive to even the smallest changes in the microstructure of the neurone. According to a recent study, DTI characteristics assessed in the sciatic or tibial nerve were strongly correlated with the existence and seriousness of DSPN in T1D individuals. Patients suffering from T1D with or without DSPN did not vary from healthy volunteers in terms of the tibial to sciatic DTI parameter ratio (Bönhof et al. 2019).

25.3.17 Functional Magnetic Resonance Imaging (fMRI)

The non-invasive, radiation-free technique known as fMRI can be used to scan large samples or to do many scans on the same person. Most importantly, it establishes a direct connection between external stimuli and functional alterations in specific cortical regions and central nuclei in real time. The amount of oxygen utilised locally in various areas of the CNS in response to external stimuli may be precisely recorded by fMRI using the ways that blood with and without O_2 have different magnetic effects. By seeing how the brain responds to different stimuli, blood oxygenation level-dependent (BOLD) imaging helps us better comprehend the connection between DPN and the central nervous system. In DPN imaging, fMRI is emerging as a novel and useful method. According to certain research that described the efficacy of fMRI in evaluating brain activity in DPN patients, the primary sensory cortex, frontal lobe, thalamus, and cerebellum were demonstrated to be considerably active in these individuals (Li et al. 2018). According to Selvarajah et al. (2019), there is a connection between functional alterations in the somatosensory cortex and brain volume that is correlated with the degree of peripheral neuropathy as determined by fMRI. A wider S1 functional representation of the foot and thigh and a greater reduction in sensory cortical thickness were found in patients with the most severe neuropathy, as indicated by NCS. This suggests that deafferentation affects sensory neurones by recruiting nearby functioning neurones (Selvarajah et al. 2019).

25.4 Bedside Tests

25.4.1 NC-Stat DPNCheck

When assessing the function of peripheral large-nerve fibres, NCS is the most objective diagnostic method. The DPN-Check is a portable point-of-care instrument that may be used to measure the intensity and velocity of conductivity of the sural nerve without the need for costly instruments or the knowledge of a neuroelectrophysiologist. The effectiveness and validity in diabetics with DSPN have been confirmed (Brown et al. 2017). According to a recent study, the DPN-check outperforms clinical tests and laser-Doppler "FLARE" imaging in terms of inter- and intra-ratter accuracy, with a 71% selectivity and 95% sensitivity, in detecting early DPN during along the feet, and ocular assessments (Yang et al. 2020). According to a more recent study, NC-stat DPNCheck is a useful diagnostic tool in resource-constrained contexts. In order to identify and triage abnormal NCV, Sharma et al., suggest using the NC-Stat DPN Check before to doing confirmatory NCS using standard electrophysiological tools (Sharma et al. 2014). However, the NC-Stat DPN Check is not free, and any additional NCS referrals require highly qualified staff and specialist clinics.

25.4.2 Vibratip

The tiny, disposable Vibratip is made out of an on/off switch, a 128 Hz electric motor, a plastic case, and a battery. Its promise in medical research has been demonstrated by its remarkable accuracy in diagnosing severe neuropathy when evaluated against clinical NDS and vibration perception thresholds (VPTs) (Bongaerts et al. 2013).

25.4.3 NeuroQuick (NQ)

The NQ is a transportable POC gadget that combines two intersecting laser lights which collide at a pre-set position away from the skin with a fan that blasts cool air. Some studies suggest that the NQ is an analysing instrument that might be used to quantify small-fibre failure. It has been demonstrated to be more accurate than both sophisticated heat testing and screening tools like the tuning fork in detecting early cases of delayed sensory paraPN (Ziegler et al. 2005).

25.4.4 Multifrequency Vibrometry

Similar to audiometry, which determines hearing thresholds for certain frequencies, multifrequency vibrometry may be used to provide VPTs at different vibration frequencies. Basically, different skin mechanoreceptors react within a certain frequency range, and their best reaction happens at a maximum frequency. In those with diabetes, VPTs at the lowest frequencies (8–32 Hz) were shown to be reduced (Nelander et al. 2012).

25.4.5 Skin Autofluorescence

Diabetic microvascular problems, such as DN, are influenced by the build-up of AGEs. The AGE Reader is a noninvasive, safe, standardised, and clinically practical method of detecting skin autofluorescence (SAF) that was developed to evaluate the accumulation of AGEs in the skin. Numerous research investigations have documented favourable correlations between SAF and diabetes and its sequelae; however, there is conflicting information about separate connections with DN, retinopathy, or nephropathy (Yoshioka 2018).

25.4.6 Sudomotor Function Tests

Sweat secretory gland activity is referred to as sudomotor function. Cholinergic nerve fibres that are sudomotor and unmyelinated innervate the sweat glands in the foot. DPN can cause impairments to these nerve fibres. Dry feet are a result of sudomotor dysfunction and are linked to a higher incidence of foot ulcers. These methods include the quantitative direct and indirect reflex test, the quantitative sudomotor axon reflex test, and the thermoregulatory sweat test, which are used to evaluate sudomotor function in DPN patients. Two straightforward techniques for evaluating sudomotor function are the Sudoscan and the Neuropad indication test. The skin is covered with a simple patch as part of the Neuropad visual test, which measures sudomotor function by undergoing a chemical process that turns it from blue to pink. Its sensitivity for diagnosing DPN is between 86 and 95%, whereas its specificity is only between 45 and 69.8% (Yang et al. 2020).

The FDA-approved non-invasive device Sudoscan, which measures electrochemical skin conductance (ESC) in the hands and feet using reverse iontophoresis, is used to diagnose DPN. Foot ESC is inversely correlated with autonomic failure, QST, and neuropathic impairments, and reduced ESC in diabetics is associated with DSPN. The tool's repeatability in controls is outstanding, and its accuracy and precision are 78% and 92%, respectively (Casellini et al. 2013). In comparison to VPT and clinical evaluation, ESC, as measured by Sudoscan, is the most sensitive test for early DPN detection, according to a clinical trial. It corresponds with other metrics such as clinical neuropathy grading systems, QST, autonomic function tests, and NCS parameters and has diagnostic usefulness equivalent to skin biopsy and IENFD examination (Yang et al. 2020).

The field of information and communication technology has advanced rapidly in recent years, allowing us to develop, create, and manufacture small mobile devices that are efficient in terms of processing power and battery life. Depending on their intended usage, wearable electronics are made of different materials and have specialised technological features. Applications of wearable technology are expanding quickly in several research fields, including social networking, sports medicine, healthcare, education, and the military, because of their affordability and viability. Peripheral neuropathy (PN) has been automatically diagnosed in a number of studies utilising a single sensor or data collection technique (Talha et al. 2023). These techniques include thermal image processing, nerve conduction studies or electromyography, foot plantar pressure, and gait analysis. However, a constant flow of data is required for testing and tracking the course of PN. Image-based techniques could provide a useful way to identify PN. Many research initiatives employ cameras to assess a subject's gait, however, it is not possible to collect data continuously considering the camera's small circle of focus, which increases the likelihood of missing data points (Goodvin et al. 2006). It is also possible to detect pressure neuro injury using non-wearable technology including thermal imaging and vibration or perception-based methods. These methods are not appropriate for gathering persistent information, which is necessary for identifying the cause and course of PN.

25.5 Utilising Single-Sensor Technology in Intelligent Wearable Systems to Diagnose Peripheral Neuropathy

The best approach to gather data continuously is to use wearable sensors, as these devices are made to be directly linked to the human body and worn by patients with ease. Additionally, they offer a mechanism to gather and save patient activity data for later use.

25.5.1 Wearable Inertial Sensor-Based Intelligent Systems for the Diagnosis of PN

Because wearable inertial measuring devices are reliable and accurate at evaluating human motion, they are mostly used for activity detection tasks like postural postures or human gait, which may have uses in health and wellness. Human gait may be recorded and analysed using a variety of inertial sensor combinations in order to facilitate diagnosis. The human body's leg, foot, or waist are where inertial sensors are typically affixed. When examining human mobility, accelerometers are the most often utilised sensor in most research. Recently, several researchers have also started examining human motion for a variety of purposes by combining an accelerometer, gyroscope, or inertial measuring device (Goodvin et al. 2006).

Chen et al. use a wearable inertial device, like an accelerometer, to screen diabetes patients for peripheral neuropathy early. Post-traumatic neuritis (PN) is characterised by a slower gait, shorter steps, and decreased knee and ankle mobility. The experimental study included 106 participants (52 males and 54 females) ranging in age from 38 to 83; 30 of them had been identified as confirmed instances of diabetic peripheral neuropathic pain, while the remaining 76 were healthy individuals (Chen et al. 2019). Diagnosing peripheral neuropathy in diabetic patients by analysing gait with a single IMU was done by Esser et al. The IMU was placed on the lower back, and the gyroscope and accelerometer data were recorded at a 100 Hz sample rate for processing. In addition to 17 subjects (14 males and 3 females) with DPN, the experimental setup comprised 52 healthy people, 30 of whom were male and 12 of whom were female, all of whom were around 63.2 ± 9.2 years old. A diabetic podiatrist with specific skills confirmed the PN patients by doing a monofilament test. The study found that, other from stride length, all spatiotemporal measures between PN patients and healthy subjects exhibited significant differences (Esser et al. 2018).

25.5.2 Pressure Sensor-Based Intelligent Wearable System for Diagnosis of PN

Foot heel pressure is a force field spread across the feet and the ground. It significantly affects the diagnosis of peripheral neuropathy. Due to damage to their foot nerves, people with diabetes frequently lack the ability to sense the appropriate amount of plantar pressure required for smooth walking. Nerve damage causes the foot to become less sensitive. As a result, walking will always require more effort from those who suffer from in the legs below the knee than from those in good condition. Because, wearable technology may be used both remotely and in clinical settings, it is possible to use a wearable pressure sensor to make an initial diagnosis of peripheral neuropathy (PN). Although there are several varieties of pressure measurement devices, they may be broadly categorised into two groups: (1) in-shoe devices and (2) platform devices.

Cao et al. introduced a method for diagnosing PN that is based on the distribution of plantar pressure in the foot. The study's main objective was to quantify and assess changes in the distribution of the stress in the feet can help older diabetics avoid ulcers. The foot plantar pressure data set for this study includes 20 male and 9 female healthy subjects (aged 65.2 ± 5.4 years), 9 male and 9 female diabetic patients with peripheral neuropathy (DPN) aged 65.7 ± 2.4 years, and 9 male and 8 female diabetic patients without DN aged 65.2 ± 6.8 years. The data was collected using a 300 Hz sample frequency insole wireless plantar pressure monitoring device manufactured by Medilogic in the USA. The results clearly showed that DPN patients tended to have higher pressure in each segmental area of the foot in contrast to diabetic or healthy people without DPN. Peak pressure readings in patients with DPN and those without were assessed (Cao et al. 2021).

25.5.3 ECG-Based Intelligent Wearable Systems for Diagnosis of PN

PN is mostly caused by diabetes, which some experts believe affects about one-third of people with acute myocardial infarction. Injured to these peripheral nerves that control the heart results in heart failure known as cardiac autonomic neuropathy (CAN). With the use of electrocardiography (ECG), May, this results from harm to the peripheral nerves necessary for the heart's proper function, may be identified early and painlessly (Voulgari et al. 2011).

Sharanya and Sridhar proposed a convolutional neural network (CNN)-based method for CAN diagnostics in order to make prediction easier. This study aimed to classify individuals as either CAN-positive or CAN-negative using ECG signal analysis. Thirteen male and six female participants participated in the experiments. Nine of them received confirmed positive findings for CAN, and ten were categorised as CAN-negative class due to their good health. At a sample frequency of 400 Hz, the ECG signals were captured for 20 min in a lead II arrangement. All things considered, the results show that CNN can manage the examination of the complex ECG features required for CAN early detection. 95.42% was the study's overall CAN diagnosis accuracy (Sharanya and Sridhar 2021).

25.5.4 Intelligent Wearable Multisensory Systems for Diagnosis of PN

Due to the numerous benefits that handheld sensors offer over conventional equipment, the clinical and medical fields have expressed a significant lot of interest in them (Sejdić et al. 2013). Acquired clinically significant data for Parkinson's disease and PN diagnosis from an acceleration signal by Khandakar et al. and determined gait characteristics from the raw accelerometer signal (Khandakar et al. 2022). Recommended a method for identifying diabetes patients' foot ulcers, which are brought on by peripheral neuropathy from high blood glucose levels. Their primary method for identifying any potential for PN-related foot ulcers was tracking changes in pressure and temperature in the diabetic foot. To capture data from both feet, a flexible thermistor and a force-sensitive resistor were employed as smart insole pressure and temperature sensors. Bigorra-Sempere et al. (Sempere-Bigorra et al. 2023). Additionally, shown how to monitor and detect diabetic foot ulcers brought on by PN using a wearable inertial sensor. A single IMU in the lumbar region was used to simulate the gait; however, additional data collecting and diagnostic methods, including as vibration and sensory tests, were used in this investigation to detect PN. A technique for the early diagnosis of PN induced by systemic autoimmune rheumatic disorders (SARDs) was presented by Z. Veličković et al. The authors concentrated on the screening method for the diagnosis of PN associated with SARDs that uses wearable sensors in conjunction with machine learning (Veličković et al. 2023).

25.6 Conclusion

Peripheral neuropathy is mostly caused by the global incidence of diabetes. Peripheral neuropathy is becoming more common; thus, it is necessary to develop effective strategies based on machine learning and current technologies. Early detection of DPN enables pre-emptive multifactorial management to slow the advancement of nerve injury. While there is information available regarding the use of many basic diagnostic modalities, the screening methods used today identify advanced illness, for which there are no viable therapies.

It is not possible to identify subclinical DPN with NCS and symptoms alone. Both peripheral and central nervous system dysfunction cannot be distinguished by the subjective QST test. The gold standard for detecting tiny fibre neuropathy is believed to be skin biopsy. Nevertheless, widespread screening and recurring biopsies are impractical. Patients were grateful that the current use of screening tests to detect small fibre pathology in diabetics in addition to retinal screening during a one-stop microvascular screening session allowed for new DPN diagnosis. Thus, a novel method for detecting early DPN alterations should be implemented during crucial stages of patient assessment. The rigorous procedures of skin biopsy and CCM will still be employed in the expert context for scientific reasons. They can increase our understanding of DPN's early natural history, how it is monitored following therapeutic therapies, and other related topics. On the other hand, the indication test The Neuropod is a sophisticated and thoroughly researched screening instrument that is highly reproducible, user-friendly, and supportive of self-evaluation. It is very dependable in the exclusion of DPN and has a high sensitivity and adverse predictive value. It could also have helped with the early diagnosis. Even if the diagnosis is getting better with the techniques available today, longer-term studies are required to.

References

Bongaerts BW, Rathmann W, Heier M, Kowall B, Herder C, Stöckl D, et al. Older subjects with diabetes and prediabetes are frequently unaware of having distal sensorimotor polyneuropathy: the KORA F4 study. Diabetes Care. 2013;36(5):1141–6.

Bönhof GJ, Herder C, Strom A, Papanas N, Roden M, Ziegler D. Emerging biomarkers, tools, and treatments for diabetic polyneuropathy. Endocr Rev. 2019;40(1):153–92.

Brown JJ, Pribesh SL, Baskette KG, Vinik AI, Colberg SR. A comparison of screening tools for the early detection of peripheral neuropathy in adults with and without type 2 diabetes. J Diabetes Res. 2017;2017(1):1467213.

Burgess J, Frank B, Marshall A, Khalil RS, Ponirakis G, Petropoulos IN, et al. Early detection of diabetic peripheral neuropathy: a focus on small nerve fibres. Diagnostics. 2021;11(2):165.

Callaghan BC, Little AA, Feldman EL, Hughes RA. Enhanced glucose control for preventing and treating diabetic neuropathy. Cochrane Database Syst Rev. 2012;2012(6):CD007543.

Cao Z, Wang F, He Y, Zhang Y, Zhang J. Analysis of plantar pressure in elderly diabetic patients with peripheral neuropathy. In: 2021 international conference on public health and data science (ICPHDS). IEEE; 2021, July. p. 184–7.

Carmichael J, Fadavi H, Ishibashi F, Shore AC, Tavakoli M. Advances in screening, early diagnosis and accurate staging of diabetic neuropathy. Front Endocrinol. 2021;12:671257.

Casellini CM, Parson HK, Richardson MS, Nevoret ML, Vinik AI. Sudoscan, a noninvasive tool for detecting diabetic small fiber neuropathy and autonomic dysfunction. Diabetes Technol Ther. 2013;15(11):948–53.

Chen S, Kang L, Lu Y, Wang N, Lu Y, Lo B, Yang GZ. Discriminative information added by wearable sensors for early screening-a case study on diabetic peripheral neuropathy. In: 2019 IEEE 16th international conference on wearable and implantable body sensor networks (BSN). IEEE; 2019, May. p. 1–4.

Cheng WY, Jiang YD, Chuang LM, Huang CN, Heng LT, Wu HP, et al. Quantitative sensory testing and risk factors of diabetic sensory neuropathy. J Neurol. 1999;246:394–8.

Divisova S, Vlckova E, Srotova I, Kincová S, Skorna M, Dusek L, et al. Intraepidermal nerve-fibre density as a biomarker of the course of neuropathy in patients with Type 2 diabetes mellitus. Diabet Med. 2016;33(5):650–4.

Esser P, Collett J, Maynard K, Steins D, Hillier A, Buckingham J, et al. Single sensor gait analysis to detect diabetic peripheral neuropathy: a proof of principle study. Diabetes Metab J. 2018;42(1):82–6.

Fan Q, Smith AG. Recent updates in the treatment of diabetic polyneuropathy. Fac Rev. 2022;11:30.

Goodvin C, Park EJ, Huang K, Sakaki K. Development of a real-time three-dimensional spinal motion measurement system for clinical practice. Med Biol Eng Comput. 2006;44:1061–75.

Iqbal Z, Azmi S, Yadav R, Ferdousi M, Kumar M, Cuthbertson DJ, et al. Diabetic peripheral neuropathy: epidemiology, diagnosis, and pharmacotherapy. Clin Ther. 2018;40(6):828–49.

Jende JM, Groener JB, Oikonomou D, Heiland S, Kopf S, Pham M, et al. Diabetic neuropathy differs between type 1 and type 2 diabetes: insights from magnetic resonance neurography. Ann Neurol. 2018;83(3):588–98.

Khandakar A, Mahmud S, Chowdhury ME, Reaz MBI, Kiranyaz S, Mahbub ZB, et al. Design and implementation of a smart insole system to measure plantar pressure and temperature. Sensors. 2022;22(19):7599.

Kim SS, Won JC, Kwon HS, Kim CH, Lee JH, Park TS, et al. Prevalence and clinical implications of painful diabetic peripheral neuropathy in type 2 diabetes: results from a nationwide hospital-based study of diabetic neuropathy in Korea. Diabetes Res Clin Pract. 2014;103(3):522–9.

Kubasch ML, Kubasch AS, Torres Pacheco J, Buchmann SJ, Illigens BMW, Barlinn K, Siepmann T. Laser Doppler assessment of vasomotor axon reflex responsiveness to evaluate neurovascular function. Front Neurol. 2017;8:370.

Li J, Zhang W, Wang X, Yuan T, Liu P, Wang T, et al. Functional magnetic resonance imaging reveals differences in brain activation in response to thermal stimuli in diabetic patients with and without diabetic peripheral neuropathy. PLoS One. 2018;13(1):e0190699.

Narayanaswamy H, Facer P, Misra VP, Timmers M, Byttebier G, Meert T, Anand P. A longitudinal study of sensory biomarkers of progression in patients with diabetic peripheral neuropathy using skin biopsies. J Clin Neurosci. 2012;19(11):1490–6.

Neeland IJ, Patel KV. Diabetes: key markers of injury and prognosis. In: Biomarkers in cardiovascular disease. Elsevier; 2019. p. 41–51.

Nelander J, Speidel T, Björkman A, Dahlin LB. Vibration thresholds are increased at low frequencies in the sole of the foot in diabetes—a novel multi-frequency approach. Diabet Med. 2012;29(12):e449–56.

Petropoulos IN, Ponirakis G, Khan A, Almuhannadi H, Gad H, Malik RA. Diagnosing diabetic neuropathy: something old, something new. Diabetes Metab J. 2018;42(4):255.

Pritchard N, Edwards K, Russell AW, Perkins BA, Malik RA, Efron N. Corneal confocal microscopy predicts 4-year incident peripheral neuropathy in type 1 diabetes. Diabetes Care. 2015;38(4):671–5.

Sejdić E, Lowry KA, Bellanca J, Redfern MS, Brach JS. A comprehensive assessment of gait accelerometry signals in time, frequency and time-frequency domains. IEEE Trans Neural Syst Rehabil Eng. 2013;22(3):603–12.

Selvarajah D, Wilkinson ID, Maxwell M, Davies J, Sankar A, Boland E, et al. Magnetic resonance neuroimaging study of brain structural differences in diabetic peripheral neuropathy. Diabetes Care. 2014;37(6):1681–8.

Selvarajah D, Wilkinson ID, Fang F, Sankar A, Davies J, Boland E, et al. Structural and functional abnormalities of the primary somatosensory cortex in diabetic peripheral neuropathy: a multimodal MRI study. Diabetes. 2019;68(4):796–806.

Sempere-Bigorra M, Brognara L, Julian-Rochina I, Mazzotti A, Cauli O. Relationship between deep and superficial sensitivity assessments and gait analysis in diabetic foot patients. Int Wound J. 2023;20(8):3023–34.

Sharanya S, Sridhar PA. A model for early diagnosis of Cardiac Autonomic Neuropathy (CAN). J Phys Conf Ser. 2021;2089(1):012053. IOP Publishing.

Sharma S, Vas PR, Rayman G. Assessment of diabetic neuropathy using a point-of-care nerve conduction device shows significant associations with the LDIFLARE method and clinical neuropathy scoring. J Diabetes Sci Technol. 2014;9(1):123–31.

Stem MS, Hussain M, Lentz SI, Raval N, Gardner TW, Pop-Busui R, Shtein RM. Differential reduction in corneal nerve fiber length in patients with type 1 or type 2 diabetes mellitus. J Diabetes Complicat. 2014;28(5):658–61.

Talha M, Kyrarini M, Buriro EA. A survey of the diagnosis of peripheral neuropathy using intelligent and wearable systems. Technologies. 2023;11(6):163.

Tesfaye S, Boulton AJ, Dyck PJ, Freeman R, Horowitz M, Kempler P, et al. Diabetic neuropathies: update on definitions, diagnostic criteria, estimation of severity, and treatments. Diabetes Care. 2010;33(10):2285.

Veličković Z, Dolijanovic SP, Tomonjic N, Janjić S, Stojic B, Radunovic G. AB1524 a novel accelerometry-based method for early detection of peripheral neuropathy associated with systemic autoimmune rheumatic diseases. Ann Rheumat Dis. 2023;82:1994.

Voulgari C, Tentolouris N, Stefanadis C. The ECG vertigo in diabetes and cardiac autonomic neuropathy. J Diabetes Res. 2011;2011(1):687624.

Watanabe K, Hirano S, Kojima K, Nagashima K, Mukai H, Sato T, et al. Altered cerebral blood flow in the anterior cingulate cortex is associated with neuropathic pain. J Neurol Neurosurg Psychiatry. 2018;89(10):1082–7.

Yang H, Sloan G, Ye Y, Wang S, Duan B, Tesfaye S, Gao L. New perspective in diabetic neuropathy: from the periphery to the brain, a call for early detection, and precision medicine. Front Endocrinol. 2020;10:929.

Yoshioka K. Skin autofluorescence is a noninvasive surrogate marker for diabetic microvascular complications and Carotid Intima–Media thickness in Japanese patients with Type 2 diabetes: a cross-sectional Study. Diab Ther. 2018;9:75–85.

Yu Y. Gold standard for diagnosis of DPN. Front Endocrinol. 2021;12:719356.

Ziegler D, Siekierka-Kleiser E, Meyer B, Schweers M. Validation of a novel screening device (NeuroQuick) for quantitative assessment of small nerve fiber dysfunction as an early feature of diabetic polyneuropathy. Diabetes Care. 2005;28(5):1169–74.

MIX
Papier aus verantwortungsvollen Quellen
Paper from responsible sources
FSC® C105338

If you have any concerns about our products,
you can contact us on
ProductSafety@springernature.com

In case Publisher is established outside the EU,
the EU authorized representative is:
Springer Nature Customer Service Center GmbH
Europaplatz 3, 69115 Heidelberg, Germany

Printed by Libri Plureos GmbH
in Hamburg, Germany